ALFRED BENZON SYMPOSIUM 19

Gene Expression
The Translational Step and its Control

GENE EXPRESSION

THE TRANSLATIONAL STEP AND ITS CONTROL

Proceedings of the Alfred Benzon Symposium 19 held at the premises of the Royal Danish Academy of Sciences and Letters, Copenhagen 19–23 June 1983

EDITED BY

BRIAN F. C. CLARK
HANS UFFE PETERSEN

Published by
Munksgaard, Copenhagen

Distributed in Japan by
Nankodo, Tokyo

Printed in Denmark by P. J. Schmidts Bogtrykkeri, Vojens
ISBN 87-16-09561-8
ISSN 0105-3639

Distributed in Japan by
Nankodo Company Ltd.

Contents

List of Participants

C. T. CASKEY
Howard Hughes Medical Institute
De Bakey Building
Texas Medical Center
Houston, Texas 77030
U.S.A.

B. F. C. CLARK
Division of Biostructural Chemistry
Aarhus University
DK-8000 Aarhus C
Denmark

F. CRAMER
Max-Planck-Institut für
experimentelle Medizin
Abeteilung Chemie
D-3400 Göttingen
Federal Republic of Germany

J. P. EBEL
Institut de Biologie Moléculaire
et Cellulaire
15, rue René Descartes
F-67084 Strasbourg
France

V. A. ERDMANN
Freie Universität Berlin
Institut für Biochemie
D-1000 Berlin 33 (Dahlem)
Federal Republic of Germany

R. ERIKSON
Department of Pathology
University of Colorado
Health Sciences Center
Denver, Colorado 80262
U.S.A.

J. GALLANT
Department of Genetics
University of Washington
Seattle, Washington 98195
U.S.A.

R. GARRETT
Division of Biostructural Chemistry
Aarhus University
DK-8000 Aarhus C
Denmark

L. GOLD
Department of Molecular, Cellular &
Developmental Biology
University of Colorado
Boulder, Colorado 80309
U.S.A.

M. GRUNBERG-MANAGO
Institut de Biologie Physico-Chimique
13, rue Pierre et Marie Curie
F-75005 Paris
France

J. W. B. HERSHEY
Department of Biological Chemistry 2043
University of California at Davis
Davis, California 95616
U.S.A.

J. HESS THAYSEN
Department of Medicine P
Rigshospitalet
DK-2100 Copenhagen Ø
Denmark

M. Innis
Cetus Corporation
1400 Fifty-Third Street
Emeryville, California 94608
U.S.A.

J. Justesen
Institute of Molcular Biology
Aarhus University
DK-8000 Aarhus C
Denmark

Y. Kaziro
Institute of Medical Science
Minatoku, Takanawa
Tokyo 108
Japan

I. M. Kerr
Imperial Cancer Research Fund
Laboratories
Lincoln's Inn Fields
London WC2A 3PX
England

N. O. Kjeldgaard
Institute of Molecular Biology
and Plant Physiology
Aarhus University
DK-8000 Aarhus C
Denmark

A. Klug
MRC Laboratory of Molecular Biology
Hills Road
Cambridge CB2 2QH
England

C. G. Kurland
Department of Molecular Biology
Biomedical Center
S-751 24 Uppsala
Sweden

O. Maaløe
Institute of Microbiology
University of Copenhagen
DK-1353 Copenhagen K
Denmark

S. Magnusson
Institute of Molecular Biology
Aarhus University
DK-8000 Aarhus C
Denmark

K. A. Marcker
Institute of Molecular Biology
and Plant Physiology
Aarhus University
DK-8000 Aarhus C
Denmark

D. L. Miller
New York State Institute for Basic Research
in Developmental Disabilities
1050 Forest Hill Road
Staten Island – New York 10314
U.S.A.

M. Nomura
Institute for Enzyme Research
University of Wisconsin
Madison, Wisconsin 53706
U.S.A.

J. Ofengand
Roche Institute of Molecular Biology
Nutley, New Jersey 07110
U.S.A.

S. Pedersen
Institute of Microbiology
University of Copenhagen
DK-1353 Copenhagen K
Denmark

S. Pestka
Roche Institute of Molecular Biology
Nutley, New Jersey 07110
U.S.A.

P. M. Pitha
Oncology Center
The Johns Hopkins University
School of Medicine
Baltimore, Maryland 21205
U.S.A.

H. U. Petersen
Division of Biostructural Chemistry
Aarhus University
DK-8000 Aarhus C
Denmark

M. Rosenberg
Laboratory of Biochemistry
National Cancer Institute
Bethesda, Maryland 20205
U.S.A.

F. M. Rottman
Department of Microbiology
Case Western Reserve University
Cleveland, Ohio 44106
U.S.A.

B. Safer
Laboratory of Molecular Hematology
National Heart, Lung and Blood Institute
Bethesda, Maryland 20205
U.S.A.

G. Schatz
Biozentrum, Abteilung Biochemie
Universität Basel
CH-4056 Basel
Switzerland

P. Schimmel
Department of Biology
Massachusetts Institute of Technology
Cambridge, Massachusetts 02139
U.S.A.

M. Sprinzl
Laboratorium für Biochemie
Universität Bayreuth
D-8580 Bayreuth
Federal Republic of Germany

O. C. Uhlenbeck
Department of Biochemistry
University of Illinois
Urbana, Illinois 61801
U.S.A.

Preface

The tremendous advances in biochemistry during the last 30 years have been preceded by a long period of slow development. Although nucleic acids were discovered more than one hundred years ago, it was the inspiration of the double helical model for DNA which gave birth to molecular biology. This has led to the explosive increase in research on the molecular mechanisms of gene expression.

Molecular biology has, thus far, concentrated on interpreting genetic information structurally, chemically, and genetically. The intensive development in molecular biological research resulted in such breakthroughs as the elucidation of the genetic code, and genetic engineering. The elucidation of the genetic code explained the way in which genetic information encoded in DNA is transcribed into messenger RNA which is then translated into protein. The link between the gene and gene product was, thus clearly established.

Genetic engineering has arisen from a growing knowledge about the basic enzymology involved in the transfer of genetic information. The importance of genetic engineering techniques in establishing new industries is clear. This provides a further stimulus for a deeper understanding of the mechanism of gene expression.

We felt that it was timely, therefore, to bring together a group of experts on gene expression, and to concentrate on examining the translational step in depth, reviewing in open discussion new material which will also have important relevance to research work outside its immediate circle of experts. This we were able to do, thanks mainly to the good auspices of the Alfred Benzon Foundation.

Our symposium focused on the various stages of protein biosynthesis both for prokaryotic and eukaryotic cells. Much is still to be learned from the use of bacteria, depending on the type of question asked, although many may feel it more topical to concentrate in the future on the molecular biology of eukaryotic cells. Components of the biosynthetic apparatus were reviewed in some detail as a prelude to a discussion of selected control systems and post-translational modification events. This gave a window onto an attempt to understand foreign gene expression, a newly topical field of importance for genetic engineering.

Finally, the special case of eukaryotic control involving interferon, and its possible application in the medical sphere, was examined as a research field at the forefront of knowledge.

We hope that the book which results from this detailed treatment of specialised aspects of gene expression will find a welcome from both established and new researchers on gene expression and translational control.

We are happy to thank Birgit Dalgaard, Lisbeth Heilesen, Margaret Clark and Nick Coppard for help in the preparation of this volume.

Brian F. C. Clark and Hans Uffe Petersen

I. Initiation of Protein Biosynthesis

Mechanism of Initiation of Prokaryotic Translation

Hans Uffe Petersen, Friedrik P. Wikman & Brian F. C. Clark

In all living cells, protein biosynthesis can be regarded as consisting of 4 main phases: initiation, elongation, termination and finally posttranslational modification. Different aspects of the initiation phase are described in the following 4 chapters, and the other 3 phases are discussed in the rest of this volume. The general mechanisms are similar in prokaryotic and eukaryotic cells. However, at the molecular level, differences become apparent. The mechanism of initiation in eukaryotic cells is discussed by J. W. B. Hershey *et al.* and B. Safer elsewhere in this volume. Here, we summarize the prokaryotic initiation process with special reference to some of the main differences found between the 2 cell types.

MECHANISM OF INITIATION OF PROKARYOTIC TRANSLATION

Each of the main steps in protein biosynthesis involves specific interactions between proteins and nucleic acids. The initiation process can be regarded as a number of steps involving the initiator $tRNA_f^{Met}$. Fig. 1 shows a schematic representation of such a view in 6 steps.

In the first 2 steps, the initiator $tRNA_f^{Met}$ is enzymatically aminoacylated with methionine by methionyl-tRNA synthetase. These steps will be discussed in chapter III. In eukaryotic cells, the initiator $tRNA_i^{Met}$ is similarly charged with methionine. In most prokaryotic cells, the Met-$tRNA_f^{Met}$ is formylated at the α-amino group of the methionine as shown in Step 3. The formyl group is donated by N_{10}-formyl-tetrahydrofolate, and the reaction is catalyzed by N_{10}-f-THF-methionyl-tRNA formyltransferase. This enzyme has been purified to homogeneity (Kahn *et al.* 1980). It is a monomer of M_r 32000, and it has a significantly higher affinity for Met-$tRNA_f^{Met}$ than for any other tRNA species. However,

Division of Biostructural Chemistry, Department of Chemistry, Aarhus University, 8000 Aarhus C, Denmark

GENE EXPRESSION, Alfred Benzon Symposium 19.
Editors: Brian F. C. Clark & Hans Uffe Petersen, Munksgaard, Copenhagen 1984.

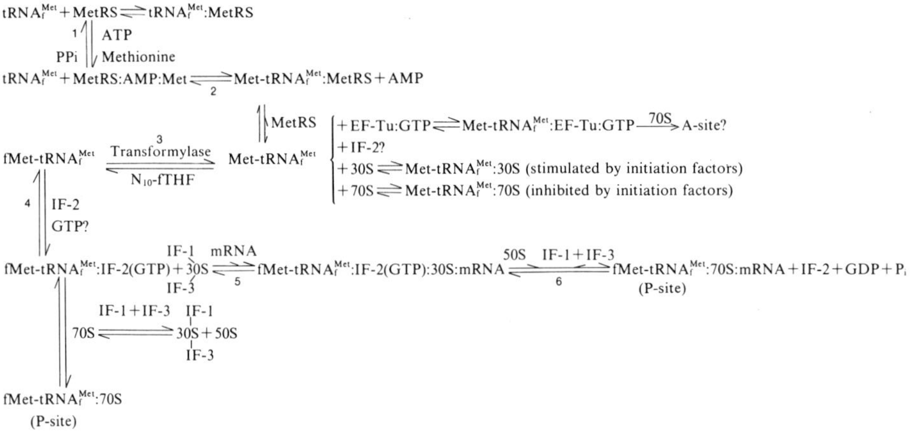

$tRNA_f^{Met} + MetRS \rightleftharpoons tRNA_f^{Met}{:}MetRS$

1 ATP / PPi Methionine

$tRNA_f^{Met} + MetRS{:}AMP{:}Met \underset{2}{\rightleftharpoons} Met\text{-}tRNA_f^{Met}{:}MetRS + AMP$

MetRS

$fMet\text{-}tRNA_f^{Met} \underset{N_{10}\text{-}fTHF}{\overset{3 \text{ Transformylase}}{\rightleftharpoons}} Met\text{-}tRNA_f^{Met}$

+ EF-Tu:GTP ⇌ $Met\text{-}tRNA_f^{Met}$:EF-Tu:GTP $\xrightarrow{70S}$ A-site?
+ IF-2?
+ 30S ⇌ $Met\text{-}tRNA_f^{Met}$:30S (stimulated by initiation factors)
+ 70S ⇌ $Met\text{-}tRNA_f^{Met}$:70S (inhibited by initiation factors)

4 IF-2 GTP?

$fMet\text{-}tRNA_f^{Met}$:IF-2(GTP) + 30S (IF-1, IF-3) $\underset{5}{\overset{mRNA}{\rightleftharpoons}}$ $fMet\text{-}tRNA_f^{Met}$:IF-2(GTP):30S:mRNA $\underset{6}{\overset{50S \quad IF\text{-}1+IF\text{-}3}{\rightleftharpoons}}$ $fMet\text{-}tRNA_f^{Met}$:70S:mRNA + IF-2 + GDP + P_i
(P-site)

IF-1 + IF-3
70S ⇌ 30S + 50S (30S with IF-1, IF-3)

$fMet\text{-}tRNA_f^{Met}$:70S
(P-site)

Fig. 1. Principal steps in the initiation of prokaryotic protein biosynthesis.

until now, only limited amounts of the pure transformylase have been available and, therefore, little is known about the interaction between this protein and the initiator tRNA. As the initiator $Met\text{-}tRNA_f^{Met}$ is the only tRNA in the cell which is formylated enzymatically, it is of great interest to elucidate details about the specific sites on the tRNA involved in this reaction. It can be expected that knowledge about the $tRNA_f^{Met}$-transformylase interaction can reveal structural elements which are unique for the initiator tRNA and different from all other tRNAs. Many exceptions to the general rule of formylation have been found in prokaryotic cells including *E. coli* (Petersen *et al.* 1976a). Moreover, and surprisingly, it was found recently that unformylated $Met\text{-}tRNA_f^{Met}$ can form a stable ternary complex with the elongation factor EF-Tu and GTP (Clark *et al.*, this volume). This raises the question whether the formylation plays a role in preventing the trapping of $Met\text{-}tRNA_f^{Met}$ in such a ternary complex. This idea might seem contradictory to the situation in eukaryotic cells, where the initiator $Met\text{-}tRNA_i^{Met}$ is not formylated. However, in higher organisms, a complex with the EF-Tu-equivalent elongation factor, EF-1α has not been demonstrated.

The formation of a pre-initiation complex between the initiator $Met\text{-}tRNA_i^{Met}$ and a protein initiation factor (eIF-2) has been demonstrated *in vitro* in eukaryotes. This is discussed in detail by B. Safer in this volume. Whether a similar complex is formed in prokaryotes has, over the past 10 years, been a puzzling problem (Petersen *et al.* 1979). The interaction between *E. coli* initiator

$tRNA_f^{Met}$ and the initiation factor IF-2 is one of the subjects discussed in detail in Chapter III of this volume. Suffice it to say that correct and efficient binding of fMet-$tRNA_f^{Met}$ to the ribosomal P-site under physiological conditions *in vitro* requires the presence of IF-2 and GTP as shown in Step 4.

The next step involves the ribosomal 30S subunit. As shown in Fig. 1, 70S ribosomes dissociate into the 2 subunits 30S and 50S. This dissociation is stimulated by the protein factors IF-1 and IF-3. IF-1 serves to increase the rate constant for the dissociation reaction, while IF-3 shifts the equilibrium strongly towards dissociation by binding to the 30S particle (Godefroy-Colburn *et al.* 1975).

It is evident that during peptide chain elongation, the correct aminoacyl-tRNA is selected by anticodon:codon interaction on the ribosome. Yet, it is not clear whether one particular sequence of binding of the initiator fMet-$tRNA_f^{Met}$ and the mRNA to the ribosome exists. At high magnesium concentration, fMet-$tRNA_f^{Met}$ can be bound to the ribosomal P-site in the absence of mRNA (puromycin-sensitive binding), which shows that the ribosomal P-site is accessible and can be selected by the initiator tRNA *in vitro* under these conditions (Petersen *et al.* 1976b). Therefore, in Fig. 1, we have shown Step 5 as the combined binding of the pre-initiation complex fMet-$tRNA_f^{Met}$:IF-2 and mRNA to the 30S ribosomal subunit.

A mechanism of selection of the correct initiation codon (in most cases AUG) in the mRNA has been proposed by Shine and Dalgarno (1974). A number of purine nucleotides varying between 3 and 9 precede the initiator codon and are complementary to nucleotides near the 3'-OH terminal of the 16S ribosomal RNA in the 30S subunit. These RNA:RNA interactions have been reviewed by Steitz (1979). In prokaryotic cells, mRNA is usually polycistronic, whereas all known eukaryotic cellular mRNAs are functionally monocistronic. Therefore, different mechanisms of initiation of translation and the regulation of this step might be involved in the 2 cell types (see below).

The final step involves the joining of the 50S ribosomal subunit and the 30S initiation complex. Before this interaction, the initiation factor IF-3 is released, while IF-1 appears to dissociate from the complex during the subunit-subunit interaction (Grunberg-Manago 1979). After the formation of the 70S complex, GTP is hydrolyzed releasing the initiation factor IF-2.

Although the initiation machinery in eukaryotic cells is similar to that just described, the cytoplasm of higher cells contains at least 9 different initiation factors as compared to the 3 in prokaryotes (Hershey *et al.*, this volume). In addition, the eukaryotic ribosome contains larger amounts of both RNA and protein than the 70S particle (Wool 1980).

As shown in Fig. 1, the pre-initiation complex can also bind directly to the 70S ribosome. It has been suggested that this interaction has physiological significance in connection with the role of the formyl group of fMet-$tRNA_f^{Met}$ where initiation takes place at internal initiation sites on polycistronic mRNAs (Petersen *et al.* 1976a, b). Such a situation can exist when a translating ribosome terminates one polypeptide chain without dissociation from the mRNA. It was shown *in vitro* that initiation on 70S ribosomes strictly requires the formylation of Met-$tRNA_f^{Met}$, whereas (as shown in Fig. 1) unformylated Met-$tRNA_f^{Met}$ can bind enzymatically to the 30S ribosomal subunit. Therefore, it could be that formylation is not strictly required for initiation at the first cistron where the mechanism involves ribosomal subunits as described. At initiation of later cistrons, formylation would normally be required. A lack of formylation, however, could be compensated by ribosomal dissociation and the use of 30S subunits for the later initiations. *In vivo* experiments involving the expression of the lactose operon support this hypothesis about the role of the formylation. Inhibition of formylation *in vivo* results in a stronger inhibition of the expression of the 5' distal cistron, thiogalactoside transacetylase, as compared to the 5' proximal β-galactocidase (Petersen *et al.* 1978). However, mutants containing ribosomes which dissociate much more easily than wild type ribosomes did not create this polarity effect upon inhibition of formylation of Met-$tRNA_f^{Met}$. Moreover, as eukaryotic cells have no polycistronic mRNAs, they would not require formylation of the initiator tRNA methionine. In fact, as mentioned, no formylation is found in the eukaryotic cytoplasmic initiation machinery.

Recent data, which support the idea that re-initiation on polycistronic mRNAs requires translation of the previous cistron (and, thus, could involve non-dissociated 70S ribosomes moving through the varying intercistronic region), is discussed later in this volume by Nomura *et al.*

In conclusion, there are at present several possible roles envisaged for formylation in the specific function of fMet-$tRNA_f^{Met}$ in prokaryotic chain initiation. Intensive research is being carried out in many laboratories, including our own, to further our understanding of this particular difference between the mechanism of initiation of prokaryotic and eukaryotic protein biosynthesis.

On pp. 41–54 we discuss some recent data on structural aspects of interactions between the prokaryotic initiator tRNA and the various proteins involved in the initiation steps, including its binding to the 70S ribosome.

ACKNOWLEDGEMENTS

The authors acknowledge financial support from the Danish Natural Science Research Council, the Niels Bohr Fellowship Fund (H.U.P.), and the Carlsberg Foundation (F.P.W.). We thank Nick Coppard for critical reading of the manuscript and Lisbeth Heilesen for help with its preparation.

REFERENCES

Godefroy-Colburn, Th., Wolf, A. D., Dondon, J., Grunberg-Manago, M., Dessen, P. & Pantaloni, D. (1975) *J. Mol. Biol. 94*, 461–478.

Grunberg-Manago, M. In: *Ribosomes, Structure, Function and Genetics* (G. Chambliss, B. R. Craven, J. Davis, K. Davis, L. Kahan & M. Nomura eds) University Park Press, Baltimore, 1980, pp. 445–477.

Kahn, D., Fromant, M., Fayat, G., Dessen, P. & Blanquet, S. (1980) *Eur. J. Biochem. 105*, 489–497.

Petersen, H. U., Danchin, A. & Grunberg-Manago, M. (1976a) *Biochemistry 15*, 1357–1362.

Petersen, H. U., Danchin, A. & Grunberg-Manago, M. (1976b) *Biochemistry 15*, 1362–1369.

Petersen, H. U., Joseph, E., Ullmann, A. & Danchin, A. (1978) *J. Bacteriology 135*, 453–459.

Petersen, H. U., Röll, T., Grunberg-Manago, M. & Clark, B. F. C. (1979) *Biochem. Biophys. Res. Commun. 91*, 1068–1074.

Shine, J. & Dalgarno, L. (1974) *Proc. Natl. Acad. Sci. 71*, 1342–1346.

Steitz, J. A. (1980) In: *Ribosomes. Structure, Function and Genetics* (G. Chambliss, B. R. Craven, J. Davis, K. Davis, L. Kahan & M. Nomura eds) University Park Press, Baltimore, pp. 479–495.

Wool, I. G. (1980) In: *Ribosomes. Structure, Function and Genetics* (G. Chambliss, B. R. Craven, J. Davis, K. Davis, L. Kahan & M. Nomura eds), University Park Press, Baltimore, pp. 797–824.

The Organization and Expression of Genes for Translational Initiation Factors

M. Grunberg-Manago, J. A. Plumbridge, C. Sacerdot, J. G. Howe, M. Springer, J. W. B. Hershey, J. F. Mayaux, G. Fayat & S. Blanquet

INTRODUCTION

A crucial step in understanding the dynamics of cell growth is the deciphering of the regulatory mechanisms for the genetic expression of the translational components. Co-ordination of the expression of these genes is likely essential for maintaining balanced cell growth. The structure of some operons encoding proteins and RNA involved in transcription and translation have been determined. The genes encoding elongation factor G (*fusA*) and one of the genes for EF-Tu (*tufA*) are located within an operon containing genes for the ribosomal proteins S7 and S12 (*rpsG* and *rpsL*). Interrelationships of this type can possibly explain the co-ordination of synthesis of elongation factors with ribosomal proteins at different cell growth rates (Nomura *et al.* 1982). On the other hand, the second gene for EF-Tu (*tufB*) is clustered with the genes for RNA polymerase subunits β and β^1 and for some tRNA molecules, indicating further levels of complexity.

The levels of initiation factors IF1, IF2 and IF3 in exponentially growing cells are co-ordinated with those of elongation factors and ribosomal proteins (Howe & Hershey 1983). However, unlike elongation factors, the genes for initiation factors are probably not located in operons whose expression is regulated at the translational level through feed-back inhibition by key ribosomal proteins. Moreover, the genes for the initiation factors are not clustered together on the *E.coli* chromosome but are grouped with genes encoding other translation or

Institut de Biologie Physico-Chimique, 13, rue Pierre et Marie Curie. 75005 Paris, and Ecole Polytechnique, 92120 Palaiseau, France

GENE EXPRESSION, Alfred Benzon Symposium 19.
Editors: Brian F. C. Clark & Hans Uffe Petersen, Munksgaard, Copenhagen 1984.

transcriptional components, i.e., *infC* the gene for IF3 is grouped with the genes for threonyl-tRNA synthetase (ThrRS) and phenylalanyl-tRNA synthetase (PheRS), and *infB* the gene for IF2 is grouped with those for ribosomal protein S15, polynucleotide phosphorylase and protein NusA which is involved in transcriptional termination.

Like most aminoacyl-tRNA synthetases, initiation factors are subject to metabolic control, while ThrRS and PheRS are also derepressed upon aminoacid starvation, permanently for PheRS or transitorily for ThrRS (Neidhardt *et al.* 1975). Less is known about the regulation of expression of the genes near *infB* but they also appear subject to multilevel control. Detailed analysis of the primary structure of such complex operons may lead to an understanding of how co-ordination of gene expression occurs in some situations and not in others. The organization of two gene clusters on the *E.coli* chromosome, one for IF3, the other for IF2, will be reported in this article. The locus of the gene for IF1 is currently unknown.

RESULTS

Gene organization of the IF3 cluster

A λ transducing phage (λp2) was isolated which complements an *E.coli* mutant producing a thermolabile IF3 (Springer *et al.* 1977). The isolation of this phage permitted the localisation of *infC,* the gene for IF3, at 38 min on the *E.coli* chromosome. This phage also carries four other genes and their promoters. The order of the genes is *thrs* for threonyl-tRNA synthetase, a $\alpha 2$ type dimeric enzyme; *infC* for IF3; *pdzA* for a protein of unknown biological function whose molecular weight is about 14,500 (previously called "P12") (Springer *et al.* 1977); *pheS* and *pheT* corresponding, respectively, to the small and large subunit of phenylalanyl-tRNA synthetase (tetrameric enzyme $\alpha 2 \beta 2$) (Springer *et al.* 1979). A 10.5 kb *E.coli* DNA fragment from λp2 was cloned in pBR322 giving plasmid pB1, which expresses these 5 genes (Plumbridge *et al.* 1980; Fayat *et al.* 1983). A restriction map of pB1 showing the localisation of the genes is given in Figure 1. All these genes are transcribed in the same direction, anticlockwise from *thrS* to *pheT* (Plumbridge & Springer 1980, Mayaux *et al.* 1983).

In order to determine the transcriptional units, different techniques have been applied, including:

1) Cloning of selected restriction fragments into different bacteriophages and plasmid vectors under conditions which reveal promoter activities.

2) DNA sequencing of most of the *E.coli* insert of pB1 DNA and identification of putative promoter and terminator sequences.

3) Mapping the *in vitro* transcription products and sequencing their 5′ ends.

1. The first method has shown that the 4 genes can be expressed on three independent transcription units, 1) *thrS,* 2) *infC,* 3) *pheS* and *pheT.* Although *thrS* and *infC* are close together and transcribed in the same direction, a promoter capable of expressing *infC in vivo* was located downstream of the *Sac*II$_1$ site (Springer *et al.* 1982a) (Fig. 1) which is within the *thrS* structural gene (Mayaux *et al.* 1983). Thus, *thrS* and *infC* can be expressed from independent promoters although the relative contributions to infC expression from the *thrS* and *infC* promoters is still not defined. The fact that *pheS* and *pheT* can be transcribed from the same promoter was shown by cloning various restriction fragments of pB1 in front of the structural genes for tetracycline resistance lacking their own promoter. No individual promoter for *pheT* was detected. Elimination of the *pheS* promoter eliminates *pheT* expression as well (Plumbridge & Springer 1980). The *pheS,T* operon can be expressed independently of the *infC* promoter since DNA downstream of the *Sac*II$_2$ site (within the *infC* structural gene) carries a promoter permitting expression of *pheS* (Springer *et al.* 1982a).

2. The DNA nucleotide sequence of a restriction fragment *Cla*I-*Bgl*I of about 5.7 kb in plasmid pB1 has been established, and the structural genes for *thrS, infC, pheS* and *pheT* have been identified by comparing the N and C terminal aminoacid sequences of ThrRS, α and *β* PheRS and IF3 with aminoacid

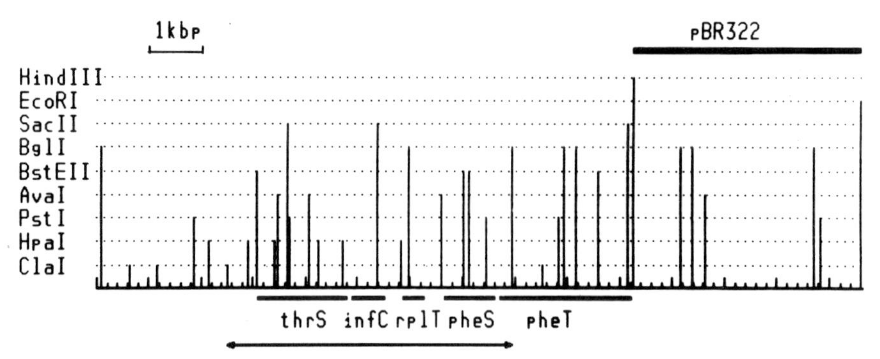

Fig. 1. Restriction enzyme cleavage sites of plasmid pB1 DNA.
pB1 carries a 10.5 kb *Eco*RI$_2$-*Hind*III fragment of *E.coli* DNA derived from the λ transducing phage λ p2. The fragment sequenced by the method of Maxam and Gilbert (1980) is indicated (←→).

sequences predicted by the DNA sequence (Sacerdot *et al.* 1982, Springer *et al.* 1982b, Mayaux *et al.* 1983). The location of these genes is consistent with the genetic data. In addition, an open reading frame which starts 350 nucleotides downstream from the end of *infC* and accounts for a protein of a molecular weight 14,543 was observed (Fayat *et al.* 1983). This initiator codon of this open reading frame is preceded by a sequence characteristic of a very strong translational initiation signal. This protein is probably the "P12" of Springer *et*

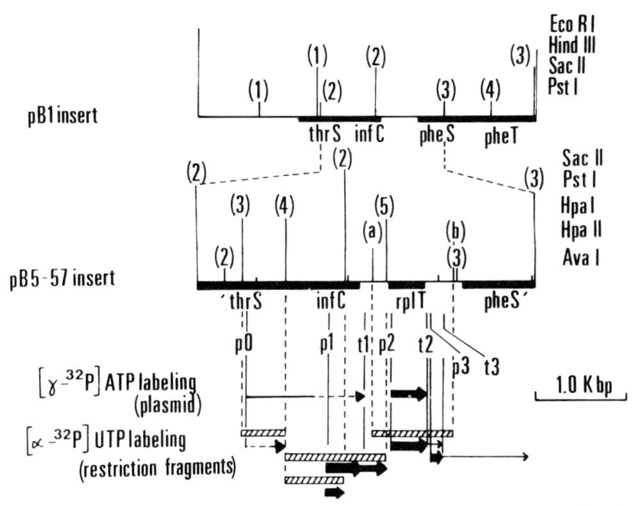

Fig. 2. Restriction maps of the pB1 chromosomal insert and of the pB5-57 plasmid with the location of the transcription signals and products.

For the transcription experiments using intact plasmid pB5-57 as a template, the origins of the transcription products were identified by 5′ end sequencing of (γ-^{32}P)ATP-labelled RNAs. Two transcripts from *E.coli* DNA with lengths of 700 and 400 nucleotides indicated on the figure by horizontal arrows (with thickness proportional to their relative synthesis) were characterized. Two start sites of transcription initiating with ATP were identified: p0 and p2. For the experiments using restriction fragments: *Hpa*I$_3$-*Hpa*I$_4$-HpaI$_4$-*Hpa*I$_5$ and *Hpa*I$_4$-*Sac*II$_2$+*Sac*II$_2$-*Hpa*I$_5$ (located on the figure by hatched bars, the subfragment *Sac*II$_2$-*Hpa*I$_5$ is not represented (α-^{32}P)UTP-labelled products were mapped. The mapped RNA products are symbolized by arrows. The experiments with the *Hpa*I$_4$-*Hpa*I$_5$ and *Hpa*I$_4$-*Sac*II$_2$ templates identify the p1 promoter within *infC*. Using *Hpa*II fragment (not numbered but marked a and b since there are numerous *Hpa*II sites) as template two main transcripts, the 400 mRNA and 150 mRNA, were identified. Two other RNA products were synthesized at a lower level, 235 mRNA and 600 mRNA. Using (γ^{32}P)-labelled RNA the sequence of the 5′ ends of 400 mRNA and 600 mRNA identified their start at the p2 site. The 5′ and 3′ ends of the 150 mRNA and 235 mRNA were identified by S1 nuclease protection studies using two different DNA probes labelled at the 5′ or 3′ ends with ^{32}P (for details see Mayaux *et al.*, in press). The two transcripts start at the site marked p3. Also shown on the figure are the transcription terminators (t1 and t2) after *infC* and *pdzA*, respectively, and the transcription terminator t3 involved in the control of the expression of the *pheS-pheT* operon.

al. (1979) which was located between *infC* and *pheS*. This P12 protein is strongly expressed in UV-irradiated cells infected by λp2 (Springer *et al.* 1977), and in a cell-free transcription-translation system programmed by pB1 (Lestienne *et al.* 1982).

The structural part of *thrS* is found upstream and on the same DNA coding strand as that of *infC* (Mayaux *et al.* in press). There are only 3 nucleotides between the stop codon of *thrS* and the atypical AUU initiator codon of the *infC* structural gene. The ribosome binding site of *infC* (Shine & Dalgarno sequence, Shine & Dalgarno 1974) is located within the *thrS* structural gene. The only potential transcriptional terminator structure is 55 bp downstream from the *infC* coding sequence. This implies that *thrS* and *infC* are co-transcribed.

The genetic data discussed above indicates that *infC* has its own promoter. The DNA sequence between *Ava*I_2 (internal to *thrS* structural gene) and the start of *infC* was, therefore, examined for the presence of sequences homologous to the prototype sequence of bacterial promoters (Rosenberg & Court 1979). The analysis revealed two sequences exhibiting more than 80% homology; the first one, p0, covers that *Hpa*I_3 restriction site (Fig. 2), the second, p0′, is found 50 bases downstream from the *Hpa*I_4 site (Fayat *et al.* in press) (Table I).

3. To assess the activity of promoter sites in this DNA region a third method, *in vitro* transcription, was applied by using as a template an intact plasmid pB5-57 obtained by subcloning the *Pst*I_2-*Pst*I_3 fragment of pB1 in pBR322. Two origins of transcription could be identified on the *E.coli* insert of the plasmid by S1 mapping and/or sequencing the 5′ ends of the transcripts. One RNA, 400 nucleotides long (representing 30% of the synthesized RNA) was recognized as a

Table I

Identified promoter sequences within the E.coli *insert of pB1*

	−35 binding regions	Pribnow's box
consensus	gTTGaca	TAtaaTPu
p0	GTTAACG	TATGATA
pl	AACGCCG	TATAATG
P2	GTTAACG	TAGAATA
P3	ATTGACT	TTCAATA

The consensus sequence is taken from Rosenberg and Court (1979). Capital letters denote greater than 75% and small letters greater than 50% agreement among promoter sequences. Pu indicates a purine.

transcript originating from the promoter p2, located in front of the *pdzA* gene downstream from *infC* (Fig. 2) (as discussed later). The second RNA (15%) was 700 nucleotides long and the 5′ terminal sequence located the start site for this transcript 33 bp downstream from *Hpa*I₃, i.e., consistent with the position of the putative p0 promoter within the structural gene for *thrS*. No promoter other than p0 could be identified with intact pB5-57 using γ^{32}P-labelled ATP or GTP. With α^{32}P UTP several RNA products longer than 1000 bp were obtained but could not be resolved by polyacrylamide gel electrophoresis. Using shorter restriction fragments as templates, no RNA was found to be expressed from p0′. This suggests that the putative p0′ site is probably not functional at least *in vitro*. This work confirms the close linkage of *thrS* and *infC* genes on the *E.coli* chromosome and the existence of individual promoters predicted by the genetic studies of *infC* (Springer *et al.* 1982).

Examination of *in vitro* transcription products from the *Hpa*I₄-*Hpa*I₅ DNA fragment revealed a major product of about 640 nucleotides (Fayat *et al.* in press). This transcript initiates at the p1 site within the *infC* gene and extends at least up to the *Hpa*I₅ extremity of the fragment (Fig. 2). Since this transcript is the major product (80%) it can be inferred that, at least *in vitro*, the terminator structure, t1, located at the end of *infC* does not terminate efficiently. The next possible terminator structure, t2, is located after the *pdzA* gene. This raises the possibility that *pdzA*, the gene immediately downstream from infC, can also be expressed from mRNA synthesized either from the *thrS* or *infC* promoter. It is worthwile noting that transcripts from p1 could not be identified using intact plasmid pB5-57 DNA and λ^{32}P ATP or GTP. This suggests either that it initiates transcription with UTP, or CTP, or that it is less active in the presence of transcription from p0. However, a strong promoter, p2, has been identified downstream from p1 just in front of *pdzA*. From this promoter a 400 bp mRNA is synthesized which (as discussed above) is also observed using pB5-57 as template The 400 bp mRNA ends in the region of transcription terminator t2. The structural part of the *pdzA* gene ends within the G+C rich region of dyad symmetry, composing the t2 terminator (Fayat *et al.* submitted). The *in vitro* transcription experiments revealed additional transcripts, among which was a 600 nucleotide product which was also shown to originate at the p2 promoter, indicating that *in vitro* transcription from p2 could extend to the next terminator t3, located 200 nucleotides downstream from t2.

The promoter, p3, for the third transcriptional unit, the *pheS* and *pheT* operon, has also been identified. It starts 336 bases upstream from the *pheS*

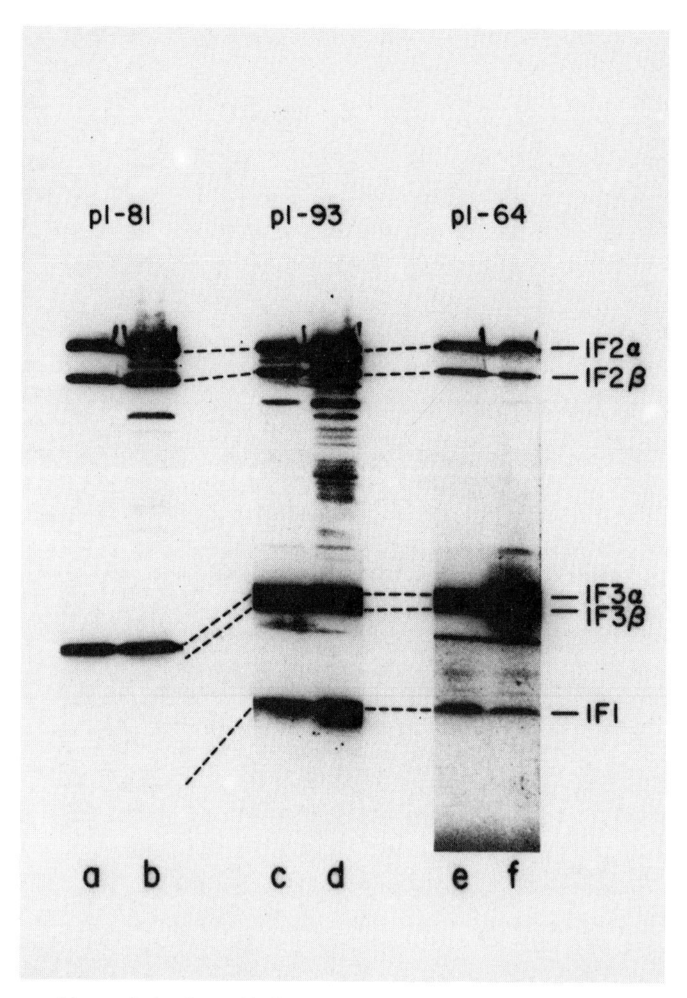

Fig. 3. Immunoblot analysis of cosmid clones.

Cell lysates of cosmid-containing bacteria were electrophoresed on 10–18% gradient SDS polyacrylamide gels blotted onto nitrocellulose paper, and incubated with anti-IF2 and anti-IF3 together (upper two-thirds of paper), or with anti-IF1 alone (the lower third). After subsequent incubation with ^{125}I-labelled *S.aureus* protein A, the nitrocellulose papers were dried and autoradiographed for 6 h. The 3 cosmids causing overproduction of IF (p1-81, p1-93 and p1-64) are analyzed as indicated in lanes b, d and f, respectively. Lanes a, c, e, are random adjacent cosmid clones typical of non-overproducers. Locations of IF are shown in the margin (from Plumbridge *et al.* 1982).

structural gene. P3 is the most proximal promoter to the *pheS* gene responsible for the expression of the *pheS* gene. The t3 terminator site is situated at the end of the attenuator controlled by the level of Phe-tRNA (Springer *et al.* 1982b). A mRNA transcript of 150 nucleotides is expressed from p3 and stops at the t3 site, but this RNA can elongate into the structural part of the *pheS-pheT* operon since longer transcripts are also observed (Fayat *et al.* 1983).

In conclusion, the *thrS, infC, pdzA, pheS* and *pheT* genes encoding ThrRS, IF3, P12, and the small and large subunits of PheRS, respectively, are located in a cluster at 38 min on the *E.coli* chromosome. In addition to the putative *thrS* promoter, 4 promoters have been identified in this region: p0 located in the structural gene of threonyl-tRNA synthetase, p1 located in the structural gene of IF3, p2 located in front of the P12 structural gene and p3, the promoter of the

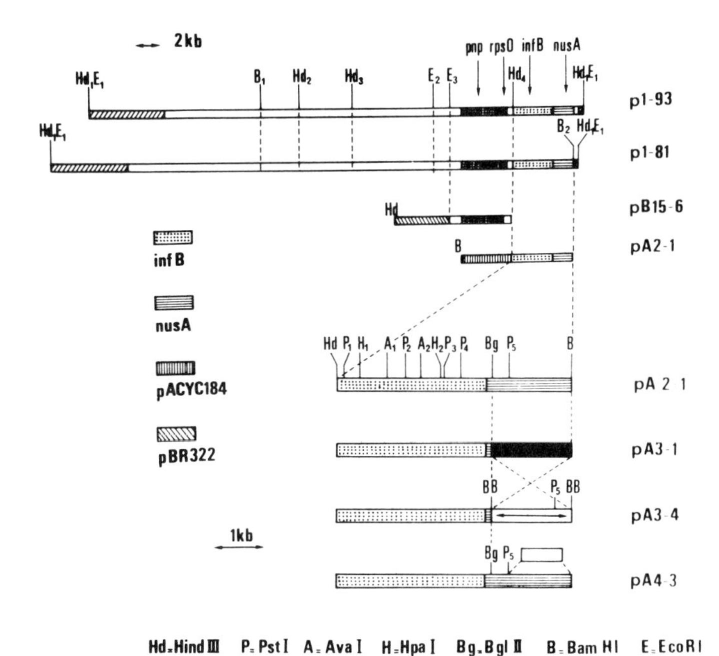

Fig. 4. Structures and maxicell analysis of the cosmids p1-81 and 1-93 and major subclones.
a) The positions of the identified genes are indicated. DNA of the various vectors used is indicated by rightward hatching, pHC79; leftward hatching, pBR322; vertical hatching, pACYC184, p1-81* is a derivative of p1-81 carrying a spontaneous insertion located between *Hind*III$_4$ and *Bam*HI$_2$. The *Hind*III$_4$-*Bam*HI$_2$ fragment of p1-81* clones in pACYC184 gives pA4-3. Restriction sites are numbered sequentially within pA2-1. *Ava*I$_2$ is a *Sma*I site. BB in pA4-3 indicates the hybrid *Bam*HI-*Bgl*II sites formed by deletion of the *Bgl*II-*Bam*HI fragment of pA2-1.

pheS-pheT operon far upstream from *pheS.* No terminator structure was found between *thrS* and *infC* which are only separated by 3 bp. Only a weak terminator, t1, was found between *infC* and the next structural gene, *pdzA.* A terminator t2 exists after *pdzA,* and the third terminator t3 was identified as that of the attenuator for the *pheS-pheT* operon.

As the coding sequence of *pdzA* ends within the t2 terminator site an optimal translation rate of *pdzA* could result in termination rclicf at t2. This read-through transcription would extend to the leader region of the *pheS-pheT*

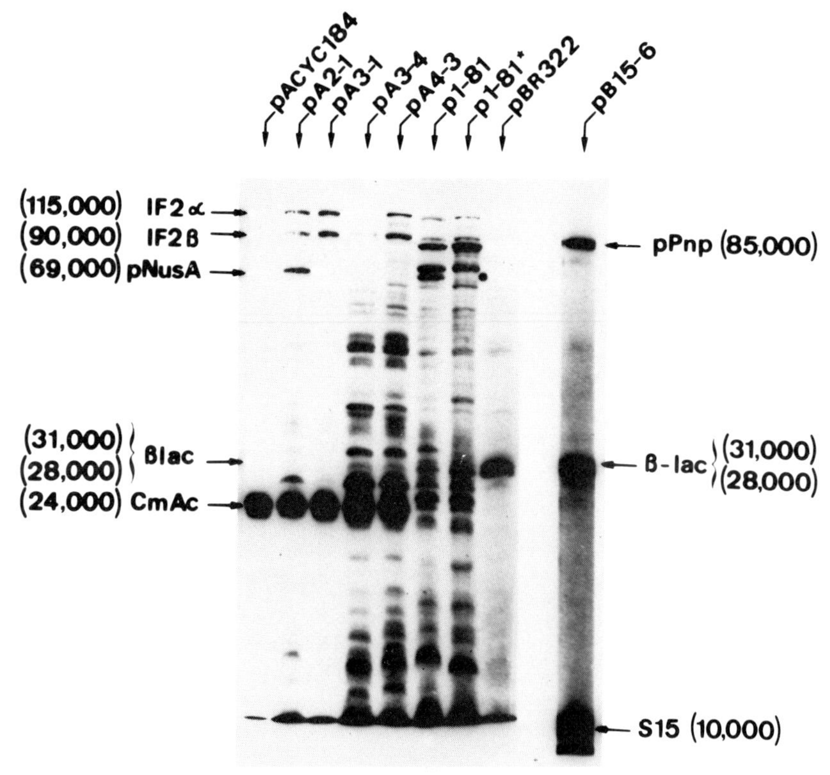

b) Proteins coded by various plasmids were labelled with (^{35}S) methionine (10 μCi/ml) as described (Plumbridge *et al.* 1982). Labelled extracts were analysed on 12.5% polyacrylamide, sodium dodecyl sulphate treated for fluorography and exposed to X-ray film. The equivalent of 0.3 A_{650} units of the exponentially growing cells were loaded onto the gels. Radioactivity loaded on each lane was about 10^5 cpm. Identified proteins are indicated, translational initiation factor IF2α and β (Mr 115,000 and 90,000); pNusA (Mr 69,000); pPnp (polynucleotide phosphorylase, Mr 85,000), and ribosomal protein S15 (Mr 10,000. β-lac=β-lactamase (Mr 31,000 and 28,000 for precursor and secreted forms), and CmAc=chloramphenicol acetylase (24,000) are plasmid vector-encoded proteins.

operon. Consequently, it is possible that mRNAs expressing *pheS-pheT* could originate upstream from *pdzA* at p2, p1, p0 or even from the promoter for *thrS*. In turn, loose translational-transcriptional coupling in the expression of *pdzA* would favor termination at t2, and *pheS-pheT* transcription would have to start from the p3 promoter site. This raises the possibility of an interaction between the expression of *pheS-pheT* and the 3 genes upstream.

In addition, an overlapping of the *infC* and *thrS* translational regulatory signals on a multicistronic mRNA should be capable of co-ordinating gene expression at the translation level, beyond the co-ordination already achieved by cotranscription. In particular, a ribosome translating *thrS* could recognize the weak translational initiation signal, the atypical AUU start codon of *infC*, and ensure equimolar synthesis of the *thrS* and *infC* products. Apparently, this is not the only mode of IF3 synthesis, since in haploid strains, IF3 is present at a concentration at least 3 to 4-fold over that of threonyl-tRNA synthetase (Lestienne *et al.* 1982, Howe & Hershey 1983).

Gene organization of the IF2 gene cluster

The gene for initiation factor IF2 (*infB*) was isolated from a cosmid library of *E. coli* DNA because the presence of the *infB* gene on a multicopy cosmid increased the cellular concentration of IF2. This increase was detectable using a sensitive, quantitative immunoblotting technique previously developed to measure changes in the levels of initiation factors in crude cell lysates (Howe & Hershey 1981, 1982). Using this method to analyse lysates of the cosmid library, clones carrying the genes for IF2 and IF3 were identified. Two cosmids, p1-81 and p1-93, were shown to have relatively increased amounts of IF2 and one cosmid, p1-64, to have enhanced IF3 levels (Fig. 3). Initiation factor IF2 is present in two forms in the cell, IF2α (M.W. 115,000) and IF2β (90,000). Both forms are overproduced. The gene products of p1-81 and p1-93 were analysed in the maxicell system. Immune precipitation of the labelled material with antiserum specific for IF2 showed IF2α, IF2β and a third band probably corresponding to an *in vitro* degradation product (Plumbridge *et al.* 1982).

Subcloning fragments in plasmids from the original cosmid located the *infB* gene to a 4.8 kb *Hind*III-*Bam*HI fragment of p1-81[1] (Fig. 4). This fragment was inserted into an integration deficient recombinant λ phage which can only lysogenize by homology. Mapping the point of lysogenization in the *E.coli* chromosome, *infβ* was located at 68 min, quite close to *argG, nusA, rpsO,* and *pnp* (genes coding, respectively, for argino succinate synthetase, a protein involved in transcription termination), ribosomal protein S15 and polynucleotide phosphorylase (Plumbridge *et al.* 1982). The gene *argG* is not carried by either cosmid, whereas the other three proteins could be identified. Maxicell analysis revealed proteins of the correct size encoded by cosmids and some subclones. The presence of *nusA* was confirmed by complementation of the $nusA_1$ mutation, polynucleotide phosphorylase by an activity test and S15 by running the maxicell-labelled extract on a 2D ribosomal protein gel. IF2α and *β* and pNusA are expressed from the same 4.8 kb fragment (plasmid pA2-1), and polynucleotide phosphorylase and S15 from the adjacent 5.2 kb $EcoRI_3$-$HindIII_4$ fragment (plasmid pB15-6) (Fig. 4b). The order of *rpsO* and *pnp* within the *EcoRI-Hind*III fragment has already been established by Portier (1982, Portier *et al.* 1981). The orientation of *infB* and *nusA* was established by deletion mapping (pA3-1; pA3-4) and isolation of a spontaneous insertion mutant which eliminated *nusA* expression (p1-81*; pA4-3) (Fig. 4). Thus, the order of the genes on the *E.coli* chromosomes is *argG, nusA, infB, rpsO, pnp* (Plumbridge & Springer 1983). The 3.1 kb *Hind*III-*Bgl*II fragment of pA3-1 is just of sufficient size to code for a protein of M.W. 115,000 but both IF2α and *β* are expressed. Thus, IF2α and IF2*β* are coded by a single gene.

Recently Kawihara and Nakamura (1983) cloned an *Eco*RI restriction fragment of *E.coli* DNA which can complement an *argG* mutation and which also complements the $nusA_1$ mutation. They showed that transcription from *argG* to *pnp* occurs anticlockwise in the chromosome. The results of these authors are consistent with ours. In addition, they localized a gene between *argG* and *nusA,* coding for a protein of 21 K and a gene between *infB* and *rpsO* coding for a protein of 15 K. By sequencing parts of this region of the chromosome, Imamoto's group found an open-reading frame corresponding to a protein of 15 K just upstream of *nusA*. The identity of these small proteins is not known, nor is their relevance to the expression of their larger neighbours. In addition, Imamoto's group found a gene coding for a minor form of initiator fMet-$tRNA^{fMet}$ (*metY*) between *argG* and the 15 K protein (personal communication). The order of the following genes is certain: *argG, metY,* 15 K, *nusA, infB, rpsO, pnp,* with possibly one or two other small genes interspaced. The direction of transcription for all the genes analysed so far is anticlockwise from *argG* to *pnp* on the *E.coli* chromosome. For *nusA* and *infB* this was determined by

[1] The *Bam*HI site of p1-81 does not exist on the *E.coli* chromosome and is a reconstructed site as a result of the *Sau3A* cloning into a *Bam*HI site.

hybridization of pulse-labelled mRNA with the separated single strands of a λ phage carrying the two genes (Plumbridge & Springer 1983), for *metY, p15K* and *nusA* (Imamoto, personal communication) and *pnp* and *rpsO* (Portier, in preparation) by DNA sequence.

We are currently determining the nucleotide sequence of the IF2 gene. The DNA coding for the N-terminal extremity of IF2α was identified by comparison with the N-terminal of the aminoacid sequence of IF2α (unpublished results). The initiator codon is AUG and 5 bases upstream from this initiation codon, a Shine and Dalgarno (1974) sequence (AAGGA) is observed (Fig. 5). Upstream from this AUG, separated by 21 nucleotides there is an open-reading frame which can be identified as the end of *nusA* by comparison with the sequence of *nusA* established by Imamoto's group (personal communication).

The transducing phage carrying the 4.8 kb *Hind*III$_4$-*Bam*HI$_2$ fragment expresses 3 high M.W. proteins IF2α, IF2β and pNusA under lysogenic conditions (in a U.V. irradiated cell system) (Fig. 6). Promoters capable of expressing *infB* and *nusA,* therefore, exist on this 4.8 kb fragment. An indication

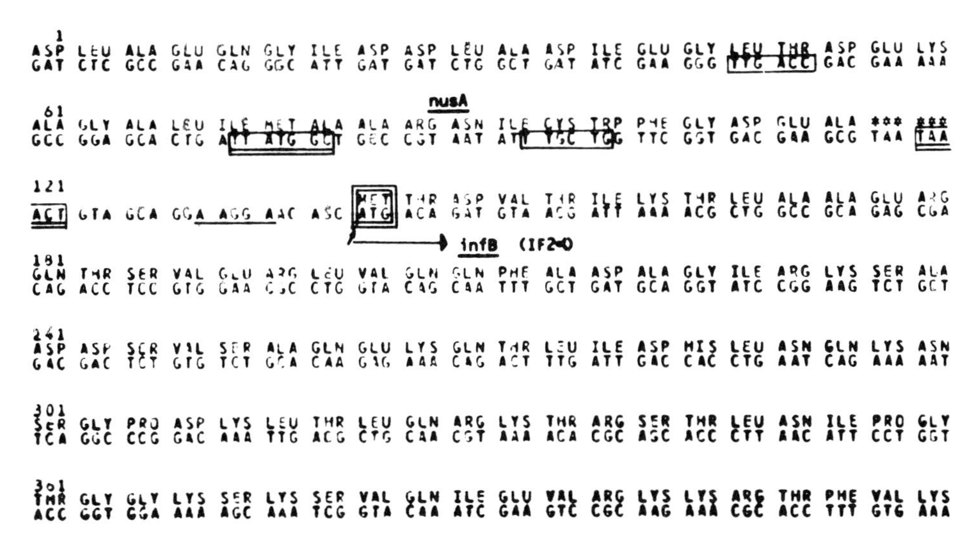

Fig. 5. The nucleotide sequence of the region containing the 3′ end of nusA and the 5′ end of infB. The sequence was determined by the method of Maxam and Gilbert (1980). The initiator codon, as shown by comparison with the IF2α protein N-terminal sequence, is indicated in a double box and the corresponding Shine and Dalgarno sequence underlined. Two possible Pribnow boxes (underlined boxes), and their corresponding – 35 regions (boxed), within the end of the *nusA* coding sequence are indicated.

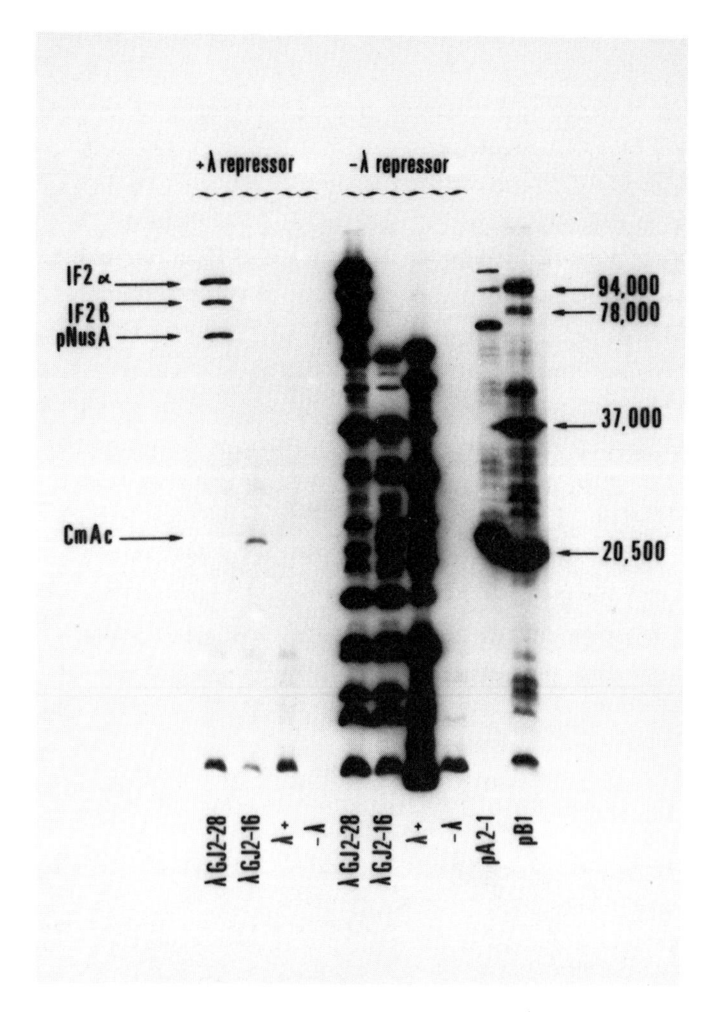

Fig. 6. Infection of UV-irradiated cells with λGJ2-28, λGJ2-16 and λ⁺ phages.

Strains 159 and 159 (λind⁻) were grown and irradiated as described in Springer *et al.* (1977). A m.o.i. of 10 was used and the cells transferred to glucose minimal medium. Newly synthesized proteins were labelled with 15 μCi/ml (^{35}S) methionine, for 30 min. Radioactive incorporation was stopped with excess unlabelled methionine and the cells were harvested by centrifugation. Cells were lysed by boiling in sodium dodecyl sulphate sample buffer and samples equivalent to 0.1 A_{650} unit of bacteria analysed by electrophoresis on 10% sodium dodecyl sulphate polyacrylamide gels The gel was stained with Coomassie blue and then treated for fluorography. The molecular weight of characterized gene products is indicated. λGJ2-28 carries the 4.8 kb *Hind*III-*Bam*HI fragment of p1-81, λGJ2-16 carries the *Hind*III-*Bam*HI fragment of pACYC184.

that this fragment carries independent promoters for *infB* and *nusA* was suggested by experiments cloning different *Pst* fragments in a plasmid which has a unique *Pst*I site situated immediately before the structural gene coding for resistance to tetracycline and lacking a functional promoter. A promoter activity was detected in the *Pst*I fragment covering the end of *nusA* and the beginning of *infB*. Furthermore, the analysis of our DNA sequence shows the presence of 2 possible promoters for the *infB* gene located inside the end of the *nusA* gene (Fig. 5). However, no sequence homologous to the consensus sequence for a rho-independent terminator was observed, neither in the small intergenic region between *nusA* and *infB*, nor inside the beginning of the *infB* gene. And moreover, several observations suggest that read-through from *nusA* into *infB* is possible. A foreign promoter (e.g., the plasmid tetracyclineresistant promoter) can enhance the expression of *nusA* and *infB in vivo*. This overproduction is eliminated by the presence of an insertion in *nusA*. Another observation which could be interpreted as suggesting that read-through transcription from the *nusA* promoter is important in *infB* expression is the considerable decrease in IF2 synthesis in maxicells carrying the plasmid pA3-4, as compared with pA2-1 (Fig. 4). In this plasmid the inversion of the *Bgl*II-*Bam*HI fragment of pA_2-1, containing most of the *nusA* gene and presumably its promoter, would eliminate any *nusA* promoter contribution to *infB* expression. This would argue in favor of a majority of transcription starting from *nusA* and continuing into *infB*.

To summarize, it has been shown: (1) that the closely linked genes *nusA* and *infB* are carried on a 4.8 kb DNA fragment; (2) that they are separated by only 21 bp; and (3) that the same fragment carries one or more promoters capable of bringing about the synthesis of IF2α-IF2*β* and pNusA. The sequence data, as well as several other observations, suggest a polycistronic expression of *nusA* and *infB*. This eliminates neither the possibility of an independent, maybe weak, expression of IF2, nor that *nusA* could also be expressed polycistronically from promoters situated in front of upstream genes, protein p15K and *metY* (Imamoto, personal communication).

CONCLUSION

The gene organization of the IF3 and IF2 clusters appears to be complex and to have many general common characteristics. Both initiation factors are located contiguously to other genes involved in translation/transcription, and an extremely tight linkage between IF3 and Thr-tRNA synthetase has been established in the IF3 cluster as well as between *nusA* and IF2 in the IF2 cluster.

In addition, both clusters carry genes for small proteins of unknown function whose transcription seems to be tightly coupled with that of the initiation factors, *nusA* and *thrS*. Co-transcription of aminoacyl-tRNA synthetase and IF3, or of *nusA* and IF2, couples their regulation and might explain a co-ordinated response to changes in growth rate. However the absolute levels of the proteins in the cell expressed by the various genes are different. Moreover, the regulation of the levels of aminoacyl-tRNA synthetase and initiation factors was shown to be unco-ordinated under a number of conditions: IF3 shows a stringent response (Lestienne *et al.* 1982) while ThrRS and PheRS are derepressed by depletion of the aminoacids they activate (Neidhardt *et al.* 1975). Beside the major promoter responsible for polycistronic transcription, promoters internal to the operon may be needed to accomplish unco-ordinated regulation of operon genes. The efficiency of translation of each cistron may depend on mRNA structure and length, i.e., on what promoter is used to transcribe the gene. Regulation may also be achieved post-transcriptionally. Different rates of translational elongation might affect the read-through of transcriptional terminators and an attenuation mechanism has been demonstrated for the synthesis of phenylalanyl-tRNA synthetase (Springer *et al.* 1982b, Fayat *et al.* 1983, Springer *et al.* 1983).

It is now important to know why these genes are organized together as an operon, and what the situation is *in vivo* under normal conditions, during changes of environmental conditions, or in the presence of mutations. We are currently using S1 mapping to define *in vivo* transcripts and to determine the extent of coupling between the different genes. It has been suggested that non-translating ribosomes prevent further biosynthesis and accumulation of ribosome particles (Nomura *et al.* 1982). Initiation factors play an important role in determining ribosomal activity. It is not unreasonable to speculate that the tight linkage between initiation factors, *nusA* and aminoacyl-tRNA synthetases is necessary to couple their synthesis with that of ribosomes.

ACKNOWLEDGEMENTS

This work was supported by the following grants: Centre National de la Recherche Scientifique (G.R. n° 18 and A.T.P. “Organisation et Expression du Génome”), Délégation Générale à la Recherche Scientifique et Technique (Convention 80.E.0872), Institut National de la Santé et de la Recherche Médicale (Contrat Libre n° 823008), Fondation pour la Recherche Médicale and Dupont de Nemours.

REFERENCES

Fayat, G., Sacerdot, C., Mayaux, J. F., Fromant, M., Dessen, P., Springer, M., Grunberg-Manago, M. & Blanquet S., *J. Mol. Biol.* (in press).

Howe, J. G. & Hershey, J. W. B. (1981) *J. Biol. Chem. 250,* 12836–12839.

Howe, J. G. & Hershey, J. W. B. (1983) *J. Biol. Chem. 258,* 1954–1959.

Kurihara, T. & Nakamura, Y. (1983) *Molec. Gen. Genet. 190,* 189–195.

Lestienne, P., Dondon, J., Plumbridge, J. A., Howe, J. G., Mayaux, J. F., Springer, M., Blanquet, S., Hershey, J. W. B. & Grunberg-Manago, M. (1982) *Eur. J. Biochem. 123,* 483–488.

Maxam, A. M. & Gilbert, W. (1980) *Methods Enzymol. 65,* 499–560.

Mayaux, J. F., Fayat, G., Fromant, M., Springer, M., Grumberg-Manago, M. & Blanquet, S., *Proc. Natl. Acad. Sci. USA* (in press).

Mayaux, J. F., Fayaut, G., Springer, M., Grunberg-Manago, M. & Blanquet, S. *J. Mol. Biol.* (in press).

Neidhardt, F. C., Parker, J. & Mc Keever, W. (1975) *Ann. Rev. Microbiol. 29,* 215–250.

Nomura, M., Jinks-Robertson, S. & Miura, A. (1982) *Developments in Biochemistry 24,* 91–104.

Nomura, M., Jinks-Robertson, S. & Miura, A. In: *Interaction of translational and transcriptional controls in the regulation of gene expression.* pp. 91–104. Elsevier Biomedical, New-York.

Plumbridge, J. A. & Springer M. (1980) *J. Mol. Biol. 144,* 595–600.

Plumbridge, J. A., Springer, M., Graffe, M., Goursot, R. & Grunberg-Manago, M. (1980) *Gene 11,* 33–42.

Plumbridge, J. A., Howe, J. G., Springer, M., Touati-Schwartz, D., Hershey, J. W. B. & Grunberg-Manago, M. (1982) *Proc. Natl. Acad. Sci. USA 79,* 5033–5037.

Plumbridge, J. A. & Springer M. (1983) *J. Mol. Biol. 167,* 227–243.

Portier, C., Migot, C. & Grunberg-Manago, M. (1981) *Molec Gen. Genet. 183.* 298–305.

Portier, C. (1982) *Gene 18,* 261–265.

Rosenberg, M. & Court, D. (1979) *Ann. Rev. Genet. 13,* 319–353.

Sacerdot, C., Fayat, G., Dessen, P., Springer, M., Plumbridge, J. A., Grunberg-Manago, M. & Blanquet, S. (1982) *EMBO J. 1,* 311–315.

Shine, J. & Dalgarno, L. (1974) *Proc. Natl. Acad. Sci. USA 71,* 1342–1346.

Springer, M., Graffe, M. & Hennecke, H. (1977) *Proc. Natl. Acad. Sci. USA 74,* 3970–3974.

Springer, M., Graffe, M. & Grunberg-Manago, M. (1979) *Molec. Gen. Genet. 169,* 337–343.

Springer, M., Plumbridge, J. A., Trudel, M., Graffe, M. & Grunberg-Manago, M. (1982a) *Molec. Gen. Genet. 186,* 247–252.

Springer, M., Plumbridge, J. A., Trudel, M., Grunberg-Manago, M., Fayat, G., Mayaux, J. F., Sacerdot, C., Dessen, P., Fromant, P. & Blanquet, S. (1982b) In: *Interaction of translational and transcriptional controls in the regulation of gene expression,* pp. 25–47. Elsevier Biomedical, New-York.

DISCUSSION

SCHIMMEL: As far as the organization is concerned I am struck by the fact that IF3 and threonyl-tRNA synthetase genes are in frame with each other and wonder if in fact an ancestral protein might be composed by both of these polypeptide sequences as one continuous polypeptide. The reason for the interest in this is related to the size diversity of aminoacyl-tRNA synthetases, and the question which arises, is whether a lot of this heterogeneity of length is due to other functional domains being associated with aa-tRNA synthetases, other than just the core enzyme which carries out charging of tRNA. It is more an observation than a question, but it does raise the question as to whether in fact IF3 could have been at one time part of the same polypeptide as threonyl-tRNA synthetase and whether in fact one could use a suppressor. One could make a hybrid protein and see whether both activities would be contained in such a protein?

GRUNBERG-MANAGO: Yes, it is one of the experiments we are planning to do. In fact, we have made a hybrid protein by deleting DNA restriction fragments in the cluster containing the infC gene. By Cloning 2 $SacII_1$-$SacII_2$ fragments of pH_1 (Fig. 1) in tandem in a λ phage, we obtained a recombinant phage which expresses, in UV-irradiated lysogenic cells, a 70 kdal protein. This protein is precipitated by both anti-IF3 and anti-ThrRS immune sera. This protein starts in the N-terminal with infC in the left $SacII_1$-$SacII_2$ fragment and ends with *thrS* within the right $SacII_1$-$SacII_2$ fragment at the C-terminal. Several fusions between ThrRS and the small or large subunits of PheRS were also obtained. One plasmid, comprising 2 PstI fragments of pB_1 (Pst_1-Pst_2 Pst_4-Pst_5) joined together, causes a slight overproduction of a 78 kdal protein. This protein is immunoprecipitated by ThrRS and by anti-PheRS (large subunit) immune sera. The 78 kdal protein is a hybrid protein made from the NH_2 extremity of ThrRS and the COOH terminal end of the large subunit of PheRS. The reaction with the specific antisera indicates that the tertiary structure of the different domains is preserved, and it is quite possible that the different domains have conserved some of the activities of the synthetases or initiation factor IF3.

CASKEY: Have you used these probes to look at the organization of these genes and other bacterial strains?

GRUNBERG-MANAGO: Not yet. We are now investigating if the organization is changed in the different mutants: ThrRS, PheRS or IF3 that we have isolated.

ROSENBERG: In the IF3 situation, is there any evidence for translational coupling between the threonyl-tRNA synthetase gene and the IF3 gene? The positioning is such that you might find IF3 translational expressions depending directly on the translational expression of the first gene.

GRUNBERG-MANAGO: We have only very preliminary results. There exist some mutants overproducing ThrRS, the regulation of expression of this enzyme appears at first sight to be at the translational level, we are looking at whether the overproduction of ThrRS affects the synthesis of IF3. We have, vice versa, isolated some mutants which overproduce or underproduce IF3, and we are looking at what are the levels of ThrRS. It is too early to have conclusive data; however, in the preliminary experiments it does not appear that the levels of ThrRS and IF3 are related. In the case of the IF2 cluster, the regulation of nusA gene expression seems to affect the expression of IF2, and the fact that these genes could be transcribed from a unique promotor is probably biologically significant. We are currently using S_1 mapping to define *in vivo* transcripts and to determine the extent of coupling between different genes. In preliminary studies, J. Plumbridge has found that *in vivo* there are transcripts containing both pNusA and IF2 gene copies, indicating that these 2 genes can be transcribed polycistronically.

As far as IF3 is concerned, our *in vitro* studies suggest that IF3 is transcribed either from its own promotor or from ThrS promotor. There is some indication that the situation *in vivo* might not be drastically different from *in vitro*, but this is a very important question, and now the main focus of our work is to find out what is the situation *in vivo*, and why the gene organization is so complicated. One interesting possibility is that the multiple mode of transcription is necessary for controlling the expression because different controls operate at different promotors.

In addition, it is possible that the translation rate could affect the transcriptional termination and may explain the metabolic regulation. The rplT coding sequence ends only 3 nucleotides before the region of dyad symmetry forming the t_2 terminator structure, consequently under normal physiological conditions in which transcription and translation are tightly coupled, a ribosome translating the C terminal extremity of rplT could hinder the formation of the terminator hairpin and stimulate the readthrough at terminator t_2. This readthrough transcription could extend into the leader region of the pheS, T operon.

Interaction between Methionine-accepting tRNAs and Proteins during Initiation of Prokaryotic Translation

Hans Uffe Petersen, Friedrik P. Wikman, Gunhild E. Siboska, Hanne Worm-Leonhard & Brian F. C. Clark

Cellular protein synthesis invokes a number of specific RNA and protein interactions. Although, as yet, little is known about the chemical nature of these interactions, recent developments in fast RNA-sequencing methods will, hopefully, enable researchers to elucidate some of the underlying molecular mechanism involved.

The purpose of this paper is to discuss structural aspects of the interaction between tRNA and protein occurring during the different initiation steps involved in prokaryotic translation. In particular, we have been interested in studying the nature of protein-binding sites on tRNA molecules. The principal reactions in the initiation mechanism are described in pp. 17–21 of this volume. A summary of the RNA:protein complexes involving the methionine accepting $tRNA_f^{Met}$ or $tRNA_m^{Met}$ is shown in Table I, which also indicates the complexes which are further discussed in this chapter. Before discussion of the experimental results, we describe the method in detail.

THE FOOTPRINTING METHOD

The method referred to here as the footprinting method is a combination of several known techniques including radioactive end-labelling of RNA, amino-acylation of tRNA, N-formylation of aa-tRNA, RNA:protein complex formation, gel filtration, partial enzymatic hydrolysis of RNA, high resolution polyacrylamide gel electrophoresis (PAGE) and autoradiography. The foot-

---

Division of Biostructural Chemistry, Department of Chemistry, Aarhus University, 8000 Aarhus C, Denmark

GENE EXPRESSION, Alfred Benzon Symposium 19.
Editors: Brian F. C. Clark & Hans Uffe Petersen, Munksgaard, Copenhagen 1984.

Table I

Complexes involving methionine-accepting tRNAs and proteins in the prokaryotic translation

	MetRS	Transfor-mylase	IF-2	EF-Tu	30S	70S A-site	70S P-site
$tRNA_m^{Met}$	⊕						
Met-$tRNA_m^{Met}$	+			+[1)]		+	
$tRNA_f^{Met}$	⊕						
Met-$tRNA_f^{Met}$	+	+		+[1)]	+		+[2)]
fMet-$tRNA_f^{Met}$			⊕		+		⊕

+ indicates that a complex can be formed

⊕ discussed in this paper

1) discussed in detail by Clark *et al.*, Chapter II

2) only in the absence of initiation factors

printing method has the potential for locating the exact positions in the tRNA molecule where splitting occurs during RNA digestion, and for comparing the digestion patterns of free tRNA and tRNA complexed to a protein. This comparison reveals the sites in the tRNA molecule which are exposed in the tRNA:protein complex and the sites which are protected against nuclease digestion. The various steps in the procedure are described in Fig. 1.

Steps in the FOOTPRINTING analysis

1. Radioactive end-labelling of tRNA (3'- or 5'-end ^{32}P incorporation)
   A. 3'-labelling:
      i) Treatment with snake venom phosphodiesterase to remove terminal A
      ii) Repair by nucleotidyltransferase using α-^{32}P-ATP
      iii) Purification of labelled tRNA by PAGE
2. Aminoacylation (and for $tRNA_f^{Met}$, formylation)
3. Complex formation (tRNA:protein)
   - isolation of complexed tRNA by gel filtration
4. Partial hydrolysis by ribonucleases (T_1, T_2, Panc., CV)
5. Analysis of RNA fragments by PAGE and autoradiography
6. Comparison of digestion pattern of complexed and uncomplexed tRNA.

Fig. 1. 1) The $tRNA_f^{Met}$ or $tRNA_m^{Met}$ are labelled with ^{32}P, either at the 3'- or the 5'- terminus. The results discussed in this article are all obtained from experiments with 3'-labelled tRNA. Radioactive labelling of the 3'-end occurs after removal of the terminal AMP with snake venom phosphodiesterase and subsequent repair of the CCA-end with nucleotidyl transferase using α-^{32}P-ATP (Boutorin *et al.* 1981). The labelled tRNA is purified by PAGE, extracted from the gel and precipitated several times with cold ethanol before further treatment. It should be emphasized that the only modification of the tRNA molecule using this labelling method is a replacement of a normal

INTERACTION WITH METHIONYL-tRNA SYNTHETASE (MetRS)

Since Clark and Marcker (1966a) separated two species of methionine-specific tRNAs, namely $tRNA^{Met}_{f}$ and $tRNA^{Met}_{m}$, the structural elements of the two tRNAs, which are responsible for the functional differences, have been the subject of intense study. For many years, these differences have been included in the theory of a functional mechanism which involves the formylation of the α-amino group only in Met-$tRNA^{Met}_{f}$, and the function of fMet-$tRNA^{Met}_{f}$ as initiatior tRNA (Clark & Marcker 1966b). Only Met-$tRNA^{Met}_{m}$ was believed to form a ternary complex with the elongation factor Tu and GTP (Ono *et al.*) and, thus, to function as an elongator tRNA, carrying methionine for internal

---

^{31}P-phosphate with a radioactive ^{32}P-phosphate resulting in a chemically and biologically native tRNA which is fully active in protein biosynthesis. 2) Where indicated, the tRNA is aminoacylated using purified methionyl-tRNA synthetase (Fig. 3) and formulated enzymatically as previously described (Petersen *et al.* 1981). 3) The footprinting method requires a high degree of complex formation as unbound tRNA results in increased background on the gel (see below), which makes the interpretation of the autoradiograms difficult or impossible. Therefore, in most cases, the tRNA:protein complex is isolated by gel filtration after its formation. 4) Partial hydrolysis (by the ribonucleases T_1, T_2, pancreatic and cobra venom RNase (from *Naja oxiana*)) is performed at 0°C, which favours complex stability. Keeping in mind that the reaction tRNA+protein⇌complex is a dynamic equilibrium, the digestion time is kept well below the half-life of the complex with respect to the first order dissociation rate constant (in the cases where these are known). To minimize the risk of 2 cuts in the same tRNA molecule, the ribonuclease:tRNA ratio is chosen such that only a minor fraction of the tRNA molecules are cut (approximately 0.1–0.2 cuts per tRNA molecule). The kinetics of the RNase reactions are studied, using increasing amounts of enzyme or by increasing the reaction time. This approach reveals non-enzymatic cuts in the tRNA, provides a test for contaminated samples and will in some cases show when secondary cuts have occurred (Wikman *et al.* 1982). 5) The radioactive fragments are separated by high voltage electrophoresis on polyacrylamide slab gels. To ensure that the same amount of radioactivity is applied to each lane, the samples are Cerenkov counted prior to loading (approximately 25000 dpm in each lane). Knowing the nucleotide sequence of the tRNA molecules, each cut can be identified on the gel by reference to a random hydrolysis ladder (obtained by limited hydrolysis in formamide) on which the bands are numbered according to the positions of the guanosine residues (obtained by treatment of denatured tRNA with ribonuclease T_1) as shown in Fig. 2 (Boutorin *et al.* 1981). 6) Fragments obtained by digesting free tRNA were loaded adjacent to those obtained by digesting complexed tRNA, making a direct comparison of corresponding bands possible. When a protein binds to a certain region of the tRNA molecule and, thereby, hinders the accessibility of the ribonuclease for this particular site, the result is a decrease of the amount of radioactivity in the corresponding band of the gel. Such an effect is illustrated on the cloverleaf structure of $tRNA^{Met}_{m}$ in Fig. 2, at the bands marked with a "P". In this case (Met-tRNA synthetase), a protection is found in the anticodon arm of the molecule. When different ribonucleases give the same result in a certain region, we interpret the protection as a region of close contact between the protein and the tRNA molecule.

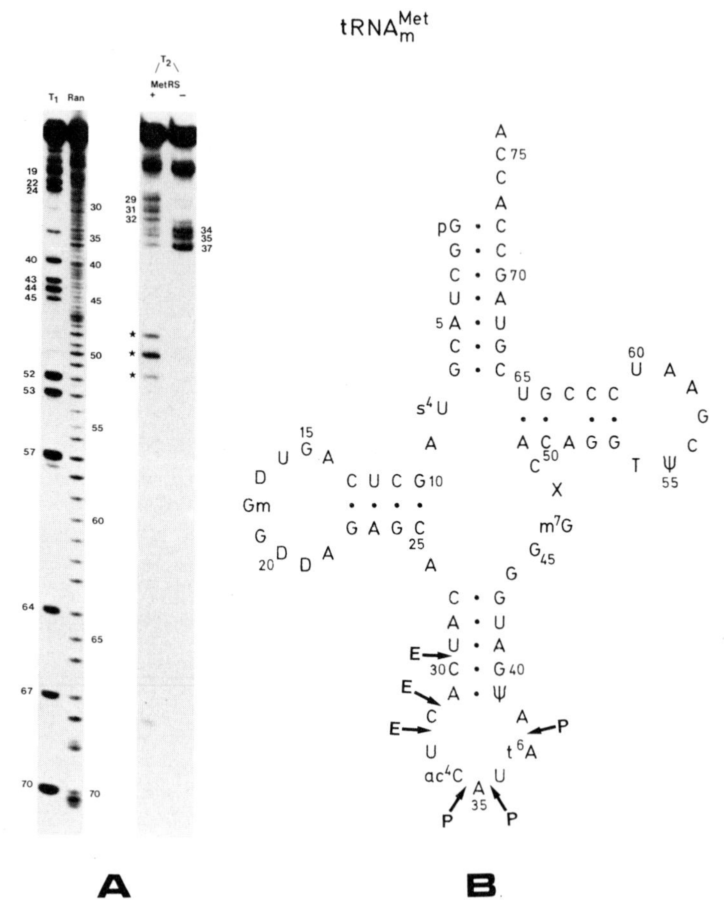

Fig. 2. Example of an autoradiogram of a footprinting gel and determination of cutting positions on the cloverleaf structure. A. Random alkaline hydrolysis of tRNA$_m^{Met}$ (Ran) and limited ribonuclease T_1 digestion of denatured tRNA$_m^{Met}$ (T_1). Limited digestion of native tRNA$_m^{Met}$ by ribonuclease T_2 in the absence (−) and in the presence (+) of MetRS. The bands are numbered according to the positions of guanosine residues as described in legend to Fig. 1 pt. 5. B. The T_2 digestion transferred to the cloverleaf structure of tRNA$_m^{Met}$ where cuts are marked by arrows. Positions which correspond to bands of lower intensity in tRNA$_m^{Met}$ complexed to MetRS as compared to free tRNA$_m^{Met}$ are marked P, and positions of bands which are more intense in complexed tRNA are marked E. The bands of positions 48–52 (*) have been shown to be artefacts (Wikman *et al.* 1982).

positions in the polypeptide chain. However, it was found recently that unformylated Met-tRNA$_f^{Met}$ can form a similar ternary complex (Tanada *et al.* 1981), and many examples have been found of cells which are able to grow

Methionine + ATP + MetRS ⇌ Methionyladenylate:MetRS + PP_i	(1)
Methionyladenylate:MetRS + $tRNA^{Met}_{f/m}$ ⇌ Met-$tRNA^{Met}_{f/m}$ + MetRS + AMP	(2)

Fig. 3. Reaction scheme for the aminoacylation of $tRNA^{Met}_f$ and $tRNA^{Met}_m$.

without formylation of Met-$tRNA^{Met}_f$ (Chapter I). These findings have made the question of structural differences and similarities between the 2 tRNAs even more important.

The methionyl-tRNA synthetase provides a particular opportunity to study this question as this enzyme is the only protein which binds specifically to the uncharged form of both tRNA species. Native methionyl-tRNA synthetase is an enzyme of the type α_2, a dimer of 2 identical subunits each of M_r 76000 (S. Blanquet, pers. comm.). Trypsin treatment of the monomer results in a fragment of M_r 64000. This fragment retains catalytic activity (Dessen *et al.* 1982), and its structure was recently determined by low resolution X-ray crystallography (Zelwer *et al.* 1982).

The aminoacylation of both tRNAs is believed to follow the reaction mechanism shown in Fig. 3. Although, as shown, the tRNA-enzyme interaction *in vivo* takes place after the binding of methionyl-adenylate to the enzyme (which may change the conformation of the enzyme), a stable complex between $tRNA^{Met}_f$ or $tRNA^{Met}_m$ and MetRS can be formed and isolated *in vitro* in the absence of other ligands.

A number of methods have been employed to study which parts of the tRNA molecules are involved in the interactions with MetRS. These include covalent crosslinking (Rosa *et al.* 1979), fluorescence spectroscopy (Blanquet *et al.* 1973), neutron scattering (Dessen *et al.* 1978) and ribonuclease treatment (Yamashiro-Matsumura & Kawata 1981). However, none of these methods could reveal the effect of the protein-binding in detail at all possible nucleotide (or phospho-diester) positions in the tRNA molecules in one experiment. The "footprinting" method described above, has been used for studying the effect of methionyl-tRNA synthetase on partial ribonuclease digestion of $tRNA^{Met}_f$ and $tRNA^{Met}_m$.

A summary of the results from experiments on the binary complex enzyme: tRNA with each of the 2 tRNAs and different ribonucleases is shown in the cloverleaf representations of the tRNAs in Figs. 4A and B (Petersen *et al.* 1984).

In both tRNAs, the 3'-side of the anticodon loop is markedly protected by MetRS. Differences are seen in the 5'-side of the anticodon stem, which is

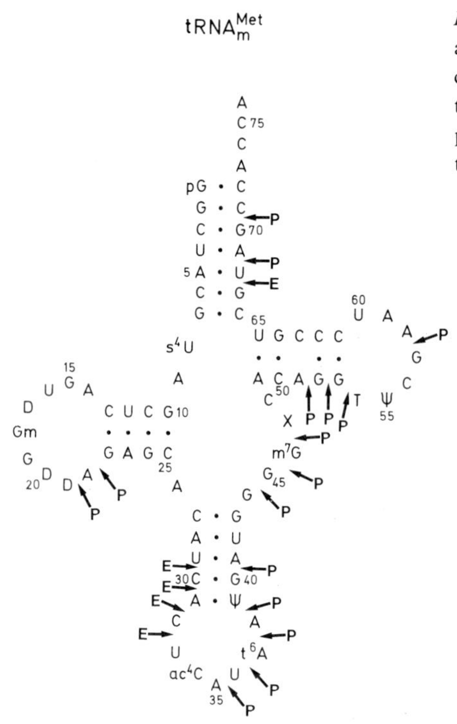

Fig. 4A. Cloverleaf structure of tRNA$^{Met}_{m}$. The arrows indicate positions cleaved by ribonucleases (T_1, T_2 or cobra venom RNase). Positions protected by MetRS are marked P, and positions cut more intensely in complexed tRNA are marked E (from Petersen *et al.* 1984).

protected in tRNA$^{Met}_{f}$, but cut more intensely in tRNA$^{Met}_{m}$ when complexed to MetRS. Alternatively, the D-loop and the extra-loop are protected in complexed tRNA$^{Met}_{m}$, whereas both these regions are cut more intensely by RNases in tRNA$^{Met}_{f}$ as a result of complex formation.

It is known that cobra venom ribonuclease cuts tRNA at many positions in the acceptor region (Boutorin *et al.* 1981). MetRS shows only weak protection in the amino acid region. This supports the idea that (at least in the absence of the other substrates of the aminoacylation reaction), the acceptor region of tRNA is not strongly bound at the surface of the enzyme (Jacques & Blanquet 1977). There are indications that also in the presence of methionine and ATP, no strong protection against cobra venom RNase digestion in the acceptor region takes place (Petersen, Siboska & Blanquet, unpublished). This result is in good agreement with the effect of MetRS on the spontaneous hydrolysis of the aminoacyl ester bound in fMet-tRNA$^{Met}_{f}$ in the presence of the initiation factor IF-2, which is discussed in the next section.

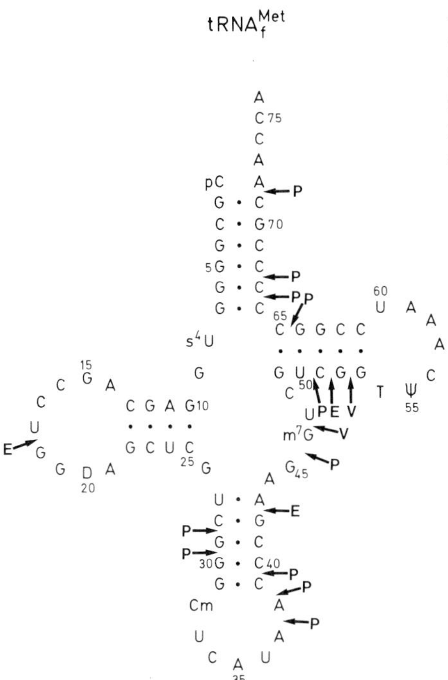

Fig. 4B. Cloverleaf structure of tRNA$_f^{Met}$. Effect of MetRS on ribonuclease digestion as described in Fig. 4A except that Fig. 4B also indicates positions where both protection and enhancement is observed (V), depending on the ribonuclease used (from Petersen *et al.* 1984).

The tertiary structure models (Fig. 5A, B) show the locations of the various protected and exposed regions within the 2 tRNAs upon binding to MetRS.

In conclusion, the interactions with the synthetase seem to be similar in the extreme parts (aminoacids end and anticodon) of the tRNAs, whereas only tRNA$_m^{Met}$ seems to bind the protein in the central parts of the molecules (the extra loop).

INTERACTION WITH THE INITIATION FACTOR IF2

The initiation factor IF2 exists in 2 forms – IF2α (M_r 115000) and IF2β (M_r 90000), which are coded for by the same gene. The DNA sequence of this gene is presently being determined in the laboratory of Dr. M. Grunberg-Manago. No information exists on the higher order structure of these proteins. Genetic aspects of these factors are discussed by Grunberg-Manago *et al.* in the preceding chapter.

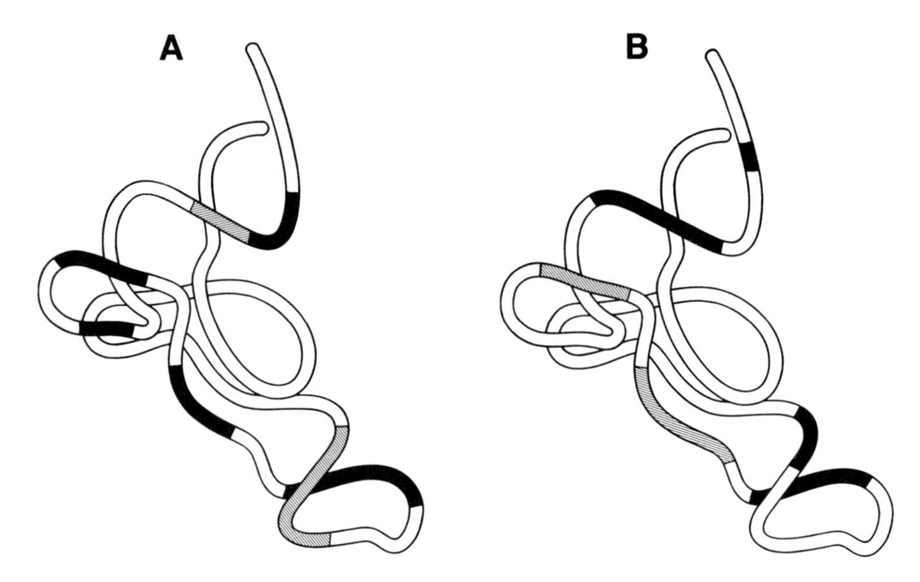

Fig. 5. Tertiary structure model of tRNA$^{Phe}_{yeast}$ (based on Robertus *et al.* 1974) A. Results transferred from Fig. 4A to illustrate regions within tRNA$^{Met}_{m}$, which are protected (■) or cut more intensely by ribonucleases (▨) when complexed to MetRS. B. Results are transferred from Fig. 4B to illustrate regions within tRNA$^{Met}_{f}$, which are protected (■) or cut more intensely (▨) when complexed to MetRS.

Whilst it is well established that the protein chain elongation factor EF-Tu functions as an aminoacyl-tRNA carrier protein in a ternary complex:EF-Tu:GTP:aa-tRNA during the elongation step of prokaryotic translation, the question of whether a similar complex is formed between the initiation factor IF2, GTP and fMet-tRNA$^{Met}_{f}$ during chain initiation has been the subject of intense debate for more than a decade. Although many different chemical and physical methods have been applied in such investigations, a complex between native IF2, fMet-tRNA$^{Met}_{f}$ and GTP has never been isolated (Petersen *et al.* 1979).

Previous studies have shown that IF2 invariably interacts with fMet-tRNA$^{Met}_{f}$. However, the extent of interaction with unformylated Met-tRNA$^{Met}_{f}$ varied from no detectable interaction to almost the same level as for the formylated species. No experiments have shown that GTP is required for IF2 to interact with the initiator tRNA. In addition, the ionic requirements for the formation of a binary complex IF2:fMet-tRNA$^{Met}_{f}$ varied considerably. A general feature of all earlier experiments is the attempt to isolate a macromolecular complex. However, such a complex may dissociate during the preparation, and we have, therefore, chosen methods which do not require the isolation of a complex. Results from experiments based on the use of 2 such methods are discussed below: 1) Effect of IF2 (and of MetRS) on the spontaneous chemical hydrolysis of the aminoacyl-ester bond in fMet-$tRNA^{Met}_f$ (Petersen *et al.* 1979), and 2) protection by IF2 against limited digestion of fMet-tRNA net by cobra venom ribonuclease (foot-printing) (Petersen *et al.* 1981).

The effect of IF2 in the first method is used as a measure of the accessibility of the ester linkage, and it is expected that binding of IF2 at, or near, the 3'-terminal CCA of the tRNA will result in a protection against hydrolysis and, thereby, increase the half-life of the esterbond.

Fig. 6. shows the rate of hydrolysis of fMet-$tRNA^{Met}_f$, free and in the presence of MetRS, IF2 or both proteins. This shows that MetRS has no effect itself on the rate of hydrolysis. The lack of protective effect is an indication that MetRS does not bind strongly at the aminoacyl-linkage. It also shows that MetRS (in the absence of AMP) does not catalyze the de-aminoacylation. In the presence of IF2, complete protection is observed. When increasing amounts of MetRS are

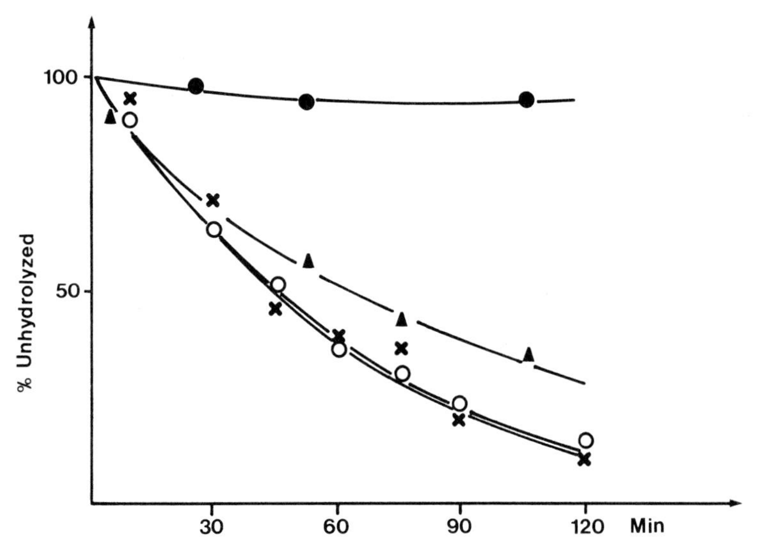

Fig. 6. Kinetics of the non-enzymatic hydrolysis of the aminoacyl ester bond of fMet-$tRNA^{Met}_f$ at 37°C in the absence (○) and in the presence of Initiation Factor IF2 at 8 times molar excess (●), MetRS at 16 times molar excess (×) or both IF2 and MetRS at 8 and 16 times molar excess respectively (▲). Results with IF2 are from Petersen *et al.* (1979).

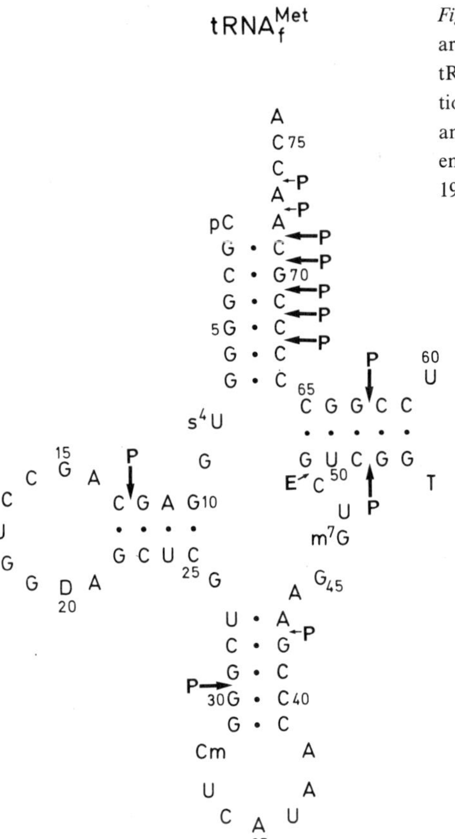

Fig. 7. Cloverleaf structure of $tRNA_f^{Met}$. The arrows indicate cleavage positions of fMet-$tRNA_f^{Met}$ by cobra venom ribonuclease. Positions strongly protected by IF2 are marked P, and cleavage positions intensified in the presence of IF2 are marked E (from Petersen et al. 1981).

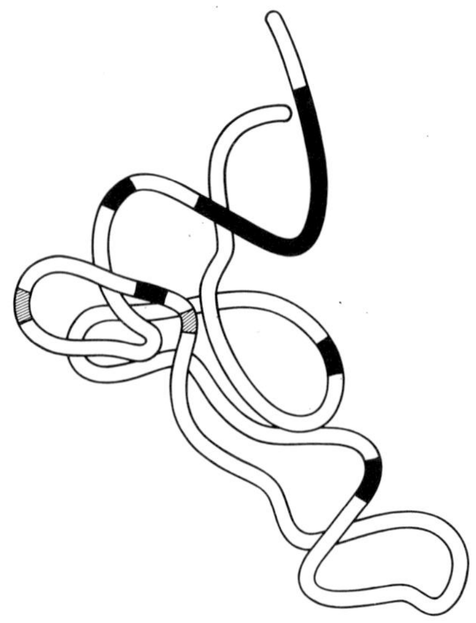

Fig. 8. Tertiary structure model of yeast-$tRNA^{Phe}$ based on Robertus *et al.* (1974). Results are transferred from Fig. 7 to illustrate regions within fMet-$tRNA_f^{Met}$ which are protected (▬) or cut more intensely (▨) by cobra venom RNase in the presence of IF2.

added in the presence of sufficient amount of IF2 to obtain such complete protection, the effect of IF2 disappears. This is taken as an indication that MetRS and IF2 have overlapping binding sites on the tRNA molecule and, thus, in this experiment, compete for the binding to this site. However, it is also clear that the binding site of MetRS is not the 3'-terminal part of fMet-$tRNA^{Met}_{f}$, which, on the other hand, seems to be one of the important binding sites for IF2.

This question has been further investigated by footprinting experiments. Figs. 7 and 8 show the positions within the initiator $tRNA^{Met}_{f}$, which are cut by cobra venom ribonuclease. The effect of IF2 on limited digestion of fMet-$tRNA^{Met}_{f}$ is indicated by the same symbolism as used in Figs. 4B and 5B. These protected regions include the 3'-end, both sides of the T-stem, the anticodon stem (in particular the 5'-side) and the D-stem.

It seems that the protected regions are mainly located at the extreme parts of the L-shaped tRNA molecule, and no protection is found in the extra loop. This result is very similar to the one found with MetRS except that here, the protein seems to be in closer contact with the 3'-end of the tRNA.

Using both methods, similar experiments have been done with unformylated initiator tRNA. In all cases, IF2 had no effect on the results obtained with free Met-$tRNA^{Met}_{f}$. We must, therefore, conclude that formylation of the initiator Met-$tRNA^{Met}_{f}$ is strictly required for the specific interaction with IF2 in the absence of ribosomes.

INTERACTION WITH ELONGATION FACTOR Tu

Apart from following the pathway through formylation and complex formation with IF2, the aminoacylated initiator Met-$tRNA^{Met}_{f}$ can form a complex with EF-Tu:GTP (p. 18, Fig. 1). Although the biological significance of this complex is unclear, the existence of such a complex has given us an opportunity to compare footprinting studies on the Met-$tRNA^{Met}_{f}$ complexed to EF-Tu:GTP with similar studies on aminoacylated elongator tRNAs. This is discussed by Clark *et al.* in Chapter II. Whether the Met-$tRNA^{Met}_{f}$ can be transferred to the ribosomal A-site is presently under investigation.

INTERACTION WITH THE 70S RIBOSOMAL P-SITE

The *in vivo* prerequisite for prokaryotic polypeptide synthesis is the 70S initiation complex consisting of the 70S ribosome with fMet-$tRNA^{Met}_{f}$ and an initiation codon of a messenger RNA positioned at the ribosomal P-site. The series of events leading to the formation of this complex is discussed in the first chapter of this volume. *In vitro* experiments have shown that the fMet-tRNA$_f^{Met}$ can bind non-enzymatically to the ribosomal P-site (Petersen *et al.* 1976a) at 15–20 mM magnesium ion concentration. The complex described here was formed in this way in the absence of initiation factors, using a poly(A,U,G) RNA chain as a messenger. In order to ensure that no unbound tRNA is present in the footprinting study, the complex was isolated on a Sepharose 6B column prior to enzymic digestion. Fig. 9 shows the cleavage points in the tRNA$_f^{Met}$, which are protected against ribonuclease digestion in the initiation complex. These are seen to be located in the aminoacyl-stem, the variable loop and in the part of the anticodon stem close to the variable loop.

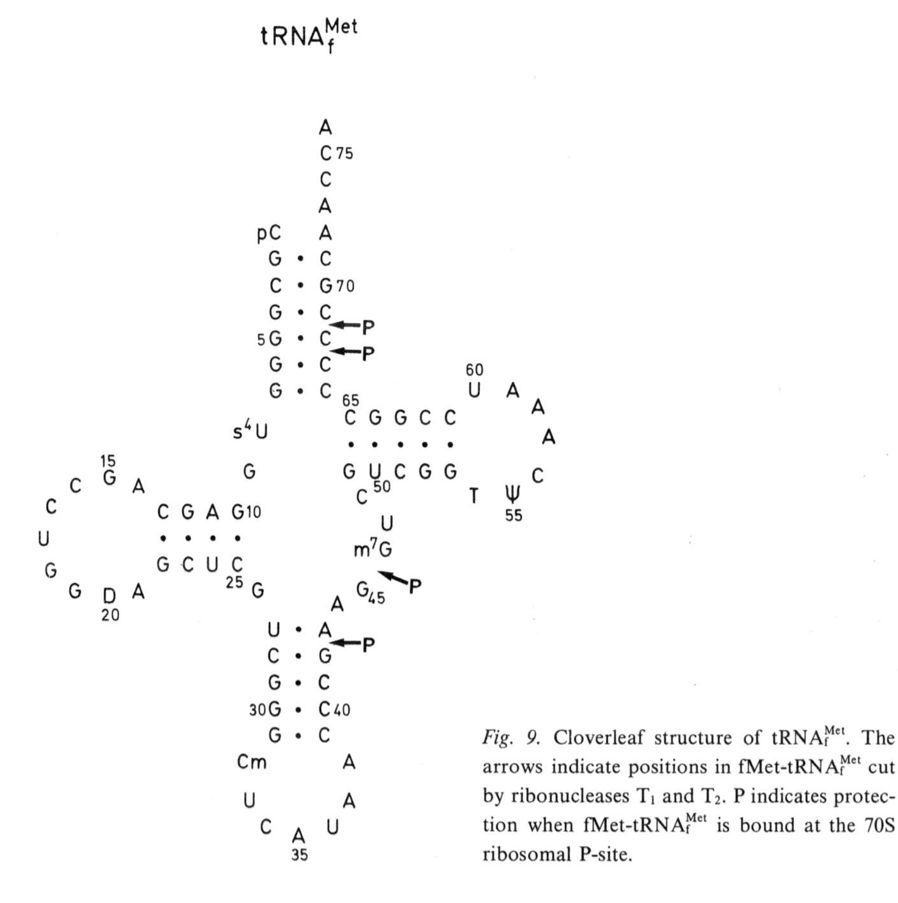

Fig. 9. Cloverleaf structure of tRNA$_f^{Met}$. The arrows indicate positions in fMet-tRNA$_f^{Met}$ cut by ribonucleases T_1 and T_2. P indicates protection when fMet-tRNA$_f^{Met}$ is bound at the 70S ribosomal P-site.

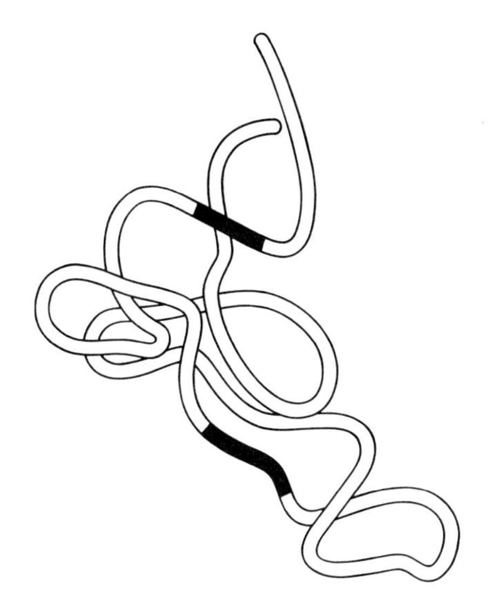

Fig. 10. Tertiary structure model of $tRNA^{Phe}_{yeast}$ based on Robertus *et al.* (1974). Results are transferred from Fig. 9 to illustrate the regions within fMet-$tRNA^{Met}_{f}$ protected (■) by the 70S ribosome when bound at the P-site.

Transferring these results to the folded representation of $tRNA^{Phe}_{yeast}$ in Fig. 10, it is observed that the protection occurs at one side of the tRNA molecule, the side exposing the extra loop.

When compared to the studies with MetRS and IF2, we observe here for the first time a strong protection of $tRNA^{Met}_{f}$ in the extra loop. This might indicate that this central part of the initiator tRNA plays an important role in the interaction with the ribosomal P-site, at least at elevated magnesium concentration.

ACKNOWLEDGEMENTS

This work was supported by grants from the Danish Natural Science Research Council, the Niels Bohr Fellowship Fund (HUP) and the Carlsberg Fund (FPW). The authors thank Helge Petersen for designing the figures of the tRNA tertiary structure and Orla Jensen for photographic assistance. We also thank Nick Coppard for critical reading of the manuscript and Lisbeth Heilesen for help with its preparation.

REFERENCES

Blanquet, S., Iwatsubo, M. & Waller, J.-P. (1973) *Eur. J. Biochem. 36,* 213–226.

Boutorin, A. S., Clark, B. F. C., Ebel, J. P., Kruse, T., Petersen, H. U., Remy, P. & Vassilenko, S. (1981) *J. Mol. Biol. 152,* 593–608.

Clark, B. F. C. & Marcker, K. A. (1966a) *J. Mol. Biol. 17,* 394–406.

Clark, B. F. C. & Marcker, K. A. (1966b) *Nature 211,* 378–380.

Dessen, P., Blanquet, S., Zaccai, G. & Jacrot, B. (1978) *J. Mol. Biol. 126,* 293–313.

Dessen, P., Fayat, G., Zaccai, G. & Blanquet, S. (1982) *J. Mol. Biol. 154,* 603–613.

Jacques, Y. & Blanquet, S. (1977) *Eur. J. Biochem. 79,* 433–441.

Ono, Y., Skoultchi, A., Klein, A. & Lengyel, P. (1968) *Nature 220,* 1304–1307.

Petersen, H. U., Danchin, A. & Grunberg-Manago, M. (1976) *Biochemistry 15,* 1357–1362.

Petersen, H. U., Røll, T., Grunberg-Manago, M. & Clark, B. F. C. (1979) *Biochem. Biophys. Res. Commun. 91,* 1068–1074.

Petersen, H. U., Kruse, T., Worm-Leonhard, H., Siboska, G. E., Clark, B. F. C., Boutorin, A. S., Remy, P., Ebel, J. P., Dondon, J. & Grunberg-Manago, M. (1981) *FEBS Letts 128,* 161–165.

Petersen, H. U., Siboska, G. E., Clark, B. F. C., Buckingham, R. H., Hountondji, C. & Blanquet, S. (1984) *FEBS Letts,* (in press).

Robertus, J. D., Ladner, J. E., Finch, J. T., Rhodes, D., Brown, R. S., Clark, B. F. C. & Klug, A. (1974) *Nature 250*, 546–551.

Rosa, J. J., Rosa, M. A. & Sigler, P. B. (1979) *Biochemistry 18,* 637.

Tanada, S., Kawakami, M. & Takemura, S. (1981) In: *Nucleic Acids Symp. Ser. no. 10,* pp. 165–168 (ed. A. E. Pritchard) IRL Press, London.

Vlassov, V. V., Kern, D., Romby, P., Giége, R. & Ebel, J.-P. (1983) *Eur. J. Biochem.*

Wikman, F. P., Siboska, G. E., Petersen, H. U. & Clark, B. F. C. (1982) *EMBO J. 1,* 1095–1100.

Yamashiro-Matsumura, S. & Kawata, M. (1981) *J. Biol. Chem. 256,* 9308.

Zelwer, C., Risler, J. L. & Brunie, S. (1982) *J. Mol. Biol. 155,* 63–81.

DISCUSSION

CRAMER: I noticed in your footprinting experiments that in some cases you had protections and very close to that enhancements, which is difficult to explain by purely structural or topographical protection.

One should keep in mind that in this kind of protection there could always be 2 effects: one effect is a general structural stabilization or destabilization, which could be long-ranged. An attachment to one end of the molecule might cause a structural rearrangement of chains, which in some other part of the molecule is visible as a protection or enhancement. This I would call a structural influence of the binding. The second effect is a direct topographical protection or deprotection and, at first look, one cannot differentiate between the 2 in such experiments.

PETERSEN: Yes, I agree on that in principle as long as you look at the effect of one particular ribonuclease. To minimize the possibility of what you call the structural influence I think it is very important that one uses many different ribonucleases, because I think that a small alteration of the structure of the tRNA caused by the binding to a protein can result in increased or decreased cutting by different nucleases. The affinity of a nuclease for one particular site may increase or decrease, if the structure changes. But I think if one uses a number of ribonucleases, and only looks at the overlapping results, one should be able to see the real protection by the protein.

EBEL: Could you comment on the interaction you described in your introduction between non-formulated methionyl-tRNA$_f^{Met}$ and the elongation factor EF-Tu.

PETERSEN: I think whether it occurs *in vivo* is an interesting question. We don't know, but it is possible to isolate a ternary complex between Met-tRNA$_f^{Met}$, EF-Tu and GTP, and actually the binding constant is about 10^6 M^{-1}, which is of the same order of magnitude as for other aminoacyl-tRNAs. But it remains a question, what happens in the living cell in the case where you inhibit the formylation.

MARCKER: But there are instances of that, aren't there, where you have bacterial mutants which can initiate without formylation.

PETERSEN: Yes, there are many examples of cells which can grow in the absence of formylation of Met-tRNA $_f^{Met}$. *S. faecalis R* does not synthesize the formyle donor *de novo* and *H. cutirubrum* does not synthesize the transformulase enzyme. Nevertheless, both organisms can grow without formylation of the initiator tRNA menthionine. There has also been found a number of *E. coli* mutants which are able to grow at conditions where the formylation is blocked by trimethoprin. In such cases it is an open question, whether some of the initiator tRNA is used for incorporating methionine internally in the polypeptide chain.

SCHIMMEL: I have a question of a somewhat technical nature. Usually, when foot-print experiments are done, you only want to cut the molecule once. I gather from your response to a previous question that the digestion lasted for 10 min. Could there be multiple hits, and could some of the enhanced cleavage be a second cut, and not the first cut that is made in the molecule, in which case you are dealing with a perturbed structure? The question is: were the experiments done under conditions where you have a single cut on average?

PETERSEN: Yes, it is only a minor part of the tRNA which is cut. About 90% of the tRNA molecules are still intact after 10 min digestion, and as I said we have done kinetics studies with the different ribonucleases. The only possibility of secondary cuts in our experiments is in the case of slow primary cutting kinetics followed by an immediate structural rearrangement of the tRNA and a fast secondary cut. This we will only be able to see by using 5′ labelled tRNA and we are just now ready to do these experiments.

GARRETT: From our studies on 5 S RNA it is clear that secondary cuts can only be detected by using both 5′ and 3′-end labelled RNA molecules. A secondary cut, by definition, occurs as a result of a primary cut, at a site that is inaccessible in the native RNA structure. It can be distinguished from a primary cut because it is detectable on gels only when it lies between the primary cut and the labelled end of the molecule as illustrated in the figure.

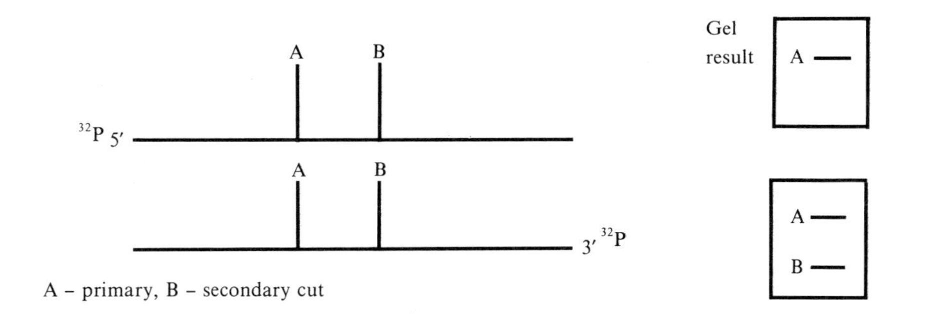

A - primary, B - secondary cut

KLUG: When I look at the drawings of the cutting positions in the tRNAs, they look to me as they were much more like each other than unlike, and in fact the small differences could possibly be due to sequence differences. I don't know the sequences well enough to see what they were, but the enzymes are sequence dependent, so you would never expect it to appear quite the same.

PETERSEN: When you look at the exact cutting positions you will find variations from one tRNA to the other, which I expect to be a result of sequence differences. Therefore, I have not focused on particular nucleotides but only on regions, i.e., stems or loops, within the tRNAs. As I showed, there are similarities mainly in the aminoacyl part and in the anticodon part of the molecules, and I think that agrees with all synthetase-tRNA systems studied until now. The main difference we observed was in the extra arm, where the elongator tRNA is protected by the synthetase, whereas we did not see any protection in the case of the initiator tRNA. I think that in all known elongator tRNA-synthetase systems the extra loop is interacting with the synthetase, but that does not seem to be the case for the initiator tRNA. So it seems that the main difference between the initiator tRNA and all other tRNAs is minor, but important, structural differences in that part of the tRNA.

The Role of Mammalian Initiation Factors in Translational Control

John W. B. Hershey, Roger Duncan, Diane O. Etchison & Susan C. Milburn

INTRODUCTION

Translational control of gene expression

Numerous examples are known in which control of gene expression in mammalian cells occurs at the level of translation. Protein synthesis rates decrease in mitotic cells, or following amino acid or serum deprivation, and increase on entering interphase, or upon refeeding starved cells. Heat shock causes changes in the mRNA population being actively translated, and virus infection often leads to inhibition of cellular protein synthesis. *A priori,* the rate of protein synthesis may be regulated in two ways: by the amount of mRNA being translated, and by the efficiency of translation of the mRNA. The former involves the complex processes of transcription and processing of mRNA in the nucleus, its transport into the cytoplasm and its half-life. The latter, mRNA translational efficiency, constitutes the focus of this paper, and may be defined as the number of proteins translated from a mRNA per unit time. Under steady-state conditions, this is equivalent to the initiation rate on the mRNA. The rate of initiation is established by intrinsic qualities of the mRNA (i.e., its structure), by the level and activity of the translational machinery, and by factors which influence the initiation rate and the distribution of mRNA between active (polysomal) and inactive (mRNP) states. These features are subject to change in response to intra- and extra-cellular signals and their modulation constitutes what we call translational control. We have focussed our investigations on the initiation phase of protein synthesis because this step appears to be rate-limiting in most cells and, therefore, must control the rate of translation.

Department of Biological Chemistry, School of Medicine, University of California, Davis, California 95616, U.S.A.

GENE EXPRESSION, Alfred Benzon Symposium 19.
Editors: Brian F. C. Clark & Hans Uffe Petersen, Munksgaard, Copenhagen 1984.

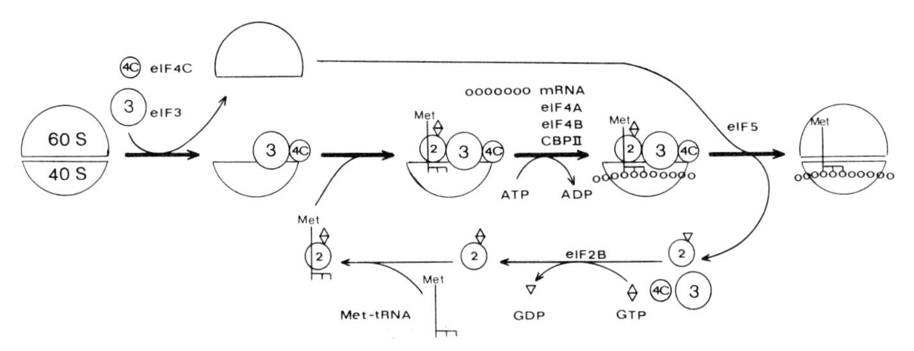

Fig. 1. Pathway of initiation of protein synthesis.

The initiation phase of protein synthesis

The general features of the initiation process in mammalian cells are shown in Fig. 1. Ribosomes dissociate into subunits, Met-tRNA and mRNA bind sequentially to the 40S subunit to form a 43S pre-initiation complex, and the 60S subunit joins it to complete formation of the 80S initiation complex. The scheme is based primarily on *in vitro* experiments with purified components (Trachsel *et al.* 1977, Benne & Hershey 1978, Jagus *et al.* 1981, Safer 1983). One or more initiation factors participate in each of the major steps (shown by heavy arrows); eIF3 and eIF4C may help dissociate ribosomes; eIF2 and eIF2B are involved in ternary complex formation and Met-tRNA binding; eIF4A, eIF4B and CBP-II promote mRNA binding and eIF5 is required for junction of the 60S subunit with the 43S pre-initiation complex. The binding of Met-tRNA is crucial for initiation on all mRNAs, and is described in detail in this volume by Dr. B. Safer. The mRNA binding step is in many ways the most interesting, since it is at this moment that a particular mRNA is selected for translation. It has been proposed (Kozak 1983) that some of the initiation factors, in conjunction with ATP hydrolysis, are involved first in altering the secondary structure of the mRNA during the cap recognition process. Following 40S binding to the mRNA, the ribosomal subunit then "scans" the mRNA, selecting (in general) the 5′-proximal AUG as the initiator codon. However, we know very little about the detailed molecular reactions of mRNA recognition and binding.

The initiation factors have been purified from rabbit reticulocytes and a variety of other mammalian cells, and their physical and biological properties have been characterized (Table I) (Schreier *et al.* 1977, Benne *et al.* 1979, Merrick 1979, Voorma *et al.* 1979). Most factors are composed of a single polypeptide chain, but eIF2, eIF2B, eIF3 and CBP-II are multi-subunit complexes containing a number of different, unrelated polypeptides. Immunoblot analyses with antisera against a number of the factors show that the purified factor

Table I
Mammalian initiation factors

Factor	Number of subunits	Mol. wt. (kDa)	Cell level (eIF/Rb)	Function
eIF-1	1	15		mRNA binding?
eIF-2	3	140	0.8	Ternary complex formation
		36		Met-tRNA binding to 40S Rb.
		48		target of HRI
		52		
eIF-2B	5	280	~0.2	Promotes GDP/GTP exchange
		26		on eIF-2
		39		
		58		
		67		
		80		
eIF-3	7	420–700	0.5	Ribosome dissociation
		170		Met-tRNA binding to 40S Rb.
		115		mRNA binding
		66		
		47		
		44		
		40		
		36		
eIF-4A	1	49	3	mRNA binding
				cap recognition
				component of CBP-II
eIF-4B	1	80	~0.4	mRNA binding
				cap recognition
eIF-4C	1	19		ribosome dissociation
				Met-tRNA binding to 40S Rb.
eIF-4D	1	17		lysine modification to hypusine
eIF-5	1	160		junction of 40S and 60S Rbs.
CBP-II	3	300		mRNA binding
p220		220		cap recognition
eIF-4A		49		site of poliovirus protease
CBP-I		24		action

proteins are indistinguishable from those in crude cell lysates (Meyer *et al.* 1982). Initial attempts to construct *in vitro* systems capable of initiating protein synthesis with the purified factors were unsatisfactory because of the poor activities obtained. More recently, with the identification of eIF2B and CBP-II and their addition to the assays, more active systems have been made which approach the efficiency of the reticulocyte lysate, and which are capable of mimicking translation control mechanisms (Voorma & Amesz 1982). Therefore, it is now possible to test directly the effects of changes in the structure of initiation components on the overall efficiency of the cell-free system.

The experimental approach

We argue, *a priori,* that a likely general mechanism for regulating protein synthesis is the reversible chemical modification of initiation factors. The phosphorylation of eIF2α is a proven case in point, and is described in detail in this volume by Dr. B. Safer. Analysis of initiation factor preparations labelled *in vivo* with (^{32}P)phosphate indicates that a number of other factor polypeptides also are labelled (Benne *et al.* 1978, Hathaway *et al.* 1979). Rather than isolate or prepare modified initiation factors, demonstrate a change in activity *in vitro,* and then seek a physiological role for the modification, we chose a different approach. We proposed to compare extracts of cells which differed in their overall rates of initiation of protein synthesis, examining the initiation factors in order to correlate changes in factor structures with changes of *in vivo* protein synthesis rates. Once such correlations are made, we intend to study the phenomenon in detail using the purified *in vitro* system.

General methods were needed to rapidly examine initiation factor structures in crude cell lysates, since approaches which involve factor purification may lead to changes in the extent of modification or to losses which are difficult to evaluate. We have chosen to use immunoblotting with polyvalent antisera or with affinity-purified antibodies specific for initiation factors. Immunoblots of cell lysates fractionated on 1-dimensional SDS polyacrylamide gels can provide quantitative comparisons of the levels and molecular weights of the factors (Howe & Hershey 1981), whereas, immunoblots of 2-dimensional isoelectric focussing/sodium dodecyl sulfate (IEF/SDS) polyacrylamide gels can show changes in the charge forms of factor proteins and, thus, can measure the extent of modifications such as phosphorylation (Duncan & Hershey 1983). We have applied this approach to studies of HeLa cells in 3 different conditions, following heat shock, serum deprivation and infection by poliovirus. In all cases, evidence for structural modifications of initiation factors is found as described below.

INITIATION FACTOR CHANGES FOLLOWING HEAT SHOCK AND SERUM DEPRIVATION

When cultured mammalian cells are subjected to elevated temperature, dramatic alterations in gene expression occur (Schlesinger *et al.* 1982). Normal cellular RNA synthesis is inhibited while a few specific heat shock genes are actively transcribed. At the level of translation, heat shock usually inhibits protein synthesis from all mRNAs. Repressed mRNAs are converted to mRNPs but are not degraded. When the cells are returned to lower temperatures, protein synthesis activity resumes and the cells eventually revert to normal patterns of gene expression. Thus, the changes exhibited by heat shock are both rapid and reversible. The precise response to heat shock depends on the cell species studied, and different mechanisms for selective repression of protein synthesis may exist. However, these are as yet poorly understood.

IEF/SDS-PAGE analysis of cell lysates

The rapid and reversible characteristics of the heat shock response suggest that covalent modification of the translational machinery may be involved. Therefore, we wished to determine whether we could detect and correlate changes in factor molecular forms, such as isoelectric variants due to phosphorylation. We employed the technique of IEF/SDS-PAGE to examine a number of the initiation factors in HeLa cell lysates (Duncan & Hershey 1983). HeLa cells were grown on the surface of plastic dishes and labelled 4 to 24 h during exponential phase with either (^{35}S)methionine or a mixture of (^{14}C)amino acids. The cells (ca. 1×10^5) were washed and then rapidly lysed into electrophoresis sample buffer and subjected immediately to IEF/SDS-PAGE. The rapid processing of intact, growing cells into denatured lysates is important for avoiding possible changes in extents of modification during work-up. Following electrophoresis, the gels were either dried for autoradiography or their proteins were electrotransferred to nitrocellulose for immunoblotting with antisera against the initiation factors. An example of a gel autoradiogram is shown in Fig. 2, and spots corresponding to initiation factors are labelled. These assignments were made on the basis of comigration of lysate and purified factor preparations, immunoblotting, and comparison of partial proteolytic degradation patterns of the labelled spots and

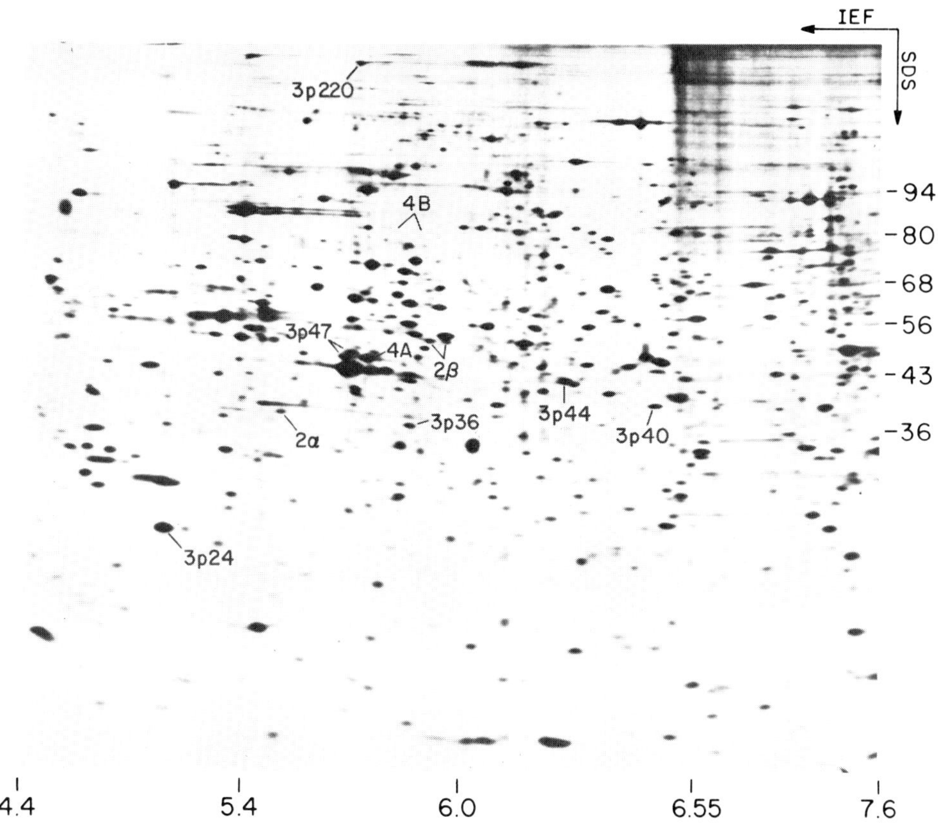

Fig. 2. Initiation factors identified in two-dimensional gel patterns of HeLa cell lysate proteins. Exponentially growing HeLa cells were labelled 4 h with (^{35}S) methionine and the lysate was analyzed by IEF/SDS-PAGE. The figure is a photograph of the fluorogram of the dried gel. Factors were identified as described in the text and elsewhere (Duncan & Hershey 1983).

the pure factor (Duncan & Hershey 1983). We have identified eIF2α, eIF2*β*, 5 subunits of eIF3, eIF4A and eIF4B. eIF2*γ* is too basic for analysis in this gel system, but has been located in lysate patterns using NEPHGE/SDS-PAGE. We observed unique molecular forms for all of these factor proteins except for eIF2*β* and eIF2*γ*, which occur as 2 and sometimes 3 spots, and eIF4B, which appears to comprise a large number of isoelectric and size variants, all of which appear to correspond to the true factor. The initiation factors are moderately abundant proteins in these lysates and their levels were determined by excising spots and measuring radioactivity. The values, reported in Table I, indicate that

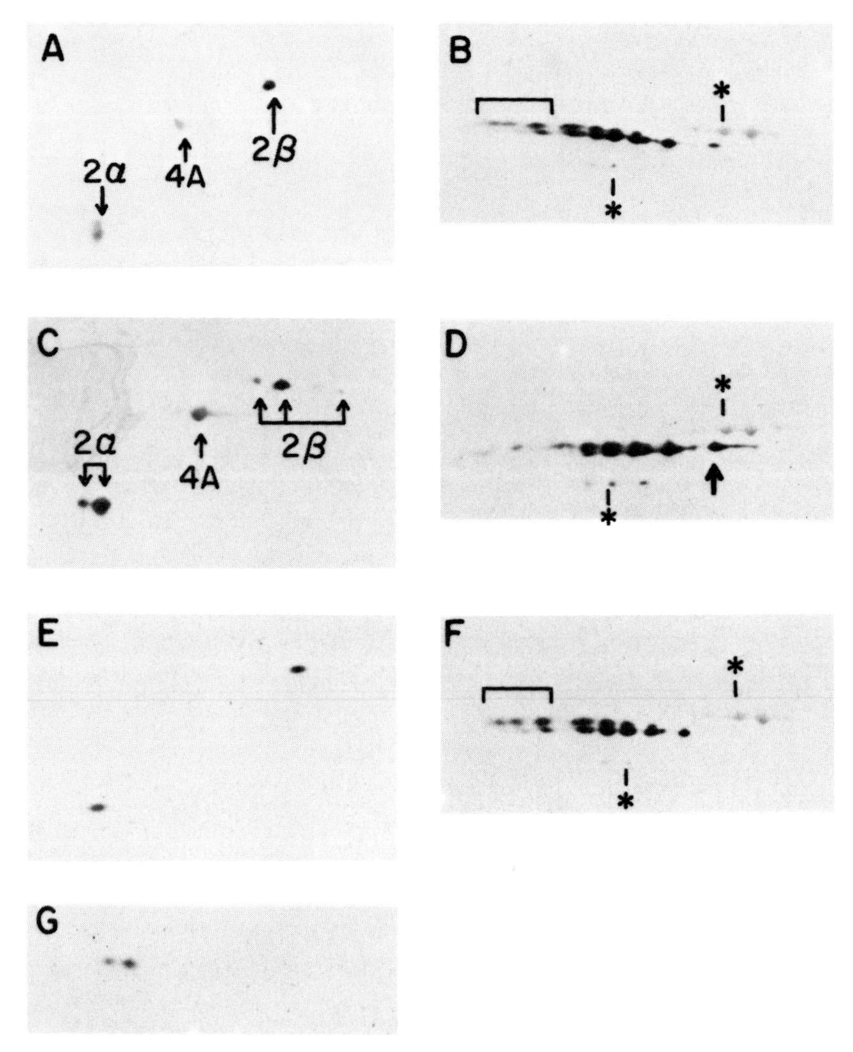

Fig. 3. Immunoblot analyses of initiation factors in lysates of heat shocked and unshocked HeLa cells. HeLa cells growing in monolayer cultures at 37° (panels A and B) were raised to 45° for 20 min (panels C, D, G), and then allowed to recover at 37° for 2 h (panels E, F). Lysates were fractionated by IEF/SDS-PAGE, proteins were electrotransferred to nitrocellulose, a small region of the blot was cut out which contained either eIF2α, eIF2β and eIF4A (left panels), or eIF4B (right panels). These sections were incubated with affinity-purified anti-eIF2α (panels A, C, E, G), anti-eIF2β (panels A, C, E), anti-eIF4A (panels A, C) and anti-eIF4B (panels B, D, F). Antibodies bound to the blots were located by incubation with ^{125}I-labelled antibody against the specific antibodies, and by autoradiography. The figure shows photographs of the autoradiograms; spots corresponding to factors are labelled appropriately.

each initiation factor protein comprises about 0.1% of the cell protein; factor/ribosome ratios are generally about 0.5 except for eIF4A which is much more abundant, occurring as 3 molecules per ribosome.

Heat shocked HeLa cells

Having established the methodology for quantitatively observing these initiation factors, we then compared lysate patterns from normal and heat shocked cells. HeLa cells growing exponentially in monolayer cultures were raised from 37° to 45° for 20 min. Under these conditions, protein synthesis is inhibited greater than 95%, but recovers to normal rates within 2 h of being returned to 37°. Analysis of the protein products during the recovery phase showed the classical pattern of the induced heat shock proteins. When initiation factors in heat shocked (45°, 20 min.) and non-shocked (37°) cell lysates were compared by IEF/SDS-PAGE and immunoblotting (Fig. 3), clear differences in some of the factors were detected. Heat shock induces a second, more acidic form of eIF2α, several new minor forms of eIF2β appear which are shifted to both the basic and the acidic side of the major eIF2β spot, and the pattern of eIF4B is shifted towards more basic forms. No change in the single form of eIF4A was seen. When heat shocked cells were returned to 37° and analyzed after 2 h when protein synthesis rates were normal, the induced changes were no longer found (Fig. 3, panels E,F) and factor patterns resembled those from non-treated cell lysates.

We then investigated the possibility that the changes in isoelectric forms induced by heat shock were due to phosphorylation or dephosphorylation of the factors. HeLa cells were incubated 20 min. in (^{32}P)phosphate medium at either 37° or 45°, and cell lysates were prepared and analyzed by IEF/SDS-PAGE (Fig. 4). Numerous enhancements as well as depressions of specific protein phosphorylation are caused by heat shock. In the 45° lysates, a new radioactive spot was found to migrate in the same position as the more acidic isoelectric variant of eIF2α, indicating that the variant is a phosphorylated form of eIF2α. Quantitative analysis of the immunoblots showed that the phosphorylated form of eIF2α comprises 25 to 30% of the factor's mass in the heat shocked state, but only 0–5% in non-heated cells. The cluster of spots corresponding to eIF4B is labelled with (^{32}P)phosphate, indicating that this factor also is phosphorylated and suggesting that phosphorylation may be responsible for the isoelectric variants. In the heat shocked state, no radioactivity is detected in the most acidic region corresponding to some of the eIF4B variants of normal cells. Thus, the

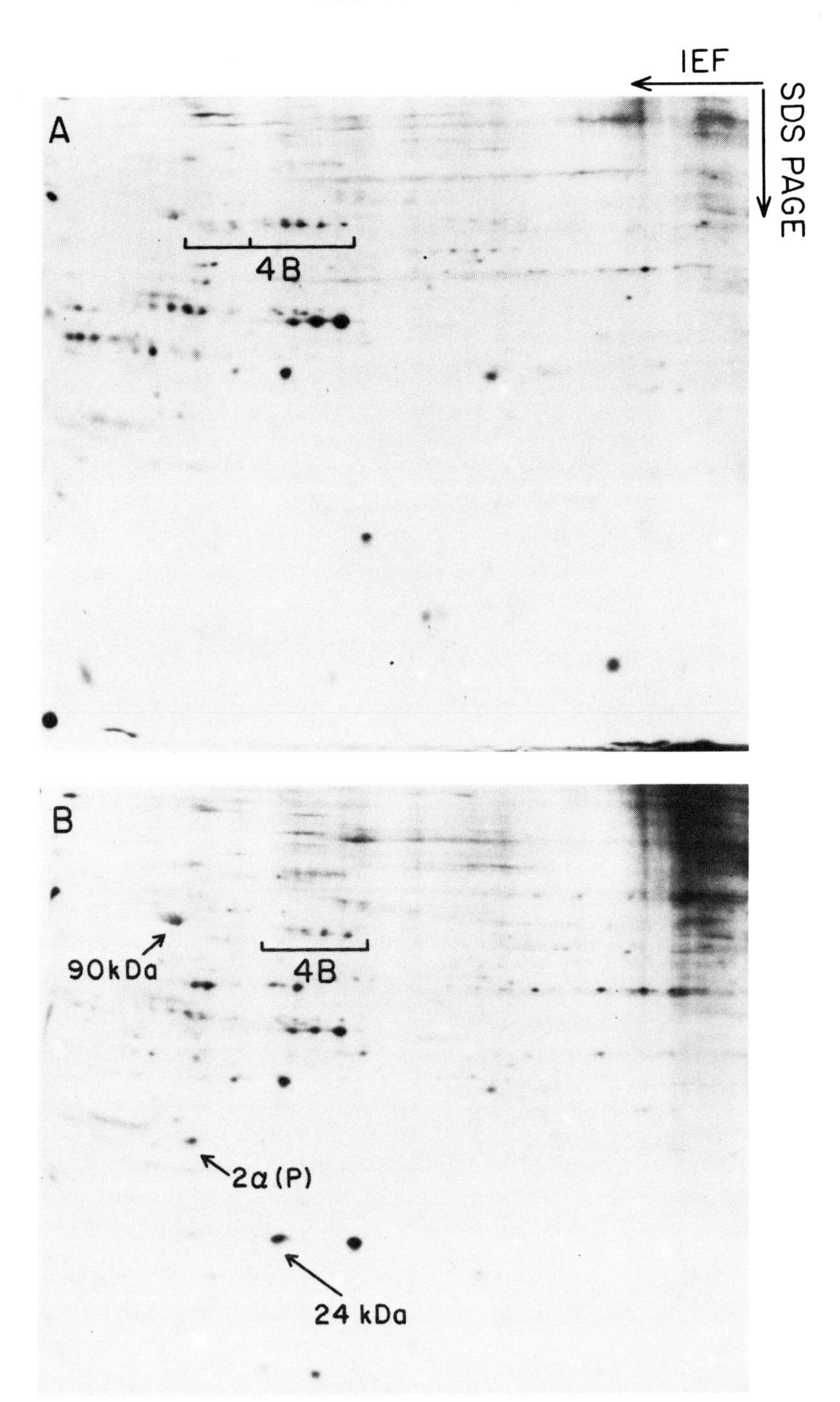
IEF
SDS PAGE
A
4B
B
90kDa
4B
2α(P)
24 kDa

^{32}P-labelling confirms the immunoblot analyses and shows a general shifting of intensity of spots toward the more basic forms. This is consistent with a partial dephosphorylation of eIF4B in the heat shocked state.

Although there is a clear correlation of heat shock inhibition with eIF2α phosphorylation, eIF2β modification, and eIF4B dephosphorylation, proof is lacking that these changes are responsible for the reduced rate of protein synthesis. We have examined normal and heat shocked cell lysates for their ability to support *in vitro* protein synthesis, but find that both types of lysates fail to initiate protein synthesis. As an alternative approach, we compared the activities of the ribosomal high-salt wash fraction of either cell type to stimulate a fractionated protein synthesis assay. The heat shock wash fraction was inactive, but could be stimulated when supplemented with normal A cut (0–40% saturated ammonium sulfate fraction of the high-salt wash), normal B cut (40–50%), or maximally by normal A+B cuts, but not at all by C cut (50–70%). This result is consistent with the view that eIF2 (in B cut) and eIF4B (in A cut) are inactive due to the modifications observed. We are currently attempting to define more precisely the initiation factors responsible for restoring activity to the heat shocked high-salt wash fraction, and to determine the mechanism of their inactivation in heat shocked cells.

Serum-deprived cells

Immunoblotting of cell lysates was also used to detect changes in initiation factor forms in cells deprived of serum. When HeLa cells in spinner culture are grown without replenishing serum and the growth medium, the rate of protein synthesis progressively declines. A similar repression of translation occurs over a period of several days when HeLa cells growing in monolayer cultures are deprived of fresh serum. In serumstarved cells greatly inhibited in protein synthesis, we observed the same changes in eIF2 and eIF4B which we observed in heat shock-inhibited cells. eIF2α was phosphorylated, eIF2β was modified and eIF4B was apparently dephosphorylated. These findings suggest that the changes in factor modification observed may be generally correlated with inhibition of protein synthesis.

Fig. 4. ^{32}P-labelled proteins in lysates of heat shocked and unshocked HeLa cells. HeLa cells in monolayer culture were grown 20 min at 37° (panel A), or 45° (panel B) in medium containing ^{32}P-phosphate. Lysates (100,000 cpm of TCA precipitable proteins) were fractionated by IEF/SDS-PAGE, and the gels were dried and subjected to autoradiography. Besides eIF-4B and eIF2α, two heat shock proteins are identified in panel B: 90 kDa and 24 kDa.

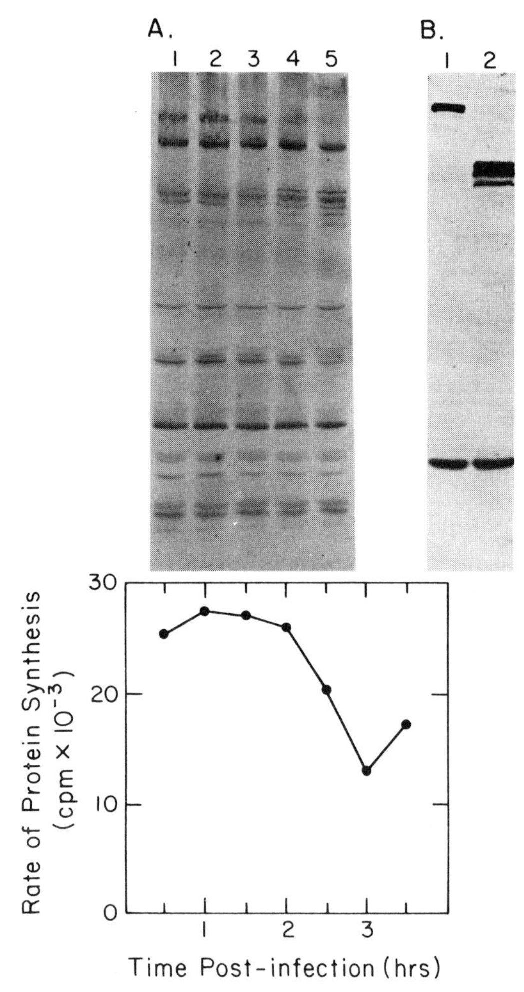

Fig. 5. The time course of p220 degradation in poliovirus infected HeLa cells. HeLa cells were infected at a m.o.i. of 10, and aliquots of cells (2×10^6) were removed and analyzed for rates of protein synthesis by incubation with (^{35}S) methionine. At the same time, cells (2.4×10^7) were also removed, lysed and analyzed by SDS-PAGE. The gels were immunoblotted with anti-eIF3 serum and are shown in panel A directly above the appropriate times indicated in the panel below. In panel A, p220 corresponds to the slowest moving band in lanes 1–4, and is nearly undetected in lane 5. In panel B, lysates from uninfected (lane 1) and poliovirus-infected (lane 2) HeLa cells were analyzed as in panel A, except that affinity-purified anti-p220 was used.

THE MECHANISM OF HOST CELL SHUT-OFF OF PROTEIN SYNTHESIS BY POLIOVIRUS INFECTION

Infection of HeLa cells by poliovirus results in a dramatic permanent inhibition of translation of capped cellular mRNAs, whereas uncapped poliovirus mRNAs which accumulate during late infection are translated efficiently (Ehrenfeld 1982). Cellular mRNA in infected cells is not degraded, but can be extracted and translated in cell-free systems. However, initiation on capped mRNAs is inhibited after about 1 to 2 h (see Fig. 5), presumably through modification of the translational machinery. Evidence that initiation factors might be involved came from showing that high-salt ribosomal washes (which contain the factors) do not promote translation of capped mRNAs *in vitro.*

Two approaches were taken to identify the "defective" component in the infected HeLa cell. In collaboration with E. Ehrenfeld (Utah), we tried to purify the initiation factors and found early in the fractionation procedures active eIF4B, but inactive eIF3 preparations (Helentjaris *et al.* 1979). Alternatively, Rose *et al.* (1978) attempted to restore translation of capped mRNA in a cell-free lysate system prepared from infected cells by supplementing it with purified initiation factors, and showed that an eIF4B preparation possessed "restoring activity". Subsequent work revealed that "restoring activity" was related to preparations of CBP-I (Trachsel *et al.* 1980), and finally to the high molecular weight complex, CBP-II (Tahara *et al.* 1981). These apparently conflicting results were resolved when it was realized that the new factor, CBP-II, tends to contaminate preparation of eIF3, eIF4B and CBP-I. The experiments indicate that translation of capped mRNAs is specifically inhibited by altering the activity of CBP-II.

Initiation factor CBP-II is degraded

In order to detect possible changes in initiation factors following poliovirus infection, we tested uninfected and infected cell lysates by SDS polyacrylamide gel electrophoresis and immunoblotting. We used antisera against eIF2, eIF3, eIF4A and eIF4B, as most of these factors are involved in mRNA binding (see Fig. 1). Our analyses with anti-eIF4A and anti-eIF4B indicated no change in levels or charge forms with these factors (Etchison *et al.* 1983). With anti-eIF3, however, we observed the disappearance of an immuno-reactive band of 220 kilodaltons, which we call p220 (Etchison *et al.* 1982). The level of p220 versus time following infection was determined, and its cleavage was observed to correlate with the inhibition of protein synthesis (Fig. 5). Anti-p220 antibodies were affinity purified and were shown to recognize the p220 component of purified CBP-II, and two proteins in uninfected cell lysates: p220 and p36 (Fig. 5). In poliovirus infected cells, the p220 protein is absent, but is replaced by a series of presumably degradation products at 100–130 kilodaltons. The p36 protein remains unaltered throughout infection; its role or relationship to other components of translation is unknown.

The results provide an explanation for how poliovirus inhibits the translation of capped mRNAs. We propose that CBP-II is required for the translation of all capped mRNAs and that intact p220 is necessary for CBP-II activity. Following poliovirus infection, a proteolytic enzymic activity is induced which specifically cleaves p220 and inactivates this factor. Remaining problems are to explain the detailed mechanism of action of CBP-II, to isolate and test *in vitro* the cleaved factor, and to explain whether, or how, cleaved CBP-II participates in the translation of polio mRNA. The recent purification of CBP-II by cap affinity columns (Tahara *et al.* 1981), or by classical techniques (Grifo *et al.* 1983), should provide sufficient material for answering these questions experimentally.

Another interesting aspect of the shut-off phenomenon is to determine the source of the protease which cleaves p220. *A priori,* the protease could be either a cellular protein activated by infection or a polio-coded protein synthesized *de novo* during the infection. In collaboration with Drs. Hanecak and Wimmer (Stoney Brook, N.Y.), we showed that a HeLa lysate programmed for translation with polio RNA (but not other RNAs) resulted in partial cleavage of p220 into 100–130 kilodalton products. When antiserum against the polio protease, 7C, was added to the *in vitro* system, this cleavage was reduced, suggesting that the polio-coded protease is responsible for the cutting of p220. Therefore, we postulate that this highly specific polio protease used to cleave the polio polyprotein also recognizes the p220 subunit of CBP-II and, thereby, inhibits the translation of cellular mRNAs.

CONCLUSIONS

Phosphorylation of the α-subunit of eIF2 and proteolytic cleavage of the p220 subunit of CBP-II lead to inhibition of initiation of protein synthesis. We provide correlative evidence that modification of the β-subunit of eIF2 and dephosphorylation of eIF4B may also play regulatory roles. Finally, the modification of eIF4D by conversion of a lysine residue to hypusine, correlates with the stimulation of protein synthesis (Cooper *et al.* 1983). Taken together, the results indicate strongly that protein synthesis rates may be regulated by covalent modification of initiation factors. Since the tools are at hand for studying the mechanism of action of the modified and unmodified factors, we can anticipate a rapid detailed elucidation of these translational control mechanisms at the molecular level.

ACKNOWLEDGEMENTS

This work was supported by a grant from the U.S. Public Health Services (NIH), GM 22135. We thank Martine Jouannaud for expert typing of the manuscript.

REFERENCES

Benne, R., Brown-Luedi, M. L. & Hershey, J. W. B. (1979) *Methods Enzymol. 60,* 15–35.

Benne, R., Edman, J., Traut, R. R. & Hershey, J. W. B. (1978) *Proc. Natl. Acad. Sci. USA 75,* 108–112.

Benne, R. & Hershey, J. W. B. (1978) *J. Biol. Chem. 253,* 3078–3087.

Cooper, H. L., Park, M. H., Folk, J. E., Safer, B. & Braverman, R. (1983) *Proc. Natl. Acad. Sci. USA 80,* 1854–1857.

Duncan, R. & Hershey, J. W. B. (1983) *J. Biol. Chem. 258,* 7228–7235.

Ehrenfeld, E. (1982) *Cell 28,* 435–436.

Etchison, D., Duncan, R. & Hershey, J. W. B. (1983) *J. Biol. Chem. 258,* 7236–7239.

Etchison, D., Milburn, S. C., Edery, I., Sonenberg, N. & Hershey, J. W. B. (1982) *J. Biol. Chem. 257,* 14806–14810.

Grifo, J. A., Tahara, S. M., Morgan, M. A., Shatkin, A. J. & Merrick, W. C. (1983) *J. Biol. Chem. 258,* 5804–5810.

Hathaway, G. M., Lundak, T. S., Tahara, S. M. & Traugh, J. A. (1979) *Methods Enzymol. 60,* 495–511.

Helentjaris, T., Ehrenfeld, E., Brown-Luedi, M. L. & Hershey, J. W. B. (1978) *J. Biol. Chem. 254,* 10973–10978.

Howe, J. G. & Hershey, J. W. B. (1981) *J. Biol. Chem. 256,* 12836–12838.

Jagus, R., Anderson, W. F., & Safer, B. (1981) *Prog. Nucl. Acids Res. Mol. Biol. 25,* 127–185.

Kozak, M. (1983) *Microbiol. Rev. 47,* 1–45.

Merrick, W. C. (1979) *Methods Enzymol. 60,* 101–108.

Meyer, L. J., Milburn, S. C. & Hershey, J. W. B. (1982) *Biochemistry 21,* 4206–4212.

Rose, J. K., Trachsel, H., Leong, K. & Baltimore, D. (1978) *Proc. Natl. Acad. Sci. USA 75,* 2732–2736.

Safer, B. (1983) *Cell 33,* 7–8.

Schreier, M. H., Erni, B. & Staehelin, T. (1977) *J. Mol. Biol. 116,* 727–754.

Schlesinger, M. J., Ashburner, M. & Tissières, A. (eds.) (1982) *Heat Shock: From Bacteria to Man,* Cold Spring Harbor Laboratory Press, New-York.

Tahara, S. M., Morgan, M. A. & Shatkin, A. J. (1981) *J. Biol. Chem. 256,* 7691–7694.

Trachsel, H., Erni, B., Schreier, M. H. & Staehelin, T. (1977) *J. Mol. Biol. 116,* 755–767.

Trachsel, H., Sonenberg, N., Shatkin, A. J., Rose, J. K., Leong, K., Bergman, J. E., Gordon, J. & Baltimore, D. (1980) *Proc. Natl. Acad. Sci. USA 77,* 770–774.

Voorma, H. O. & Amesz, H. (1982) In: *Interaction of translational and transcriptional controls in the regulation of gene expression* (Grunberg-Manago, M. & Safer, B., eds.), pp. 311–325, Elsevier/North-Holland, New-York.

Voorma, H. O., Thomas, A., Goumans, H., Amesz, H. & van der Mast, C. (1979) *Methods Enzymol. 60,* 124–135.

DISCUSSION

MARCKER: The discussion here I think should centre on 2 aspects. The first one should be a global one. That which is very interesting and still unknown in eukaryotic initiation is actually how the message is selected and the first AUG is found. So, I would like first of all to have questions pertaining to how you see this, and then we should proceed to talk about the modifications of the factors and their influence on initiation rates.

KURLAND: I was thinking of an even more global question than our Chairman, which is the issue which you raised in the introductory remarks. You said that the regulation of protein synthesis is initiation-limited. By this do you mean that in the eukaryotic system, if only you increase the initition rate, the rate of synthesis will go up? Or do you have a different sense of limitation by initiation?

HERSHEY: In this talk I did not distinguish different kinds of initiation events, but I would like to do so at this moment. I think that there are basically 2 kinds of initiation events which differ mechanistically in some unknown fashion. The 2 kinds are the very first initiation event on a messenger RNA, i.e., what I would call the mobilization of RNP into polysomes, in other words the very first ime a ribosome binds to a messenger RNA. The second kind are all subsequent initiation events which build and maintain the polysomal state of that messenger RNA. I think there is good evidence that these 2 kinds of initiation events are in fact different, at least the rates at which they occur are different. Analysis of a specific mRNA following sucrose gradient centrifugation usually shows a bell-shaped distribution around the average polysome size commensurate with the mRNA's size, very low amounts of mRNA in monosomes and disomes, but much higher amounts in the RNP fraction. This indicates that the rate of mobilization is slower than the subsequent initiation events on the mRNA. A good example is found during early development of sea-urchin embryos. At the moment of fertilization, less than 1% of ribosomes and mRNA are in polysomes, but that part which is active occurs as full-sized polysomes, not very small ones. Following fertilization, the rates of re-initiation and elongation on polysomes do not change more than 2- to 3-fold, whereas the 30-fold activation of protein synthesis is due primarily to an increase in the rate of mobilization (i.e., the number of mRNAs being translated). Mammalian cells also regulate overall translation rates by modulating either the mobilization or the re-initiation/elongation rates.

KURLAND: Do you believe that the ribosome is the rate-limiting component, and that regulation is the traffic control for getting the messenger into that limiting component?

HERSHEY: There is certainly an excess of mRNA in the form of RNP, whereas in rapidly growing cells about 90% of the ribosomes are found in polysomes. However, under some conditions one finds substantial amounts of 80S ribosomes which apparently are not active at that moment. The cell appears to regulate the number of mRNAs that are actually being translated, keeping the number in reasonable balance with the number of ribosomes available so that ribosomes do not become excessively rate-limiting. However, I have no knowledge of the mechanisms by which the cell senses ribosomes' availability or regulates the number of mRNAs in polysomes.

SCHATZ: I would like to expand on your interesting discussion. I have always been intrigued by the possibility that some of the components for initiation and, for that matter, protein synthesis itself, may be compartmentalized in the cell. Some of the "soluble" initiation factors might really attach to something, this attachment could be regulated by phosphorylation or dephosphorylation. What is your view on this?

HERSHEY: The question is whether there is a kind of cellular compartmentalization that might contribute to translational control. I would say yes, I suspect so, but I am not sure. The compartmentalization of which you surely are thinking, or at least of which I am thinking, is the attachment of messenger RNA (i.e., polysomes) to the cytoskeletal framework as shown by Penman and co-workers. No one knows the molecular nature of the interaction, or whether it is, in fact, truly to the cytoskeleton as we generally call it. Another piece of evidence is from Hans Trachsel's laboratory in Basel. He prepared a monoclonal antibody against cap-binding proteins, which also binds to intermediate filaments of the cytoskeleton. This is a difficult result to interpret since we cannot be certain that the immunostaining antibody is actually binding exclusively to cap-binding proteins. We have also analyzed cytoskeletal fractions prepared according to Penman, by solubilizing cells very gently with detergent, and found that most of the initiation factors are in the cytoskeletal fraction. I understand that Hans Trachsel does not get the same results, so this may be controversial. The factors appear not to be attached to the mRNA, but rather to the ribosomes. Thus, there are a number of people pursuing the idea that the cytoskeleton is involved in protein synthesis, but is has been very difficult to generate compelling evidence.

GOLD: As the Chairman is interested in messenger selection, I would like to add one comment, which is that the most conserved nucleotide in prokaryotic messengers, aside from the nucleotides in the initiation codon, turns out to be an A at −3 upstream from the initiator codon. A highly conserved A occurs at the same position in eukaryotic messages. Marilyn Kozak recently reported an experiment in which she changed that A to a C and changed a message which is beautifully translated into a message which is not translated at all. So, one begins to recognize further signals between the 5′-cap and the initiation codon.

KLUG: As I understood you, the poliovirus-induced proteolytic cleavage of the p220 means that there is a change in the specificity of translation. Is it so simple? Can one thereby deduce that this particular factor simply recognizes the cap structure? Does the cap structure bind to the protein?

HERSHEY: It is difficult to answer. CBP-II or (eIF-4F) is clearly involved in the cap recognition process, I think, although it is not clear exactly how. There are other initiation factors that are also involved. When crude cell fractions and ATP are mixed with oxidized capped mRNAs, one sees crosslinking to CBP-I, eIF-4A and eIF-4B. Both CBP-I and eIF-4A are components of CBP-II, along with p220, yet no crosslinking of caps to p220 has been reported. Perhaps p220 is involved in organizing the other proteins in a form active for cap recognition and binding, and that proteolytic cleavage of p220 prevents complex formation. Whether or how the altered CBP-II recognizes uncapped mRNA is an open question.

KERR: I am trying to get back to what is happening *in vivo* in the intact poliovirus-infected cell. I seem to remember 10–15 years ago people did experiments with amino acid analogues, whereby they could inhibit the cleavage of the precursor polio protein which would prevent the production of the polio protease. I am trying to remember whether or not host protein synthesis was shut off in those cells.

HERSHEY: This is a very interesting question. I am not aware of any experiments which properly test this idea in poliovirus-infected HeLa cells. In those experiments which demonstrated the presence of the full-length polio polyprotein, the amino acid analogues were added a number of hours after infection, when host cell protein synthesis was fully inhibited and essentially only polio protein was being synthesized. However, in a closely related system, mengovirus infection of L cells, high concentrations of p-fluorophenyl-alanine prevented the normal inhibition of host cell protein synthesis as well as polyprotein cleavage (Baltimore, D., Franklin, R. & Callendar, J. (1963) Mengovirus-induced inhibition of host ribonucleic acid and protein synthesis. *Biochem. Biophys. Acta 76,* 425–430).

Regulation of eIF-2B-Mediated Guanine Nucleotide Exchange by Limited Phosphorylation of eIF-2α

Brian Safer

INTRODUCTION

During eukaryotic protein synthesis initiation, binding of Met-tRNA$_i$ to native 43S ribosomal subunits is mediated by formation of an eIF-2 · GTP · Met-tRNA$_i$ ternary complex. As the first step in an extremely endergonic process, ternary complex formation is tightly regulated by two distinct, though related mechanisms. The first is the direct regulation of eIF-2 activity by the energy charge of the guanine nucleotide pool. Ternary complex formation occurs by an ordered sequential process, in which eIF-2 is first activated by the initial formation of an eIF-2 · GTP binary complex (Safter *et al.* 1975, Siekierka *et al.* 1983). Binding of GDP, however, prevents binding of Met-tRNA$_i$, and consequently all other steps of translational initiation (Jagus *et al.* 1981). As the K_D^{GDP} of eIF-2 is 3.1×10^{-8} M, while the K_D^{GTP} is 2.5×10^{6} M (Walton & Gill 1975, Safer *et al.* 1982), binding of Met-tRNA$_i$ at physiologic concentrations of guanine nucleotides is directly regulated by small changes in the GTP/GDP ratio, and, thereby, the overall metabolic status of the cell. In addition, since hydrolysis of GTP present in the ternary complex upon completion of the 80S initiation complex is required for release of eIF-2 and other initiation factors (Peterson *et al.* 1979, Merrick 1979), eIF-2 is thought to be released as an inactive eIF-2 · GDP binary complex. Reactivation of eIF-2 at rates sufficient to support the translational requirements of the cell, equivalent to the exchange of bound GDP for GTP (Jagus *et al.* 1981, 1982, Clemens *et al.* 1982), requires the

Laboratory of Molecular Hematology, National Heart, Lung, and Blood Institute, Bethesda, Maryland 20205, USA

GENE EXPRESSION, Alfred Benzon Symposium 19.
Editors: Brian F. C. Clark & Hans Uffe Petersen, Munksgaard, Copenhagen 1984.

participation of an ancillary factor designated eIF-2B. Modulation of their interaction, required to promote guanine nucleotide exchange, is the second major process by which eIF-2 activity is regulated, though indirectly, by the energy charge of the guanine nucleotide pool. The fundamental basis of this regulation, studied intensively in the hemin-dependent rabbit reticulocyte system, now appears to be understood.

REGULATION OF TRANSLATION BY HEMIN-AVAILABILITY AND eIF-2α KINASE

When rabbit reticulocyte lysate is supplemented with hemin, the rate of translation closely approaches that found in the intact reticulocyte. In the absence of hemin, however, normal rates of protein synthesis are maintained only for 4–6 min, and concomitant with a loss of polysomes, translation becomes inhibited by greater than 90%. As the onset of translational inhibition is preceded by activation of a cAMP-independent kinase specific for a single phosphorylation site on the α subunit of eIF-2, phosphorylation of eIF-2 is accompanied by decreased Met-$tRNA_i$ binding to initiating 43S ribosomal subunits, and the addition of exogenous eIF-2 to inhibited lysate transiently restores active translation, phosphorylation of eIF-2α was thought to be directly responsible for the observed inhibition of protein synthesis in hemin-deficient lysates (Farrell *et al.* 1977, Levin *et al.* 1981, Safer & Jagus 1981). To answer the question, however, as to whether phosphorylation is necessary and sufficient for translational inhibition, the following questions were asked: 1) What is the extent of eIF-2α phosphorylation? 2) Does the extent of phosphorylation correlate with the degree of translational inhibition observed? 3) Does phosphorylation inhibit eIF-2 activity directly? 4) What is the normal mechanism for eIF-2 reactivation that allows its catalytic recycling during protein synthesis initiation? 5) How is this affected by eIF-2α phosphorylation?

THE PHOSPHORYLATION STATE OF eIF-2 IN RABBIT RETICULOCYTE LYSATE

The phosphorylation state of eIF-2 was determined in hemin-supplemented lysate undergoing active translation and in hemin-deficient lysate after the onset of translational inhibition. eIF-2 was first purified in the presence of 50 mM NaF and 10 mM EDTA, and tracer amounts of eIF-2($\alpha^{32}P$, $\beta^{32}P$) to monitor for possible phosphatase activity, and measure the recovery of eIF-2 during its

Table I
Phosphorylation state of eIF-2

Condition	pmol phosphate/pmol subunit		
	α	β	γ
None	$\leq$0.05	1.9	0
eIF-2 Phosphatase		0	0
eIF-2α Kinase	0.9–1.1	$\leq$0.05	0
eIF-2β Kinase	$\leq$0.05	1.8–2.0	0
eIF-2 (hemin-supplemented lysate)	0.1	1.9	0
eIF-2 (hemin-deficient lysate)	0.3	2.0	0

[1] Chemical determination
[2] ^{32}P incorporation into dephosphorylated eIF-2
[3] Derived from ^{32}P incorporation in eIF-2 purified in the presence of 10mM EDTA and 50 mM NaF.

purification. No residual enzymatic activity could be detected and the recovery of eIF-2 was approximately 80%. The α, β and γ subunits of eIF-2 were then separated by reverse phase chromatography on C_{18} μBondapak using a Waters HPLC system. In both hemin-deficient and hemin-supplemented lysates, the two phosphorylation sites on the β subunit are almost totally occupied (Table I). This is in agreement with a low turnover rate for phosphate on the β subunit, resulting from the association of eIF-2 with either or both Met-$tRNA_i$ and eIF-2B (Crouch & Safer 1983). In contrast, phosphate on the single α subunit site phosphorylated by the hemin-regulated kinase has a halflife in lysate of 15–20 sec, and is accessible to both kinase and phosphatase activity in lysate (Safer & Jagus 1979). In hemin-supplemented lysate, less than 10% of the α subunit sites are phosphorylated. In hemin-deficient lysates, the extent of eIF-2α phosphorylation increases to 25% (Safer *et al.* 1981, Jagus *et al.* 1982). The low extent of eIF-2α phosphorylation appears to result from a high level of eIF-2 phosphatase activity in lysate and its continued accessibility to the α site in hemin-deficient lysate. Thus, the extent of eIF-2α phosphorylation in hemin-deficient lysates, approximately 25–30%, is low relative to the greater than 95% inhibition of translation observed. The major question posed, then, is why does not the remaining 70–75% of nonphosphorylated eIF-2 function normally during protein synthesis initiation.

DOES PHOSPHORYLATION OF eIF-2α INHIBIT ITS ACTIVITY?

To examine the *in vitro* effects of eIF-2 phosphorylation, eIF-2 having defined phosphorylation states was first prepared. Following an initial dephosphorylation by incubation with its purified phosphatase and repurification, dephosphorylated eIF-2 was incubated with either or both kinases specific for sites on its α and β subunits. Two mol phosphate are incorporated into its β subunit sites, and one mol into its α subunit site by eIF-2β kinase (casein kinase II) and eIF-2α kinase (HRI), respectively (Table I). When either Met-tRNA$_i$ or GDP binding was examined, no significant differences were observed between dephosphorylated, and both α and/or β phosphorylated preparations. All preparations of eIF-2 were 60–70% active in forming eIF-2 · GTP · Met-tRNA$_i$ ternary complexes, and almost totally active in binding GDP.

The ability of phosphorylated and non-phosphorylated eIF-2 to restore translational activity in hemin-deficient lysates was also compared. Both preparations of eIF-2 produced an immediate resumption of translational activity following their addition to inhibited hemin-deficient lysates. When the extent of this rescue was quantitated, both non-phosphorylated and phosphorylated eIF-2 were found to promote only one initiation event per eIF-2 added. This is in contrast to the highly catalytic utilization of eIF-2, and other initiation factors normally observed. Thus, eIF-2 does not appear to undergo any inactivation as a direct consequence of its phosphorylation, when only its stoichiometric function is observed. Rather, the inhibition of translation in hemin-deficient lysates following the partial phosphorylation of the eIF-2 pool is the result of an inability to utilize eIF-2 catalytically.

PURIFICATION AND CHARACTERIZATION OF THE eIF-2 RECYCLING COMPLEX eIF-2B

At physiologic ionic strengths, eIF-2 exists as either the free αβγ trimer, or as part of much larger complex designated eIF-2 · eIF-2B (Konieczny & Safer 1983, Siekierka *et al.* 1981, Amesz *et al.* 1979, Gupta *et al.* 1982, Matts *et al.* 1983). Following partial purification by phosphocellulose and DEAE chromatography, free eIF-2 and the eIF-2 · eIF-2B complex can be readily resolved by glycerol gradient centrifugation (Fig. 1). Gradient fractions were assayed for both GTP-dependent Met-tRNA$_i$ binding and the ability to restore translation in inhibited hemin-deficient lysates. The polypeptide composition of each fraction was also analyzed by SDS PAGE. In the 6S region of the gradient (fractions 7–8), Met-tRNA$_i$ binding activity coincides with the distribution of free eIF-2 seen in the stained gel. A much smaller peak of Met-tRNA$_i$ binding activity is also seen, however, in the 12S region (fractions 15–17). When assayed for the ability to stimulate [^{14}C]valine incorporation into globin, free eIF-2 has

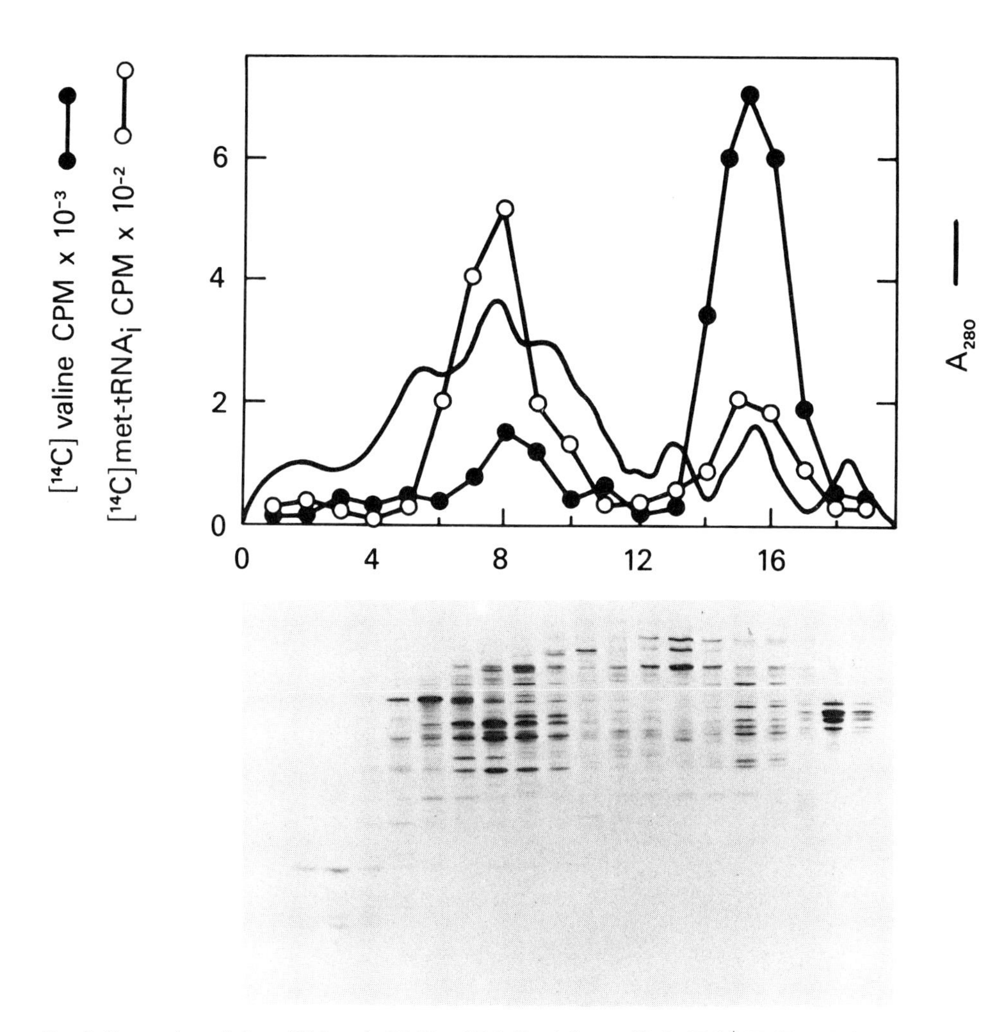

Fig. 1. Separation of free eIF-2 and eIF-2B · eIF-2. Partially purified eIF-2/eIF-2B · eIF-2 was resolved by centrifugation on 15–40% glycerol gradients at 40,000 rpm for 32 h. A. The A_{280} profile (—), GTP-dependent [^{14}C]Met-tRNA$_i$ binding activity (○—○), and the stimulation of [^{14}C]valine incorporation into globin in hemin-deficient rabbit reticulocyte lysate (●—●) are shown. B. SDS PAGE analyses of gradient fractions showing the bimodal distribution of eIF-2 subunits (●) in free eIF-2 and eIF-2B · eIF-2.

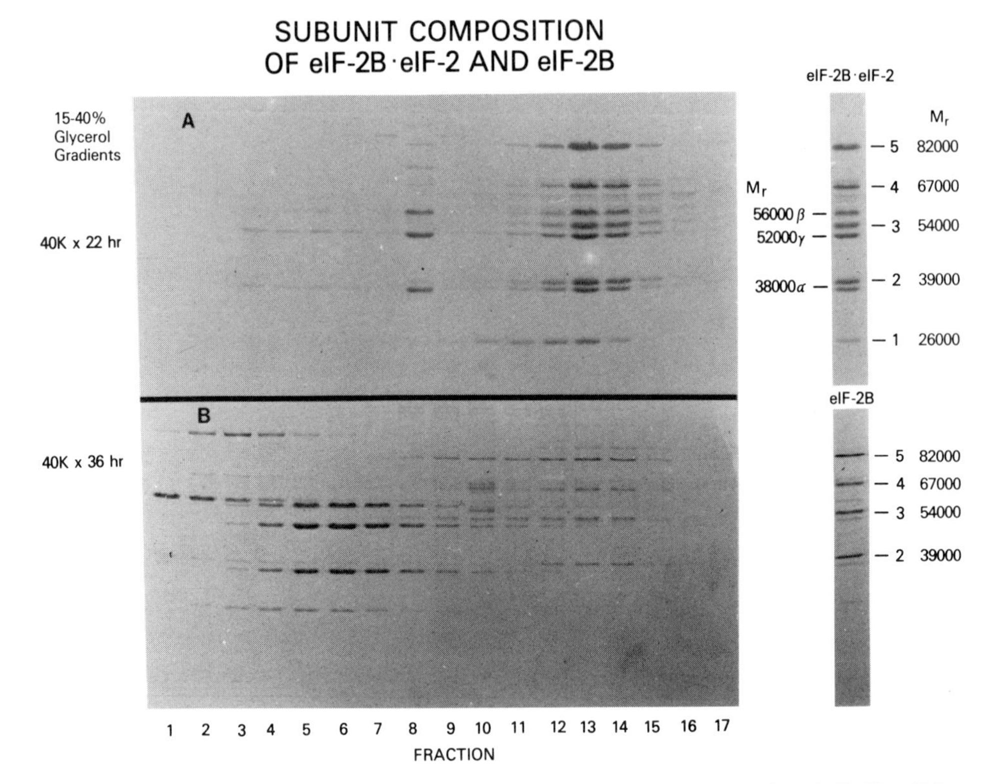

Fig. 2. SDS PAGE analyses of eIF-2B · eIF-2 and free eIF-2B. A. Final purification of eIF-2B · eIF-2 on 15–50% glycerol gradients. 2 μg eIF-2 have been added to fraction 8 as a marker. B. Separation of eIF-2B · eIF-2 into free eIF-2 and eIF-2B by glycerol gradient centrifugation at 500 mM KCl. The identification of the subunit composition and M_r are presented to the right of (A) and (B).

only 10% of the activity seen at 12S, which by SDS PAGE analysis also contains the α, β and γ subunits of eIF-2. While the Met-tRNA$_i$ binding activity in these 2 regions of the gradient appeared to be proportional to the relative amounts of free and complexed eIF-2, restoration of translational activity was markedly more efficient in the latter. The final purification of this complex, designated eIF-2 · eIF-2B is shown in Figure 2. In A, the polypeptide composition of fractions following recentrifugation on 15–40% glycerol gradients is shown. The eIF-2 · eIF-2B complex consists of 8 subunits present in equimolar amounts, 3 of which are the α, β and γ subunits of eIF-2, and 5 of which compose the eIF-2B component. The native M_r of this complex is 393,000 daltons, consistent with a

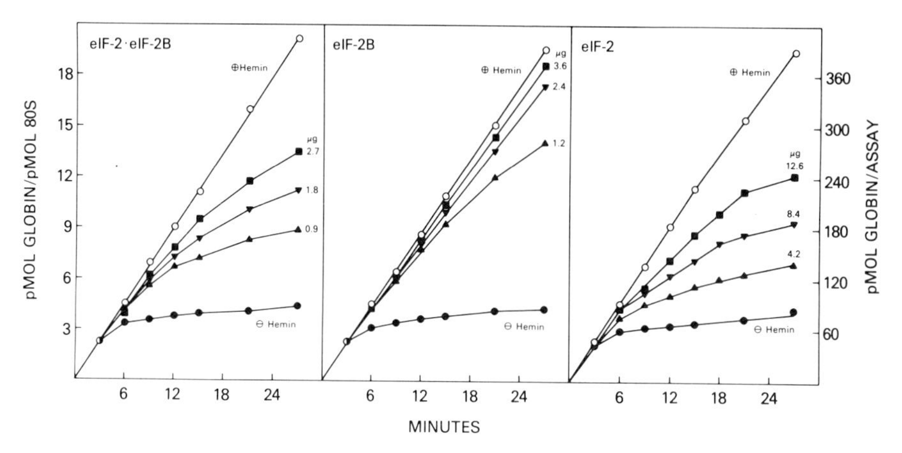

Fig. 3. Comparison of eIF-2B · eIF-2, eIF-2B, and eIF-2 activity in hemin-deficient reticulocyte lysate. The ratio of globin synthesized to pmol of added factor, after 20 min incubation, shows catalytic stimulation by eIF-2B · eIF-2 and free eIF-2B, but not free eIF-2.

$S_{20,w}=12$. At high salt concentrations or upon phosphorylation of eIF-2α at physiologic ionic strengths, eIF-2 · eIF-2B is dissociated into free eIF-2 and eIF-2B (5 subunits; $M_r=26$, 39, 58, 67 and 82 KD).

eIF-2B RESTORES CATALYTIC ACTIVITY OF eIF-2 IN HEMIN-DEFICIENT LYSATES

The relative abilities of free eIF-2, the eIF-2 · eIF-2B complex, and free eIF-2B to restore translational activity in hemin-deficient lysates was examined (Fig. 3). In the absence of any additions, biphasic kinetics of translational inhibition in hemin-deficient lysates are observed. The addition of all 3 preparations to hemin-deficient lysates is seen to rescue translational activity. For each pmol of eIF-2 added, however, only 1 to 2 pmols of globin are synthesized. In contrast, eIF-2 · eIF-2B restores translational activity in a highly catalytic fashion, and for each pmol added, 30 to 40 pmol globin are synthesized. The ability to restore translation catalytically results from the eIF-2B moiety of this complex, since identical results are obtained with free eIF-2B.

The ability of eIF-2B to maintain translational activity in hemin-deficient lysates appears to result from its effect on ternary complex formation at physiologic concentrations of guanine nucleotides. Table II shows that in the

Table II

Additions	met-tRNA$_i$ bound			met-tRNA$_i$/eIF-2	
	eIF-2	eIF-2(αP)	eIF-2B	eIF-2	eIF-2(αP)
130 μM GTP	9.0	8.6	0	0.7	0.7
125 μM GTP 5 μM GDP	1.8	1.9	0	0.2	0.2
125 μM GTP 5 μM GDP 0.4 μg eIF-2 · eIF-2B	6.7	2.3	0.6	0.6	0.2
125 μM GTP 5 μM GDP 0.2 μg eIF-2B	6.8	2.3	0	0.6	0.2

presence of GTP alone, purified eIF-2 is 70% active in ternary complex formation. Phosphorylation of the α subunit does not affect either the rate or extent of Met-tRNA$_i$ binding. Free eIF-2B does not bind Met-tRNA$_i$ directly. At physiologic GTP and GDP concentrations, however, Met-tRNA$_i$ binding by free eIF-2 is inhibited by 80%. This is a consequence of the unfavorable K_D^{GTP} of free eIF-2 relative to its K_D^{GDP}. At physiologic concentrations of GTP and GDP, 130 and 5 μM, respectively, 95% of the total eIF-2 pool (50 pmol/ml lysate) would exist as the inactive eIF-2 · GDP complex. This inhibition of ternary complex can be prevented, however, by either eIF-2 · eIF-2B or free eIF-2B. In the presence of both GTP and GDP, the addition of eIF-2 · eIF-2B promotes binding of an additional 4.9 pmol Met-tRNA$_i$. The small amount of Met-tRNA$_i$ binding seen with eIF-2B · eIF-2 alone results from the eIF-2 moiety of the complex. Similarly, the addition of 0.5 pmol free eIF-2B stimulates the binding of an additional 5 pmol Met-tRNA$_i$. This catalytic promotion of ternary complex formation at physiologic guanine nucleotide concentrations (eIF-2 · eIF-2B and free eIF-2B stimulate Met-tRNA$_i$ binding 5-10 fold) is totally inhibited, however, by phosphorylation of eIF-2α (Safer *et al.* 1982, Clemens *et al.* 1982, Siekierka *et al.* 1982, Voorma & Amesz 1982).

EFFECT OF eIF-2B ON GTP AND GDP BINDING

To examine the mechanism by which eIF-2B promoted ternary complex formation at physiologic GTP and GDP concentrations, the K_D of both free and

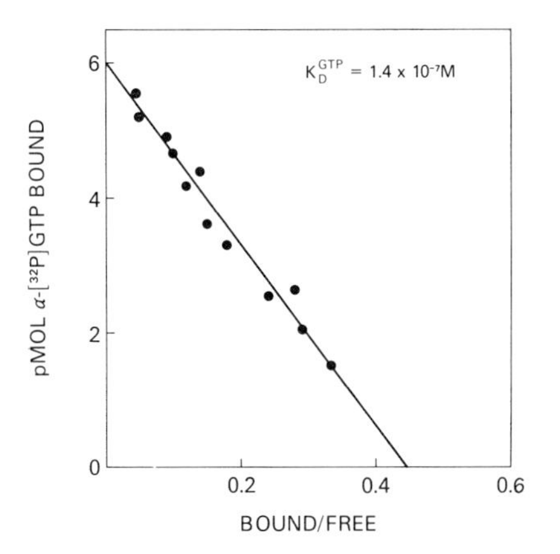

Fig. 4. Scatchard analysis of GTP binding by eIF-2B · eIF-2. 5 μg of eIF-2B · eIF-2 and [8-^{3}H]GTP (1–400 μM) were incubated for 3 min at 30° in 50 mM KCl, 2 mM $MgCl_2$, 10 mM MES, pH 7.1, and 2 mM DTT. GTP binding to eIF-2B · eIF-2 was measured by nitrocellulose binding and analyzed according to Scatchard, 1949.

eIF-2B-complexed eIF-2 for guanine nucleotides were determined by Scatchard analysis. In agreement with the data of Walton and Gill (1975), a dissociation constant of 3.1×10^{-8} M for eIF-2 · GDP was found. Phosphorylation of eIF-2α had no effect on the K_D^{GDP} of free eIF-2. (Konieczny & Safer 1983). The K_D^{GTP} of eIF-2 could not be determined directly due to the inability to form a stable eIF-2-GTP binary complex. When eIF-2 is complexed with eIF-2B, however, the K_D^{GDP} is increased to 1.8×10^{-7} M and a stable eIF-2 · GTP · eIF-2B complex can be isolated having a $K_D^{GTP} = 1.7 \times 10^{-7}$ M (Fig. 4). At physiologic guanine nucleotide concentrations, therefore, the percent of eIF-2 present as the active GTP complex is increased to approximately 90%. The exact value would depend on the relative sizes of the free eIF-2 and eIF-2B pools. The association of eIF-2 with eIF-2B, therefore, appears to be required to effect GDP:GTP exchange, analogous to the prokaryotic EF-Tu/EF-Ts system (Kaziro 1979). The requirement for an enzymatic guanine nucleotide exchange system can be bypassed, however, by increasing the GTP/GDP ratio in reticulocyte lysate (Fig. 5). The addition of 1 mM GTP · Mg^{2+} to hemin-supplemented lysates has no effect, but immediately restores control rates of protein synthesis in hemin-deficient lysates. This resumption of translational activity is not accompanied by eIF-2α dephosphorylation, which only occurs after several minutes have elapsed.

To investigate further the mechanism by which eIF-2 association with eIF-2B promotes guanine nucleotide exchange and, thereby, allows catalytic utilization of eIF-2, the subunit specificity of 8-azidoguanosine triphosphate crosslinking to either eIF-2 · eIF-2B and/or free eIF-2 was examined (Fig. 6). Following incubation with [^{32}P]-8-azido GTP, alone or in the presence of GTP, GDP or ATP, the subunit localization of guanine nucleotide binding sites was examined by autoradiography following SDS PAGE. The position of the 8 eIF-2 · eIF-2B subunits are shown in the photograph of the Coomassie blue stained gel; the α and β subunits of eIF-2 are indicated at M_r=38,000 and 56,000 respectively, in the autoradiogram. It is important to note that the amount of free eIF-2 utilized in these experiments was 30 fold higher than the eIF-2 · eIF-2B complex. With either free eIF-2 or the eIF-2 · eIF-2B complex, the α and β subunits are weakly labelled upon activation of the GTP photo-affinity probe. In the presence of both, α-[^{32}P]-8-azido GTP crosslinks strongly to the α and β subunits of eIF-2. These data indicate that both the α and β subunits may participate in guanine nucleotide exchange in the eIF-2B · eIF-2B complex, and that association of eIF-2 with this complex is required for binding of GTP. In addition, crosslinking of 8-azido GTP is specific for the α and β subunits of eIF-2B · eIF-2. The data also supports the rapid exchange of free eIF-2 in and out of the complex, independent of Met-tRNA$_i$; that is, release of active eIF-2 · GTP from the

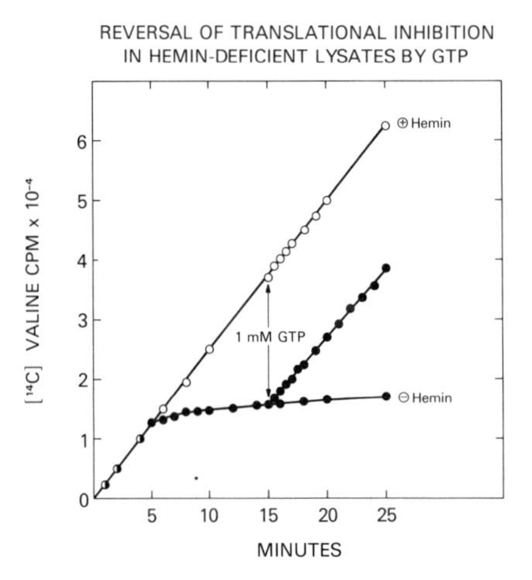

Fig. 5. GTP produces an immediate resumption of protein synthesis in inhibited hemin-deficient lysate. 1 mM GTP · Mg was added after 10 min incubation to both hemin-supplemented and hemin-deficient rabbit reticulocyte lysate.

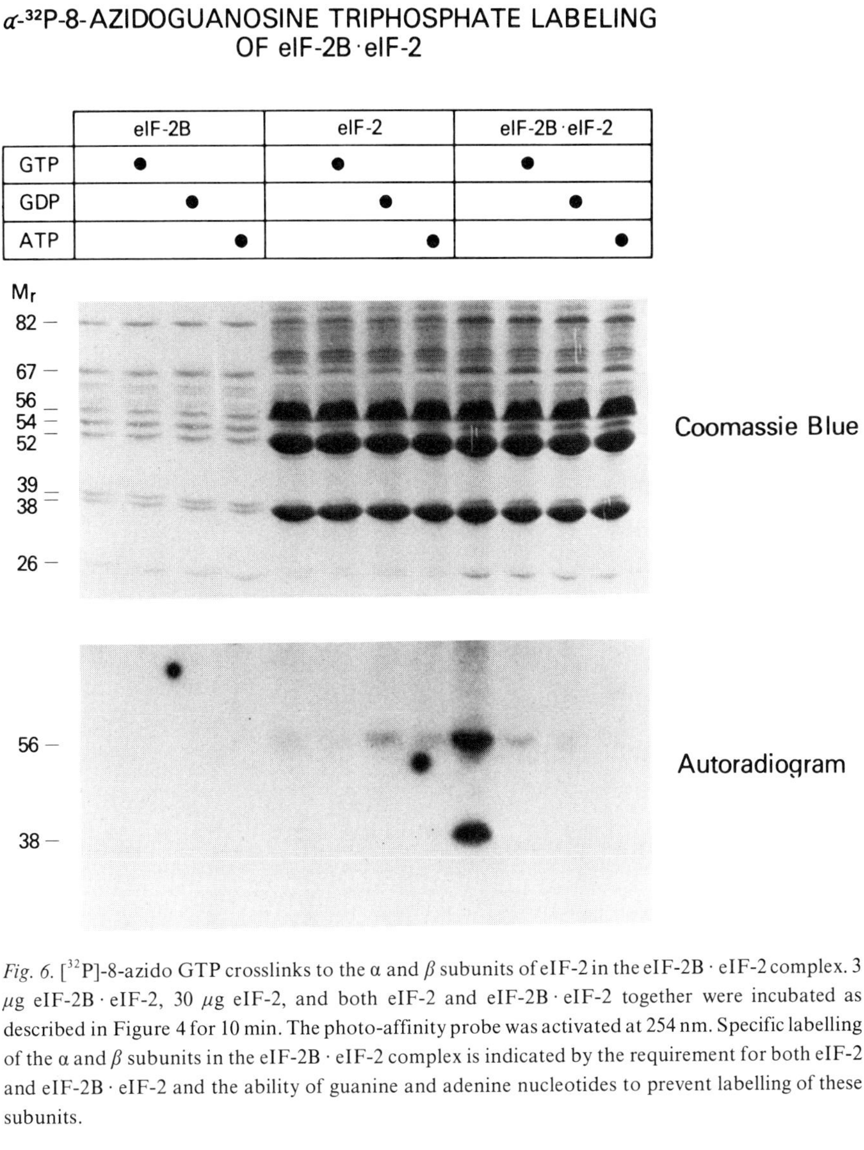

α-^{32}P-8-AZIDOGUANOSINE TRIPHOSPHATE LABELING OF eIF-2B · eIF-2

	eIF-2B			eIF-2			eIF-2B · eIF-2		
GTP	●			●			●		
GDP		●			●			●	
ATP			●			●			●

M_r
82 —
67 —
56 —
54 —
52 —
39 —
38 —
26 —

Coomassie Blue

56 —
38 —

Autoradiogram

Fig. 6. [^{32}P]-8-azido GTP crosslinks to the α and β subunits of eIF-2 in the eIF-2B · eIF-2 complex. 3 μg eIF-2B · eIF-2, 30 μg eIF-2, and both eIF-2 and eIF-2B · eIF-2 together were incubated as described in Figure 4 for 10 min. The photo-affinity probe was activated at 254 nm. Specific labelling of the α and β subunits in the eIF-2B · eIF-2 complex is indicated by the requirement for both eIF-2 and eIF-2B · eIF-2 and the ability of guanine and adenine nucleotides to prevent labelling of these subunits.

complex does not require prior formation of the eIF-2 · GTP · Met-tRNA$_i$ ternary complex.

Additional support for a role of the eIF-2 α and β subunits in guanine nucleotide exchange comes from initial studies (in collaboration with Joel Moss and Drucilla Burns) on the ability of turkey erythrocyte enzyme to ADP ribosylate the α and β subunits of free eIF-2 and eIF-2 · eIF-2B. This and other related ADP-ribosylation enzymes such as cholera and pertussis toxins appear to selectively derivatize argenine residues at guanine nucleotide binding sites in a variety of systems including EF-2, and the adenylate cyclase and insulin receptor regulatory subunits (Burns *et al.* 1982, Cooper *et al.* 1981). While EF-2 is readily ADP-ribosylated by cholera toxin, only the turkey enzyme is able to ADP-ribosylate eIF-2. The basis of this selectivity and its effects on the guanine nucleotide exchange process are currently being investigated.

TWO POSSIBLE MODELS OF TRANSLATIONAL INHIBITION IN HEMIN-DEFICIENT LYSATES

The regulation of eIF-2 activity by guanine nucleotide binding, the effect of eIF-2B association on the stability of guanine nucleotide binding by eIF-2, and the ability of exogenous eIF-2B to effect a catalytic rescue of translational activity in hemin-deficient lysates support a major role for eIF-2B in the mechanism of inhibition associated with eIF-2α phosphorylation. From these and other studies, it does not appear that the defect in hemin-deficient lysates results from a direct inactivation of eIF-2 upon phosphorylation. Rather, the inhibition of translational initiation appears to result from an inability to utilize phosphorylated eIF-2 catalytically at physiologic concentrations of guanine nucleotides (Safer & Jagus 1982). It is important to recall that translational activity is immediately restored by artificially increasing the GTP concentration to 1 mM. Since this would require the immediate utilization of endogenous eIF-2, without dephosphorylation, it is apparent that no direct inactivation of eIF-2 in this system has occurred.

Two models can be proposed based on the apparent inhibition of eIF-2B-mediated guanine nucleotide exchange in hemin-deficient lysates. In the first, the association of eIF-2 with eIF-2B would be directly inhibited by eIF-2α phosphorylation (Voorma & Amesz 1982, Siekierka 1981). In the second model, eIF-2B would be sequestered by eIF-2(αP) into a complex unable either to exchange guanine nucleotides, or release eIF-2 · GTP upon exchange for free eIF-2 · GDP (Safer *et al.* 1982, Konieczny & Safer 1983, Siekierka *et al.* 1983). A major problem with the first model is that in hemin-deficient lysates, only 25–30 percent of eIF-2α is phosphorylated. Therefore, why is not the remaining non-phosphorylated eIF-2 used normally?

GUANINE NUCLEOTIDES STABILIZE THE eIF-2(αP) · eIF-2B COMPLEX

In the absence of guanine nucleotides, eIF-2α phosphorylation inhibits the exchange of free eIF-2(α^{32}P) for eIF-2 in the eIF-2 · eIF-2B complex (Fig. 7A). In contrast, exchange of eIF-2(β^{32}P) is unaffected (Fig. 7B). Such data would appear, at first, to favor the model of translational inhibition based on eIF-2 · eIF-2B dissociation. However, re-examination of the effect of eIF-2α phosphorylation was performed at physiologic concentrations of GTP and GDP, since the major function of the eIF-2 · eIF-2B complex concerned the exchange of bound guanine nucleotides from its eIF-2 moiety. Figure 7C shows

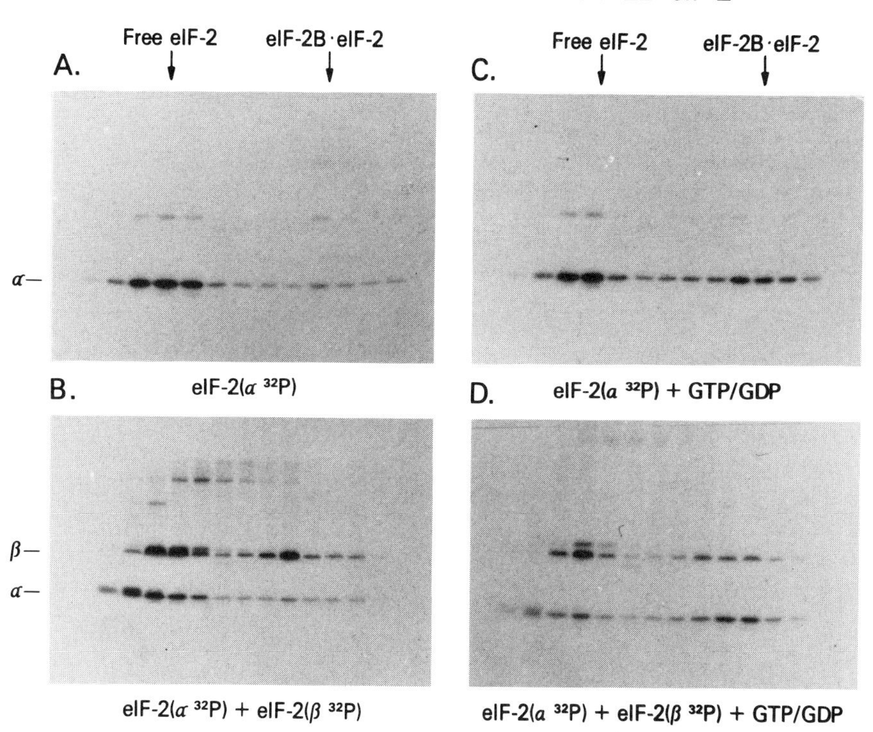

Fig. 7. Guanine nucleotides stabilize the eIF-2B · eIF-2(α^{32}P) complex. A. eIF-2(α^{32}P) exchanges poorly into eIF-2B · eIF-2. B. eIF-2(β^{32}P) exchanges readily into eIF-2B · eIF-2. In the presence of 130 μM GTP and 6 μM GDP, however, (C) eIF-2(α^{32}P) readily exchanges into eIF-2B · eIF-2 and (D) exchanges in preference to eIF-2(β^{32}P). The distribution of eIF-2(α^{32}P) and eIF-2(β^{32}P) were analyzed by autoradiography of glycerol gradient fractions following SDS PAGE.

that when exchange of free eIF-2(α^{32}P) for complexed eIF-2 is examined in the presence of guanine nucleotides, eIF-2(α^{32}P) readily exchanges for eIF-2 in the eIF-2 · eIF-2B complex. Furthermore, when eIF-2 · eIF-2B is incubated with eIF-2(α^{32}P) and eIF-2(β^{32}P), guanine nucleotides promote a more stable association of eIF-2(α^{32}P) · eIF-2B than eIF-2(β^{32}P) · eIF-2B (Fig. 7D). It is possible, therefore, that phosphorylation of eIF-2α, by promoting the formation of this complex, effectively sequesters eIF-2B into an inactive complex incapable of guanine nucleotide exchange, or one that is unable to exchange eIF-2 · GTP from the complex for inactive eIF-2 · GDP.

RELATIVE SIZE OF THE eIF-2 AND eIF-2B POOLS

The previous results are compatible with the model of translational inhibition involving sequestration of eIF-2B into a complex incapable of effective guanine nucleotide exchange. Since phosphorylation of eIF-2α only occurs to a limited extent in hemin-deficient lysate (25–30 percent), it is important to determine the relative sizes of the eIF-2 and eIF-2B pools. Figure 8 shows the results of a glycerol gradient analysis of the total eIF-2B and eIF-2 · eIF-2B pool obtained by prior phosphocellulose and DEAE chromatography of rabbit reticulocyte lysate. In panel A, the position of free eIF-2 (peak fraction=4) and of the eIF-2 · eIF-2B complex (peak fraction=12) is located. Panel B is an autoradiogram of gradient fractions corresponding to A, following the addition of [γ-^{32}P]ATP to each fraction. The autophosphorylation of eIF-2α kinase is seen, as well as phosphorylation of the free eIF-2α subunit present in these fractions. To obtain the distribution of eIF-2 throughout the gradient, each fraction was, therefore, supplemented with [γ-^{32}P]ATP and exogenous eIF-2α kinase. Figure 9C shows that eIF-2α is now radiolabelled and is localized to those regions of the gradient corresponding to free eIF-2 and the eIF-2 · eIF-2B complex. By quantitation of alkali labile ^{32}P released from the eIF-2α subunit in these regions, a eIF-2:eIF-2B ratio of 6:1 is obtained. Identical results are obtained by Western blot analysis

---

Fig. 8. The relative sizes of the free eIF-2 and eIF-2B · eIF-2 pools was determined by glycerol gradient fractionation of the partially purified eIF-2/eIF-2B pool. (A) Coomassie blue stain of glycerol gradient fractions analyzed by SDS PAGE. (B) Autoradiogram of (A) following the addition of [γ-^{32}P]ATP to gradient fraction aliquots. (C) Autoradiogram of (A) following the addition of [γ-^{32o}]ATP and eIF-2α kinase to gradient fraction aliquots.

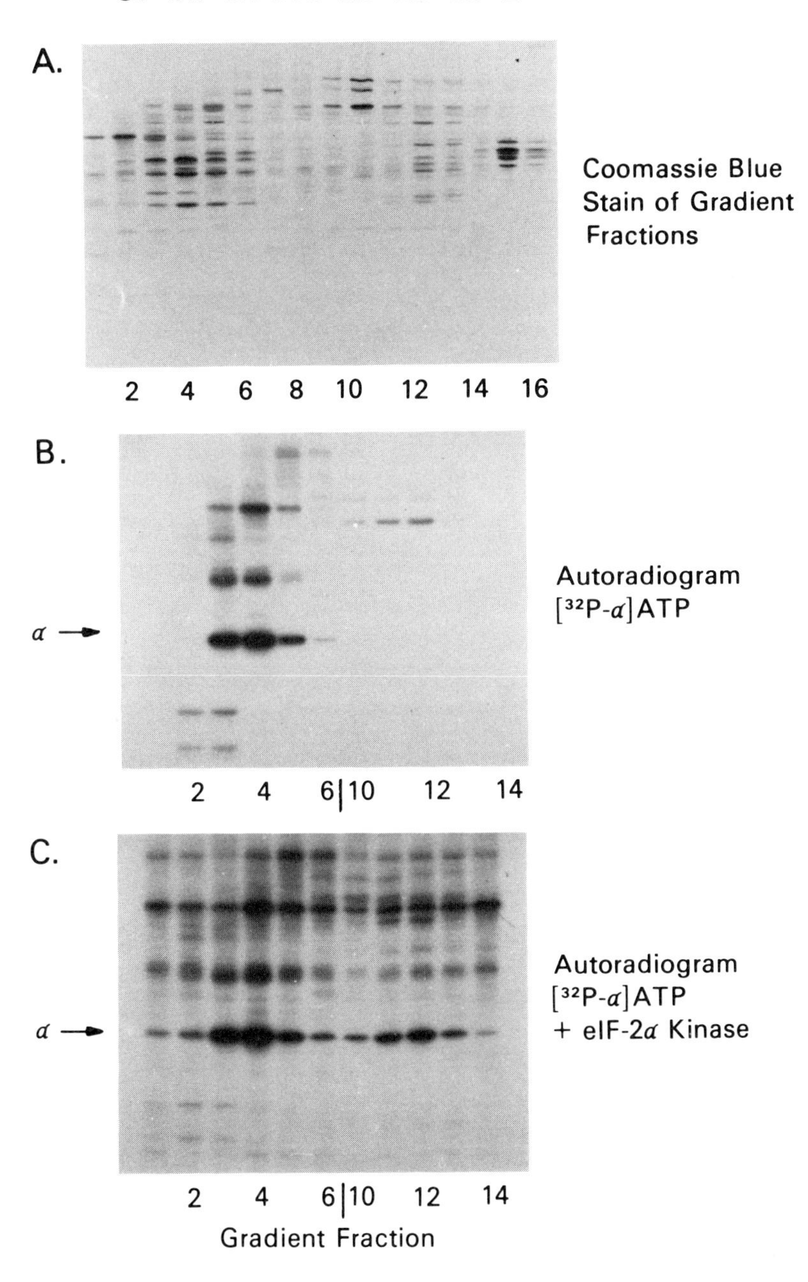
GLYCEROL GRADIENT ANALYSIS
OF THE RELATIVE POOL SIZES
OF eIF-2 AND eIF-2B·eIF-2
A.
Coomassie Blue
Stain of Gradient
Fractions
2 4 6 8 10 12 14 16
B.
α
Autoradiogram
[32P-α]ATP
2 4 6|10 12 14
C.
α
Autoradiogram
[32P-α]ATP
+ eIF-2α Kinase
2 4 6|10 12 14
Gradient Fraction

using anti-eIF-2β IgG and [^{125}I]protein A. These data agree well with the relative amounts of eIF-2 and eIF-2B purified from rabbit reticulocyte lysate. Partial phosphorylation of the free eIF-2 pool would, therefore, generate sufficient eIF-2(αP) to sequester most, if not all, of the eIF-2B pool.

MECHANISM OF HEMIN-REGULATED TRANSLATIONAL INHIBITION

The normal interaction of eIF-2 with other translational components during protein synthesis initiation, its association with eIF-2B which permits its catalytic recycling, and the mechanism of inhibition upon activation of eIF-2α kinase are summarized in Fig. 9. Following formation of the eIF-2 · GTP · Met-tRNA$_i$ ternary complex, binding to native 43S ribosomal subunits containing eIF-3 and eIF-4C occurs to form a 43S pre-initiation complex. This is followed by the ATP-dependent binding of mRNA, mediated by a large protein complex (CBP II) containing several polypeptides including 24,000 dalton cap binding protein (Tahara *et al.* 1981), and the 48,000 dalton protein formerly identified as eIF-4A (Griffo *et al.* 1982). This forms the second pre-initiation complex which now sediments at 48S. Upon 60S ribosomal joining and in the presence of eIF-5 (Peterson *et al.* 1979, Merrick *et al.* 1979), hydrolysis of GTP releases all ribosomal-bound initiation factors. Because of its high affinity for GDP, eIF-2

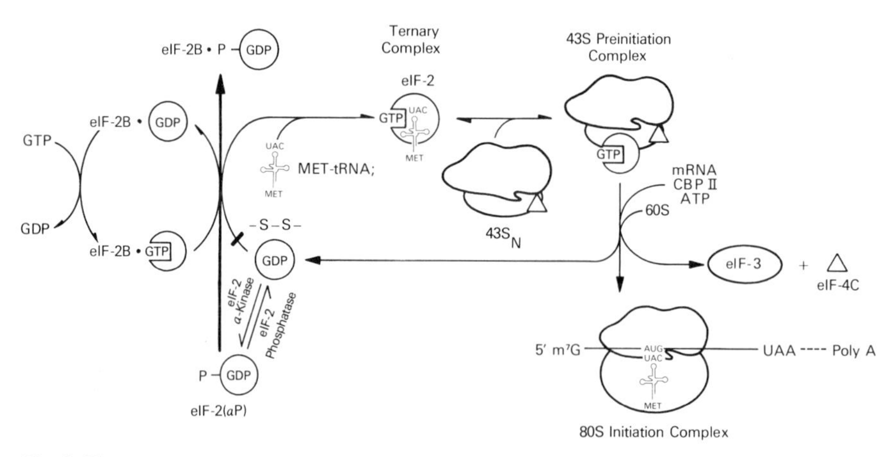

Fig. 9. The mechanism of eIF-2 recycling during protein synthesis initiation and its inhibition in hemin-deficient lysate.

is thought to be released as an eIF-2 · GDP binary complex. This is supported indirectly by the presence of equimolar GDP in preparations of eIF-2 (Siekierka *et al.* 1983). The eIF-2 · GDP binary complex is inactive and reutilization of eIF-2 requires the exchange of GTP for bound GTP. Since the K_D for GDP and GTP (3.1×10^{-8} M and 2.5×10^{-6} M, respectively) are unfavorable for rapid exchange at physiologic concentrations of guanine nucleotides, this exchange is mediated by a second factor designated eIF-2B. A complex of eIF-2B and eIF-2 has now been isolated by several laboratories (Amesz *et al.* 1979, Konieczny & Safer 1983, Siekierka *et al.* 1982, Matts *et al.* 1983, Gupta *et al.* 1982). Under physiologic conditions optimal for protein synthesis, eIF-2B exists in a stable complex with eIF-2. While dissociation of eIF-2 · eIF-2B is unlikely, free eIF-2 can readily exchange for bound eIF-2 in the complex (Safer *et al.* 1982). Although eIF-2 does not form a stable binary complex with GTP, a stable eIF-2 · GTP · eIF-2B complex can be isolated and the K_D^{GTP} determined (1.7×10^{-7} M). It is likely, therefore, that the guanine nucleotide exchange mediated by eIF-2B occurs by an exchange of bound eIF-2 · GTP for free eIF-2 · GDP. The similarity of the exchange process to the prokaryotic EF-Tu/EF-Ts system (Kaziro 1978) has been noted (Safer *et al.* 1982, Konieczny & Safer 1983, Siekierka *et al.* 1982). The molecular basis of this exchange reaction has not been determined to date, but no direct binding of guanine nucleotides by the eIF-2B moiety of the complex has been found, either by direct binding studies or crosslinking of 8-azido GTP. While the initial analysis of the number of guanine nucleotide binding sites indicates a single site, [^{32}P]-8-azido GTP crosslinks to both the α and β subunits of eIF-2 while in the eIF-2 · eIF-2B complex. Preparations of eIF-2 consisting of only the α and γ subunits, however, bind Met-tRNA$_i$ (Stringer *et al.* 1979), and are able to transiently rescue hemin-deficient lysate (unpublished observations, this laboratory). This may indicate that the β subunit may be involved only in guanine nucleotide exchange, while the α subunit contains the primary GTP binding site in the eIF-2 · GTP · met-tRNA$_i$ complex. Studies of α and β phosphorylation site accessibility to both kinase and phosphatase support an altered eIF-2 conformation resulting from its association with eIF-2B (Crouch & Safer 1983), or guanine nucleotides (Safer & Jagus 1979). eIF-2, therefore, is likely to be an allosteric protein whose Met-tRNA binding activity is modulated by structural alterations induced by binding of guanine nucleotides and/or eIF-2B.

In hemin-deficient lysates, eIF-2 phosphorylation interferes with the catalytic regeneration of eIF-2 · GTP. Since phosphorylated eIF-2 is able to participate normally in many, if not all of the partial reactions of protein synthesis initiation, and increased GTP levels allow normal utilization of endogenous eIF-2 without its prior dephosphorylation, the GTP/GDP exchange process mediated by eIF-2B is the critical process affected by eIF-2α phosphorylation. The mechanism of the inhibition appears to be the sequestration of eIF-2B into an inactive eIF-2B · eIF-2(αP) complex that is either unable to exchange bound GDP for GTP or is unable to release eIF-2(αP) · GTP from the complex in exchange for eIF-2 · GDP (Voorma *et al.* 1982, Siekierka *et al.* 1982, Safer *et al.* 1982, Clemens *et al.* 1982). Biphasic kinetics of inhibition are seen in lysate because phosphatase activity for eIF-2(αP) is high, and eIF-2α kinase activation only increases the steady-state level of eIF-2 phosphorylation from <10 to 30%. During each cycle of initiation, however, a proportion of eIF-2B is sequestered from the active pool into an inactive eIF-2B · eIF-2(αP) · GDP complex. As the pool size of eIF-2B is only 10–20% of eIF-2, the partial phosphorylation of eIF-2 can still generate sufficient eIF-2(αP) to exceed the amount of eIF-2B present in lysate (Konieczny & Safer 1983, Siekierka *et al.* 1983, Matts *et al.* 1983). When the pool of active eIF-2B is diminished to the point at which it is no longer able to regenerate eIF-2 · GTP at a rate sufficient to support initiation, translational inhibition is the result.

ACKNOWLEDGEMENTS

This work summarizes the creative and diligent efforts of my associates: Rosemary Jagus, Andrej Konieczny and Deborah Crouch, with whom it has been my pleasure to work. Appreciation is also given to Kirsten Cook for her excellent preparation of this manuscript.

REFERENCES

Amesz, H., Goumans, H. J., Haubrich-Moree, T., Voorma, H. O. & Benne, R. (1979) *Eur. J. Biochem. 98,* 513–520.

Burns, D. L., Moss, J., & Vaughan, M. (1982) *J. Biol. Chem. 257,* 32–34.

Cooper, D. M. F., Jagus, R., Somers, R. L. & Rodbell, M. (1981) *Biochem. Biophys. Res. Commun. 101,* 1179–1185.

Clemens, M. J., Pain, V. M., Wong, S.-T. & Henshaw, E. C. (1982) *Nature 296,* 93–95.

Crouch, D. & Safer, B. (1983) *Fed. Proc. 42,* 224.

Farrell, P. J., Balkow, K., Hunt, T., Jackson, R. J. & Trachsel, H. (1977) *Cell 11,* 187–200.

Gupta, N. K., Grace, M., Baherjee, A. C. & Bagchi, M. K. (1982) In: *Interaction of Translational and Transcriptional Controls in the Regulation of Gene Expression.* (M. Grunberg-Manago & B. Safer, eds.), pp. 339–358. Elsevier, New York.

Jagus, R., Anderson, W. F. & Safer, B. (1983) *Prog. Nucleic Acid Res. Mol. Biol. 25,* 127–15.

Jagus, R., Crouch, D., Konieczny, A. & Safer, B. (1982) *Curr. Top. Cellular Reg. 21,* 35–63.

Kaziro, Y. (1978) *Biochem. Biophys. Acta 505,* 95–127.

Levin, D., Ernst, V., Leroux, A., Petryshyn, R., Fagard, R. & London, I. M. (1981) In: *Protein Phosphorylation and Bioregulation,* (G. Thomas, E. J. Podesta, & J. Gordon, eds.), pp. 142–50, Karger, Basel.

Konieczny, A. & Safer, B. (1981) *J. Biol. Chem. 258,* 3402–3408.

Merrick, W. C. (1979) *J. Biol. Chem. 254,* 3708–3711.

Matts, R. L., Levin, D. H. & London, I. M. (1983) *Proc. Natl. Acad. Sci. USA 80,* 2559–2563.

Peterson, D. T., Safer, B. & Merrick, W. C. (1979) *J. Biol. Chem 254,* 7730–7735.

Safer, B., Anderson, W. F. & Merrick, W. C. (1975) *J. Biol. Chem. 250,* 9067–9075.

Safer, B. & Jagus, R. (1979) *Proc. Natl. Acad. Sci. USA 76,* 1094–1098.

Safer, B., Jagus, R. (1981) *Biochimie 63,* 709–717.

Safer, B., Jagus, R., Konieczny, A. & Crouch, D. (1982) In: *Interaction of Translational and Transcriptional Controls in the Regulation of Gene Expression.* (M. Grunberg-Manago & B. Safer, eds). pp. 311–325, Elsevier, New York.

Safer, B., Jagus, R., Crouch, D. & Kemper, W. (1981) In: *Protein Phosphorylation and Bioregulation.* (G. Thomas, E. J. Podesta & J. Gordon, eds.), pp. 142–153, Karger, Basel.

Scatchard, G. (1949) *Ann. N. Y. Acad. Sci. 51,* 660–686.

Siekierka, J., Manne, V., Mauser, L. & Ochoa, S. (1983) *Proc. Natl. Acad. Sci. USA 80,* 1323–1235.

Siekierka, J., Mauser, L. & Ochoa, S. (1982) *Proc. Natl. Acad. Sci. USA 79,* 2537–2540.

Siekierka, J., Mitsui, K.-I. & Ochoa, S. (1981) *Proc. Natl. Acad. Sci. USA 78,* 220–223.

Stringer, E. A., Chaudhuri, A. & Maitra, U. (1979) *J. Biol. Chem. 254,* 6845–6848.

Tahara, S. M., Morgan, M. A. & Shatkin, A. J. (1981) *J. Biol. Chem. 256,* 7691–7694.

Voorma, H. O. & Amesz, H. (1982) In: *Interaction of Translational and Transcriptional Controls in the Regulation of Gene Expression.* (M. Grunberg-Manago & B. Safer, eds.), pp. 297–309, Elsevier, New York.

Walton, G. M. & Gill, G. N. (1976) *Biochim. Biophys. Acta 447,* 11–19.

DISCUSSION

MARCKER: I seem to remember there is also a double-stranded RNA-mediated phosphorylation. Is that not correct?

SAFER: Yes. It is the same phosphorylation site on the eIF-2α subunit, but a different kinase.

KURLAND: There is a very interesting parallel to this mechanism in *E. coli.* If I understand you correctly, one possible interpretation of the kinetics you have would be that the catalytic use of eIF-2 is being slowed down by its phosphorylation. The system is being stopped by the inability to exchange GTP for GDP in the inactive eIF-2 · GDP complex. In fact, we have been studying the effect of magic spot on the cycling of the GDP · Tu complex. We can show that magic spot does a similar thing, and under physiological conditions this is probably the dominant effect on protein synthesis via the elongation factors. Magic spot traps Ts in a complex with Tu and prevents or slows down the other Tu · GDP complexes in their exchange of GTP for GDP. It is quite simple, but effective, because there is a large excess of Tu over Ts in the system.

SAFER: We would like to know the protein chemistry involved, and what the association constant is for eIF-2 · eIF-2B and the effect of eIF-2α phosphorylation of this. Obviously, this model relies quite heavily on the work of Dr. Kaziro with recycling of elongation factor Tu.

GRUNBERG-MANAGO: What is the effect of GTP? You say that it releases inhibition. Does it help to dissociate eIF-2 from eIF-2B, or how does it work?

SAFER: Under physiologic conditions the association between eIF-2 and eIF-2B appears to be very stable. You have to raise the potassium concentration quite high to dissociate the two. Guanine nucleotides have no effect on the dissociation of the complex. Rather than a dissociation of the complex, there is an exchange of eIF-2 coming into the complex for eIF-2 that was previously in the complex. There does not appear to be any free eIF-2B that we can detect under normal conditions. It is always in a complex with eIF-2.

HERSHEY: Do you have any evidence that indicates that all of the subunits of eIF-2B are required for the activity?

SAFER: In fact the evidence that we have indicates that only 4 out of 5 eIF-2B subunits are required. Phosphorylation of the eIF-2α in the absence of guanine nucleotides dissociates the complex, so it releases the 3- subunit eIF-2 and 5-subunit eIF-2B. High salt releases eIF-2B containing only the 4 largest subunits. Both will restore translation and exchange guanine nucleotides quite nicely. One can speculate that small subunit eIF-2B is in fact Gupta's eIF-2A which is about the same molecular weight.

KERR: This is really a joint question to Dr. Hershey and Dr. Safer. It is of general importance that this mechanism has been beautifully worked out for the *in vitro* system, but what evidence is there, that it actually operates in cells other than the reticulocyte? I was also very interested in the heat shock data that you presented, but it was my impression that in HeLa cells one of the differences from, say, drosophila, is that while you get a switch-on of the heat shock proteins you do not get a dramatic inhibition of protein synthesis in general. I wondered how good the correlation really is between the phosphorylation state of eIF-2α and the activity of protein synthesis. Also, does it actually mean that the heat shock proteins are specifically escaping translational inhibition or not? Is there any mRNA specificity involved, in other words?

HERSHEY: I can comment on some of that. HeLa cells growing in monolayer cultures must be heated to 45°C to see any effect on the rate of protein synthesis, i.e., at 43°C you see very little change at all. These are higher temperatures than are required to get inhibition in suspension culture, but we do not know the basis for the difference in response.

KERR: What happens to the phosphorylation of eIF-2α and the synthesis of heat shock proteins between 43° and 45°C?

HERSHEY: We don't see the phosphorylation of eIF-2α until we see the inhibition of protein synthesis, i.e., at 45°C. The correlation should be regarded as preliminary; further work is required to establish the correlation more precisely. We do know, however, that in the inhibited state at 45°C, protein synthesis is essentially totally inhibited, and the heat shock proteins do not escape the inhibition. You see their synthesis only in the recovery phase at 37°C following heat shock at 45°C.

SAFER: The phosphorylation of eIF-2α is really an on/off survival switch for the cell. In terms of it being universal people have found eIF-2α kinases in other cells. It is likely that in cells other than reticulocytes the regulation of the kinase does not occur by the availability of hemin but by alternative mechanisms. Conditions of oxyen deprivation, substrate starvation, lowered availability of NADH, or decreased energy charge of the cell would tend to activate the kinase, and basically bring protein synthesis to a halt. This is the kind of feedback system you would like since protein synthesis is a very energy demanding system. For the cell's survival, you would like to shut off protein synthesis temporarily. So I believe that the mechanism probably exists generally in other cells, and that it acts as an on/off switch rather than as a mechanism to alter the specificity of mRNA translation.

II. Elongation and Termination of Protein Biosynthesis

In *Escherichia coli* Individual Genes are Translated with Different Rates *in vivo*

Steen Pedersen

INTRODUCTION

The disproportionate use of synonymous codons in mRNAs has been evident since the first structural genes were sequenced (Fiers *et al.* 1976). In particular, genes for ribosomal proteins were observed to use some codons very frequently, and others very rarely (Post *et al.* 1979). As first found for the *lacI* gene by Farabaugh (1978), genes coding for less abundant proteins have a broader spectrum of codon usage.

Differences in codon usage can be correlated with the concentration of the cognate tRNAs in the cell (Ikemura 1981), and/or with the codon-anticodon affinity (Grosjean & Fiers 1982), but no consequence of the presence of rare codons in a given mRNA species has been demonstrated previously.

It has been proposed (Grosjean & Fiers 1982, Gouy & Gautier 1982) that rare codons might be translated more slowly, but no direct evidence for this has hitherto been obtained.

The translation time for gene products from *rpsA*, *fus*, *tuf*, *tsf*, *lacI* and *bla* was compared in experiments, demonstrating that the presence of relatively many rare codons in the *bla* and *lacI* genes correlates with a 30% increase in the average codon translation time.

MATERIALS AND METHODS

Strains:

NF929 *thr, leu, his, argH, pyrE, thi, relA*$^+$, *relC*$^+$, *spoT*$^+$ (Fiil *et al.* 1977). Maxicell CSR603 (Sancar *et al.* 1979).

University of Copenhagen, Institute of Microbiology, Øster Farimagsgade 2A, DK-1353, Copenhagen. Denmark

GENE EXPRESSION, Alfred Benzon Symposium 19.
Editors: Brian F. C. Clark & Hans Uffe Petersen, Munksgaard, Copenhagen 1984.

Plasmids:
pBR322 (Bolivar *et al.* 1977). pR2172: pBR322 with a partial HindII fragment, containing the *lacI* gene with the $lacI^q$ promoter, obtained from J. Hays, University of Maryland, Baltimore, USA.

Growth of cells:
Maxicell experiments were performed as described by Christiansen & Pedersen (1981). Steptime experiments were carried out at 24.5°C±0.2°C in the A+B medium (Clark & Maaløe 1967), supplemented with 0.2 per cent glucose, 50 μg/ml of the required amino acids and 0.5 μg/ml thiamine. The growth rate was 0.34 doublings per h, a reduction by a factor 3.15 from 37°C. All cultures had been growing exponentially for more than four generations at 24.5°C before the experiment started.

Labelling:
Reference cells were labelled for three generations with ^{3}H-leucine (20 μCi/ml, 10 μg/ml), and harvested at OD_{436}=1.0.

Pulse-labelling:
To 13 ml culture at OD_{436}=0.7, 300 μCi carrierfree ^{35}S-methionine (specific activity approximately 1000 Ci/mmol) were added. Six to eight sec later non-radioactive methionine was added to 200 μg/ml. At frequent intervals during the next 3 min 12 one-ml samples were harvested into ice-cold tubes with chloramphenicol (300 μg/ml). Ten μl were TCA-precipitated, and the incorporated radioactivity determined. The remaining sample was centrifuged for 2 min in the cold and the pellet frozen at −80°C. The complete harvesting procedure lasted 6 min.

2D-gel analysis:
To each sample ^{3}H reference cells were added to give a $^{3}H/^{35}S$ ratio of about 5, and the cells were opened by sonication and treated as described by Pedersen *et al.* (1976). We used a 3–10 ampholine mixture in the first dimension, and a 10% polyacrylamide gel in the second dimension.

Spots were located by autoradiography, cut out from the gels, digested by H_2O_2 as described by Pedersen *et al.* (1976), and the radioactivity was determined. Results are given as the $^{35}S/^{3}H$ ratio in a spot divided by the $^{35}S/^{3}H$ ratio of the total cell extract. The resultant curve, therefore, gives the accumulation of radioactivity in arbitrary units for each protein (Reeh *et al.* 1976). EFG, EFTu, EFTs, and S1 were located on the 2D-gel spot pattern (Pedersen *et al.* 1978). To locate β-lactamase and the *lac* repressor, the plasmids pBR322 and pR2172 were transformed into the maxicell and a maxicell experiment performed. The positions of labelled *lac* repressor and β-lactamase were then located relative to the other spots by staining (Pedersen *et al.* 1976), and a comparison of an autoradiogram with the stained gel. To verify these positions NF929, NF929/pBR322 and NF929/pR2172 were pulse-labelled during exponential growth and analyzed on the 2D-gels. Both *lac* repressor and β-lactamase are uncontaminated by other proteins and has the molecular weight and isoelectric point (40,000, pI about 8.0, and 28,000, pI about 6.5, respectively) as expected from the amino acid sequence.

The DNA sequences of these 6 proteins have been established (Sutcliffe 1978; An *et al.* 1981, An & Friesen 1980, Yokota *et al.* 1980, Farabaugh 1978, Schnier *et al.* 1982) with the exception of EFG that appears to have a molecular weight of 83,000 on SDS gels (Pedersen *et al.* 1978), and, therefore, should have about 750 amino acids. The N-terminal end of EFG has been sequenced (Post & Nomura 1980), and is rich in methionine. In the calculation of codon translation times the number of amino acids from the first methionine in the molecule to the end of the molecule has been: 220 for β-lactamase; 282 for EFTs; 302 for EFTu; 360 for *lac* repressor; 556 for S1; and 750 for EFG.

RESULTS

The method used was devised by Bremer & Yuan (1968). Labelling with a pulse (^{35}S-methionine) much shorter than the synthesis time for the peptide was followed by a chase. Samples were taken at various times, and the kinetics of the appearance of radioactivity in finished polypeptides measured. Theoretically, the radioactivity in finished polypeptide chains will increase with time until the first methionine in the molecule is located in the completed peptide. After this time the radioactivity in the protein will remain constant. Knowing the amino acid sequence, this permits a direct calculation of the ribosomal steptime on this mRNA. The theoretical shape of the accumulation curve is dependent on the location of methionine residues in the peptide and is shown in Fig. 1 for 2 of the proteins that have been analyzed. In practice more smooth curves were observed (see Fig. 3).

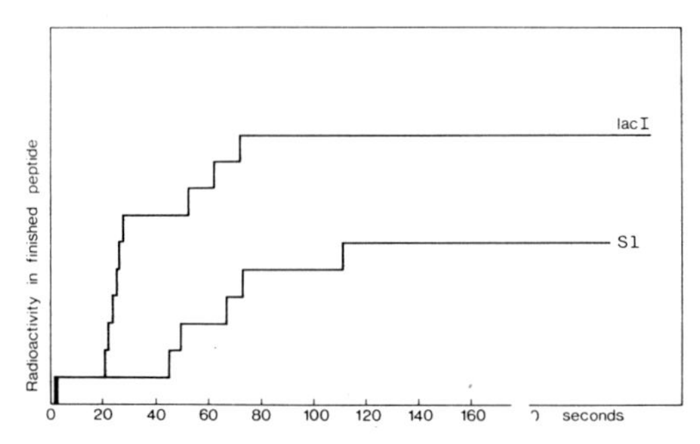

Fig. 1. Theoretical incorporation curves for *lacI* and S1. The curves show the expected effect of the distribution of methionine residues on the accumulation of radioactivity in finished polypeptides. The translation steptime is assumed to be 0.2 sec./codon, and the pulse of ^{35}S-methionine in the pulse-chase experiment infinitely short.

The incorporation of radioactivity into total TCA-precipitable material is shown in Fig. 2. It is seen that the chase after 5 sec is effective in blocking further incorporation. It is important that no protein synthesis takes place during the harvesting procedure. This condition is met with, because the synthesis curves for the individual proteins extrapolate back to (0.0) (see Fig. 3).

Fig. 3 shows the kinetics of labelling the EFTs, EFTu, S1, EFG, *lac* repressor and β-lactamase peptides. The average codon translation time is calculated to

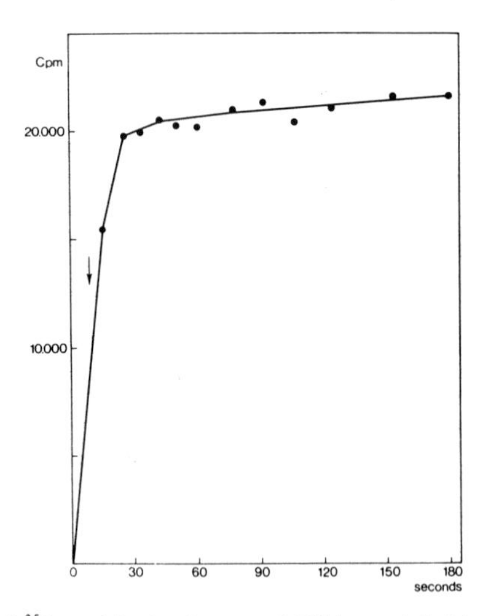

Fig. 2. Incorporation of ^{35}S-methionine into total TCA precipitable material in the pulse-chase experiment. The arrow indicates time for addition of unlabelled methionine at 7 sec.

0.18, 0.21, 0.23, 0.20, 0.29, and 0.29 sec per amino acid, respectively. The observed codon translation time on *lacI* and *bla* mRNA is seen to be about 30%

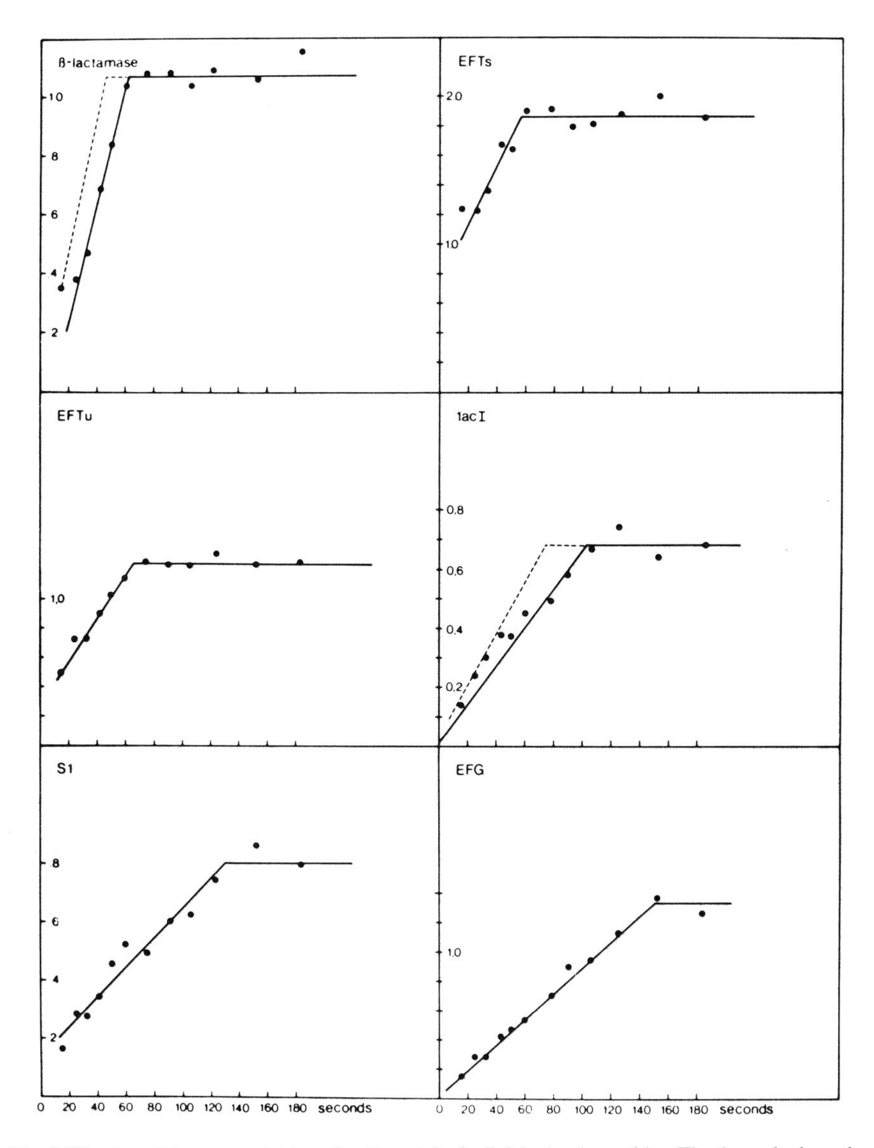

Fig. 3. Kinetics of the accumulation of radioactivity in finished polypeptides. The 6 panels show the accumulation of radioactivity in arbitrary units for: β-lactamase, EFTs, EFTu, *lacI* gene product, ribosomal protein S1, and EFG. The dotted curves show the results for β-lactamase and *lac* repressor if they had had the same steptime as determined for the 4 other proteins in this experiment.

longer than on the 4 ribosomal genes. This result has been obtained consistently in 5 experiments.

DISCUSSION

Before interpreting the measured rates of peptide synthesis, it must be ascertained that the position of the spots on the 2D-gels is not influenced by post-translational modifications. In the case of β-lactamase such a complication might exist, due to signal peptide cleavage (Sutcliffe 1978), but it presented no problem in this experiment, because only one spot for β-lactamase was observed. The removal of the signal peptide must take place during synthesis of the *bla* gene product, because this spot comigrates with the long-term ^3H label from the reference culture. The N-terminal of EFTu protein is acetylated (Arai *et al.* 1980), but no change in spot position for pulse-labelled versus continuously labelled EFTu was observed. For the other 4 proteins no such complications exist, and calculation of the codon translation time is straightforward. The scattering of the data is minimal, and the difference in translation time is consistently observed. Codon translation time, thus, depends on sequences on the mRNA. *In vivo* experiments cannot determine the reason for this slow translation. It is, however, possible to correlate the slow translation with rare codon usage on both *lacI* and *bla.*

I have chosen to name the following 22 codons as "rare": TTA, TTG, CTT, CTC, CTA, ATA, GTC, TCA, TCG, CCT, CCC, ACA, ACG, GCC, AAT, CGA, CGG, AGT, AGA, AGG, GGA, and GGG. These codons are used in ribosomal protein genes with a frequency of less than 10% of the number of codons for the same amino acid (Grosjean & Fiers 1982). In the *tuf, tsf, rpsA* and *fus* genes, these codons occur with a frequency of 5%, whereas the frequency in *lacI* and *bla* is 23 and 28%, respectively. If the slow translation were caused only by the presence of rare codons, the translation time on each rare codon would be about 0.5 sec per amino acid at 24.5°C, i.e., 2.5 times slower than the translation time of a non-rare codon. If the slow translation were caused by a subset of the rare codons, translation would be even slower.

The differences in translation time I have demonstrated here could be envisaged to play a role in 1) generating signals, designed for export through membranes; 2) decreasing of the translation of mRNA; 3) being of importance for protein folding into its tertiary structure; 4) generating signals for transcriptional polarity or mRNA folding. Interestingly enough, one cluster of rare codons in the *bla* gene is positioned within the signal peptide region.

ACKNOWLEDGEMENTS

The expert help from Marit Warrer has been vital to this work, which was supported by grants from the Danish Natural Science Research Council (no. 11-3771), and from the NOVO's Foundation.

REFERENCES

An, G. & Friesen, J. D. (1980) *Gene 12,* 25–31.

An, G., Bendiak, D. S., Mamelak, L. A. & Friesen, J. D. (1981) *Nucleic Acid. Res. 9,* 4162–4172.

Arai, K., Clark, B. F. C., Duffy, L., Jones, M. D., Kaziro, Y., Laursen, R. A., L'Italien, J., Miller, D. L., Nagarkatti, S., Nakamura, S., Nielsen, K. M., Petersen, T. E., Takahashi, K. & Wade, K. (1980) *Proc. Natl. Acad. Sci USA 77,* 1326–1330.

Bolivar, F., Rodriguez, R. L., Greene, P. J., Betlach, M. C., Heyneker, H. L., Boyer, H. W., Crosa, J. H. & Falkow, S. (1977) *Gene 2,* 95–113.

Bremer, H. & Yuan, D. (1968) *J. Mol. Biol. 34,* 527–540.

Christiansen, L. & Pedersen, S. (1981) *Mol. Gen. Genet. 181,* 548–551.

Clark, D. J. & Maaløe, O. (1967) *J. Mol. Biol. 23,* 99–112.

Farabaugh, P. J. (1978) *Nature 274,* 765–767.

Fiers, W., Contreras, R., Duerinck, F., Haegeman, G., Iserentant, D. & Merregaert, J. (1976) *Nature 260*, 500–507.

Fiil, N. P., Willumsen, B. M., Friesen, J. D. & von Meyenburg, K. (1977) *Mol. Gen. Genet. 150,* 87–101.

Gouy, M. & Gautier, C. (1982) *Nucleic Acids Res. 10,* 7055–7074.

Grosjean,H. & Fiers, W. (1982) *Gene 18,* 199–209.

Ikemura, T. (1981) *J. Mol. Biol. 146,* 1–21.

Pedersen, S., Bloch, P. L., Reeh, S. & Neidhardt, F. C. (1978) *Cell 14,* 179–190.

Pedersen, S., Reeh, S., Parker, J., Watson, R., Friesen, J. & Fiil, N. P. (1976) *Mol. Gen. Genet. 144,* 339–343.

Post, L. E. & Nomura, J. D. (1980) *J. Biol. Chem. 255,* 4660–4666.

Post, L. E., Strycharz, G. D., Nomura, M., Lewis, H. & Dennis, P. P. (1979) *Proc. Natl. Acad. Sci USA 76,* 1697–1701.

Reeh, S., Pedersen, S. & Friesen, J. D. (1976) *Mol. Gen. Genet. 149,* 279–289.

Sancar, A., Hack, A. M. & Rupp, W. D. (1979) *J. Bacteriol. 137,* 692–693.

Schnier, J., Kimura, M., Foulaki, K., Subramanian, A.-R., Isono, K. & Wittmann-Liebold, B. (1982) *Proc. Natl. Acad. Sci. USA 79,* 1008–1011.

Sutcliffe, J. G. (1978) *Proc. Natl. Acad. Sci. USA 75,* 3737–3741.

Yokota, T., Sugisaki, H., Takanami, M. & Kaziro, Y. (1980) *Gene 12,* 25–31.

DISCUSSION

SCHATZ: One of the aspects you raised (which for technical reasons you could not investigate), was translational "pausing". Linda Randall from The University of Washington at Pullman took logarithmically growing *E. coli* cells, pulse-labelled them for short times, killed the cells, extracted them with SDS and immunoprecipitated specific proteins, destined to be exported from the cell. Instead of a continuous smear from the position of the complete protein to the bottom of the SDS gel she saw definite bands which might indicate such translational "pauses". It would be interesting to look at her data and to compare the sizes of these fragments with the position of rare codons in the gene.

PEDERSEN: I agree. I was aware of this study. There are other experiments mentioned in the literature, but you can raise many technical objections against them. For the experiments you mention, you would have to know that the antibody really would precipitate all uncompleted chains.

SCHATZ: In the field of transmembrane movement of proteins, there have been many discussions as to whether proteins go through the *E. coli* membrane co-translationally or post-translationally, and whether domains had to fold before a protein can get through a membrane. One of the interesting, but often ignored possibilities, has been that translational pauses, perhaps caused by rare codons, may facilitate the transient formation of domains which then go through the membrane in a folded state. Thus, the protein does not go through completely folded or completely extended. This could be controlled by the phenomena you alluded to.

KURLAND: We have looked at the problem from another point of view. Our *in vitro* system is working pretty close to the *in vivo* rates, and you could ask if you get a rate change if you use another messenger, say alternating poly UG instead of poly U. The answer is no, in both cases you get approximately 10 peptide bonds per second per active ribosome. That suggests that if there are different rates they are not at the Kcat of the ribosome. However, if you go back to the *in vivo* situation and look at the average tRNA for more constants, we would say that ribosomes are 90–95% saturated in terms of their Km's *in vivo*. So, they are running at their Kcat. That is for your average tRNA. The rare tRNA's would have tremendous problems getting the rates up. That leads to a critical question.

To what extent have you engineered this difference? If you increase the drain on the rare tRNA's by increasing the amount of mRNA's with rare codons you might be creating a rate effect, which *E. coli* does not experience.

PEDERSEN: I was going to investigate this question just before the meeting by doing the same experiment in a strain where the lacI gene was on an episome and not on a high copy number plasmid. This should reduce the artificial high drain on the rare tRNA's by a factor of say 50. The experiment has not been done, and, so far, I think I have just shown a principle which is suitable for hypothesis.

KLUG: I wonder if there is a real biological problem here. The experiments are very nice, but there is only a factor of 3. If you take Ikemura's results to mean that there has already been an adjustment in the total number of tRNA's present for rare codons, then how delicate can the machinery be? It seems to me that you pay a price, perhaps in time; and a factor of 2 or 3 does not seem to me to be enough on an average to make some significance in, say, the general export problem, unless you have a very specific pause.

PEDERSEN: I think that the factor 3 would be more like a factor 10 if you put the effect on the codons with rare tRNA's.

HERSHEY: Another way to interpret different transit times for different messenger RNA's would be the aspect of the secondary structure of the messenger RNA, and this problem might be more important at 25°C than at 37°. I wonder if you have taken this possibility into account.

PEDERSEN: I can only tell you that ribosomal protein Sl mRNA has many possibilities for very stable structures and this mRNA is still translated fast. As I have already said, *in vivo* experiments can never really reveal details. They can only establish correlation, and the correlation with rare codon usage might break down, if I counted more proteins.

INNIS: I would like to make a couple of comments. First, the 2 genes you have chosen to study both have disastrous ribosomal binding sites, and I suspect that for them the initiation would be very slow. In fact, we have modified the binding sites for β- lactamase and increased the level of expression more than 10-fold. Second, is it not true that these rare codons are regulated, and that, for example, in our case we are capable of overproducing genes with disastrous "rare codon usage" to the level of 25–35% of total *E. coli* protein without any apparent effect on rate.

PEDERSEN: I do not know that the rare tRNAs are overproduced in strains with elevated levels of mRNAs using the rare codons. With respect to the first comment, my experiment measures the time from formation of the first peptide bond to the time of formation of the last one. Any period the ribosome spends finding the binding site does not show in the curves I presented.

INNIS: Finally, I would like to raise the question that instead of rare codon utilization *per se*, it is the context of the positioning of those rare codons that might affect the translation. There are genes which make functional messenger RNA which will translate in the reticulocyte system but not in yeast, so that it is a steric hindrance between adjacent rare codons that prevents them from being used, i.e., it is the context, rather than the abundance *per se*, that affects the rates of translation.

GOLD: Do you think it is the minimum step time you measure in the *in vivo* experiment? Is it not the transit time for the RNA polymerase you are really measuring?

PEDERSEN: From measurements of the capacity for protein synthesis after addition of rifampicin or streptolygidin you get the result that about half of the messengers are completed.

GOLD: I guess I am really asking whether you depend most on the data points close to the plateau, to determine the step time.

SPRINZL: It is not known exactly how many of each iso-acceptor *E. coli* has. We have now developed an affinity chromatography method by which one group of iso-acceptors can be isolated. We see many more tRNAs in *E. coli* than expected. In context of the translational regulation one should not only measure concentration of tRNA but also see how well this tRNA will complex with EF-Tu. There are 1–2 orders of magnitude differences in the binding constants of different tRNAs.

ERDMANN: Did you consider the possibility of a correlation between the rare codons used and modification of the proteins during protein synthesis? Is this assumption possible?

PEDERSEN: Yes, but with respect to signal-peptide cleavage from β-lactamase and acetylation of EF-Tu, they happen so fast that I do not see them, simply because I am looking only at the finished products.

Structure, Expression, and its Regulation of the Genes Coding for Polypeptide Chain Elongation Factor Tu

Y. Kaziro, J. Mizushima-Sugano, S. Nagata, Y. Tsunetsugu-Yokota & A. Naito

The polypeptide chain elongation factor Tu (EF-Tu) promotes a GTP-dependent binding of an aminoacyl-tRNA to the A-site of ribosomes (Kaziro 1978). EF-Tu from *E. coli* consists of a single polypeptide chain with Mr 43,000, and the primary structure comprising 393 amino acid residues has been established (Arai *et al.* 1980). The protein binds one mole of GTP/GDP per mol of protein and alters its conformation and reactivity with the ligand change from GTP to GDP (or *vice versa*). This property is of prime importance for the function of EF-Tu, i.e., the energy-dependent transfer of an aminoacyl-tRNA to ribosomes, and the common mechanism of energy utilization has been proposed for other translational factors requiring GTP (Kaziro 1978). EF-Tu may be regarded as a simplest form of the energy transducing protein. Many proteins with similar functions have since been found to be involved in various systems; including DNA replication, self-assembly systems of microtubules and micro-filament, and signal transmission in hormonal receptor and photoreceptor GTPases. The basic mechanism underlying these reactions appears to be very much analogous (Kaziro 1980, 1983).

E. coli EF-Tu is encoded by 2 nearly identical unlinked genes (Jaskunas *et al.* 1975), *tufA* at 73 min and *tufB* at 89 min on the recalibrated genetic map of *E. coli* (Bachmann 1983). Both *tufA* (Shibuya *et al.* 1979) and *tufB* (Miyajima *et al.* 1979) have been cloned and their sequences determined (Yokota *et al.* 1980, An & Friesen 1980). The sequences of *tufA* and *tufB* are nearly homologous and

---

Institute of Medical Science, University of Tokyo, Minatoku, Takanawa, Tokyo 108, Japan.

GENE EXPRESSION, Alfred Benzon Symposium 19.
Editors: Brian F. C. Clark & Hans Uffe Petersen, Munksgaard, Copenhagen 1984.

differ only in 13 positions. The gene products, EF-TuA and EF-TuB, are identical except for the C-terminal amino acid which is glycine (GGC) for *tufA,* and serine (AGC) for *tufB.*

In eukaryotic systems, the counterpart of prokaryotic EF-Tu, designated as EF-1α and consisting of a single polypeptide chain of Mr 47,000–53,000, has been purified from various sources (Iwasaki & Kaziro 1979). In addition to EF-1α, which functions in the cytoplasmic fractions in conjunction with 80S ribosomes, eukaryotic cells possess mitochondrial EF-Tu (designated as mtEF-Tu) that functions in the mitochondrial translational apparatus (Richter & Lipmann 1970, Piechulla & Küntzel 1983). Translational factors as well as ribosomal proteins in the mitochondria are encoded by nuclear genes, synthesized in cytoplasmic fractions, and transported into mitochondria (Richter 1971). The translational machineries of mitochondria have been thought to be closer to the prokaryotic ones than to the machinery present in eukaryotic cytoplasm (Borst 1972).

In this paper, we describe first the structure of the genes coding for *E. coli* EF-Tu and the expression of *tufB* in the cell-free system, especially the regulation of the synthesis of *tufB* mRNA by ppGpp. We then describe the cloning of the nuclear gene of yeast coding for mtEF-Tu. Isolation of the gene, *tufM,* is based on the cross-hybridization of *tufM* with the *E. coli tufB* gene under low stringency conditions.

STRUCTURE AND EXPRESSION OF THE GENES CODING FOR *E. coli* EF-Tu

Cloning of tufA *and* tufB *genes*

The *EcoR*I fragments carrying *tufA* and *tufB* were obtained from the transducing phages $\lambda fus3$ and λrif^{d} 18, respectively, and cloned on a ColEl derivative plasmid RSF2124 to yield 4 plasmids, pTUA1 and pTUA2, and pTUB1 and pTUB2 (Shibuya *et al.* 1979, Miyajima *et al.* 1979) (Fig. 1). The 2 sets of the plasmids contain the DNA fragment carrying *tufA* and *tufB* in an opposite orientation with respect to the vector DNA. The *tufA* gene in pTUA1 is not expressed as it has no promoter for its transcription; *tufA* in *E. coli* chromosome is co-transcribed with its upstream genes for S12, S7, and EF-G in the *str* operon (Lindahl *et al.* 1979). The *tufA* gene in pTUA2 is expressed presumably by read-through from the *ColEl* promoter on the vector plasmid.

On the other hand, *tufB* in both pTUB1 and pTUB2 is well expressed indicating that its natural promoter is present in the cloned 8.6 kb *EcoR*I

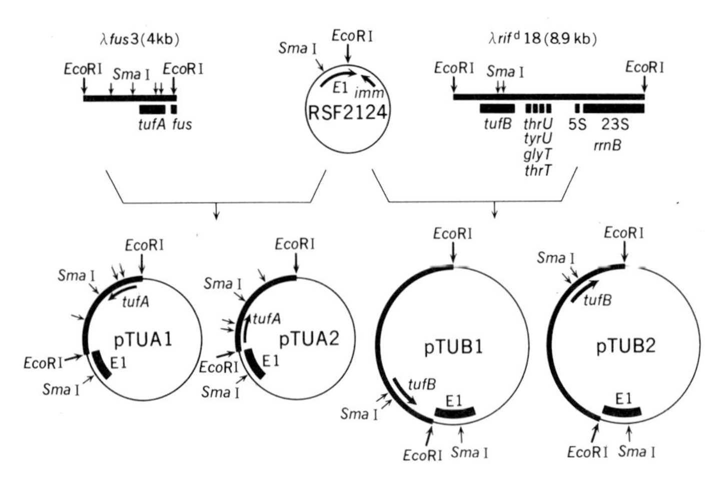

Fig. 1. Cloning of the *E. coli tufA* and *tufB* genes. The 4 kb *EcoR*I fragment carrying *tufA* and the 8.9 kb *EcoR*I fragment were obtained from λ *fus*3 DNA and λ *rif*[d]18 DNA, respectively. They were inserted in the *EcoR*I site of the vector plasmid RSF2124. For details, see Shibuya *et al.* (1979) and Miyajima *et al.* (1979).

fragment. The fragment contained the genes for part of 23S rRNA and 5S rRNA *(rrnB), rts,* 4 tRNAs (*thrU, tyrU, glyT,* and *thrT*), and EF-Tu (*tufB*), and downstream of *tufB,* the genes for unknown protein "U", and part of L11 (*rplK*). By comparing the RNA nucleotide sequence of the *in vitro* transcripts with the DNA sequence determined by An and Friesen (1980), we have shown that the transcription of *tufB* mRNA is initiated at the promoter located upstream of *thrU,* and *tufB* is cotranscribed with its adjacent 4 tRNA genes (Miyajima *et al.* 1981). This conclusion was confirmed independently by Lee *et al.* (1981) and Hudson *et al.* (1981) from the different experimental approach.

Expression of the tufB *gene and its regulation*

The studies in our laboratory and several other laboratories have shown that the rate of synthesis of elongation factors is proportional to the growth rate under balanced growth conditions, and is subject to the stringent control under amino-acid starvation (Miyajima & Kaziro 1978). We have studied the expression of the *tufB* gene in the DNA-dependent cell-free transcription-translation coupled system using λ*rif*[d]18 DNA or pTUB1 DNA as template (Shibuya & Kaziro 1979). In this system, the synthesis of EF-Tu was found to be greatly suppressed by the addition of 0.1 to 0.2 mM ppGpp. Further studies have revealed that the transcription of the *thrU-tufB* operon by purified RNA polymerase holoenzyme was strongly inhibited by low concentrations of ppGpp (Miyajima *et al.* 1981). Then we studied the effect of ppGpp on the transcription of *tufB* further in detail using the *recA* promoter as a reference promoter that is not subject to stringent control (Mizushima-Sugano *et al.* 1983). As a source of the *tufB* template, we utilized plasmid pTUB2, rather than pTUB1, in which 3 out of 4 tRNA genes

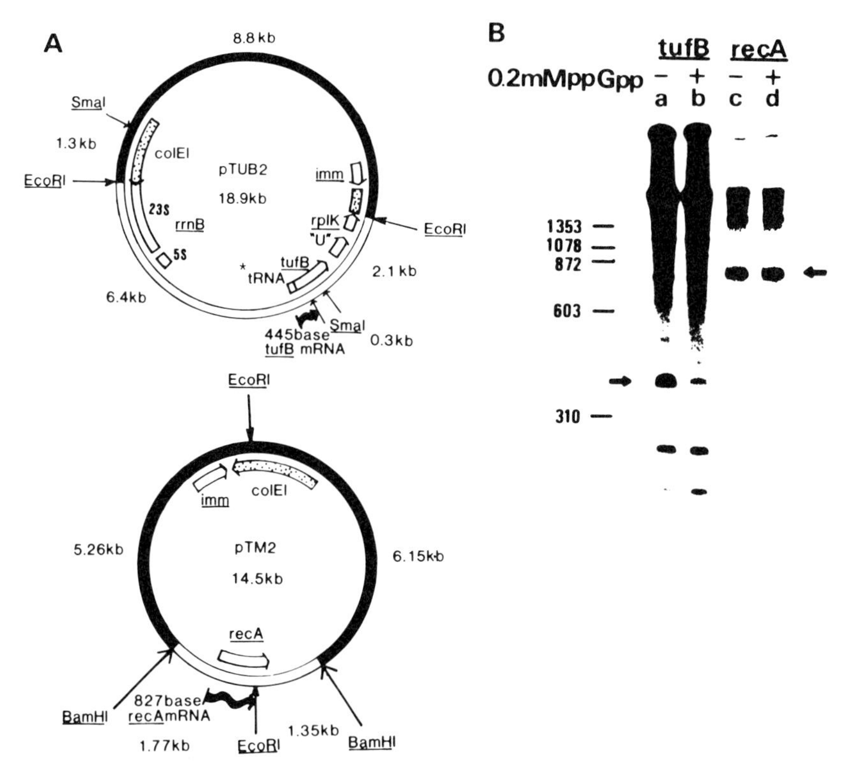

Fig. 2. Transcription of the truncated *tufB* and *recA* templates. A, Physical and functional maps of pTUB2 and pTM2 plasmids. The black and white regions in the outside circle represent the vector and host genomes, respectively. The initiation site, size, and the direction of reading of truncated *tufB* and *recA* mRNAs are indicated by black wavy lines with arrows. The *tRNA gene in pTUB2 is a hybrid *thrU-thrT* gene produced by a homologous recombination in the *thrU-tyrU-glyT-thrT* region of pTUB1 (Miyajima *et al.* 1983). B, *tufB* and *recA* transcripts synthesized by RNA polymerase holoenzyme and the effect of ppGpp. The *Sma*I digested pTUB2 (lanes *a* and *b*), and the *Eco*RI digested pTM2 (lanes *c* and *d*) were transcribed in the absence (lanes *a* and *c*) or presence (lanes *b* and *d*) of 0.2 mM ppGpp. For details, see Mizushima-Sugano *et al.* (1983).

Pribnow Box HhaI +1 TaqI EcoRI

CTCGCATGTCTCCATAGAATGCGCGCTACTTGATGCCGACTTAGCTCAGTAGGTAGAGCGCACCCTTGGTAAGGGTGAGGTCGGCAGTTCGAATTC

CTCGCATGTCTCCATAGAATGCGCGCTACTTGATGCCGACGGAATTC Δ56

CTCGCATGTCTCCATAGAATGCGCGCTACTTGATGGGAATTC Δ53

CTCGCATGTCTCCATAGAATGCGCGCTACGGAATTC Δ39

CTCGCATGGAATTC Δ41

GGAATTC Δ44

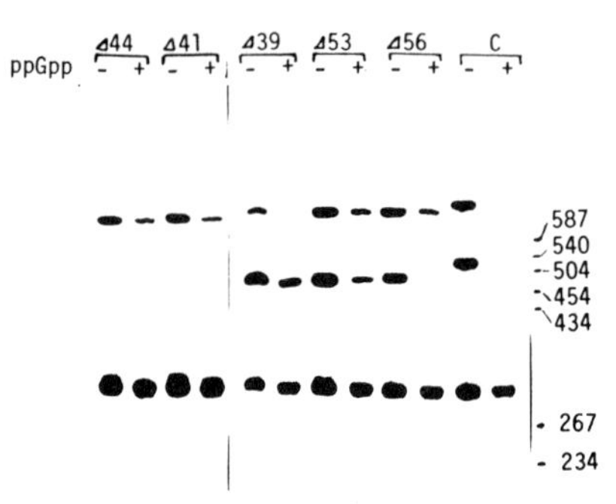

Fig. 3. Construction of deletion mutants of the *thrU-tufB* promoter region and the transcription of the mutant DNAs. Plasmid pTUB2-mt8 was cleaved at the *EcoR*I site and digested with enzyme Bal31 for various times. *EcoR*I linkers were inserted to the digested ends and plasmids were re-ligated. Lengths of the individual deletions were determined by DNA sequencing, and the structure of the pTUB2-mt8 and its deletion mutants are shown in the top of the figure. Black bars indicate the sequence of pBR327 and the framed sequences are the Pribnow box and the sequence of *EcoR*I linkers. At the bottom, radioautograms of transcripts of pTUB2-mt8 (lane *c*) and its various deletion mutants in the presence and absence of 0.5 mM ppGpp are shown.

were deleted due to a spontaneous fusion between homologous regions of *thrU* and *thrT* (Miyajima *et al.* 1983, see Fig. 2).

Plasmid pTUB2 was digested with restriction endonuclease *Sma*I to cleave the structural gene for *tufB* at the position corresponding to the codon for Pro-82. The transcription on this truncated DNA template would yield a fragment of *tufB* mRNA of the size of 445 nucleotides. On the other hand, the *EcoR*I digested plasmid pTM2 (Ogawa *et al.* 1978) would give rise to a *recA* mRNA fragment of 827 nucleotides. As shown in Fig. 2, distinct bands corresponding to the expected size of *tufB* and *recA* mRNA fragments were observed on the polyacrylamide gel electrophoresis. The formation of *tufB* mRNA was almost completely abolished by the addition of 0.2 mM ppGpp, whereas the transcription from the *recA* promoter is insensitive to ppGpp (Mizushima-Sugano *et al.* 1983).

As these results were obtained in the system consisting of pure RNA polymerase holoenzyme, the question arises as to how RNA polymerase could distinguish between stringently and non-stringently controlled promoters. Our current hypothesis is that when RNA polymerase is bound to a stringent promoter, its conformation may be altered to the form to which ppGpp is accessible to exert its inhibitory effect on transcription. Alternatively, the conformation of RNA polymerase which is bound to a non-stringent promoter is refractory to ppGpp. The other possibility which is not strictly ruled out is that an unknown factor required for transcription of stringently controlled operons may still be present in our polymerase preparation. These points are now currently under investigation.

More recently, we attempted to locate more precisely the nucleotide sequence which is necessary for the stringent control in the *tufB* operon. As the transcription of the *thrU-tufB* operon is inhibited at the step of initiation (Mizushima-Sugano *et al.* 1983), these sequences must be located near the promoter and upstream of *thrU.* We prepared smaller DNA restriction fragments containing the *thrU-tufB* promoter, and tested for ppGpp sensitivity of its transcription. It was found that the transcription using the *Bgl*II-*Cla*I fragment (about 730 bp), and the *Mbo*II-*Taq*I fragment (160 bp) containing the *thrU-tufB* promoter was as sensitive to ppGpp as when the *Sma*I-digested pTUB2 was used as a template. The 160 bp *Mbo*II-*Taq*I fragment was then cloned on pBR327 at the *Ava*I/*EcoR*I sites to yield plasmid pTUB2-mt8. The hybrid plasmid was then cleaved with *EcoR*I and various deletion mutants were constructed by digesting with Bal31. As shown in Fig. 3, digestion of plasmid pTUB2-mt8 from its *EcoR*I site (position +64) up to the position +2 did not destroy the sensitivity to ppGpp. The conclusion drawn from this experiment at present stage is that the sequence required for the stringent control resides within the DNA stretch of 100 bp ranging from position −98 to +2. Further attempt to locate the canonical sequence for stringent control is now under way.

CLONING AND SEQUENCE DETERMINATION OF THE NUCLEAR GENE CODING FOR YEAST MITOCHONDRIAL EF-Tu

Isolation of a sequence homologous to E. coli tufB *from yeast DNA*

The presence of a sequence homologous to *E. coli tufB* in *Saccharomyces*

```
                                                                                          -220                 -200
                                                                       (tufM) TTTCATATTTTTAAGGATTTTGTTTTAGCACCCATCCGA
 -180           -160            -140            -120            -100            -80             -60
CCTCAGTCAATATATCCTTTCGCGACCAGGCTTTCCTCCCTTTTGCTGCTAACTGGTTACAGATTTTCCTATTTTTGGTCATTTTTATCTTTGAAACTGATTAAGCTGAAAAAATTTGAGCTTCTTTGTTGTAAACTATTTTG
 -40            -20          -1 1                                30                                    60
TGCTTTCAGTTTTATTCTAGCTCGACAAAGGTAACAGACAAAA ATG TCA GCT TTA TTA CCA AGA TTA CTC ACA AGA ACA GCT TTT AAA GCT TCT GGG AAA CTT CTG AGG CTC TCT TCA
                                            Met Ser Ala Leu Leu Pro Arg Leu Leu Thr Arg Thr Ala Phe Lys Ala Ser Gly Lys Leu Leu Arg Leu Ser Ser
                                             1                               10                                      20

                 90                                      120                             150                                     180
GTA ATT TCT AGG ACC TTT TCT CAA ACT ACT ACT TCC TAT GCA GCT GCT TTT GAT CGT TCC AAA CCG CAT GTA AAT ATA GGT ACG ATC GGC CAT GTT GAT CAT GGG AAG
Val Ile Ser Arg Thr Phe Ser Gln Thr Thr Thr Ser Tyr Ala Ala Ala Phe Asp Arg Ser Lys Pro His Val Asn Ile Gly Thr Ile Gly His Val Asp His Gly Lys
                 30                                  40                                      50                                      60
                                                 1                               10                              20
                                                Ser Lys Glu Lys Phe Glu Arg Thr Lys Pro His Val Asn Val Gly Thr Ile Gly His Val Asp His Gly Lys
                                     (tufA) GTG TCT AAA GAA AAA TTT GAA CGT ACA AAA CCG CAC GTT AAC GTT GGT ACT ATC GGC CAC GTT GAC CAC GGT AAA
                                             1                                   30                                  60

                                 210                         240                                 270
ACA ACT TTA ACC GCA GCC ATT ACG AAA ACG TTA GCC GCA AAA GGT GGT GCC AAC TTC TTG GAC TAT GCT GCC ATC GAT AAG GCT CCG GAA GAA AGA GCT CGT GGT ATT
Thr Thr Leu Thr Ala Ala Ile Thr Lys Thr Leu Ala Ala Lys Gly Gly Ala Asn Phe Leu Asp Tyr Ala Ala Ile Asp Lys Ala Pro Glu Glu Arg Ala Arg Gly Ile
                     30          70                     40          80                      50          90                      60
Thr Thr Leu Thr Ala Ala Ile Thr Thr Val Ile Ala Lys Thr Tyr Gly Gly Ala Ala Arg Ala Phe Asp Gln Ile Asp Asn Ala Pro Glu Glu Lys Ala Arg Gly Ile
ACT ACT CTG ACC GCT GCA ATC ACC ACC GTA CTG GCT AAA ACC TAC GGC GGT GCT GCT CGT GCA TTC GAC CAG ATC GAT AAC GCG CCG GAA GAA AAA GCT CGT GGT ATC
                 90                                  120                         150                                     180

        300                             330                             360                                 390
ACA ATT TCT ACT GCA CAC GTG GAA TAC GAA ACG GCC AAG AGA CAT TAT TCT CAC GTC GAC TGT CCA GGC CAC GCT GAT TAC ATC AAG AAT ATG ATT ACC GGT GCT GCT
Thr Ile Ser Thr Ala His Val Glu Tyr Glu Thr Ala Lys Arg His Tyr Ser His Val Asp Cys Pro Gly His Ala Asp Tyr Ile Lys Asn Met Ile Thr Gly Ala Ala
        100                             70          110                     80          120                     90          130
Thr Ile Asn Thr Ser His Val Glu Tyr Asp Thr Pro Thr Arg His Tyr Ala His Val Asp Cys Pro Gly His Ala Asp Tyr Val Lys Asn Met Ile Thr Gly Ala Ala
ACC ATC AAC ACT TCT CAC GTT GAA TAC GAC ACC CCG ACC CGT CAC TAC GCA CAC GTA GAC TGC CCG GGG CAC GCC GAC TAT GTT AAA AAC ATG ATC ACC GGT GCT GCT
                                210                             240                             270

                        420                             450                             480
CAA ATG GAT GGT GCT ATC ATT GTT GTA GCT GCT ACC GAT GGA CAA ATG CCC CAA ACT AGA GAA CAT TTA CTT TTG GCC AGA CAA GTT GGT GTC CAA CAT ATT GTC GTT
Gln Met Asp Gly Ala Ile Ile Val Val Ala Ala Thr Asp Gly Gln Met Pro Gln Thr Arg Glu His Leu Leu Leu Ala Arg Gln Val Gly Val Gln His Ile Val Val
        100             140                         110     150                         120         160                         130
Gln Met Asp Gly Ala Ile Leu Val Val Ala Ala Thr Asp Gly Pro Met Pro Gln Thr Arg Glu His Ile Leu Leu Gly Arg Gln Val Gly Val Pro Tyr Ile Ile Val
CAG ATG GAC GGC GCG ATC CTG GTA GTT GCT GCG ACT GAC GGC CCG ATG CCG CAG ACT CGT GAG CAC ATC CTG CTG GGT CGT CAG GTA GGC GTT CCG TAC ATC ATC GTG
    300                                     330                             360                                     390

510                                 540                             570                                     600
TTT GTT AAC AAG GTT GAT ACC ATT GAT GAT CCA GAA ATG TTA GAG TTA GTC GAA ATG GAA ATG AGA GAA CTT TTA AAC GAA TAT GGG TTT GAC GGT GAT AAT GCT CCA
Phe Val Asn Lys Val Asp Thr Ile Asp Asp Pro Glu Met Leu Glu Leu Val Glu Met Glu Met Arg Glu Leu Leu Asn Glu Tyr Gly Phe Asp Gly Asp Asn Ala Pro
170                         140         180                     150         190                         160         200
Phe Leu Asn Lys Cys Asp Met Val Asp Asp Glu Glu Leu Leu Glu Leu Val Glu Met Glu Val Arg Glu Leu Leu Ser Gln Tyr Asp Phe Pro Gly Asp Asp Thr Pro
TTC CTG AAC AAA TGC GAC ATG GTT GAT GAC GAA GAG CTG CTG GAA CTG GTT GAA ATG GAA GTT CGT GAA CTT CTG TCT CAG TAC GAC TTC CCG GGC GAC GAC ACT CCG
                        420                             450                             480

            630                             660                             690                                 720
ATT ATC ATG GGT TCT GCC CTT TGC GCT TTG GAA GGT CGC CAA CCT GAA ATT GGG GAG CAG GCC ATC ATG AAA CTT TTG GAT GCA GTG GAT GAG TAT ATT CCT ACA CCT
Ile Ile Met Gly Ser Ala Leu Cys Ala Leu Glu Gly Arg Gln Pro Glu Ile Gly Glu Gln Ala Ile Met Lys Leu Leu Asp Ala Val Asp Glu Tyr Ile Pro Thr Pro
    170         210                         180         220                         190 230                             200 240
Ile Val Arg Gly Ser Ala Leu Lys Ala Leu Glu Gly Asp Ala     Glu     Trp Glu Ala Lys Ile Leu Glu Leu Ala Gly Phe Leu Asp Ser Tyr Ile Pro Glu Pro
ATC GTT CGT GGT TCT GCT CTG AAA GCG CTG GAA GGC GAC GCA     GAG     TGG GAA GCG AAA ATC CTG GAA CTG GCT GGC TTC CTG GAT TCT TAT ATT CCG GAA CCA
                                                                                570                                     600
```

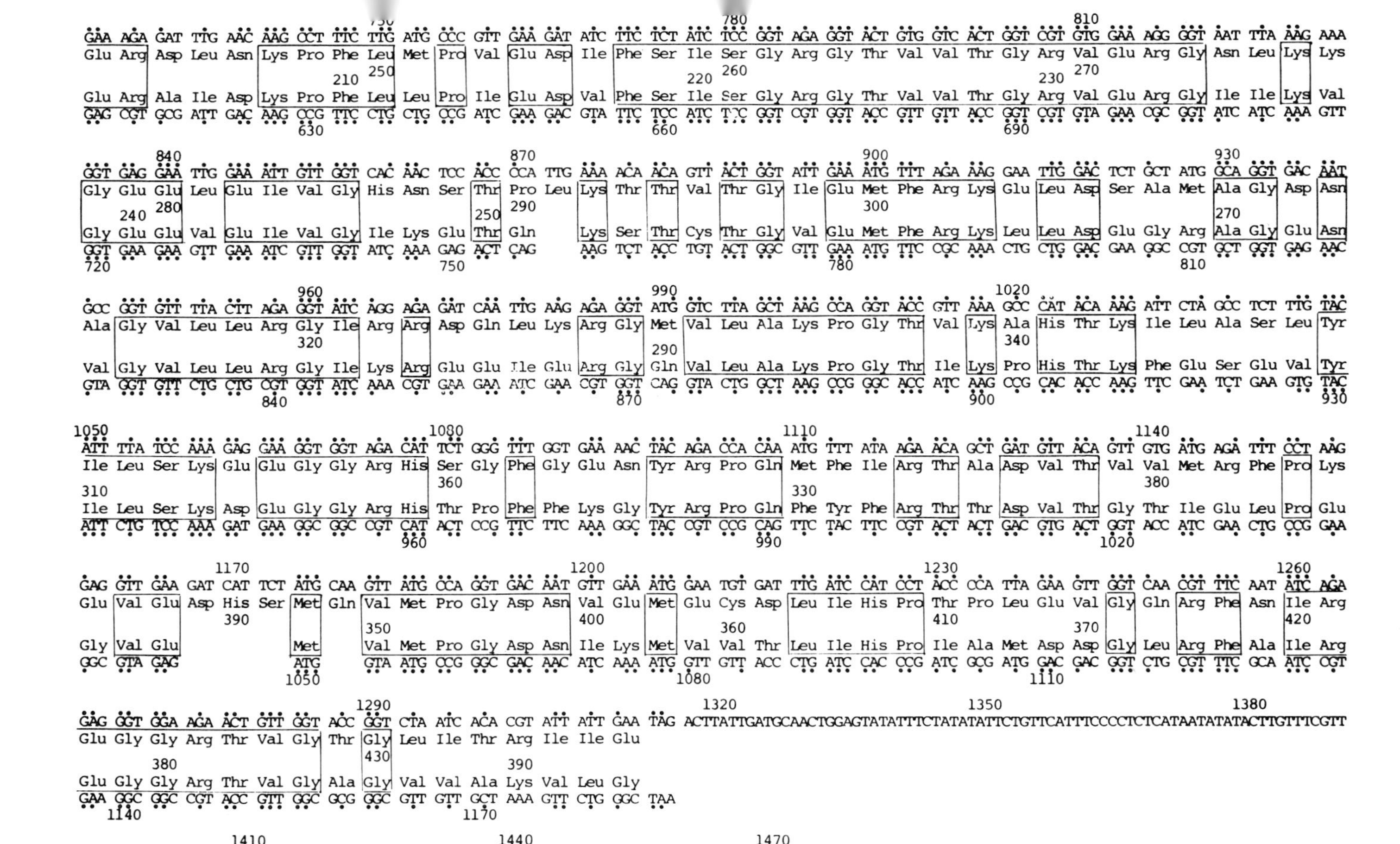

Fig. 4. Comparison of the nucleotide sequences of yeast *tufM* (Nagata *et al.* 1983) and *E. coli tufA* (Yokota *et al.* 1980) and the deduced amino acid sequences. The sequences of *tufM* and *tufA* are aligned to give maximal homology by introducing several gaps in *tufA*. Identical amino acids are framed and identical nucleotides are marked by dots.

cerevisiae DNA was surveyed utilizing Southern hybridization under low stringency conditions (Nagata *et al.* 1983). For a probe, we constructed plasmid pYT1 by subcloning the 1.5 kb *Hpa*I fragment of pTUB1 into pBR322, as plasmids pTUB1 and pTUB2 contained several other *E. coli* genes besides *tufB*. The 1.5 kb *Hpa*I fragment covered almost the entire coding sequence of *tufB* together with about 300 nucleotides of its 3'-flanking region. The results indicated that the restriction fragments of *E. coli* DNA gave 2 bands hybridizing with the *tufB* probe due to the presence of 2 genes for EF-Tu in *E. coli* (Jaskunas *et al.* 1975). However, a single DNA band was detected with the fragments of yeast DNA indicating that at least one sequence homologous to the *E. coli tufB* gene is present in yeast.

DNA from *Saccharomyces cerevisiae* 106A was then digested with *Bgl*II, and fragments (3.1 kb in length) hybridizing with pYT1 were isolated and ligated to pBR327 at the *BamH*I site. The hybrid plasmids were used to transform *E. coli* SK1592 to yield 768 ampicilin-resistant and tetracycline-sensitive clones. Clones containing the yeast *tufM* gene were identified by Southern hybridization of plasmid DNA with a *tufB* probe (obtained from pTUB311, an R1-derivative plasmid containing *tufB*) under low stringency conditions. A clone which gave a positive result was isolated and designated as pYYB.

Nucleotide sequence analysis

The Southern hybridization analysis of restriction fragments of pYYB with *E. coli tufB* showed that most of the yeast sequence homologous to *tufB* resides within the 1.0 kb *Pvu*II fragment. We therefore sequenced this region as well as the regions flanking it and the nucleotide sequence (1712 bp) thus obtained, is shown in Fig. 4, together with the sequence of the *E. coli tufA* gene (Yokota *et al.* 1980) aligned to obtain the maximal homology. The translation initiation site for yeast *tufM* was assigned to the methionine codon AUG at nucleotide positions 1 to 3, and the termination codon to UAG at 1318 to 1320 in the same reading frame. Because we found no intron in the coding sequence of *tufM*, the gene codes for a protein of 437 amino acids including the NH_2-terminal methionine with a calculated molecular weight of 47,980, in a single open reading frame of 1311 nucleotides.

A comparison of the sequence of yeast *tufM* with that of *E. coli tufA* revealed that the homology is 60% and 66% for the nucleotide and amino acid sequences, respectively. To obtain maximum homology, the initiator methionine for *tufA* was aligned to Ser-37 of *tufM,* and a limited number of gaps were introduced into the sequence of *tufA.* A most remarkable homology was found in amino acid residues 74–118 of *tufA* and 111–155 of *tufM,* where 41 of 44 amino acid residues were identical (91 % homology). As this region is supposed to be an active site for interaction with aminoacyl-tRNA (Kaziro 1978), the sequence conservation of this region might be due to a functional requirement. Other homologies were found in amino acid sequences 5–32, 208–234, and 371–386 of *tufA,* with the corresponding sequences of *tufM.* Whether or not these conserved regions constitute functional domains of the protein remains to be determined.

Yeast tufM *encodes for mitochondrial EF-Tu*

Plasmid YRpYB was constructed by cloning of the 2.5 kb *EcoR*I fragment of pYYB carrying *tufM* into a yeast vector YRp-7, and, an mRNA hybridizable with *tufM* was isolated from *Saccharomyces cerevisiae* D13–1A transformed with YRpYB. On translation in the reticulocyte lysate, the mRNA could direct the synthesis of a protein with Mr 48,000 which was immunoprecipitatable with an anti-*E. coli* EF-Tu antibody but not with an antibody against yeast cytoplasmic EF-1α (kindly donated by Dr. M. Miyazaki). The results indicate that the *tufM* gene is a nuclear gene coding for the yeast mtEF-Tu.

In general, nuclear-coded mitochondrial proteins are synthesized as precursors in the cytoplasm, and transported into mitochondria (Shatz & Butow 1983). It is noteworthy that the *tufM* gene codes for a protein 37 amino acids longer than *E. coli* EF-Tu at the amino terminal end. The 37 amino acid peptide is strongly basic having 4 arginine, and 2 lysine residues but no acidic amino acids. It is also rich in threonine (6 residues) and serine (7 residues). This sequence may serve as a signal for translocation across the mitochondrial membrane.

Recently, the structures of the proteolipid subunit of mitochondrial ATP synthase (Viebrock *et al.* 1982), and cytochrome *c* peroxidase (Kaput *et al.* 1982) have been published. The signal sequences of these proteins are 68 and 66 amino acids long, respectively, and possess properties similar to the above N-terminal sequence. As the precursor of cytochrome *c* peroxidase is transported into the inner membrane space of the mitochondria, with the first 18 amino acid residues of the signal sequence facing the mitochondrial matrix, we have compared the structure of the signal sequence of cytochrome *c* peroxidase with that of the mitochondrial EF-Tu. As shown in Fig. 5, the first 20 amino acid residues of mtEF-Tu have a distinct homology with that of cytochrome *c* peroxidase, which

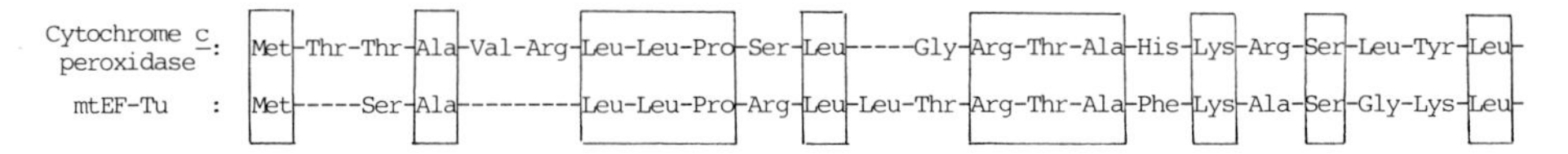

Fig. 5. Comparison of the amino terminal sequence of the precursor of yeast mitochondrial EF-Tu and the signal sequence of yeast cytochrome *c* peroxidase. The signal sequence of yeast cytochrome *c* peroxidase is from Kaput *et al.* (1982). The boxed amino acids indicate identical amino acids; and dashes show gaps introduced to obtain the maximal homology. From Nagata *et al.* (1983).

may suggest that a homologous sequence is important for proteins made in the cytoplasm that are to be imported into the mitochondria.

In closing, we would like to point out that the cloning procedure used in the present study may be applicable for isolation of other eukaryotic nuclear genes encoding for the mitochondrial proteins, especially the genes encoding for mitochondrial translational factors and ribosomal proteins. Through these studies, we may be able to throw more light on the organization of the genes for mitochondrial proteins in nuclear chromosomes, their expression and its regulation, and the structure of the signal peptides and the mechanism of import of proteins into mitochondria.

REFERENCES

An, G. & Friesen, J. D. (1980) *Gene 12,* 33–39.

Arai, K., Clark, B. F. C., Duffy, L., Jones, M. D., Kaziro, Y., Laursen, R. A., L'Italien, J., Miller, D. L., Nagarkatti, S., Nakamura, S., Nielsen, K. M., Petersen, T. E., Takahashi, K. & Wade, M. (1980) *Proc. Natl. Acad. Sci. U.S.A. 77,* 1326–1330.

Bachmann, B. J. (1983) *Microbiol. Rev. 47,* 180–230.

Borst, P. (1972) *Annu. Rev. Biochem. 41,* 333–376.

Hudson, L., Rossi, J. & Landy, A. (1981) *Nature 294,* 422–427.

Iwasaki, K. & Kaziro, Y. (1979) In: "*Methods in Enzymol. 60,*" pp. 657–676. Academic Press, New York.

Jaskunas, S. R., Lindahl, L., Nomura, M. & Burgess, R. R. (1975) *Nature 257,* 458–462.

Kaput, J., Goltz, S. & Blobel, G. (1982) *J. Biol. Chem. 257,* 15054–15058.

Kaziro, Y. (1978) *Biochim. Biophys. Acta 505,* 95–127.

Kaziro, Y. (1983) In: "*The Future of Nucleic Acid Research*" in press, Academic Press, New York.

Lee, J. S., An, G., Friesen, J. D. & Fiil, N. P. (1981) *Cell 25,* 251–261.

Lindahl, L., Post, L., Zengel, J., Gilbert, S. F., Strycharz, W. A. & Nomura, M. (1977) *J. Biol. Chem. 252,* 7365–7383.

Miyajima, A. & Kaziro, Y. (1978) *J. Biochem.* (Tokyo) *83,* 453–462.

Miyajima, A., Shibuya, M. & Kaziro, Y. (1979) *FEBS Lett. 102,* 207–210.

Miyajima, A., Shibuya, M., Kuchino, Y. & Kaziro, Y. (1981) *Mol. Gen. Genet. 183,* 13–19.
Miyajima, A., Yokota, T., Takebe, Y., Nakamura, M. & Kaziro, Y. (1983) *J. Biochem. (Tokyo), 93,* 1101–1108.
Mizushima-Sugano, J., Miyajima, A. & Kaziro, Y. (1983) *Mol. Gen. Genet. 189,* 185–192.
Nagata, S., Tsunetsugu-Yokota, Y., Naito, A. & Kaziro, Y. (1983) *Proc. Natl. Acad. Sci. U.S.A. 80,* 6192–6196.
Ogawa, T., Wabiko, H., Tsujimoto, T., Matsukata, H. & Ogawa, H. (1978) *Cold Spring Harbor Symp. Quant. Biol. 43,* 909–915.
Piechulla, B. & Küntzel, H. (1983) *Eur. J. Biochem. 132,* 235–240.
Richter, D. & Lipmann, F. (1970) *Biochemistry 9,* 5065–5070.
Richter, D. (1971) *Biochemistry 10,* 4422–4425.
Schatz, G. & Butow, R. A. (1983) *Cell 32,* 316–318.
Shibuya, M. & Kaziro, Y. (1979) *J. Biochem. (Tokyo), 86,* 403–411.
Shibuya, M., Nashimoto, H. & Kaziro, Y. (1979) *Mol. Gen. Genet. 170,* 231–234.
Viebrock, A., Perz, A. & Sebald, W. (1982) *EMBO J. 1,* 565–571.
Yokota, T., Sugisaki, H., Takanami, M. & Kaziro, Y. (1980) *Gene 12,* 25–31.

DISCUSSION

KURLAND: When you did your assay for the magic spot effect on the transcription of *tufB,* was Tu present, i.e., was it a coupled assay, or was it a pure transcriptional assay?

KAZIRO: Pure transcriptional assay, so there is no Tu in the reaction mixture.

SCHATZ: From our nucleotide sequence of the gene for mitochondrial Tu, you infer quite correctly that the extra sequence that you find, which is not present in the 2 bacterial genes, probably codes for a cleaved "leader" or "signal" sequence. In fact, unpublished experiments which were done partly in collaboration with Dr. Clark, have yielded some additional evidence for this. When we made an antibody against EF-Tu from *E. coli* and used this antibody to immunoplot an SDS gel of whole yeast cells, we found 2 cross-reactive bands. One cross-reacted strongly; we presumed it to be mitochondrial Tu. One cross-reacted weakly; we presumed it to be cytoplasmic Tu. When we grew the cells, under conditions which lead to an accumulation of precursors to mitochondrial proteins, the strongly cross-reactive band split in 2: the one which we normally saw and a band of higher molecular weight. This larger protein is almost certainly the precursor of mitochondrial Tu.

KAZIRO: How much larger was it?

SCHATZ: By gels, it is about 3000 daltons bigger.

KAZIRO: In the *tufM* sequence, we found a 36 amino acids stretch, which is sticking out beyond the corresponding N terminal of *tufA.*

SCHATZ: If you osmotically shock *E. coli* cells you obtain a soluble periplasmic shock fluid and a cellular residue. The proteins in the shock fluid are the periplasmic proteins described by Leon Heppel and the proteins in the cellular residue are all the cytoplasmic proteins. The notable exception to this is elongation factor Tu much of which appears in the shock fluid. For some reason EF-Tu is realeased by an osmotic shock, in preference to other cytosolic proteins of *E. coli.*

PEDERSEN: I can add that I repeated this experiment with a mutant in one of the Tu genes with a changed electrophoretic mobility. Both gene products are released by osmotic shock by this procedure.

KAZIRO: Regarding the question whether mitochondrial EF-Tu precursor is actually processed or not, we have to analyze the N-terminal amino acid sequences of mitochondrial Tu and compare it with the DNA sequence. This has to be done.

MILLER: I am very much surprised at how much homology there is between the 2. You might expect, since this divergence occurred billions of years ago, that you would have practically no base sequence homology. Is it possible to calculate an expected extent of homology from the relative evolution rates in 2 different types of organisms?

KAZIRO: I do not know. Mitochondria are supposed to be a prokaryotic parasite in the eukaryotic cells. The genes in mitochondria may be transferred to the nuclei by some kind of unknown mechanism. And the structures may have been conserved as they carry out the same function.

MILLER: My other question is, are the important sites conserved between the 2 organisms? For example the histidines have been thought to be important.

KAZIRO: They are preserved, yes.

MILLER: And how about the sulphydryl group?

KAZIRO: SH_2 (essential for aminoacyl-tRNA binding) is preserved, but there is no SH_1 (essential for nucleotide interaction) in mitochondrial Tu. But SH groups may not be very important, for example, EF-Tu from *T. thermophilus* has no SH.

MAGNUSSON: As you have 2 very similar Tu's in bacteria, and you have another Tu in mitochondria, which still has the same function, it could appear that the 2 forms in *E. coli* are very similar, not because of evolutionary pressure to stay similar, but because it is a recent evolutionary event. That raises the question whether the gene duplication affected only the actual coding region or whether it also affects some of the flanking regions. I wonder if you have looked at the nucleotide sequence to see whether there is any evidence that there is a longer stretch duplicated?

KAZIRO: Flanking regions of *tufA* and *tufB* have no homology. The nucleotide displacements which are in 13 positions, are found mostly in the N-terminal or C-terminal portions. Therefore, the middle part of the coding region is more conserved between *tufA* and *tufB*.

CLARK: I would like to expand a little on what Dr. Schatz was saying. We have looked at the immune cross-reactivity between prokaryotic and eukaryotic EF-Tu, and of course we find only a cross-reactivity from the anti-coli-Tu with the mitochondrial Tu, not with cytoplasmic EF-1α's. We have also done the reverse experiments with similar results. This, I suppose, corresponds with the well-known fact that if you look at *in vitro* systems you cannot substitute EF-1α for EF-Tu.

KAZIRO: It is well known that neither mitochondrial Tu nor *E. coli* EF-Tu acts with the 80S ribosomal system. On the other hand, Dr. Kunzel in Göttingen has shown that mitochondrial Tu can replace *E. coli* Tu in the poly (U)-directed poly-phenylalanine synthesizing system with the 70S *E. coli* ribosome.

Structure of Bacterial Elongation Factor EF-Tu and its Interaction with Aminoacyl-tRNA

B. F. C. Clark, T. F. M. la Cour, K. M. Nielsen, J. Nyborg, H. U. Petersen, G. E. Siboska & F. P. Wikman

INTRODUCTION

The interactive recognition of nucleic acids by proteins is a central process in the regulation of gene expression. To gain insight into such interactions requires a knowledge of the appropriate three-dimensional structures and biochemical functional information. Few biological macromolecular complexes are currently accessible to such detailed investigations. One convenient object for study is the strong and specific interaction between aminoacyl-tRNA (aa-tRNA) and the bacterial elongation factor EF-Tu, in a ternary complex, which also contains GTP. The ternary complex is a central feature in carrying aa-tRNA to the ribosomal A-site for decoding the mRNA and, thereby, placing the amino acid in the correct order during its incorporation into a polypeptide chain (Clark 1980). A knowledge of the recognitory interaction of aa-tRNA with EF-Tu should illuminate other types of nucleic acid/protein interactions and, hopefully, allow general principles to be proposed concerning regulatory mechanisms for gene expression.

Clearly, the direct way of obtaining information about the nucleic acid/protein recognition in a ternary complex would be to solve the three-dimensional structure to high resolution (about 0.25 nm), using X-ray crystallography. Since no suitable crystals large enough for X-ray studies exist at present, we decided on an indirect approach to the problem. We have used what one of us (BFCC) has, rather glibly, called the Crick "Rule of Thumb" approach. This approach, to be

Department of Chemistry, Aarhus University, 8000 Aarhus C, Denmark.

GENE EXPRESSION, Alfred Benzon Symposium 19.
Editors: Brian F. C. Clark & Hans Uffe Petersen, Munksgaard, Copenhagen 1984.

used when no suitable crystals of a macromolecular complex are available, aims to solve each macromolecular component structure and fit the components together by model building using all available information, both biochemical and structural.

As, the three-dimensional structure for yeast $tRNA^{Phe}$ is known and is thought to be a good general model for tRNAs, and, as there is good evidence that the structure of tRNA does not change significantly on charging to aa-tRNA, we decided to study the structure of EF-Tu by X-ray crystallography, and the ternary complex by biochemical and chemical methods as a prelude to model-building studies for the ternary complex.

Here we wish to report the current progress in these studies. Firstly, we shall describe how a high resolution electron density map (0.29 nm) of EF-Tu has enabled us to locate functionally important residues in the three-dimensional structure, conveniently illustrated using a 0.5 nm model. Secondly, we shall discuss the parts of the protein which interact with aa-tRNA in the ternary complex. Thirdly, we shall report how footprinting studies of aa-tRNA in the ternary complex have enabled us to propose a convenient model in terms of a 0.5 nm model of $tRNA^{Phe}$ for which parts of the tRNA are in contact with the protein. Lastly, we shall discuss the special case of the interaction of initiator tRNA with EF-Tu:GTP, using results from footprinting studies.

STRUCTURE OF EF-Tu

EF-Tu is the most abundant protein in *Escherichia coli,* constituting up to 6% of the total protein, depending on the growth conditions (Furano 1975). Interestingly, this relatively high concentration (~0.3 mM) in the bacterial cell is about the same as the aa-tRNA concentration. Perhaps therefore, the role of EF-Tu is to sequester the aa-tRNA during protein biosynthesis to prevent the easy chemical hydrolysis of the bond between the amino acid and the tRNA, which is labile at physiological pH. This action would conserve energy in the form of ATP, the hydrolysis of which is needed to reform aa-tRNA.

The ratio of the amounts of elongation factors and ribosomal proteins, earlier thought to be unity, has been found to vary somewhat with cellular growth condition (Reeh & Pedersen 1978). In rapidly multiplying cells, the ratio of EF-Tu:EF-Ts:EF-G:ribosomal proteins is about 6:1:1:1, reflecting the high concentration of EF-Tu in the cells. The situation is complicated by the fact that EF-Tu is a unique bacterial protein being coded for by 2 different genes, *tufA* and

Primary Structural Features of EF-Tu

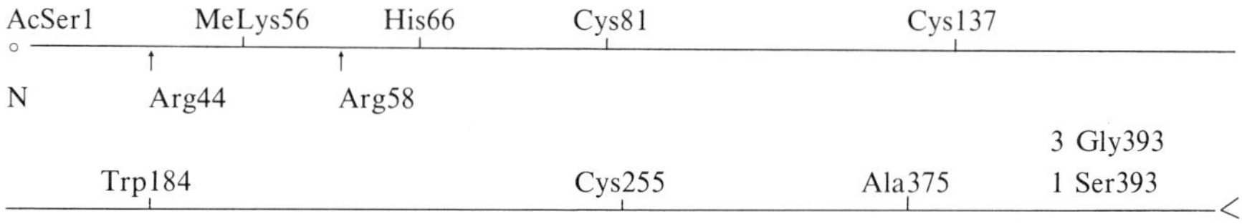

AcSer1 MeLys56 His66 Cys81 Cys137

N Arg44 Arg58

3 Gly393

Trp184 Cys255 Ala375 1 Ser393

Fig. 1. Summary of primary structure of *E. coli* EF-Tu (Jones *et al.* 1980) shown in highlighted form. The standard abbreviations for amino acids are shown. The location of two amino acids at the C-terminal position 393 shown in the ratio 3/1 for Gly/Ser is due to the determination being carried out on a mixture of two gene products.

tufB, located at 72′ and 88′, respectively, on the *E. coli* genetic map. When we decided to determine the three-dimensional structure of EF-Tu, no primary structure was known, so we completed the primary structure with the the co-operation of Dr. S. Magnusson's laboratory at Aarhus University's Department of Molecular Biology (Jones *et al.* 1980). We were rather fortunate in that it was possible to elucidate the structure of the protein isolated as a mixture of 2 gene products. It was known that the *tufA* gene was expressed about 3.5 times as much as the *tufB* gene (Reeh & Pedersen 1978), and we found only one ambiguity in the peptides produced after the protein's degradation, in that the C-terminal Gly was found in three times the amount of C-terminal Ser. We proposed, therefore, that the Gly containing product arose from the *tufA* gene and the Ser-containing product from the *tufB* gene. The determined DNA sequences of the *tufA* gene (Yokota *et al.* 1980) and the *tufB* gene (An & Friesen 1980) confirmed this situation. Actually, the 2 genes contained 12 other base differences, which were not expressed as different amino acids. The problem of differential gene expression and its regulation has not so far been explained from the knowledge of the gene sequences and their flanking regions.

As a background to the three-dimensional structure, some highlights of the primary structure of *E. coli* EF-Tu (Jones *et al.* 1980, Arai *et al.* 1980) are shown in Fig. 1. EF-Tu is 393 amino acids long of molecular weight about 43,000 daltons, starting with a blocked amino acid, acetyl Ser, and ending with either Gly or Ser, depending on which gene the product arises from, as described above. Interestingly, Lys 56 was methylated to about 60% to give monomethyl Lys, depending on the cellular growth conditions. Trp 184 is the only Trp in the protein, so it could be useful as a probe for physical chemical studies. There are 3 Cys residues: Cys 81 is thought to be concerned with aa-tRNA binding, Cys 137 with GDP/GTP binding, and Cys 255 is buried. His 66 is also thought to be concerned with the aa-tRNA binding site (see later). Ala 375 is the residue which is replaced in a kirromycin resistant strain (see later). During our EF-Tu crystallization studies, in the production of the most useful form discovered so far for data collection (Morikawa *et al.* 1978), the protein becomes nicked at Arg 58 (also perhaps later at Arg 44) when good crystals are produced. We have not completely excluded the possibility that minute traces of proteinase remain active in the crystallization dish to produce cuts at Arg 44 as well as Arg 58 (Arai *et al.* 1976). However, we prefer at present the explanation based on weak points in the polypeptide backbone being split due to restrictions in the spatial structure.

The binary complex EF-Tu:GDP, either intact or enzymically digested, crystallized readily in a variety of different forms. So far, no useful crystals of EF-Tu:GTP have been obtained. In our laboratory, we have experienced that a spontaneously nicked form of EF-Tu:GDP gives the most useful crystals for data collection and heavy metal derivation. The preliminary results of our studies with this orthorhombic crystal form describing data collection to 0.26 nm and a low resolution model (0.5 nm) were published previously (Morikawa *et al.* 1978).

Other workers have reported the results of low resolution X-ray diffraction studies on the trypsin-cleaved binary complex EF-Tu:GDP (Kabsch *et al.* 1977, Jurnak *et al.* 1980). There were some differences in the way in which the protein was modified prior to and during crystallization and in the molecular packing within the different crystal forms, but all 3 structures were similar at least at the low resolution described.

When we produced the first high-resolution electron density map, we were not able to correlate secondary structure with sequences, because only part of the primary structure was known. However, we were able to locate the bound GDP with respect to elements of secondary structure. Improvements in the method of analysis and the determination of the primary structure enabled us to locate about 70% of the structure. The range of data was reduced to 0.29 nm because of lack of isomorphism of the heavy atom derivatives. (Rubin *et al.* 1981). With the revised map, it was possible to describe the structure conveniently in terms of 3 domains I, II, and III, and locate the amino acid residues in each domain (Clark *et al.* 1982). The domains are also called the tight, the loose, and the floppy for I, II, and III, respectively, based on the degree of ordered structural elements. A simplified drawing is shown in Fig. 2, which also places some highlighted

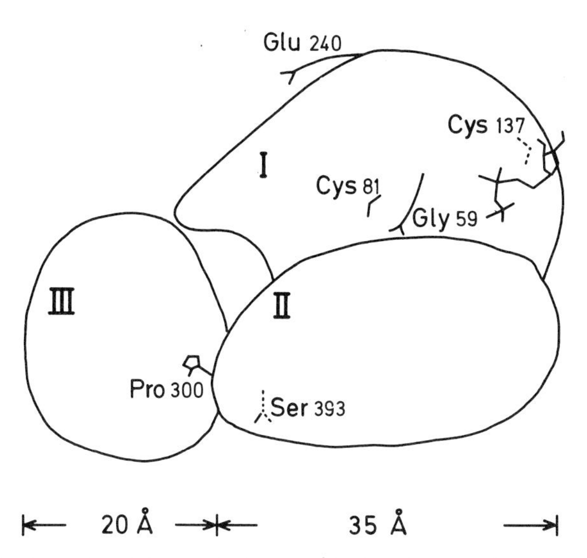

Fig. 2. A simplified drawing of the overall shapes of the domains of EF-Tu:GDP. The position of GDP is shown by the skeletal structure. Connecting amino acids and biochemically important positions are indicated. The domains I, II, and III are also referred to as the tight, the loose, and the floppy.

residues in the spatial structure. An end-on view of this simplified spatial structure is shown in Fig. 3. Progress in amino acid assignments in the spatial structure has come from refinement of the structure, and from molecular-model building in a molecular-graphics system due to collaboration of the Molecular Structure Group at Uppsala University's Biomedical Centre.

With our revised map, it has been possible to assign amino acids 60 to 240 to the tight domain I, which contains a high amount of secondary structure and contains the GDP binding site, shown by the skeletal structure in Figs. 2 and 3. The correctness of the map assignment is supported by location of the heavy atoms Hg, Pt and Pb at the expected amino acid residues, Cys, Met and acidic amino acids, respectively. The folding pattern of domain I (Rubin *et al.* 1981) shows a 5-stranded parallel β-sheet, which is extended by a sixth anti-parallel strand. The β-strands are connected by 6 α-helices via loops that are short except in one case. The β-sheet forms a central hydrophobic core with a relatively large left-handed twist. Surrounding this core, the α-helices provide an efficient interface with the solvent in a compact arrangement.

The positions of Cys 81 and Cys 137 associated with co-factor and aa-tRNA

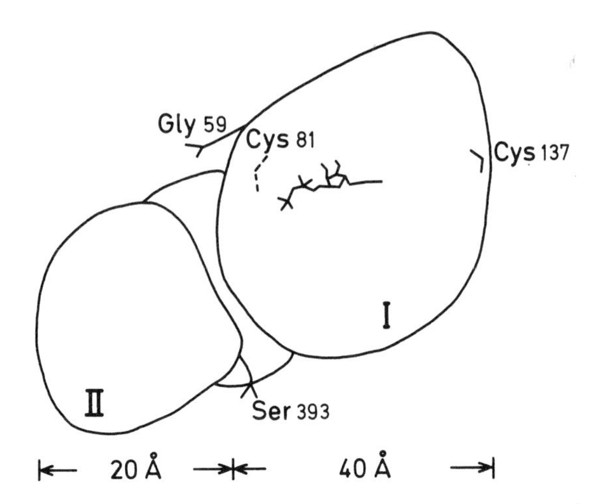

Fig. 3. A drawing of the overall shapes of the domains I, II, and III of EF-Tu: GDP viewed at 90° from the GDP binding end to the view shown in Fig. 2. Amino acids and GDP are indicated as in Fig. 2.

binding are also shown in Figs. 2 and 3. Domain II contains about 100 residues (approximately 300–393), which form a number of β-strands but no α-helices. This loose domain is in contact with helices I and II of the tight domain (Rubin *et al.* 1981). The location of the C-terminal amino acid (in this case Ser 393) is shown in the cleft between domains I and II (Figs. 2, 3). Residue 59 in Figs. 2 and 3, shows where the chain tracing begins in our model building. The electron density in domain III is still difficult to trace due to the apparent lack of secondary structure guidelines, which we could use for model building. This domain contains some less well-defined electron density for residues 1–58 and 240–300. The connection between domains II and III is round about residue 300 shown in Fig. 2.

There are indications that the interface between domains I and II is involved in allosteric changes, which are observed when EF-Tu exerts some of its functions. Some mutants of EF-Tu have been shown to be resistant towards the antibiotic kirromycin, which is an inhibitor of protein biosynthesis. One of these mutants has been shown to contain Thr in position 375 instead of Ala 375 in the wild type. (Duisterwinkel *et al.* 1981). In our preliminary interpretation of the electron density near the C-terminal end of the polypeptide chain, we have located residue 375 next to the cleft formed between domains I and II, and not far from the putative position of strand 44–58. Binding of the antibiotic causes a significant change of the affinities of the protein towards both GDP and GTP and alters the GDP binary complex to bind aa-tRNA. This allows us more confidently to use the structure of EF-Tu:GDP in our model-building studies for the ternary complex. Although normally, EF-Tu:GDP does not bind aa-tRNA, under the special conditions described above, it can. Thus, it seems likely that the disposition of the domains will change, going from the GDP to the GTP form of EF-Tu, but structures within the domains will probably remain the same. Work is continuing to improve the interpretability of our electron density map, especially to locate the atomic environment and bonding for the bound cofactor GDP.

INTERACTION OF EF-Tu:GTP WITH aa-tRNA

EF-Tu provides a good contrast in the problem of specific recognition of aa-tRNAs to an aa-tRNA synthetase, because EF-Tu recognizes all elongator tRNAs, whereas a synthetase recognizes at most a set of isoaccepting species. We are attempting to find out what determines how the ternary complex binds to the ribosome. The structure of EF-Tu:GDP itself may be important in deciding how EF-Tu is released from the ribosome after the hydrolysis of GTP.

Quite often, without keeping in mind the relative sizes of the macromolecules concerned, mistaken ideas of how EF-Tu might recognize aa-tRNA can arise. Fig. 4 shows, for comparative purposes, 0.5 nm Balsa wood models of both EF-Tu:GDP and uncharged yeast $tRNA^{Phe}$. For reasons already given, we believe these structures are relevant in a general way to the problem of the structure of the ternary complex. The models shown depict low-resolution structures to emphasize shapes, but in both cases, data for the model building was taken from high-resolution data analyses by Morikawa *et al.* (1978), and Ladner *et al.* (1975) for the EF-Tu and tRNA, respectively. Although the protein and the tRNA have very different molecular weights of about 43,000 and 25,000 daltons, the spatial sizes are rather similar, viz., $0.75 \times 0.5 \times 0.45$ nm^3 for EF-Tu and $0.8 \times 0.5 \times 0.3$ nm^3 for $tRNA^{Phe}$. Clearly, the nucleic acid has a less compact structure than the protein has.

(a) Parts of EF-Tu which interact with aa-tRNA

Studies selectively labelling essential SH-groups with N-ethylmaleimide by Laursen *et al.* (1977) indicated that Cys 81 was concerned with aa-tRNA binding, whereas Cys 137 was concerned with co-factor binding. That Cys 81 is

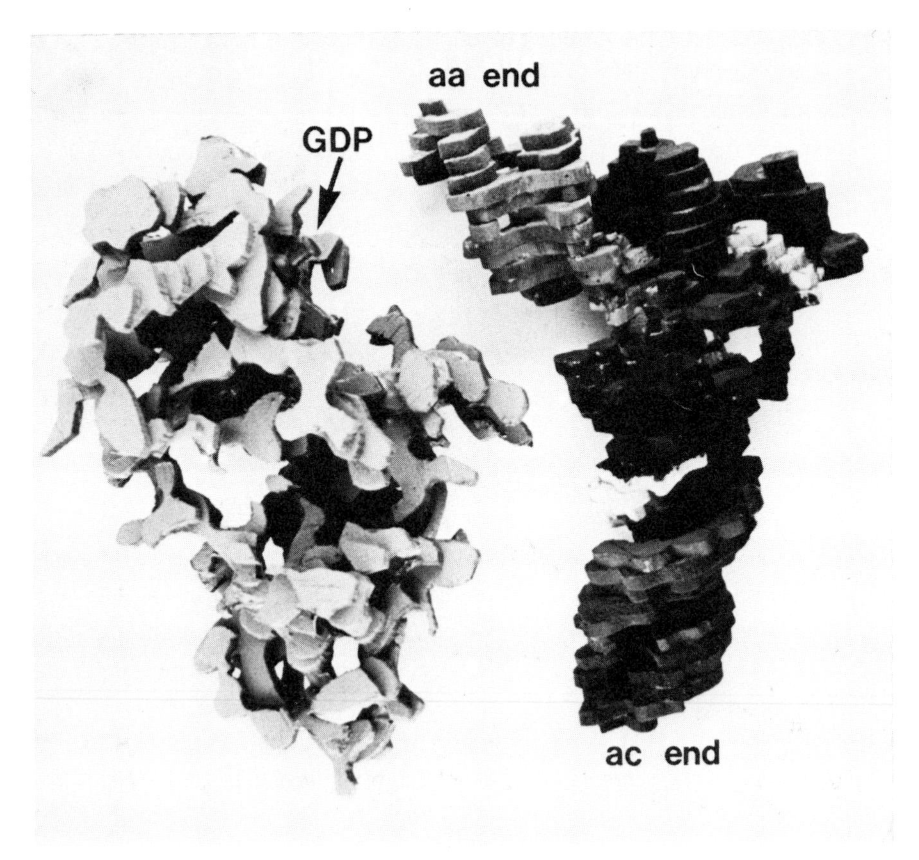

Fig. 4. Photograph of 0.5 nm Balsa wood models of nicked EF-Tu:GDP, containing essentially all of the protein polypeptide on the left and yeast $tRNA^{Phe}$ on the right.

involved with the binding of aa-tRNA was confirmed by Jonák *et al.* (1982), who used a reversible inhibitor N-tosyl-L-phenylalanyl chloromethane (TPCK) of aa-tRNA binding. Inhibition was caused by reaction of TPCK with Cys 81 (see Figs. 2, 3 for position). However, Cys 81 must be involved as a structural determinant in an indirect way, because other species of EF-Tu, for example from *Thermus thermophilus,* not containing an equivalent to Cys 81 still bind aa-tRNA (Kaziro 1978).

Other evidence for this general region of EF-Tu being concerned in the interaction with aa-tRNA comes from the biochemical studies of Johnson *et al.* (1978), who crosslinked ε-bromoacetyllysyl-tRNA to EF-Tu:GTP. The crosslink was shown to occur at His 66, which is located on β-strand 1 of the tight domain (Rubin *et al.* 1981), (with a location probably on the surface of domain I a little higher than Gly 59 in Fig. 3). Other workers (Jonák & Rychlík 1980, Nakamura & Kaziro 1981) have also shown that His residues are concerned in the ability of the factor to bind aa-tRNA, using a photo-oxidation technique specific for His residues. Recently, J. Jonák (unpublished results) has found that His 66 and His 118 of EF-Tu appear to be protected by aa-tRNA against reaction when in ternary complex. Domain I has also been implicated in binding the 3′-end of aa-tRNA in preliminary experiments by J. R. Rubin in this laboratory. By treatment of EF-Tu:GDP crystals with the analogue puromycin (which resembles tyrosyladenosine), subsequent X-ray data collection and difference Fourier analysis, changes in the electron density above helix VI (Rubin *et al.* 1981) were observed. Assuming that puromycin binds at the correct aa-tRNA binding site, this locates the 3′-end of the tRNA on the top surface of domain I approximately above the guanine base of GDP shown in Fig. 3.

Further information in the vicinity of the extra arm of the aa-tRNA should soon be available for an analysis of the crosslinked peptide position in studies where Kao *et al.* (1983) have succeeded in crosslinking X 47 of *E. coli* Phe-$tRNA^{Phe}$ with a bridging reagent to EF-Tu:GTP.

(b) Parts of aa-tRNA which interact with EF-Tu:GTP

Several observations indicate that the aminoacyl residue, and especially its α-amino group, is an essential feature for the recognition of aa-tRNA by EF-Tu:GTP. Such observations come from studies showing that neither uncharged nor N-acetylated aa-tRNA (Ravel *et al.* 1967) forms a ternary complex, and that EF-Tu:GTP prevents the chemical acylation of the α-amino group of aa-tRNA Sedláček *et al.* 1977).

That the 3′-end of the tRNA is also interacting in some recognitory way is known from the following experimental results. EF-Tu:GTP protects aa-tRNA against spontaneous chemical deacylation (Beres & Lucas-Lenard 1973) at physiological pH and the CCA-end against pancreatic ribonuclease digestion (Jekowsky *et al.* 1977). In addition, we have shown that the binding of GpGpN (N is any nucleotide) to its complementary sequence *NpCpC*pA at the 3′-end of aa-tRNA (where *N* is complementary with N) is inhibited by EF-Tu:GTP (Kruse *et al.* 1980), and that EF-Tu does not affect the tumbling of a spin label attached to the C2-position of the 3′-penultimate base of yeast-$tRNA^{Phe}$ (Sprinzl *et al.* 1978). Furthermore, aa-tRNA, in which the 3′-terminal A is replaced by the fluorescent analogue formycin, will also form a ternary complex, in which the formycin is rather accessible to reagents (P. Remy, personal communication). In summary, these foregoing observations suggest that the EF-Tu:GTP recognizes the amino group of the aminoacyl residue and the backbone, but only parts of the bases of the single-stranded 3′-region of aa-tRNA.

There were several reports in the literature which suggested that additional contact points on aa-tRNA are present in the strong recognitory interaction by EF-Tu:GTP. These concerned the types of tRNA species which would not form a ternary complex with EF-Tu:GTP. Such species included the Gly-tRNA involved in bacterial cell wall biosynthesis, denatured Leu-tRNA and the 3′-half of Val-tRNA. Originally, it was thought that the non-formylated form of the bacterial initiator tRNA, Met-$tRNA_f^{Met}$ would not form a ternary complex with EF-Tu:GTP, but recently, Tanada *et al.* (1981, 1982) have reported that Met-$tRNA_f^{Met}$ forms a ternary complex about half as stable as the ternary complex containing the elongator tRNA, Met-$tRNA_m^{Met}$. It is still unclear whether this rather controversial result has any biological meaning. Our footprinting studies on the ternary complex containing Met-$tRNA_f^{Met}$ will be described later.

So far, it has not been possible to make a general strategy to determine the chemical reactivity of an aminoacyl-tRNA complexed with EF-Tu:GTP, because the chemical reagents inactivate the protein too rapidly. Probing the structure of aa-tRNA in ternary complex has been more successfully studied by ribonuclease digestion. This nuclease digestion of the [^{32}P]-labelled aa-tRNA in ternary complex followed by gel electrophoresis gives a ladder of oligonucleotides with gaps when compared with the nuclease digestion of naked aa-tRNA. The ladder with gaps due to steric hindrance or covering by the protein gives rise to the "footprint" of the protein on the tRNA. We applied the footprinting technique first, using a double-strand specific ribonuclease, Cobra Venom ribonuclease (CV RNase) to digest 4 different aa-tRNAs in ternary complex (Boutorin *et al.* 1981). The results indicated that EF-Tu:GTP protected both the aa-stem and T-stem against CV RNase digestion. Furthermore, these tRNAs, yeast Phe-$tRNA^{Phe}$ and *E. coli* Met-$tRNA_m^{Met}$, Phe-$tRNA^{Phe}$ and Glu-$tRNA^{Glu}$ all had enhanced cutting in the anticodon stem (ac-stem) by the CV RNase in the presence of EF-Tu:GTP. These results suggested that EF-Tu:GTP interacts with other parts of the aa-tRNA in addition to the 3′-end and that no protective covering exists for the ac-stem.

There are several other pieces of evidence confirming that EF-Tu:GTP does not interact with the ac-arm of aa-tRNA. In this respect, modifications of the ac-loop have no effect on ternary complex formation (Kruse & Clark 1978, Weygand-Durasevic *et al.* 1981), EF-Tu:GTP binds to aa-tRNA when this tRNA is bound to another tRNA, having a complementary anticodon (Yamane *et al.* 1981), and the cuts in the ac-stem made by CV RNase do not prevent ternary complex formation (Boutorin *et al.* 1981).

Although EF-Tu is not thought to bind to the ac-loop, spin-labelling studies (Kruse & Clark 1981, Weygand-Durasevic *et al.* 1981), and footprinting experiments (Boutorin *et al.* 1981), indicate that a structural rearrangement of this region takes place upon ternary complex formation. Another study suggesting conformational changes, especially in the tRNA backbone on binding EF-Tu:GTP, has been reported by Riehl *et al.* (1983). They compared the phosphate accessibility towards ethylnitrosourea in yeast Phe-tRNA complexed with *E. coli* EF-Tu:GTP, suggesting that the fit between the EF-Tu:GTP and aa-tRNA is loose enough for a small reagent to penetrate between the macromolecules. However, they find a change of reactivity of some phosphate residues in the presence of EF-Tu, suggesting conformational changes in the backbone.

Recently, we have more extensively studied the structure of aa-tRNA in the ternary complex using partial ribonuclease digestion (Wikman *et al.* 1982) to locate the regions of the aa-tRNA responsible for complex formation. A similar approach was used earlier by Jekowsky *et al.* (1977), before the rapid RNA sequencing technology was available. Their experiments also were limited by the use of exhaustive digestion of the tRNA. This precludes the detection of small changes in accessibility of the various tRNA sites, and also it must be expected that the first cut alters the structure of tRNA, so changing the binding interactions with EF-Tu:GTP and exposing new sites for digestion. In summary, Jekowsky *et al.* (1977) found cuts in the D- and T-loops of tRNA and concluded that these loops do not participate in the complex formation.

We have studied the kinetics of digestion of different aa-tRNAs by the single-strand specific RNases, T_1, T_2 and A, and have also investigated the effect of different ratios of enzyme to substrate with time. In our experiments, we have, therefore, limited the amount of digested tRNA to less than 20% and used the results from the kinetic experiments to achieve 'single hit' conditions. Various controls were run to eliminate non-specific cuts (Wikman *et al.* 1982), and attempts were made to run the digestions under similar enzymic conditions. As an example, a ladder gel showing footprinting is shown in Fig. 5. The results in this figure are from an experiment where 3′ [^{32}P]-labelled *E. coli* Met-tRNA$_m^{Met}$,

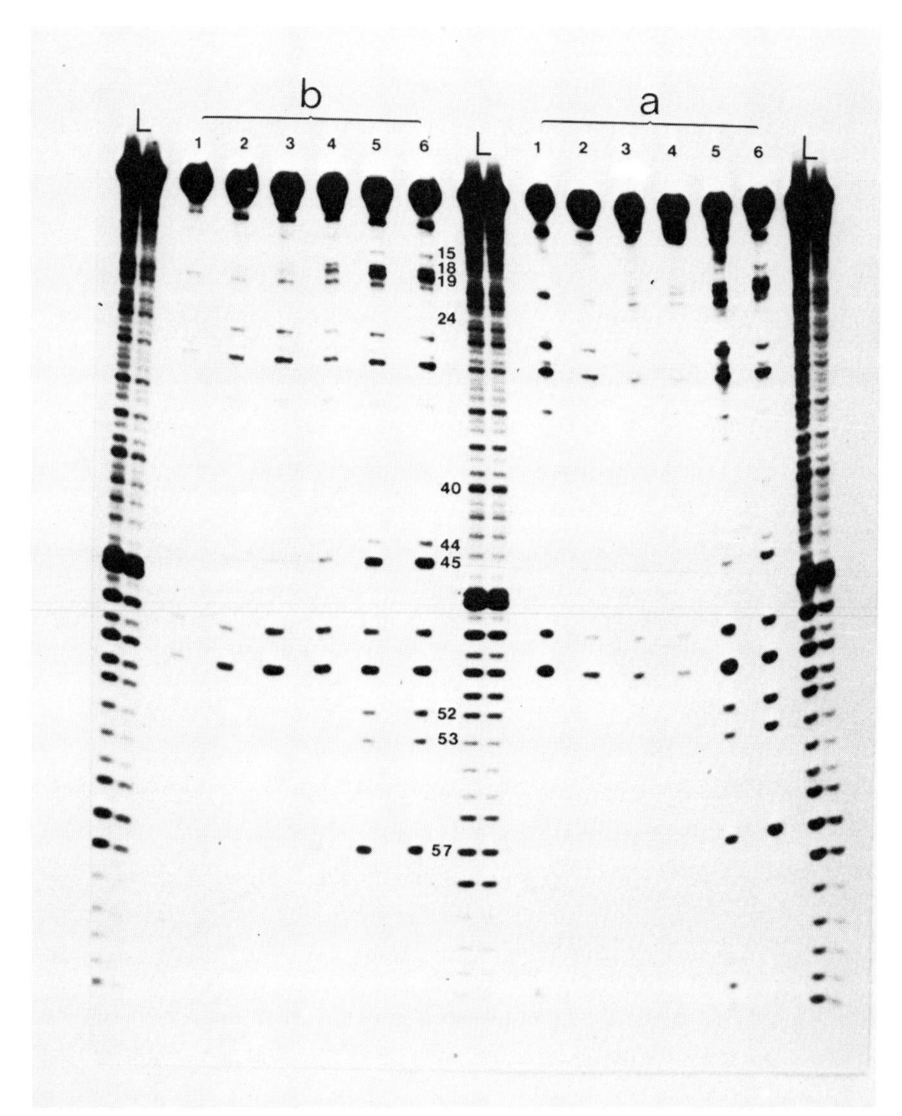

Fig. 5. Photograph of an autoradiograph of a polyacrylamide gel electrophoresis of [^{32}P] 3′-end labelled *E. coli* Met-tRNA$_m^{Met}$ digested by RNase T$_1$ (Wikman *et al.* 1982). Lane L, random digestion in formamide at 100°C for 30 min. The letters a and b denote digestion of [^{3}H]-Met-[^{32}P]-tRNA$_m^{Met}$ in ternary complex and free in solution, respectively. The numbers indicate different enzyme/tRNA ratios: 1) no enzyme added; 2) 5 ng/μg tRNA; 3) 10 ng/μg tRNA; 4) 20 ng/μg tRNA; 5) 40 ng/μg tRNA; 6) 80 ng/μg tRNA.

naked or in complex with EF-Tu:GTP, is digested by increasing amounts of RNase T_1. This figure shows that G-specific cuts increase with increasing amounts of enzyme as expected, but a number of contaminating bands of almost constant intensity not due to the enzyme are also apparent. These contaminating cuts are rejected from the control experiments and are difficult to exclude with highly radioactively labelled aa-tRNA. Most obviously, we can see the protection of the extra arm (shown by cuts 44, 45) by EF-Tu:GTP in Fig. 5. It is also obvious that these single-strand specific nucleases also cut stem regions of aa-tRNA.

A summary of our studies is shown in Fig. 6, where the results of treating *E. coli* Met-$tRNA_m^{Met}$, or Phe-$tRNA^{Phe}$ naked or in ternary complex with the three RNases mentioned above are described schematically. The positions of cuts in the aa-tRNA arranged on a standard folded structure (Ladner *et al.* 1975) are given in terms of protection, no change (neutral) and enhancements in the presence of EF-Tu:GTP. The results show protection of parts of the aa-stem, the

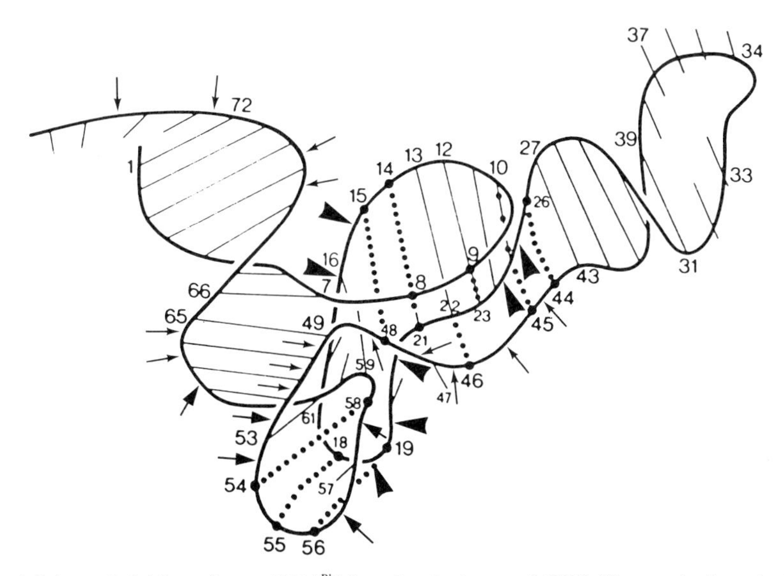

Fig. 6. Schematic folding of yeast $tRNA^{Phe}$ (based on Ladner *et al.* 1975). The standard numbering of yeast $tRNA^{Phe}$ has been used. The arrows summarize cleavage positions both for Met-$tRNA_m^{Met}$ and Phe-$tRNA^{Phe}$ (Wikman *et al.* 1982). The type of arrow indicates protection of cuts (small arrow), neutral cuts (medium-sized arrow) and enhancement of cuts (large arrowheads) due to EF-Tu:GTP in the ternary complex. Experimental details are found in Wikman *et al.* (1982). For the reasons in the text, the arrows and arrowheads are omitted in the ac-arm.

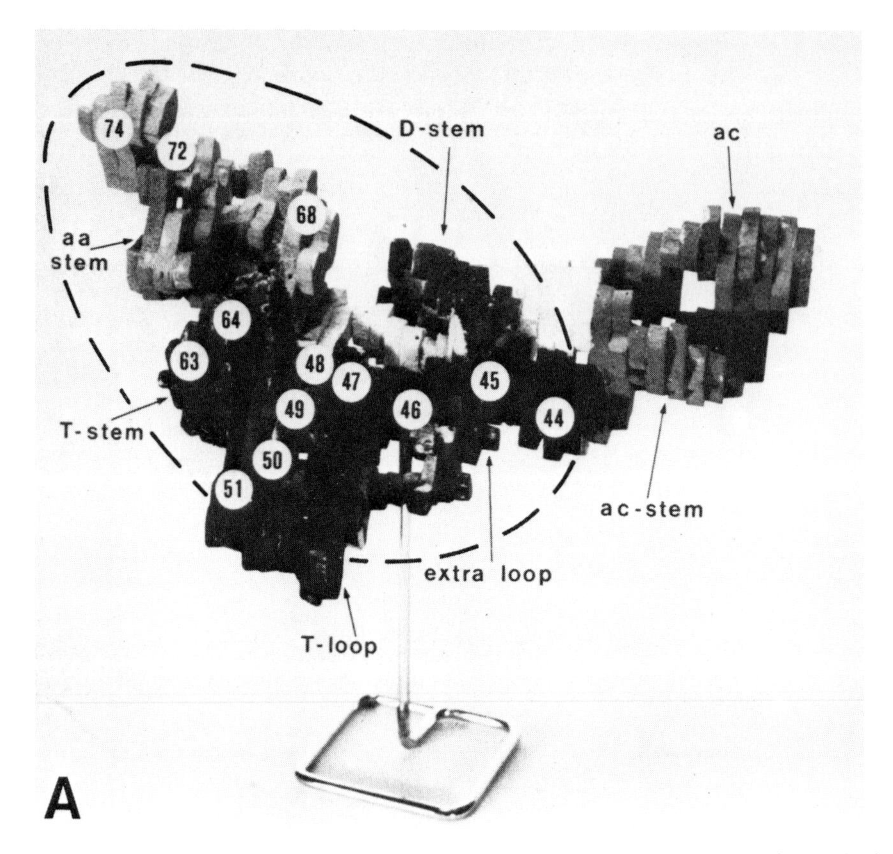

Fig. 7. Photograph of a low resolution (0.5 nm) Balsa wood model of tRNA based on the standard model of yeast $tRNA^{Phe}$. Protection sites are marked with numbered circles, neutral sites with squares, and enhancement sites with triangles. A: Omitting the cleavage sites in the ac-arm as in Fig. 6, the protection sites are observed to be clustered on the same side of the molecular structure that exposes the extra loop. B: Enhancement sites and neutral sites can be seen to be located on the opposite side of the molecule to the EF-Tu:GTP binding site, i.e., the side exposing the D-loop. Again, the cleavage sites in the ac-arm are omitted.

T-stem and the extra loop by the EF-Tu:GTP against ribonuclease digestion. Cuts in the ac-arm are not marked in the figure for the reasons given above that we do not believe that the ac-arm is in contact with EF-Tu:GTP. This folded structure shows both sides of the molecule at the same time, so has the confusing properties of an electron micrograph.

When the results of the ribonuclease digestion studies were marked on a low resolution (0.5 nm) Balsa wood model of tRNA (Fig. 7A), the picture was

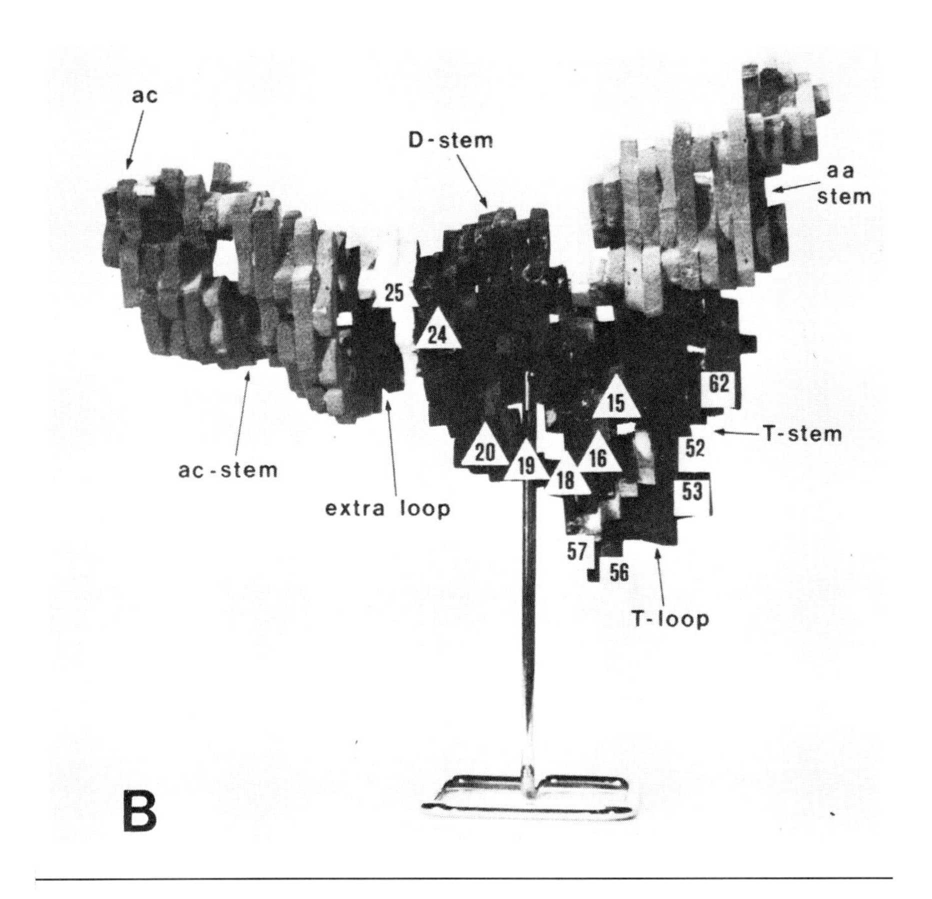

startlingly clarified. All of the protection sites against nuclease digestion lay on the same side of the model, the side of the L-shape (Rich & Kim 1977), on which the extra loop is exposed (Wikman *et al.* 1982). Enhanced cuts are located on the opposite side of the model (Fig. 7B). Parts of the T-stem and T-loop contain digestion points apparently unaffected by the presence of EF-Tu:GTP (Fig. 7B). The enhancements are most conveniently interpreted as being due to local disturbances in conformation after EF-Tu:GTP binding to the other side of the molecule.

Unexplained contradictory results have been obtained by Bertram & Wagner (1982), who proposed binding of EF-Tu:GTP to the D-arm from protection results against kethoxal modification. Also Douthwaite *et al.* (1983) found a protection by EF-Tu:GTP against a weakly modified C13 in the D-arm. These chemical modification results could be explained by steric hindrance due to local conformational changes (Riehl *et al.* 1983).

Support of our model involving the extra loop in interaction with EF-Tu:GTP comes from two other types of studies so far. Kao *et al.* (1983) have been successful in crosslinking X 47 of *E. coli* Phe-$tRNA^{Phe}$ to EF-Tu:GTP with a bridging molecule. They are currently attempting to determine the location of the crosslink of EF-Tu:GTP. In addition, NMR studies of the methyl group of m^7G46 in aa-tRNA (P. Agris, personal communications) shows a line broadening due to putative interaction in the presence of EF-Tu:GTP.

The concluding scheme for EF-Tu binding that can be deduced from our studies is shown in Fig. 7A, where the dotted line limits the region of tRNA where the EF-Tu:GTP is thought to interact. This might suggest that interaction of the ternary complex with the ribosomal A-site would involve tRNA interactions on the D-loop side.

INTERACTION OF INITIATOR tRNA WITH EF-Tu:GTP

Recently, Tanada *et al.* (1982) reported that the affinity of Met-$tRNA_f^{Met}$ for EF-Tu:GTP is much stronger (~50×) than that of fMet-$tRNA_f^{Met}$ as expected for the bacterial initiating species and unexpectedly comparable in magnitude to that of the elongator Met-$tRNA_m^{Met}$. Thus, EF-Tu:GTP in bacteria may well discriminate against only the species fMet-$tRNA_f^{Met}$. The formylation could help play a role as security against the Met-$tRNA_f^{Met}$ acting as an elongator and translating incorrectly at GUG and UUG, which the initiator species theoretically can do during initiation (Clark & Marcker 1966). What happens during eukaryotic protein biosynthesis in this connection is rather unclear, because the initiator species itself is not formylated and exists as Met-$tRNA_f^{Met}$. However, the whole selection of the initiator tRNA is more restricted in eukaryotes in the sense that more protein components are involved. Of course, during bacterial protein initiation, initiation factors must also play a significant role in placing the initiator tRNA in the correct P-site for initiation.

We have confirmed the strong interaction of *E. coli* Met-$tRNA_f^{Met}$ with EF-Tu:GTP and have carried out footprinting studies on the complex similar to those described above for elongator tRNA. Our results (Petersen *et al.* 1983) are shown in Fig. 8. Although the studies have not been so extensive, we observed a similar general pattern of protection by EF-Tu:GTP against nuclease cutting of the initiator tRNA for elongator tRNAs but with some small differences. The

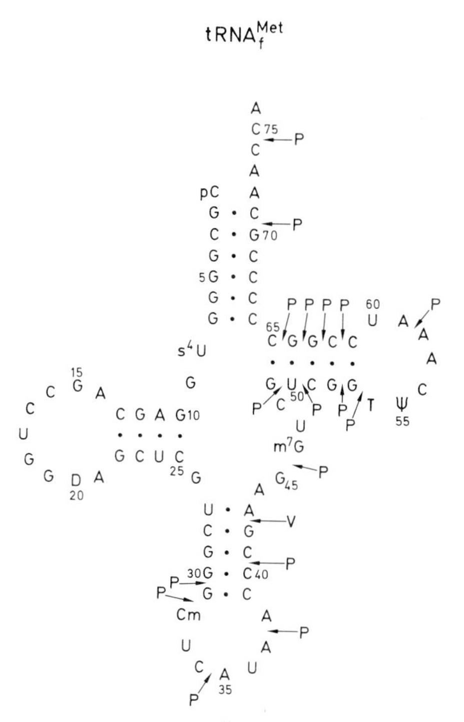

Fig. 8. Protection of cleavage sites on Met-tRNA$_f^{Met}$ (arranged in a cloverleaf) due to RNases T_1 and A. P represents protection in the presence of EF-Tu:GTP, V shows a site with variable results in different experiments. Results are taken from Petersen *et al.* 1984.

cuts are shown on a standard cloverleaf in Fig. 8, but we shall ignore the protected regions in the ac-arm for this discussion as before. Again, we see protection in the aa-stem, T-stem and extra loop. In addition, the protection runs into the T-loop on both sides where we saw no changes for the elongator tRNAs. We are not able to decide at this time whether this should be interpreted in terms of extra covering by EF-Tu:GTP of the tRNA or steric hindrance due to local conformational changes.

What appears to be a reasonable conclusion is that bacterial Met-tRNA$_f^{Met}$, the initiator tRNA in the unformylated state, can indeed form a ternary complex with EF-Tu:GTP. This ternary complex also appears almost normal in the disposition of the aa-tRNA on the EF-Tu:GTP, so it should be able to bind to the ribosomal A-site if its formation is possible *in vivo.* Whether the small differences in the T-loop binding are sufficient for making a fit poor enough in the A-site to be selected against needs more information from further experimentation.

ACKNOWLEDGEMENTS

We gratefully acknowledge financial support from the Danish Natural Science Research Council, the Niels Bohr Stipendium Committee (HUP), and the Carlsberg Foundation (FPW). We wish to thank Dr. D. L. Miller for co-operation and advice, and Anne Merete Mathiesen for technical help. In addition, the help of Orla Jensen, Arne Lindahl, and Lisbeth Heilesen in the production of the manuscript is noted.

REFERENCES

An, G. & Friesen, J. D. (1980) *Gene 12,* 33–39.

Arai, K., Clark, B. F. C., Duffy, L., Jones, M. D., Kaziro, Y., Laursen, R. A., L'Italien, J., Miller, D. L., Nagarkatti, S., Nakamura, S., Nielsen, K. M., Petersen, T. E., Takahashi, K. & Wade, M. (1980) *Proc. Natl. Acad. Sci. USA 77,* 1326–1330.

Arai, K., Nakamura, S., Arai, T., Kawakita, M. & Kaziro, Y. (1976) *Biochem. J. 79,* 69–83.

Beres, L. & Lucas-Lenard, J. (1973) *Biochemistry 12,* 3998–4005.

Bertram, S. & Wagner, R. (1982) *Biochem. Int. 4,* 117–126.

Boutorin, A. S., Clark, B. F. C., Ebel, J. P., Kruse, T. A., Petersen, H. U., Remy, P. & Vassilenko, S. (1981) *J. Mol. Biol. 152,* 593–608.

Clark, B. F. C. (1980) *Trends Biochem. Sci. 5,* 207–210.

Clark, B. F. C., la Cour, T. F. M., Fontecilla-Camps, J., Morikawa, K., Nielsen, K. M., Nyborg, J. & Rubin, J. R. (1982) In: *Cell Function and Differentiation,* Part C, pp. 59–64, FEBS Symposia, Vol. 66, Alan R. Liss Inc., New York.

Clark, B. F. C. & Marcker, K. A. (1966) *J. Mol. Biol. 17,* 394–406.

Douthwaite, S., Garrett, R. A. & Wagner, R. (1983) *Eur. J. Biochem. 131,* 261–269.

Duisterwinkel, F. J., De Graaf, J. M., Kraal, B. & Bosch, L. (1981) *FEBS Letters 131,* 89–93.

Furano, A. (1975) *Proc. Natl. Acad. Sci. USA 72,* 4780–4784.

Jekowsky, E., Schimmel, P. & Miller, D. L. (1977) *J. Mol. Biol. 114,* 451–458.

Johnson, A. E., Miller, D. L. & Cantor, C. R. (1978) *Proc. Natl. Acad. Sci. USA 75,* 3075–3079.

Jonák, J. & Rychlík, I. (1980) *FEBS letters 117,* 167–171.

Jones, M. D., Petersen, T. E., Nielsen, K. M., Magnusson, S. M., Sottrup-Jensen, L., Gausing, K. & Clark, B. F. C. (1980) *Eur. J. Biochem. 108,* 507–526.

Jurnak, F., McPherson, A., Wang, A. H. J. & Rich, A. (1980) *J. Biol. Chem. 255,* 6751–6757.

Kabsch, W., Gast, W. N., Schultz, G. E. & Leberman, R. (1977) *J. Mol. Biol. 117,* 999–1012.

Kao, T.-H., Miller, D. L., Abo, M. & Ofengand, J. (1983) *J. Mol. Biol. 166,* 383–405.

Kaziro, Y. (1978) *Biochem. Biophys. Acta 505,* 95–127.

Kruse, T. A. & Clark, B. F. C. (1978) *Nucleic Acid Res. 5,* 879–892.

Kruse, T. A., Clark, B. F. C., Appel, B. & Erdmann, V. A. (1980) *FEBS Letters 117,* 315–318.

Ladner, J. E., Jack, A., Robertus, J. D., Brown, R. S., Rhodes, D., Clark, B. F. C. & Klug, A. (1975) *Proc. Natl. Acad. Sci. USA 72,* 4414–4418.

Laursen, R. A., Nagarkatti, S. & Miller, D. L. (1977) *FEBS Letters 80,* 103–106.

Miller, D. L. (1978) In: *Gene Expression,* Vol. 43 of FEBS Symposia (Clark, B. F. C., Klenow, H. & Zeuthen, J. eds) pp. 59–68, Pergamon Press.

Morikawa, K., la Cour, T. F. M., Nyborg, J., Rasmussen, K. M., Miller, D. L. & Clark, B. F. C. (1978) *J. Mol. Biol. 125,* 325–338.

Nakamura, S. & Kaziro, Y. (1981) *J. Biochem. 90,* 1117–1124.

Petersen, H. U., Clark, B. F. C. & Hershey, J. W. (1984) in preparation.

Ravel, J. M., Shorey, R. L. & Shive, V. (1967) *Biochem. Biophys. Res. Commun. 29,* 68–73.

Reeh, S. & Pedersen, S. (1978) In: *Gene Expression,* Vol. 43 of FEBS Symposia (Clark, B. F. C., Klenow, H. & Zeuthen, J. eds) pp. 89–98, Pergamon Press.

Rich, A. & Kim, S. H. (1977) *Scient. Amer. 238,* 52–62.

Riehl, N., Giegé, R., Ebel, J. P. & Ehresmann, B. (1983) *FEBS Letters 154,* 42–46.

Rubin, J. R., Morikawa, K., Nyborg, J., la Cour, T. F. M., Clark, B. F. C. & Miller, D. L. (1981) *FEBS Letters 129,* 177–179.

Sedláček, J., Jonák, J. & Rychlík, I. (1977) In: *Translation of Natural and Synthetic Poly nucleotides* (Legocki, A. B. ed) pp. 188–191, University of Agriculture, Poznan.

Sprinzl, M., Siboska, G. E. & Petersen, J. A. (1978) *Nucleic Acids Res. 5,* 861–877.

Tanada, S., Kawakami, M., Yoneda, T. & Takemura, S. (1981) *J. Biochem. 89,* 1565–1572.

Tanada, S., Kawakami, M., Nishio, K. & Takemura, S. (1982) *J. Biochem. 91,* 291–299.

Weygand-Durasevic, I., Kruse, T. A. & Clark, B. F. C. (1981) *Eur. J. Biochem. 116,* 59–65.

Wikman, F., Siboska, G. E., Petersen, H. U. & Clark, B. F. C. (1982) *EMBO J. 1,* 1095–1100.

Yamane, T., Miller, D. L. & Hopfield, J. J. (1981) *Biochemistry 20,* 449–452.

Yokota, T., Sugisaki, H., Takanami, M. & Kaziro, Y. (1980) *Gene 12,* 25–31.

DISCUSSION

INNIS: You mentioned making changes in the nucleotide sequence and around the anticodon regions: that that had no effect on the structures that you were probing. Could you expand on that?

CLARK: I said that chemical modification had no effect on the formation of the ternary complex, because of some work we did some time ago with Mathias Sprinzl, where you have an arginine tRNA from *E. coli* which contains a thio C which can be modified. It can be alkylated, and this alkylation still allows ternary complex formation. You can also modify tRNAs which have hypermodified bases on the other side of the anticodon; modify those and they still form a ternary complex. We have not done any genetic engineering of the anticodon loop which, if you were at the recent tRNA workshop in Japan, is the "in" thing at the moment. Everybody is engineering anticodon loops. I guess we are going to hear about that from Olke Uhlenbeck.

INNIS: What effects, if any, are there due to changing the magnesium and spermidine concentrations when you do these nuclease digestion experiments?

CLARK: We try to do all the digestions under physiological conditions under very comparative situations. We don't do Sl nuclease digestions for that very reason, for example. In general, spermine and spermidine were interchangeable in the crystallization conditions during the early structure studies, except when we got really good crystals where spermine was needed. It is a kind of glue for the tRNA structure. I don't think it has any essential effect in our experiments, because you can substitute with other ions and get the same structure.

CRAMER: As the protection of the molecule on the one side of course is obvious, because the enzyme cannot approach from this side, the enhancement on the other side of course, is not so easily understood, at least not at first sight. One explanation, for me at least, would be that the molecule somehow will bend towards the side where the EF-Tu is binding and, therefore, the side, where it does not bind, is somehow strained and the phosphate backbone is, therefore, under chemical stress, and, thus, hydrolyzed easier.

CLARK: Yes, I think in fact, Dr. Ebel wants to make a note about the backbone, which may illuminate that further.

EBEL: I would like to make a short comment about recent experiments we have done using ethylnitrosourea as a modifying agent. This reagent alkylates phosphates giving rise to tri-esters. Tri-esters are very labile at alkaline pH and are therefore hydrolized. With denatured tRNA, cuts occur randomly at every phosphate. But with native tRNA, there are phosphates which are protected by the tertiary structure so that this reagent is a good probe for tertiary structure. When the reaction is performed on the complex between tRNA and aminoacyl-tRNA synthetases additional phosphates are protected. So we thought that with the EF-Tu:GTP: aminoacyl-tRNA complex we would also find protected phosphates, but to our surprise no additional protection was observed. This suggests that the interaction between the aminoacyl-tRNA and the EF-Tu factor is rather loose. On the contrary the accessibility of several phosphates is increased, suggesting a structural rearrangement of the tRNA induced by complex formation.

OFENGAND: My first comment is that we can support, at least in one area, Dr. Clark's interaction on the extra loop side of the tRNA by a cross-linking study that was done in collaboration with Dr. Miller. We found that a derivative of the base X47 cross-links very effectively to the EF-Tu molecule. It is rather interesting, perhaps, that when the cross-linker is 20Å long, the cross-linking is quite effective, but just by decreasing 6Å to 14Å, the cross-linking disappears entirely. One is tempted to conclude that, while the interaction may be on that side, it may not be in every case very close, which might be somewhat in agreement with the results of Dr. Ebel. One other comment: there are some recent measurements of the affinity of tRNA for EF-Tu which are done by equilibrium measurement in solutions using fluorescent tRNAs, and these give somewhat different numbers for some of the other species, so that (this is work done by Art Johnson) the interaction of the Met-$tRNA_f^{Met}$, that is the charged but not formylated species, was considerably lower than in the experiments that were done by Dr. Tanada. I am not sure whether you did your experiments by that same chemical protection method or by a different approach?

CLARK: Yes, as I stated previously we have just shown that you can isolate a complex for footprinting, but we did not actually measure the binding constant.

OFENGAND: I wonder if you would like to comment on the mutual exclusivity of the sites of interaction for the EF-Tu molecule and for the synthetase?

CLARK: We usually make our EF-Tu complex for isolation in a mixed charging and EF-Tu complex inculation. Thus, the EF-Tu kicks-off the synthetase very effectively, and this is the way we normally get full conversion to the ternary complcx.

PETERSEN: Based on our protection results, my suggestion would be that the synthetase and EF-Tu cannot bind at the same time. As I showed the synthetase binds especially to the extra loop on the $tRNA_m^{Met}$ and to the 5′ side of the T-stem, as we have found with EF-Tu. The difference is in the anticodon region, where EF-Tu, as far as we can see, does not bind. I can also add for the Met-tRNA synthetase, there exists a tryptic fragment of the native enzyme, which is a monomer of about 64,000 daltons. This fragment also binds to the tRNA, and we plan to do footprinting experiments with this tryptic fragment in collaboration with Dr. Blanquet in Palaiseau, hopefully that can tell if both subunits are involved in the binding to tRNA or if the entire protection can be obtained by the monomer alone.

KLUG: Could I ask Brian Clark and in fact other people who have done similar types of nuclease protection, or footprinting experiments, why they don't use chemical accessibility. When you use nucleases as a probe, you may only get a very general picture, but with chemical accessibility you actually measure a very local effect. The 2 methods are complementary to some extent.

CLARK: We have tried using the Peattie method and we found, as a number of other people, that the interaction of the protein is so fast, the complex is destroyed. We had a programme to do these studies. Going back earlier to the type of experiments done in Cambridge, David Miller tried to do such experiments about 7 years ago and also destroyed the complex right away with a kind of carbodiimide. We thought that the Peattie method, because it just touches things very lightly, would be better, and she has reported results concerning the ribosomal binding sites, but I don't know of anybody who succeeded with the synthetase, and I think the results with the EF-Tu are a mess.

Peptide Chain Termination

C. Thomas Caskey, Wayne C. Forrester & Warren Tate[1]

INTRODUCTION

Peptide-chain termination is directed by one of 3 specific codons (UAA, UAG, or UGA) and results in the release of the completed peptide from ribosomal bound tRNA. In *E. coli* 2 proteins (release factors RF1 and RF2) have different codon specificities while a third factor, which stimulates the event, recognizes GDP and GTP, but not codons. A single larger protein factor (RF) isolated from rabbit reticulocytes, functions with any of the 3 termination codons and recognizes GTP. The ribosomal peptidyltransferase is required for hydrolysis of peptidyl-tRNA at chain termination. Thus, the process of peptide-chain termination differs markedly from chain initiation and elongation since the codon recognition molecules are proteins not tRNA, and a perturbation of the peptidyltransferase leads to hydrolysis of the peptidyl-tRNA rather than the formation of peptide bonds (Caskey & Campbell 1979).

RELEASE FACTORS

The participation of protein factors in peptide-chain termination was initially proposed by Ganoza (1966), and subsequently established by Capecchi (1969). Development of a simple assay *in vitro* for peptide-chain termination led to our identification of two codon-specific protein factors in *E. coli* (1969). Using an initiation complex (f[^{3}]HMet-tRNA·AUG·ribosome) as substrate for the peptide-chain termination event, release of f[^{3}H]methionine (a peptide analogue) was found to require protein factors, termination codons, and, under certain conditions, guanine nucleotides.

Howard Hughes Medical Institute, Baylor College of Medicine, Houston, Texas U.S.A., and the [1]Department of Biochemistry, University of Dunedin, New Zealand

GENE EXPRESSION, Alfred Benzon Symposium 19.
Editors: Brian F. C. Clark & Hans Uffe Petersen, Munksgaard, Copenhagen 1984.

Two codon-specific release factors (RF1 for UAA or UAG and RF2 for UAA or UGA) have been purified to homogeneity from *E. coli.* As the factors are free of oligo or polynucleotides the codon specificity results from protein recognition mechanisms (Scolnick *et al.* 1968). These acidic proteins have been found to have common immune determinants and, therefore, probably some structural homology although they differ significantly in their physical properties (Campbell *et al.* 1979).

Recently, we have obtained recombinant plasmids which carry and express the RF2 gene. These recombinant plasmids were initially identified by means of an RIA identification of RF2 overproduction in certain isolates found in the *E. coli* recombinant Colicin E1 bank prepared by Clarke and Carbon (1976). Each recombinant was grown to stationary phase, incubated 12 h further in the presence of chloramphenicol, washed free of chloramphenicol and proteins

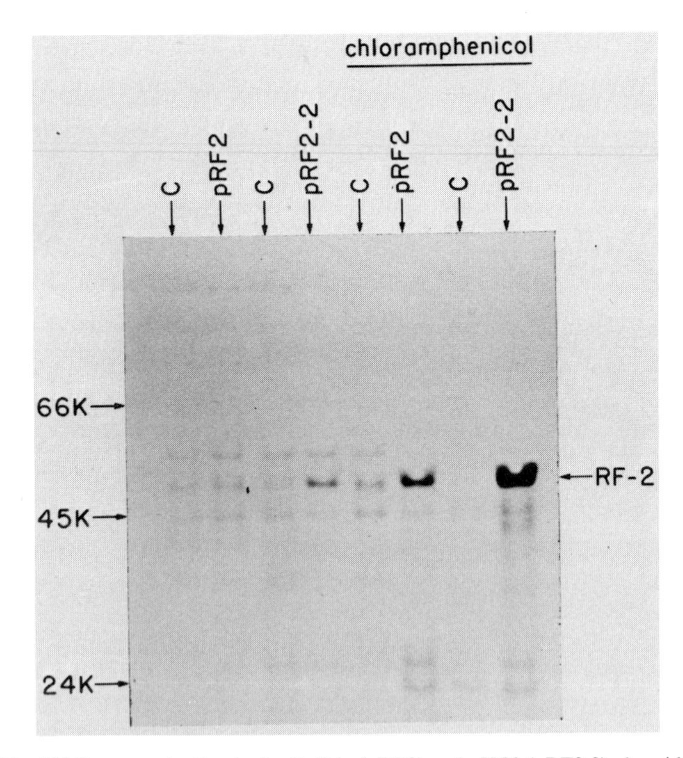

Fig. 1. RIA of RF2 overproduction in the Colicin (pRF2) and pUC9 (pRF2-2) plasmid are shown when radiolabelling was carried out with and without preceeding chloramphenicol plasmid amplification. C corresponds to plasmid containing no recombinant insert.

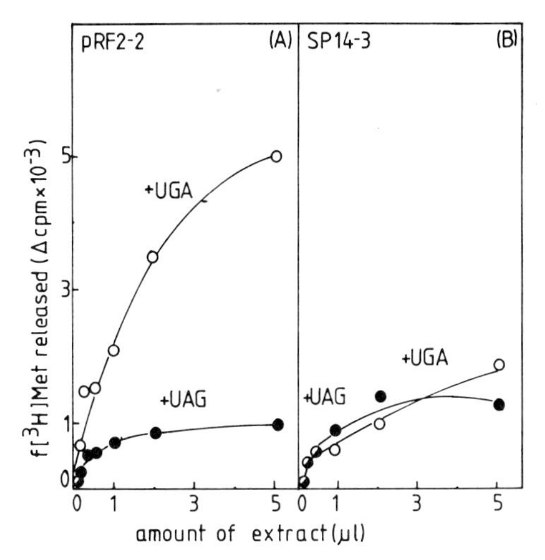

Fig. 2. Release factor activity of extracts prepared from pRF2-2 and control (SP14-3) recombinants. RF1 activity is determined by f[^{3}H]methionine release with UAG while RF2 activity is determined by UGA.

radiolabelled for 20 min in the presence of [^{35}S] methionine. The radiolabelled RF2 is then reacted with anti-RF2, immune complex precipitated by Staph A, radiolabelled product resolved by SDS PAGE, and identified by radioautography. The method is similar to that reported by Neidhardt *et al.* (1980). After study of 750 recombinants, 2 adjacent recombinants were identified which overproduced RF2 as detected by the above RIA. Having screened this number of recombinants, we had reached 0.9 probability for RF2 identification. The effects of chloramphenicol on RF2 overproduction by the ColE1 recombinant, pRF2 is shown in Fig. 1. Also shown in Fig. 1 is the level of RF2 expressed in a plasmid pUC9, originally developed by Vieira and Messing (1982). This latter plasmid is known to achieve a higher copy number under chloramphenicol amplification conditions. No RIA signal was detected with either plasmid when anti-RF1 replaced anti-RF2. The restriction endonuclease maps of pRF2 (Colicin) and pRF2-2 (pUC9) are established. The level of functional RF2 in pRF2-2 and a control plasmid without a recombinant insert is shown in Fig. 2. Since RF1 and RF2 differ in codon specificity, it is possible to measure their relative level in an unfractionated RF extract by codon-specific release of f[^{3}H]methionine from f[^{3}H]met-tRNA · AUG · ribosome complexes. In these and other studies which examine more purified fractions, RF2 activity is

Table I
Soluble Chain Termination Factors

Designation	Molecular Weight	Codon Specificity	GTP Recognition
RF1	47,000[a]	UAA and UAG	None
RF2	48,000[a], 35,000[b]	UAA and UGA	None
RF3	46,000[a]	None	Yes
Rabbit reticulocyte RF	56,500[a], 105,00[b]	UAAA, UAGA and UGAA	Yes

~5-fold elevated by the presence of pRF2-2. Thus, the data indicate that we have succeeded in identifying a plasmid recombinant containing and expressing the RF2 gene. Personal communication with John Gallant, whose paper is published in this volume, reveals he has succeeded in identifying a plasmid expressing RF1. These developments offer new approaches to the study of RF structure, function, and regulation.

During the purification of RF1 and RF2 we discovered a third release factor (RF3), which alone had no release activity in the presence of termination codons, but stimulated the activities of RF1 and RF2 (Milman *et al.* 1969). This protein factor interacts with GTP or GDP and, as will be discussed subsequently, stimulates RF1 and RF2 binding to the release from ribosomes (Caskey & Campbell 1979). The RF3 molecule has not yet been further characterized. Capecchi and Klein found that a purified fraction of EF-Tu possessed this stimulatory activity and, thus, concluded that RF3 and EF-Tu are identical (Capecchi & Klein 1969). In our laboratory, however, preparations with high RF3 activity have been separated from EF-Tu activity (Goldstein & Caskey 1970), and we conclude that RF3 is a specific protein factor involved in peptide-chain termination, not chain elongation.

Eukaryotic release factor was initially isolated from rabbit reticulocytes and, subsequently, from mammalian liver (Goldstein *et al.* 1970), and insect cells (Ilan 1973). The molecular characteristics of rabbit reticulocyte release factor are given in Table 1. The single protein factor will *in vitro* stimulate release of f[^{3}H]methionine from the reticulocyte ribosomal substrate (f[^{3}H]Met-tRNA: tetranucleotides UAAA, UAGA, or UGAA. The requirement for a tetranucleotide rather than trinucleotide is, we believe, a characteristic of the *in vitro* assay since substitution of different bases in the 3'position of the tetranucleotide had no adverse effect on recognition. Thus, it appears the 3' terminal base is not a part of the recognition process but confers an element of stability on the codon recognition complex. The reticulocyte RF interacts with GTP and GDP and promotes an active ribosomal dependent GTPase reaction (Goldstein *et al.* 1970).

TERMINATION CODON RECOGNITION

The recognition of peptide-chain termination codons has been of considerable interest as it involves a deviation from the principle of the recognition of one nucleic acid by another by base pairing. The recognition of peptide-chain termination codons on ribosomes is studied as a partial reaction, utilizing [^{3}H]-labelled terminator codons for the detection of RF-[^{3}H] terminator codon: ribosome complexes, which are retained on millipore filters (Scolnick & Caskey 1969). In this assay, RF1 binds to ribosomes in response to UAA and UAG, while RF2 binds in response to UAA and UGA. Capecchi and Klein (1969) attempted to study this recognition independent of ribosomes by equilibrium dialysis of RF and oligonucleotides containing terminator codons. Although the specificity of codon recognition was not absolute in the absence of ribosomes the studies supported the concept that RF molecules were responsible for codon recognition. More recently the recognition of specific nucleotide sequences of DNA by proteins, restriction endonucleases, has been extensively characterized (Roberts 1980).

The molecular structure of the RF "anticodon" region is still unknown. The specificity of codon recognition has, however, been investigated indirectly using modified trinucleotides (3-Me-UAH, 3-Me-UAA, 5-Me-UAG, 5-Me-UAA, Br-UAG, h-UAH and UAI). These studies revealed a specificity for RF which is remarkably similar to that of the Watson-Crick and wobble base pairing (Smut *et al.* 1970).

Dalgarno and Shine (1973) proposed, on the basis of the sequence of the 3' terminus of the 16S rRNA, that recognition of termination codons may involve these sequences. They found a high frequency of UUAOH terminal sequences as well as a UCA sequence within 8 residues of the terminus. Such sequences could accommodate base pairing with UAA, UAG, and UGA. The importance of the 3' terminus of the 16S rRNA in the recognition of termination codons is clear from our studies employing the specific nuclease cloacin DF13 (Caskey *et al.* 1977). Cleavage with cloacin of 49 nucleotides from the 3' terminus of 16S RNA in ribosomes inactivates the recognition of codons by RF, but not the RF-mediated peptidyl-tRNA hydrolysis. It is unclear if this effect of the nuclease on codon recognition is the consequence of a perturbation of the ribosomal structure. Using another approach we found the recognition of codons on *B. subtilis* ribosomes by *E. coli* RF to be qualitatively unaltered, although the 3' terminus of the 16S rRNA of *B. subtilis* does not contain sequences complementary to termination codons. Thus, the preponderance of data favors the idea that RF1 and RF2 molecules recognize directly termination codons.

A single protein factor in rabbit reticulocyte extracts recognizes all 3 termination codons, a situation significantly different from that in *E. coli* and rabbit reticulocytes are active only with the homologous ribosomes (Konecki *et al.* 1977).

RF SOLUBLE FACTOR AND GTP REQUIREMENTS

In *E. coli* the protein factor RF3 promotes binding to and release from ribosomes of RF1 and RF2 (Goldstein & Caskey 1970). RF3 reduces the Km for trinucleotide codons by 5- to 8-fold during peptide chain termination *in vitro* without affecting the Vmax of peptidyl-tRNA hydrolysis, suggesting that the stimulatory activity is related to ribosomal binding (Milman *et al.* 1969). The RF3 stimulates formation of RF1 or RF2: [^{3}H]UAA: ribosomal intermediates supporting this concept. Furthermore, addition of GDP to such recognition complexes formed in the presence of RF3 leads to their rapid disruption. We have not succeeded in demonstrating a requirement for GTP for the binding of RF1 or RF2 to ribosomes in the presence of RF3, although RF3 is known to have a ribosomal dependent GTPase activity. It is clear that EF-G, another soluble factor that has a ribosomal-dependent GTPase activity, cannot simultaneously occupy the ribosome with RF1 or RF2. These data indicate that both binding and release of RF1 and RF2 are stimulated by RF3, and suggest that GTP and its hydrolysis are involved. However, proof of this proposed cyclic series of events has not been possible *in vitro*.

In the case of peptide-chain termination using reticulocyte ribosomes and RF, the situation is clear. The release factor possesses a ribosomal-dependent GTPase (Beaudet & Caskey 1971) which is markedly stimulated by codons. The binding of RF to ribosomes is stimulated by GTP or the nonhydrolysable analog GDPCP, but not by GDP. As GTP, but not GDPCP, stimulates the catalytic behavior of RF it appears likely that GTP facilitates RF binding to the ribosome whereupon GTP hydrolysis occurs, followed by displacement of the RF and GDP from the ribosome.

PEPTIDYL-tRNA HYDROLYSIS

The role of the peptidyltransferase activity of the 50S ribosomal subunit in peptide-chain termination was initially suggested from the study of antibiotics which inhibited both the peptidyltransferase and the peptidyl-tRNA hydrolysis step of chain termination (Caskey & Beaudet 1971). This partial reaction of peptide chain termination could be studied independent of codons if RF molecules were made to bind to the ribosome by the addition of ethanol (Caskey & Campbell 1979). The antibiotics amicetin, lincocin, chloramphenicol, and sparsomycin inhibit peptidyltransferase and RF-mediated peptidyl-tRNA hydrolysis in parallel. These antibiotics do not inhibit the binding of RF to ribosomes in response to codons, indicating an inhibition of a specific partial reaction of peptide chain termination.

Peptidyltransferase has several catalytic activities which can be demonstrated *in vitro,* including hydrolysis of peptidyl-tRNA (Scolnick *et al.* 1970). It catalyses peptide-bond formation when an amino group is the nucleophilic agent attacking the aa-tRNA ester linkage; this occurs normally during peptide-chain elongation by aa-tRNA. The peptidyltransferase center will also catalyse ester formation when the nucleophilic agent is an alcohol, and hydrolysis of the nascent peptidyl-tRNA when the nucleophilic agent is water. Since RF has no esterase activity and peptidyltransferase is required for peptidyl-tRNA hydrolysis, it appears most likely that RF interacts with the peptidyltransferase to stimulate hydrolysis of the peptidyl-tRNA. It is of interest that 2 antibiotics, lincocin (with *E. coli* ribosomes) and anisomycin (with rabbit reticulocyte ribosomes) have differential effects on the peptidyltransferase activities of peptide bond and ester formation, and peptidyl-tRNA hydrolysis. Each inhibits, on the respective ribosomes, peptide bond and ester formation while stimulating markedly peptidyl-tRNA hydrolysis (Caskey & Beaudet 1971). Such modification of peptidyltransferase specificity might operate in chain termination, with RF, rather than lincocin or anisomycin, being the modifying component. Recently, using lincocin analogs the structural requirements for this effect have been determined (Tate *et al.* 1975).

RIBOSOMAL SITE OF RF BINDING

The ribosomal protein required for peptide-chain termination have been investigated with: (1) antibiotics (2) antibodies directed against specific proteins, (3) ribosomes depleted of proteins, and (4) cross-linkage of RF to ribosomal proteins. The results, collectively, give an indication of the ribosomal domain(s) of contact.

The ribosomal proteins S4 and L4 are involved in peptide-chain termination because the antibiotics streptomycin and erythromycin are inhibitors of peptide-chain termination, and mutations in these proteins lead to resistances to these antibiotics (Caskey & Beaudet 1971). Using antibodies developed to specific ribosomal proteins, the proteins L7/L12, S2 and S3 were shown to be required for RF binding, while L16 and L11 affected peptidyl-tRNA hydrolysis (Tate *et al.* 1975). These proteins have been implicated as part of the peptidyltransferase center and surrounding domain, respectively. Both L16 and L11 are required for chloramphenicol binding (Roth & Nierhaus 1975). Ribosomes depleted of L7/L12 are defective for RF binding (Brot *et al.* 1974), thus confirming the antibody studies. Depleting ribosomes of L11 did not significantly affect the peptidyltransferase activity and had minimal effect on peptide-chain termination (Armstrong & Tate 1978). Direct linkage of RF to ribosomal proteins through di-imido, bifunctional, cross-linking agents has provided direct evidence for close contact (Stöeffler *et al.* 1982). RF was found to crosslink to L2, L7/L12, L11, S17, S21 and S18. Pongs & Rossner (1975), using an affinity probe, which is a derivative of the termination codon UGA, found cross-linkage to S18 and S4 upon codon recognition. It is not surprising that RF makes contact with many of the ribosomal proteins that also interact with EF-Tu (Jose *et al.* 1976). The need for an intact 3' terminus of the 16S rRNA has been previously discussed.

Models of the 30S and 50S ribosomal subunits have been developed by Lake (1976) and Stöeffler *et al.* (1979) based upon immune electron microscopy which estimates shape and position of ribosomal proteins. The 30S ribosomal subunit has an embryo shape with S4, S18, and S21 localized to the small head, the region involved in codon recognition for aa-tRNA and interaction with IF2 and IF3 (Brimacombe *et al.* 1978). The 50S ribosomal subunit is evisaged as a "crowned seat" where L2, L7/12, and L11 are located within the seat. This binding domain for RF is a region of the interface between the two ribosomal subunits (Brimacombe *et al.* 1978).

REGULATION OF PEPTIDE-CHAIN TERMINATION CODONS

Mutations which generate chain-termination codons lead to premature chain termination and release of protein fragments. Such mutations can be phenotypically suppressed in cells that carry suppressor tRNA mutations. The mutant suppressor aa-tRNA can recognize termination codons and translate them. RF and aa-tRNAsu compete for the termination codon. Thus, regulation of mutant gene expression is achieved. Nonsense mutations occur in prokaryotes and eukaryotes, and a variety of suppressor tRNA mutations have been identified in both (Caskey & Campbell 1979).

ACKNOWLEDGEMENTS

This work was supported by a grant from the National Institutes of Health (NIGMS #GM27235), The Robert A. Welch Foundation (grant #Q-533), and the Howard Hughes Medical Institute.

REFERENCES

Armstrong, I. L. & Tate, W. P. (1978) *J. Mol. Biol. 120,* 155.

Beaudet, A. L. & Caskey, C. T. (1971) *Proc. Natl. Acad. Sci. USA 68,* 619.

Brimacombe, R., Stöeffler, G. & Wittman, H. G. (1978) *Annu. Rev. Biochem. 47,* 217.

Brot, N., Tate, W. P., Caskey, C. T. & Weissbach, H. (1974) *Proc. Natl. Acad. Sci. USA 71,* 89.

Campbell, J. M., Reusser, F. & Caskey, C. T. (1979) *Arch. Biochem. Biophys. 90,* 1032.

Capecchi, M. R. & Klein, H. A. (1969) *Cold Spring Harbor Symp. Quant. Biol. 34,* 469.

Caskey, C. T. & Beaudet, A. L. (1971) In: *Molecular Mechanisms of Antibiotic Action on Protein Biosynthesis and Membranes,* p. 326, Elsevier, Amsterdam.

Caskey, C. T., Bosch, L. & Konecki, D. S. (1977) *J. Biol. Chem. 252,* 4435.

Caskey, C. T. & Campbell, J. M. (1979) In: *Nonsense Mutations and tRNA Suppressors,* p. 81, Academic Press, New York.

Clarke, L. & Carbon, J. (1976) *Cell 9,* 91–99.

Dalgarno, L. & Shine, J. (1973) *Nature (London) New Biol. 245,* 261.

Ganoza, M. C. (1966) *Cold Spring Harbor Symp. Quant. Biol. 31,* 273.

Goldstein, J. L., Beaudet, A. L. & Caskey, C. T. (1970) *Proc. Natl. Acad. Sci. USA 67,* 99.

Goldstein, J. L. & Caskey, C. T. (1970) *Proc. Natl. Acad. Sci. USA 67,* 537.

Ilan, J. (1973) *J. Mol. Biol. 77,* 437.

Jose, C. S., Kurland, C. G. & Stöeffler, G. (1976) *FEBS Let. 71,* 133.

Konecki, D., Aune, K. C., Tate, W. & Caskey, C. T. (1977) *J. Biol. Chem. 13,* 4514.

Lake, J. A. (1976) *J. Mol. Biol. 105,* 131.

Milman, G., Goldstein, J., Scolnick, E. & Caskey, T. (1969) *Proc. Natl. Acad. Sci. USA 63,* 183.

Neidhardt, F. C., Reinhard, W., Smith, M. W. & Van Bogelen, R. (1980) *J. Bacteriol. 143,* 535–537.

Pongs, O. & Rossner, E. (1975) *Hoppe-Seyler's Z. Physiol. Chem. 356,* 1297.

Roberts, R. J. (1980) *Nucleic Acids Res. 8,* 63.

Roth, H. E. & Nierhaus, K. H. (1975) *J. Mol. Biol. 94,* 111.
Scolnick, E. M. & Caskey, C. T. (1969) *Proc. Natl. Acad. Sci. USA 64,* 1235.
Scolnick, E., Milman, G., Rosman, M. & Caskey, C. T. (1970) *Nature 225,* 152.
Scolnick, E., Tompkins, R., Caskey, T. & Nirenberg, M. (1968) *Proc. Natl. Acad. Sci. USA, 61,* 768.
Smut, J., Kemper, W., Caskey, T. & Nirenberg, M. (1970) *J. Biol. Chem. 245,* 2753.
Stöeffler, G., Tate, W. P. & Caskey, C. T. (1982) *J. Biol. Chem. 257,* 4203.
Tate, W. P., Caskey, C. T. & Stöeffler, G. (1975) *J. Mol. Biol. 93,* 375.
Tischendorf, G. W., Zeichhardt, H. & Stöeffler, G. (1979) *Mol. Gen. Genet. 134,* 209.
Vieira, J. & Messing, J. (1982) *Gene 19,* 259–268.

DISCUSSION

KURLAND: When you are overproducing the RF2, what happens to suppression frequencies?

CASKEY: Jonathan Gallant will discuss in his lecture the reversal of UAG suppression by RF1 overproduction.

GALLANT: Dr. Caskey has demonstrated that release factors compete with suppressor tRNA. We used this observation to develop a detection method for RF1 overproduction. We used a *lac*Z amber allele in a strain with a weak amber suppressor, which produced baby-blue colonies on X-GAL. My students, Bob Weiss and Jim Murphy, shotgun cloned *Sau*3a fragments into the Bam site in the tet^r gene of PBR322 derivative, selected the amp^r, tet^s transformants, and then screened them for suppression efficiency on X-GAL. Among about 2200 clones, they found 9 white colonies. One of these 9 exhibited reduced suppression of phage amber alleles as well, and was found to overproduce RF1. The RF1 gene insert reduces suppression of the *lac*Z amber by about a factor of 10. We have also transferred Dr. Caskey's RF2 plasmid into a strain carrying a *lac*Z UGA allele and a UGA suppressor, and it reduced suppression efficiency by about a factor of 5.

MILLER: Is there any interaction of the release factor with ribosomes in the absence of the release codon? Are the very high concentrations of RF2 deleterious to the cell?

CASKEY: Dr. Gallant can really comment on that. We have not looked at the recombinant growth rates.

GALLANT: I wanted to ask you about that! All we can say is that the RF1 plasmid is somewhat unstable. We have to keep it under antibiotic selection or it is lost.

CASKEY: We have not attempted to characterize instabilities. Once the cells have been chloramphenicol amplified we made lysates for biochemical studies. I have not tried to grow them on non-selective media following this amplification to determine the RF lethality.

GALLANT: Did you just do the RIA or did you look at all the protein on the gel?

CASKEY: We have looked at the protein. You can detect it. It is below several per cent.

GALLANT: Is there an excess of small proteins?

CASKEY: No. The pattern looks the same with or without the plasmid amplifications.

KURLAND: Kim Boothe, while he was in Uppsala, looked at chloramphenicol amplified cells for other reasons, and found at least one protein, in which errors accumulated with chloramphenicol stoppage. There was some weak indications in Tim's experiments that chloramphenicol was not doing it directly. During chloramphenicol inhibition you change the tRNA pools, because RNA synthesis does not stop as fast as protein synthesis, and as a consequence you upset the tRNA iso-acceptor ratios?

GALLANT: There is a short paper similar to that by Fred Neidhardt's group. After a period of chloramphenicol inhibition, even without a plasmid, the pattern of proteins synthesized is very altered by 2D-gel.

GOLD: What were the other 8 clones that you got?

GALLANT: I don't know.

III. Suppressions of Nonsense Mutations, Fidelity of Protein Biosynthesis

Anticodon Loop Substitution in tRNA

Olke C. Uhlenbeck, Lance Bare & A. Gregory Bruce[1]

INTRODUCTION

An important approach to understanding the function of different regions of a tRNA molecule has been to alter the chemical structure of particular nucleotides and study the function of the modified tRNA using one or more biochemical assays. Numerous experiments with chemically modified tRNAs have clearly demonstrated the usefulness of such an approach. For example, Schulman & Pelka (1977) have used a variety of reagents to modify nucleotides in *E. coli* $tRNA_f^{Met}$ in order to identify the regions of the molecule which interact with methionine tRNA synthetase. However, since many chemical modification procedures employ reagents which react at different positions on the RNA molecule, it is often not possible to control the reaction sufficiently, to modify a single nucleotide. It is even more difficult to modify an RNA molecule at a nucleotide of choice. The success of experiments using reactions which quantitatively alter a certain modified nucleotide at a single location in tRNA emphasize the importance of having unique modifications for structure-function studies (Kuechler & Ofengand 1980). Such site-specific reactions are particularly useful if a number of different chemical modifications can be made at the same location so that the nature of the perturbation can be better understood.

We have developed "recombinant RNA" procedures that permit the preparation of variant tRNA molecules with a variety of modifications at predetermined positions. As the single-stranded structure of the tRNA anticodon loop makes it available to reaction with enzymes, our experiments have focussed on changing nucleotides in this region of the molecule. A tRNA with a gap in the

Department of Biochemistry, University of Illinois, Urbana, and [1]Howard Hughes Medical Institute, Department of Biology, University of Utah, Salt Lake City, Utah 84112, U.S.A.

GENE EXPRESSION, Alfred Benzon Symposium 19.
Editors: Brian F. C. Clark & Hans Uffe Petersen, Munksgaard, Copenhagen 1984.

anticodon loop is prepared by enzymatic or chemical cleavage of the polynucleotide chain. T4 polynucleotide kinase and RNA ligase are then used to insert an oligonucleotide into the gap. Despite the complexity of the protocols (some require 7 or 8 steps), and the relatively low yield of products, the synthetic flexibility of this approach makes it an important addition to the available methods to modify tRNA. The broad substrate specificity of T4 RNA ligase permits the insertion of virtually any base modification and several different sugar modifications into a polynucleotide chain.

In this communication, we will briefly describe the anticodon loop substitution procedures used for yeast $tRNA^{Phe}$ and $tRNA^{Tyr}$, discuss the available data on the role of the anticodon on the aminoacylation reaction, and summarize the data demonstrating the activity of these tRNAs in prokaryotic and eukaryotic translation systems.

METHODS OF ANTICODON LOOP SUBSTITUTION

The procedure used to insert an amber anticodon into positions 34–37 of yeast $tRNA^{Phe}$ is outlined in Fig. 1 and described in detail in Bruce & Uhlenbeck (1982a). First, the hypermodifed Y nucleotide at position 37 is selectively depurinated with acid and the chain cleaved with aniline (Thiebe & Zachau

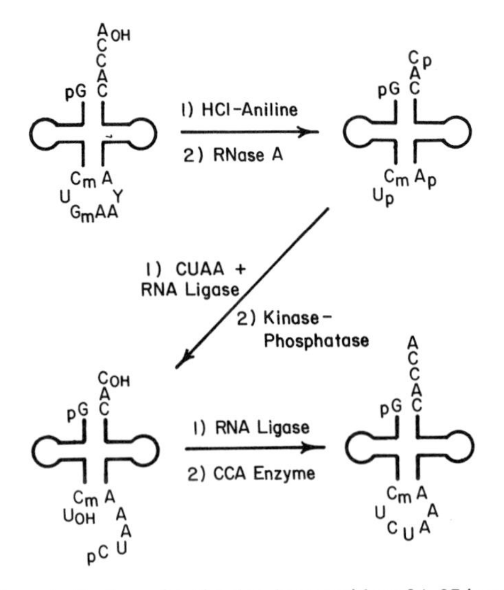

Fig. 1. Procedure for the substitution of anticodon-loop residues 34–37 in yeast $tRNA^{Phe}$ (Bruce & Uhlenbeck 1982a).

1971). The joined half molecules are then reacted with pancreatic ribonuclease at high salt and magnesium concentrations and low temperature. Under these conditions, the chain is only cleaved twice at low nuclease concentrations, resulting in half molecules with terminal phosphates that are missing the anticodon residues 34–37, and the two 3′ terminal residues 75 and 76. The combined half molecules are then reacted with the tetranucleotide CpUpApA and T4 RNA ligase, resulting in the preferential addition of the tetramer to the 5′ phosphate of the 3′ half molecule. The 5′ phosphate of the 5′ half molecule is not reactive due to the secondary structure of the acceptor stem. Next, the 3′ phosphates are removed, and a new 5′ phosphate is simultaneously introduced by the kinase and 3′phosphatase activities of T4 polynucleotide kinase. Finally, the anticodon loop is then resealed with RNA ligase and the 3′ terminal CpA is re-introduced with tRNA nucleotidyl transferase.

The 6 steps in the $tRNA^{Phe}$ anticodon-loop procedure are carried out sequentially without purification of intermediates. After the final step, the anticodon substituted $tRNA^{Phe}$ is purified on a polyacrylamide gel. The procedure gives very high yields (20 to 40%), and is quite general for the sequence inserted. More than 30 different anticodon substituted $tRNA^{Phe}$ have been prepared with different sequences and lengths varying from 3 to 15 nucleotides. In several cases, the identity of the product was confirmed by enzymatic RNA sequence determination. We will use the nomenclature $tRNA^{Phe}_{CUAA}$ for the $tRNA^{Phe}$ with CpUpApA inserted into positions 34–37.

The steps used for substitution of positions 33–35 of yeast $tRNA^{Tyr}$ to form an amber suppressing tRNA (Fig. 2) resembles the $tRNA^{Phe}$ substitution procedure in many respects. In this case, partial digestion with ribonuclease A results in cleavages on the 3′ side of cytidine-32, pseudouridine 34 and cytidine 74 to give the joined half molecules shown as intermediate *2*. This intermediate is treated with T4 *Pset* 1 polynucleotide kinase to introduce a 5′ phosphate on the 3′ half molecule. As the mutant polynucleotide kinase is missing the 3′ phosphatase activity (Soltis & Uhlenbeck 1982), the 3′ phosphates are undisturbed in this case. The oligonucleotide is then added to the 3′ half molecule as before to give intermediate *4*. The 3′ phosphates are removed and a 5′ phosphate is introduced using both activities of wild type T4 polynucleotide kinase. Finally, the half molecules are sealed and 3′ the terminus repaired as before.

Although a substantial number of different sequences have been inserted into $tRNA^{Tyr}$, it is clear that this procedure is less efficient than the $tRNA^{Phe}$ procedure. Overall yields range from 3 to 8%. The major difficulty is the

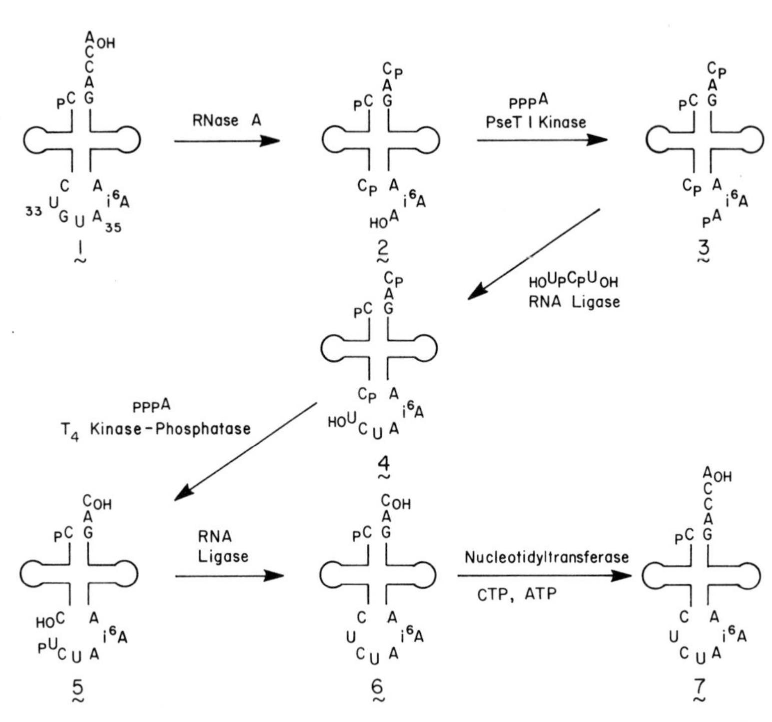

Fig. 2. Procedure for the substitution of anticodon-loop residues 33–35 in yeast $tRNA^{Tyr}$.

addition of the trinucleotide to the 3′ half molecule. Despite intensive efforts to increase the efficiency of the reaction, the yield on this step is never better than 50%. This results in production of a substantial amount of $tRNA^{Tyr}$ with an anticodon loop of 4 residues that must be separated from the correct product at the end of the synthesis. We will use the nomenclature $tRNA^{Tyr}_{UCU}$ for $tRNA^{Tyr}$ with UpCpU inserted in positions 33–35.

Enzymatic substitution of nucleotides into the anticodon loop of tRNA appears to be possible for any tRNA, provided that specific cleavages can be made. Other workers have developed similar protocols for a number of different yeast and *E. coli* tRNAs in order to study aminoacylation and tRNA modification (Carbon *et al.* 1982, Schulman *et al.* 1983). The success of the procedure primarily depends upon the availability of the anticodon loop of the two reannealed half molecules to react with enzymes. Similar strategies for substitition at other positions in the tRNA may require separation of the tRNA molecule into fragments and alteration of the sequence of the fragment before re-annealing and ligation. The secondary and tertiary structure of the tRNA may reduce the efficiency of the religation reaction.

ROLE OF THE ANTICODON IN AMINOACYLATION

Substitution of nucleotides into the anticodon of $tRNA^{Phe}$ generally results in a decrease in the rate of aminoacylation with yeast phenylalanine synthetase (Bruce & Uhlenbeck 1982b). Under reaction conditions where $tRNA^{Phe}$ saturates to more than 90% aminoacyl tRNA in a few minutes, substituted tRNAs aminoacylate more slowly and saturate at much lower levels. The lower levels of aminoacylation reflects an equilibrium between the slow forward rate and the spontaneous non-enzymatic deacylation. If more enzyme is included in the reaction, the substituted tRNAs can be fully acylated. As shown in Table I, the K_m of the aminoacylation reaction is affected much more than the V_{max}, and the change is greatest in position 34 and least in position 37. These data clearly suggest that yeast phenylalanine synthetase contacts $tRNA^{Phe}$ in the anticodon.

The above conclusion also appears to be the case for several other tRNAs. Schulman *et al.* (1983) have substituted each position in the anticodon of *E. coli* $tRNA_f^{Met}$, and found a similar decrease in the aminoacylation rate, with substitutions in positions 34 and 35 much more detrimental than position 36. As the effect on the catalytic rate was spectacularly high (about 10^6 fold), it was not possible to determine which kinetic parameters were altered. Reports of other amino-acyl tRNA synthetases which are sensitive to nucleotide changes in the anticodon at the cognate tRNA include the *E. coli* glycine, glutamine and tryptophan enzymes (Carbon & Squires 1971, Yarus *et al.* 1977).

Preliminary aminoacylation data for several anticodon substituted $tRNA^{Tyr}$

Table I
Effect of Anticodon Substitution on Aminoacylation of $tRNA^{Phe}$

Oligomer Inserted	K_m	V_{max}
G_mAAY ($tRNA^{Phe}$)	30	(1.0)
GAAG	43	1.5
GAAU	79	1.7
GAUG	71	0.98
GUAG	118	0.42
UAAG	217	1.0

(Data from Bruce & Uhlenbeck 1982b)

are shown in Fig. 3. As higher enzyme concentrations give greater levels of aminoacylation, the low levels of amino-acylation are again indicative of a slower forward rate in equilibrium with spontaneous deacylation. It is clear that the identity of nucleotides at both positions 34 and 35 are important for amino-acylation. It is interesting to note that substitution of ψ to U at position 35 reduces the aminoacylation rate. Since the ψ modification appears to be essential for active suppression (Johnson & Abelson 1983), this suggests that the poor suppression observed for the U-35 tRNA may simply be a consequence of reduced aminoacylation. In the same vein, it is interesting that the G to C change at position 34 reduces the aminoacylation rate. Although the $tRNA^{Tyr}_{UC\psi}$ has not yet been made, this result suggests that some suppressor tRNAs may amino-acylate less efficiently as a result of the nucleotide change in the anti-codon.

The available data, therefore, indicate that the correct anticodon sequence is required for a normal rate of aminoacylation. A suggestion that the anticodon sequence may play an essential role in the discrimination of the cognate tRNA is shown in Fig. 4. $tRNA^{Tyr}_{UGA}$, which has an anticodon similar to $tRNA^{Phe}$, can be aminoacylated by high concentrations of phenylalanine synthetase. As $tRNA^{Tyr}$ does not aminoacylate under these conditions, the misacylation is entirely a consequence of the ψ to A substitution at position 35. A single nucleotide change

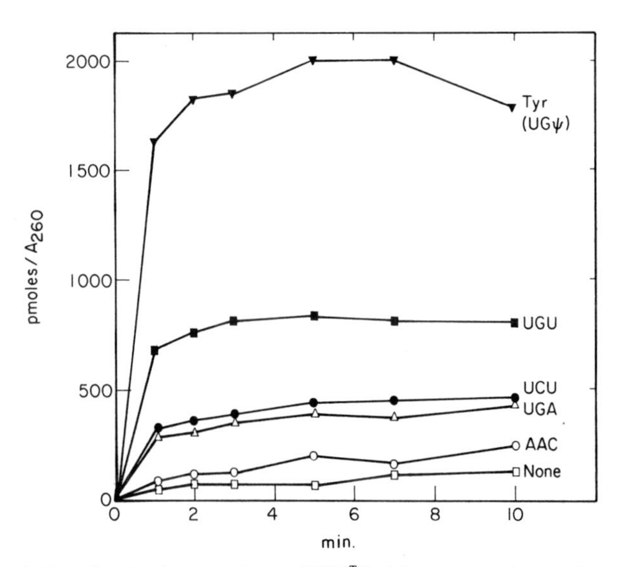

Fig. 3. Aminoacylation of anticodon substituted $tRNA^{Tyr}$ with yeast tyrosine synthetase. Sequences inserted for positions 33–35 are indicated.

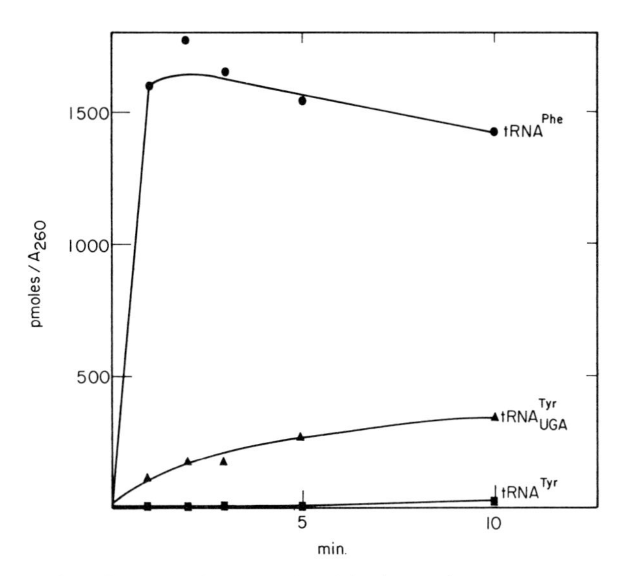

Fig. 4. Aminoacylation of tRNAs with yeast phenylalanine synthetase.

at position 35 has also been shown to alter the aminoacylation of Su_7^+ tRNA from tryptophan to glutamine (Yarus *et al.* 1977).

The importance of the anticodon in the aminoacylation reaction has been recognized for many years (Kisselev & Frolova 1964). Presumably, functional groups of the anticodon nucleotides make specific contacts with aminoacid side chains on the synthetase. A large number of experiments clearly show that these cannot be the only specific contacts. Thus, the surfaces of the 2 macromolecules must interact at a variety of locations to give the specific interaction.

TRANSLATION WITH *E. COLI* RIBOSOMES

In collaboration with T. Yamane, we have assayed the anticodon substituted tRNA^{Phe} with poly U programmed *E. coli* ribosomes in 2 different ways. In the first, the dipeptide synthesis assay of Erbe & Leder (1968) was used to measure enzymatic tRNA binding and peptidyl transfer. The reaction mixture contained poly U programmed *E. coli* ribosomes, initiation factors, unlabelled yeast N-acetyl phe-tRNA^{Phe}, and elongation factor T_u-GTP bound to the anticodon substituted [^{3}H] phe-tRNA^{Phe}. By adding the previously charged tRNA as a ternary complex, deacylation was minimized. The kinetics of formation of N-

Table II
Relative Activities of Anticodon Substituted tRNA^{Phe} on E. coli *Ribosomes*

Controls	acphe-phe	poly phe
GmAAY	(100)	(100)
GAAG	73	56
UUUG	3	0
tRNA$_2^{Leu}$ (GAG)	–	0
Size of Anticodon		
GAG	0	0
GAAG	73	56
GAAAG	5	5
Position 37		
GAAG	73	56
GAAA	58	28
GAAC	21	18
GAAU	30	16
Position 35+36		
GAAG	73	56
GAUG	12	0
GACG	12	0
GUAG	3	1
GCAG	12	2
Position 34		
GAAG	73	56
AAAG	33	9
UAAG	45	10
CAAG	30	5

(Yamane, Bruce & Uhlenbeck, manuscript in preparation)

acetyl-phe(^{3}H) phe was determined at 0°C in order to obtain linear initial rates using only a few percent of the available substrate. The rates of dipeptide formation for each tRNA$^{Phe}_{XXXX}$ relative to unmodified tRNA^{Phe} are shown in Table II.

The activity of the different anticodon substituted tRNA^{Phe} in a poly-phenylalanine synthesis reaction was determined using the method of Jelenc & Kurland (1979), with 2 modifications. First, the reaction was started with N-acetyl phe-tRNA^{Phe} and, second, the anticodon substituted tRNAs were added as a ternary complex to avoid deacylation. Incubation was at 30°C and the linear initial rates of synthesis were compared with yeast $tRNA^{Phe}$ after the minus poly U blank was subtracted (Table II).

The results of the 2 different assays are similar although the poly U system is more accurate. The $tRNA^{Phe}_{GAAG}$ is quite active on *E. coli* ribosomes, however, when a non-complementary anticodon is inserted, no activity is observed. The $tRNA^{Phe}_{GAG}$, with a 6 nucleotide anticodon loop, is totally inactive while the $tRNA^{Phe}_{GAAAG}$, shows a small amount of activity. This is consistent with the observation that +1 frameshift suppressor tRNAs with an 8- membered anticodon loop have been isolated, but corresponding −1 frameshift suppressors have not.

In agreement with the suppression data in the next section, the substitution of nucleotides at position 37 has only a modest effect on the rate of translation. Pyrimidines are less effective than purines in this position. Substitution at positions 35 and 36 give tRNAs which have greatly reduced activity in the dipeptide assay and are totally inactive in the more accurate polyphenylalanine assay. These results are consistant with many experiments which show that accurate base pairs are essential in the first 2 positions of the codon-anticodon interaction. Substitutions in the wobble nucleotide show some activity in both assays. This position in the codon-anticodon interaction is most susceptible to misreading. It is especially interesting that $tRNA^{Phe}_{AAAG}$, which can presumably form a correct AU pair, is no more active than $tRNA^{Phe}_{CAAG}$, or $tRNA^{Phe}_{UAAG}$ which cannot form the base pair. These data emphasize the relatively low importance of the third position on the codon-anticodon interaction (Mitra *et al.* 1977).

IN VITRO SUPPRESSION

In collaboration with J. Atkins, N. Wills & R. Gesteland, we have assayed $tRNA^{Phe}_{CUAA}$ and $tRNA^{Tyr}_{UCU}$ for activity in an *in vitro* suppression assay (Bruce *et al.* 1982, Bare *et al.* 1983). The mammalian translation system, described in detail in Atkins *et al.* (1975), uses ribosomes and soluble factors from Krebs II ascites cells, and initiation factors partially purified from rabbit reticulocytes. Among the mRNAs tested were BMV4 RNA and Qβ GB11 RNA, the latter having an amber mutation in the Qβ replicase gene. The ^{14}C-labelled products of the translation reactions were separated on an SDS gel and located by autoradiography. When BMV is used as a mRNA, the primary product is BMV coat protein. Including an amber suppressor tRNA in the reaction mixture results in the synthesis of a read-through protein 8 amino acids longer. Similarly, when Qβ

GB11 RNA is used, an amber peptide of the replicase is made and an amber suppressor tRNA results in the synthesis of a full-sized replicase. The relative amounts of terminated and read-through proteins could be obtained from the autoradiogram by densitometry. It was found that both $tRNA^{Phe}_{CUAA}$ and $tRNA^{Tyr}_{UCU}$ were active suppressor tRNAs, giving amounts of suppression approximately equivalent to a yeast $tRNA^{Ser}$ amber suppressor tRNA. These results clearly establish that the anticodon-loop substitution procedures can lead to functioning tRNAs.

It was somewhat unexpected that $tRNA^{Phe}_{CUAA}$ and $tRNA^{Tyr}_{UCU}$ were efficient suppressor tRNAs, as they did not contain the modified nucleotides expected to be necessary for efficient function. Thus, $tRNA^{Phe}_{CUAA}$ has an unmodified A at position 37, and modifications at that position are known to effect the codon-anticodon interaction. Mutants of yeast which do not suppress nonsense mutations are deficient in isopentenyl A at this position (Laten *et al.* 1980, Janner *et al.* 1980). The *in vitro* system was not able to insert a modification into $tRNA^{Phe}_{CUAA}$, so it is clear that the unmodified tRNA functioned as a suppressor. Similarly, the $tRNA^{Tyr}_{UCU}$ is missing the ψ at position 35 known to be important in yeast suppression (Johnson & Abelson 1983). It is possible that the disparity between the *in vivo* and *in vitro* results is simply a matter of the relative concentration of suppressor tRNA. It is clear that the yeast mutants missing ipA can grow normally, so the modification is not essential. When the yeast suppressor tRNA without the ψ is present in higher concentration in the cell, it appears to be functional. As the amount of suppression is a consequence of competition between the tRNA and release factors, an excess of tRNA or inactive release factors could account for the unexpectedly high activity of the unmodified tRNAs in the *in vitro* system.

In order to examine the relative importance of nucleotides on either side of the amber anticodon, we prepared $tRNA^{Phe}_{CUAN}$ with the 4 common nucleotides in position 37, and $tRNA^{Tyr}_{NCU}$ with the 4 common nucleotides in position 33. All 8 of these tRNAs were tested with BMV RNA for their effectiveness as amber suppressors. As summarized in Table III, nucleotide changes in both positions alter the suppression efficiency. These differences do not reflect differences in the ability of these tRNAs to aminoacylate, as the addition of yeast tRNA synthetases to the extracts does not alter the amount of suppression. If a purine is at position 37, the tRNA is a much better suppressor than if a pyrimidine is at that position. This is consistent with the fact that a modified purine is always found at position 37. This result confirms the view that the nucleotide at position

Table III
Efficiency of suppression of the BMV termination codon

Suppressor tRNA	Percent Read-Through Protein
$tRNA^{Phe}_{CUAN}$	
N=U	2
N=C	2
N=A	27
N=G	19
$tRNA^{Tyr}_{NCU}$	
N=U	61
N=C	52
N=A	38
N=G	20

0.25 μg of tRNA was used in a standard translation reaction (Bruce *et al.* 1982).

37 stabilizes the codon-anticodon interaction by increasing the stacking. The relatively small effect for substitutions at position 33 is somewhat surprising considering that the uridine at position 33 is one of the most conserved nucleotides in tRNA. Cytidine can effectively substitute for uridine and even tRNAs with a purine at position 33 are not more than three-fold less effective as suppressors. These data clearly show that the constant uridine at this position does not play an obligatory role in the protein synthesis mechanism. However, the data does not necessarily conflict with the high degree of conservation at this position. A small increase in translational efficiency due to a U at position 33 would not be detected by our assay, but would be sufficient to maintain a strong selection.

Our enzymatic substitution methods in the anticodon of tRNA nicely complement the recent elegant work of Yarus and co-workers on the modification of the anticodon loop region of the *E. coli* $tRNA^{Trp}$ Su^{+}_{7} gene (Thompson *et al.* 1982). Substitution of nucleotides in various positions in the anticodon loop result in relatively small, but clear changes in the efficiency of suppression *in vivo*. The data, therefore, suggest that although the positions flanking the anticodon are not essential for suppression, the anticodon loop is finely tuned to optimize its effectiveness in the translation process (Yarus 1982).

REFERENCES

Atkins, J. F., Lewis, J. B., Anderson, C. W. & Gesteland, R. F. (1975) *J. Biol. Chem. 250,* 5688–5695.

Bare, L., Bruce, A. G., Gesteland, R. F. & Uhlenbeck, O. C. (1983) *Nature* (in press).

Bruce, A. G., Atkins, J. F., Wills, N., Uhlenbeck, O. C. & Gesteland, R. F. (1982) *Proc. Natl. Acad. Sci. USA 79,* 7127–7131.

Bruce, A. G. & Uhlenbeck, O. C. (1982a) *Biochemistry 21,* 855–861.

Bruce, A. G. & Uhlenbeck, O. C. (1982b) *Biochemistry 21,* 3921–3926.

Carbon, J. & Squires, C. (1971) *Cancer Res. 31*, 663–666.

Carbon, P., Haumont, E., De Henau, S., Keith, G. & Grosjean, H. (1982) *Nuc. Acids. Res. 10,* 3715–3732.

Erbe, R. W. & Leder, P. (1968) *Biochem. Biophys. Res. Comm. 31,* 798–803.

Janner, F., Vogeli, G. & Fluri, R. (1980) *J. Mol. Biol. 139*, 207–219.

Jelenc, P. C. & Kurland, C. G. (1979) *Proc. Natl. Acad. Sci. USA 76,* 3174–3178.

Johnson, P. F. & Abelson, J. (1983) *Nature 302,* 681–687.

Kisselev, L. L. & Frolova, L. Y. (1964) *Biokhimiya 29,* 1177.

Kuechler, E. & Ofengand, J. (1980) In: *Transfer RNA: Structure, Properties, and Recognition,* pp. 413–444. Cold Spring Harbor Laboratories, Cold Spring Harbor, New York.

Laten, H., Gorman, J. & Bock, R. M. (1980) In: *Transfer RNA: Biological Aspects,* pp. 395–406. Cold Spring Harbor Laboratories, Cold Spring Harbor, New York.

Mitra, S. K., Lustig, F., Akesson, B., Lagerkvist, U. & Strid, L. (1977) *J. Biol. Chem. 252,* 471–478.

Schulman, L. H. & Pelka, H. (1977) *Biochemistry 16,* 4256–4265.

Schulman, L. H., Pelka, H. & Susani, M. (1983) *Nuc. Acid. Res. 11,* 1439–1446.

Soltis, D. A. & Uhlenbeck, O. C. (1982) *J. Biol. Chem. 257,* 11332–11339.

Thiebe, R. & Zachau, H. G. (1971) *Methods Enzymology 20,* 179–182.

Thompson, R. C., Cline, S. W. & Yarus, M. (1982) In: *Interaction of Translational and Transcriptional Controls in the Regulation of Gene Expression,* pp. 189–202. Elsevier Science, New York.

Yarus, M. (1982) *Science 218,* 646–652.

Yarus, M., Knowlton, R. & Söll, L. (1977) In: *Nucleic Acid – Protein Recognition,* pp. 391–408. Academic Press, New York.

DISCUSSION

GALLANT: I have a crazy question for you: the procedure means that you can make dummy tRNAs with no anticodon at all. Does the material inhibit protein synthesis *in vitro* or ternary complex formation?

UHLENBECK: I have not done that experiment. We can say that the tRNA-Tyr missing the anticodon aminoacylates very poorly.

KURLAND: A factor of 3 effect can be the whole world for a bacterium.

UHLENBECK: Yes, and very hard for a biochemist to measure.

KURLAND: Well, that's our problem. We are seeing all the time small effects *in vitro* that might make all the difference in the world *in vivo*. I agree with you that it really is a surprising result that the tRNA with an A substituted for the G at position 34 is so inactive. Could that be due to the fact that you are using a yeast tRNA which is missing a Y base in an *E. coli* system, and the tuning is not quite right?

UHLENBECK: Yes, you may be right. Grey Bruce has recently succeeded in doing anticodon loop substitution in *E. coli* $tRNA^{Phe}$, so I am sure that the homologous experiment is going to be done.

SPRINZL: Your comment on the aminoacylation was that the replacement of the anticodon shows that there is a contact with the synthetase at this position. Can you really say it so explicitly? Does it really mean that it is a contact, or could it mean that there is a fine tuning of the whole tRNA structure by the anticodon loop? A second question is, would you get complete aminoacylation, if you would add EF-Tu and avoid the hydrolysis?

UHLENBECK: In response to your first question, although the effects we measured with Phe and Tyr are rather small, they were at very low ionic strength. At higher ionic strengths the effect will be larger. La Donne Schulman observes a 5 order of magnitude effect on substitution in the same position in *E. coli* $tRNA_f^{Met}$. I generally favour the simpler idea of a synthetase contact in the anticodon. As to the second question, the levelling off is indeed due to hydrolysis. If you include Tu or add more enzyme, aminocylation goes to higher levels. The low levels are due to an equilibrium between the spontaneous hydrolysis in aminoacylation buffer and the forward aminoacylation rate.

OFENGAND: Another question is, if you used organic solvents in the aminoacylation reaction would you also get an increase in the overall level of some of thesc modified tRNAs?

UHLENBECK: We probably would, but we have not done it.

OFENGAND: In the protein synthesis experiments I am not clear on what happens with those modified tRNA species that did not charge.

UHLENBECK: It is not that they don't charge, they just charge less well. So, if you put a large excess of synthetase in, you can aminoacylate these just as well as a regular tRNA. The charged tRNA was added as a complex with an excess of Tu. As there is no yeast synthetase in the protein synthesis system, each tRNA was only used once.

EBEL: Does your mischarging experiment, when you change the anticodon, suggest that the anticodon is part of the recognition site of the synthetase? Experiments by Zachau's group showed that when the anticodon is cut out there is no great effect on aminoacylation, except on the V_{max}.

UHLENBECK: Yes, I'm aware of those experiments. Actually, the reason why we picked yeast tRNA-Phe to do anticodon loop substitution was because of the Zachau experiments. We did not want aminoacylation to be affected, as we wanted to study protein synthesis. It turns out that Zachau used a large excess. We studied the aminoacylation kinetics on the same molecules that they studied, we found about a 10-fold effect on the K_m for the molecule without the anticodon.

EBEL: Actually, many protection and crosslinking experiments showed that interaction takes place in the anticodon region, but I did not expect that substitution of the anticodon would affect the specificity of aminoacylation.

ERDMANN: I was wondering, did you use electrophoresis or HPLC for isolation of the tRNA lacking the 3 nucleotides?

UHLENBECK: All the data I presented are for tRNAs purified on sequencing gels like the one I showed. You are absolutely correct, we are beginning to use HPLC to purifiy tRNA, because of the obvious advantage of being able to put much more on columns.

GALLANT: Given the considerable residual activity you get in the di-Phe assay with mis-matches in the first position, do you get any residual activity with double mis-matches, first and second positions?

UHLENBECK: We have not tried it.

ERDMANN: Have you looked at magic spot synthesis? It would be extremely interesting to check your tRNAs in such an assay, because we have done some experiments which show that different tRNAs may exhibit different dependences on mRNA to trigger magic spot synthesis.

UHLENBECK: We have not done it, but it sounds interesting.

Evolutionary Aspects of the Accuracy of Phenylalanyl-tRNA Synthetases

Friedrich Cramer & Hans-Joachim Gabius

The phenylalanyl-tRNA synthetases from *Escherichia coli,* yeast, *Neurospora crassa,* and turkey liver, activate a number of phenylalanine analogues. Upon complexation with $tRNA^{Phe}$ the enzyme · $tRNA^{Phe}$ complexes show an increased initial discrimination of these analogues. The overall accuracy is further enhanced by proof-reading.

The strategies employed by the enzymes with respect to accuracy differ. Better initial discrimination in the aminoacylation and less elaborated proof-reading for the *E. coli* enzyme can be compared to a more efficient proof-reading by other synthetases. In this way, the poor initial amino acid recognition in the case of the yeast and *N. crassa* enzymes is balanced. A striking difference can be noted for the proof-reading mechanisms. Whereas, the enzymes from *E. coli,* yeast and *N. crassa* follow the pathway of post-transfer proof-reading, the turkey liver enzyme uses tRNA-dependent pre-transfer proof-reading. It seems that proof-reading is a late invention of evolution which becomes more refined in the higher branches of the evolutionary tree.

Phenylalanyl-tRNA synthetases from mitochondria of yeast and hen liver resemble their corresponding cytoplasmic counterparts. Whereas slight intra-species differences at the amino acid binding site are exploitable by phenyl-alanine analogues, no intraspecies difference can be noted for the strategies to achieve the high fidelity of protein synthesis. While the yeast mitochondrial enzyme follows the pathway of post-transfer proof-reading, the hen liver mitochondrial enzyme uses a tRNA-dependent pre-transfer proof-reading in the case of the natural amino acids. The accuracy of mitochondrial phenylalanyl-

---

Max-Planck-Institut für experimentelle Medizin, Abteilung Chemie, Hermann-Rein-Str. 3, D-3400 Göttingen, West Germany.

GENE EXPRESSION, Alfred Benzon Symposium 19.
Editors: Brian F. C. Clark & Hans Uffe Petersen, Munksgaard, Copenhagen 1984.

tRNA synthetases appears to be even better than the accuracy of the corresponding cytoplasmic enzymes. A similarity of mitochondrial enzymes to the phenylalanyl-tRNA synthetase from *E. coli* is not observed.

INTRODUCTION

Aminoacyl-tRNA synthetases esterify an amino acid with its cognate tRNA with an error rate of less than 10^{-4} (Loftfield & Vanderjagt 1972). In fact, a specificity of smaller than 10^{-5} is observed as a result of a proof-reading step subsequent to initial binding and activation of the amino acid (Cramer *et al.* 1979, von der Haar *et al.* 1981). For the proof-reading capacity of the aminoacyl-tRNA synthetases, different mechanistic interpretations are given in the literature. It was suggested by Hopfield that specificity is enhanced by kinetic proof-reading via the preferential dissociation of the wrong aminoacyl-adenylate (Hopfield 1974, Hopfield *et al.* 1976). This pathway has been questioned by von der Haar (1977), Fersht (1977b) and Igloi *et al.* (1978). Von der Haar and Cramer have suggested chemical proof-reading, as a corrective step after transfer of the amino acid to the tRNA which leads to enzymatic hydrolysis of the ester linkage between the tRNA and the wrong amino acid (von der Haar & Cramer 1975, 1976). A similar scheme for the discrimination between cognate and non-cognate substrates depending on the relative rates of synthesis and hydrolysis was developed by Fersht, who introduced the double-sieve model (Fersht 1977a, Tsui & Fersht 1981). Generalization of such mechanistic descriptions must be done with caution, since synthetases from specific organisms are necessarily used for these investigations. Discrepancies in the mechanistic interpretations are possible because it is not sufficiently well-established that the proof-reading mechanisms of specific synthetases from different organisms are identical. An indication of different behaviour is also provided by the example of different extent of activation of some non-protein amino acids that are produced only in certain plants (e.g. Lea & Norris 1977).

We have studied the interaction of 11 phenylalanine analogues with the phenylalanyl-tRNA synthetase from *E. coli, Saccharomyces cerevisiae* (yeast), *N. crassa* and turkey liver by analysis of the ATP/PP_i pyrophosphate exchange, amino-acylation of yeast $tRNA^{Phe}$-C-C-A and $tRNA^{Phe}$-C-C-A(3'NH_2), and AMP production during aminoacylation of $tRNA^{Phe}$-C-C-N. From the studies, a comprehensive picture emerges of how the phenylalanyl-tRNA synthetases from different organisms achieve the required fidelity. It seems that proof-reading is a late invention of evolution which becomes more refined in the higher branches of the evolutionary tree (for experimental details see H. J. Gabius *et al.* 1983a). In this connection it is of interest to extend these studies to mitochondrial phenylalanyl-tRNA synthetases. As fungal and animal mitochondria exhibit strikingly different patterns of gene organization and transcription (Gray & Doolittle 1982), both pairs of intracellular heterotopic isoenzymes, from yeast and hen liver, are, therefore, examined. Besides the description of the interaction of phenylalanine analogues with the mitochondrial enzymes and the mechanistic implications, the role of certain amino acid residues in the catalytic activity of the phenylalanyl-tRNA synthetases from *E. coli* and the 2 different intracellular compartments of yeast and hen liver is comparatively studied by chemical modification using different types of reagents (H. J. Gabius *et al.* 1983b).

SUBSTRATE DISCRIMINATION IN PHENYLALANYL-TRNA SYNTHETASES FROM *E. COLI,* YEAST AND TURKEY LIVER

Phenylalanyl-tRNA synthetases from different sources will promote ATP/PP_i pyrophosphate exchange with a wide variety of structurally dissimilar amino acids (Gabius & Cramer 1982). Neither a closed phenyl ring, nor an unsubstituted α-NH_2 group is essential. Additionally, a larger ring system can be tolerated. The lack of specificity for phenylalanyl-tRNA synthetases was noted earlier in the case of the yeast enzyme (Igloi *et al.* 1978, Igloi & Cramer 1978).

The difference of the kinetic parameters upon binding of $tRNA^{Phe}$ compared to activation, as measured by ATP/PP_i pyrophosphate exchange in the absence of tRNA, indicates an improvement in amino acid discrimination, and a probable narrowing of the binding pocket for the amino acid. This may be mediated by the enzyme via a conformational change triggered by the 3' adenosine of $tRNA^{Phe}$ (von der Haar & Gaertner 1975, Fasiolo *et al.* 1981). A similar change for the kinetic parameters was noted for the *E. coli* enzyme by Santi *et al.* (1971).

In Table I, the aminoacylation of $tRNA^{Phe}$-C-C-A(3'NH_2) is shown. In this tRNA no post-transfer proof-reading is possible because of the stability of the amide linkage. p-Fluorophenylalanine can be isolated as a stable aminoacyl-tRNA without significant species-specific differences. The planar system of 2-amino-4-methylhex-4-enoic acid, tolerated by the yeast enzyme, appears very unfavourable for the *E. coli* enzyme. As with the natural amino acids, the turkey

Table I

Substrate properties of phenylalanine analogs upon aminoacylation of $tRNA^{Phe}$-C-C-A(3'NH_2) by phenylalanyl-tRNA synthetase from E. coli (E. c.), yeast (S. c.), and turkey liver (t. l.).

	K_m (μM)			V_{Phe}/V_{analog}		
Amino Acid	E.c.	S.c.	t.l.	E.c.	S.c.	t.l.
Phenylalanine	2	9.4	13			
Tyrosine	1,700	1,000	-	5,400	415	-
Leucine	10,000	14,000	-	54,000	27,000	-
Methionine	12,000	16,000	-	33,000	23,000	-
p-Fluoro-Phe	24	26	125	13.5	2.9	14.5
2-Amino-4-methylhex-4-enoic acid	400	85	-	625	11	-
N-Benzyl-L-Phe	1,500	6,100	-	5,400	11,000	-
N-Benzyl-D-Phe	30,000	26,000	-	27,000	97,000	-

liver enzyme gives no aminoacylation of $tRNA^{Phe}$ -C-C-A(3'NH_2) or $tRNA^{Phe}$-C-C-A.

As indicated by Santi & Webster (1976), *E. coli* phenylalanyl-tRNA synthetase possesses, in addition to the phenyl-binding pocket, a second hydrophobic area near the α-NH_2 group. A similar weak hydrophobic area is also indicated by the present data, however, access to it is restricted by complexation with tRNA. So N-benzyl-L-phenylalanine containing a hydrophobic α-amino substitute is activated by all tested enzymes, however, transfer to the tRNA is only detected for the *E. coli* enzyme and to a very small extent for the yeast enzyme. No aminoacylation of $tRNA^{Phe}$-C-C-A(3'NH_2) could be measured for the turkey liver enzyme. This aspect of species-dependent differences, and its implication of pharmacological application, has been discussed by von der Haar *et al.* (1981).

From the data it appears that the initial recognition of the amino acid by the free enzyme is most accurate in the case of *E. coli.* This difference in accuracy and its subsequent effect on aminoacylation is only valid in a comparison with the yeast enzyme. The turkey liver enzyme on the other hand appears to have developed a different approach in order to obtain a high fidelity of aminoacylation. In agreement with the data from the cysteinyl-tRNA synthetase of *E. coli* (Fersht & Dingwall 1979b), a lower proof-reading capacity can be expected for the *E. coli* enzyme, owing to high initial amino acid recognition with an error rate that is close to 10^{-6} with non-cognate natural amino acids (Table I).

In this respect the results of the AMP/PP_i-independent hydrolysis of aminoacyl-tRNA are of special importance. With the quotient of the turnover numbers of amino-acylation of $tRNA^{Phe}$-C-C-A(3'NH_2)and non-stoichiometric

Table II

Aminoacylation of tRNA^{Phe}-C-C-A by phenylalanyl-tRNA synthetase with phenylalanine analogs and relation of tRNA^{Phe}-C-C-A(3'NH$_2$) aminoacylation to AMP production.

	Aminoacylation of tRNA^{Phe}-C-C-A%			$\frac{k_{cat}\ (tRNA^{Phe}(3'NH_2))}{k_{cat}\ (AMP\ production)}$		
	E.c.	S.c.	t.l.	E.c.	S.c.	t.l.
Phenylalanine	100	100	100	81	30	8,2
Tyrosine	0	0	0	9.7	6.2	–
Leucine	0	0	0	14	1.6	–
Methionine	0	0	0	13	0.7	–
p-Fluoro-Phe	45	62	57	62	38	16.5
2-Amino-4-methylhex-4-enoic acid	0	8	0	78	1.8	–
N-Benzyl-L-Phe	7	<1	0	35	0.6	–
N-Benzyl-D-Phe	0	0	0	6.5	–	–

AMP production, it can be deduced that accurate initial recognition processes correlate with less elaborate hydrolytic capacity (Table II).

The ratio of the primary synthetic and the subsequent hydrolytic step occurring during aminoacylation is significantly smaller in the case of non-cognate natural amino acids than in the case of phenylalanine for the *E. coli* and yeast enzyme, and indicates a further enhancement of accuracy after initial recognition of the amino acid by the enzyme.

As we have shown that the transfer of amino acid to tRNA^{Phe}-C-C-A (3' NH$_2$) is applicable to the normal catalytic cycle, it can be concluded from the data that the route of proof-reading differs for the enzymes. With the turkey liver enzyme, natural amino acids promote ATP hydrolysis without being covalently bound to the tRNA. Even in the presence of tRNA^{Phe}-C-C-2' dA that is not a substrate for phenylalanylation, a very efficient hydrolysis of natural non-cognate amino-acyl-adenylates takes place with the turkey liver enzyme. This is clearly a case of pre-transfer proof-reading in the terminology of Jakubowski & Fersht (1981). AMP production in the presence of modified tRNAs^{Phe} further implicates the 3' OH group of the terminal adenosine in this reaction. From the significant ATP hydrolysis, occurring non-stoichiometrically with respect to aminoacylation of tRNA^{Phe}-C-C-A(3'NH$_2$), and the exclusion of dissociation of non-cognate aminoacyl-adenylates from the enzyme, one can conclude that some chemical proof-reading occurs even in the case where an amino group exists at the adjacent 3' OH position.

MECHANISTIC CONSIDERATIONS: POST-TRANSFER VERSUS PRE-TRANSFER PROOF-READING

As suggested above, the turkey liver enzyme uses a pre-transfer proof-reading mechanism. No mischarged $tRNA^{Phe}$-C-C-A or $tRNA^{Phe}$-C-C-A(3'NH_2) can be observed in this case. This is further substantiated in a pre-steady-state experiment shown in Fig. 1.

Considering the capability of the turkey liver enzyme to use pre-transfer proof-reading for non-cognate natural amino acids and post-transfer proof-reading for synthetic analogues, it can be suggested that there is probably not a fundamental mechanistic difference between the pre-transfer and post-transfer pathway. The interplay between the active site of the enzyme, the 3′ -terminal adenosine and the aminoacyl-adenylate may give an intermediary complex that, potentially, can dissipate its energy into either pathway, dependent on the structure of the amino acid. Donor and acceptor interactions within the active site, perhaps via a carboxylate, the 3′ OH group and a histidine residue, may result in a correct orientation and necessary polarization, common to both subsequent reactions. For example, in the case of tyrosine, the amino acid is either transiently transferred to the 2′ OH group and proof-read via the influence of the 3′ OH group (*E. coli, S. cerevisiae*), or the aminoacyl moiety of the aminoacyl-adenylate is positioned for the hydrolysis before transfer (turkey liver) (Fig. 2).

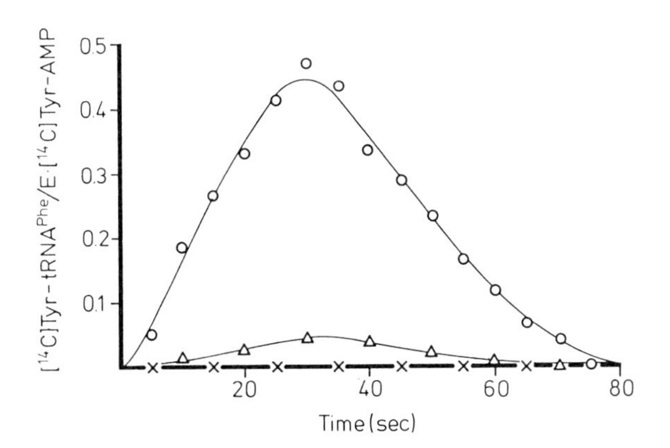

Fig. 1. Transient formation of [^{14}C]aa-$tRNA^{Phe}$-C-C-A during rejection of phenylalanine analogues by the phenylalanyl-tRNA synthetases from yeast and turkey liver. Transfer of [^{14}C]tyrosine to $tRNA^{Phe}$ by yeast (o) and turkey liver (x) phenylalanyl-tRNA synthetase. Transfer of [^{14}C]N-benzyl-L-phenylalanine to $tRNA^{Phe}$ by the turkey liver phenylalanyl-tRNA synthetase (*Δ*).

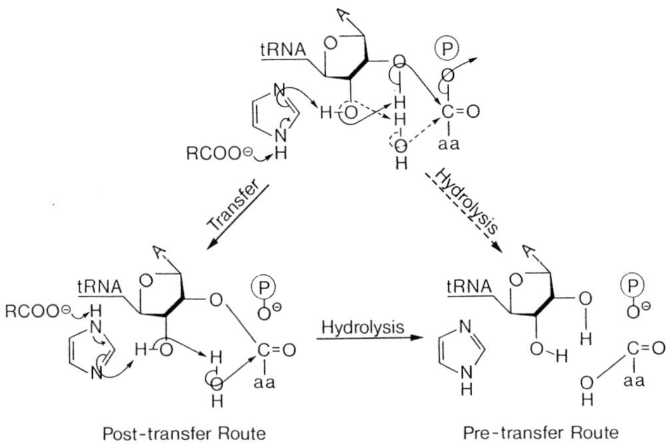

Fig. 2. Schematic mechanistic depiction of aminoacyl-adenylate processing starting with an enzyme·tRNA·aminoacyl-adenylate complex (above) leading to AMP and amino acid via immediate hydrolysis (right pathway, pre-transfer route) or via intermediate transfer to tRNA (left pathway, post-transfer route).

Thus, the different enzymes appear to use different strategies such that the overall cost of proof-reading, as calculated by Savageau & Freter (1979), will decrease until the energy by an additional decrement in proof-reading is equal to the increment in energy waste resulting from a concomitant increase in net error. The valyl-tRNA synthetases of *E. coli, B. stearothermophilus,* yeast and *Lupinus luteus,* for example, all follow primarily the post-transfer pathway (Fersht & Kaethner 1976, Igloi *et al.* 1978, Fersht & Dingwall 1979a, Jakubowski & Fersht 1981).

DO MITOCHONDRIA BEHAVE LIKE DOMESTICATED PROKARYOTES?

We have isolated the mitochondrial phenylalanyl-tRNA synthetases from yeast and hen liver, and compared their mechanisms and properties with those of the corresponding cytoplasmic synthetases (Gabius *et al.* 1983b). The results are summarized in Table III. It turns out that in every respect the mitochondrial synthetases behave like the cognate cytoplasmic enzymes. This applies also to the pattern of activity inhibition by chemical modification using different types of reagents as shown in Table IV.

Table III

Similarity in comparison of phenylalanyl-tRNA synthetase pairs from hen liver and yeast

Method		cyto/mito	cyto/procaryotic	mito/procaryotic
Amino acid composition (SΔQ)		+	–	–
Trypsin-pattern		+	–	–
Ouchterlony double diffusion		+	–	–
Immunoelectrophoresis		+	–	–
Immunotitration		+	–	–
Immunoblotting		+	–	–
Immunospotting		+	–	–
Active site mapping	ATP	+	(+)	(+)
	Phe	+	(+)	(+)
Catalytic cooperativity		+	+	+
Active site labelling		+	+	+
Proofreading mechanism		+	–	–
Chemical modification		+	–	–
Glycoprotein		–	+	–

\+ high degree of similarity
(+) low degree of similarity
– no similarity

CONCLUSIONS

Throughout the evolutionary tree, phenylalanyl-tRNA synthetases do have the same overall structure, a molecular weight of about 260,000 and an $\alpha_2\beta_2$ subunit

Table IV

Influence of chemical modification of phenylalanyl-tRNA synthetase on pyrophosphate exchange (top) or aminoacylation (bottom).

	Phenylalanyl-tRNA Synthetase from				
Reagent	E.c.	S.c.		t.l.	
(against)		cyto	mito	cyto	mito
Diethylpyrocarbonate	+	++	++	–	–
(Histidine)	++	++	++	++	++
Phenyl-phosphorodichloridate	–	+	+	–	–
(Nucleophiles)	++	+	+	++	++
3[2-Ethyl-5-isoxazolio]benzo-	– –	–	–	(+)	(+)
sulfonate (Carboxyl groups)	– –	–	–	(+)	(+)
Pyridoxalphosphate	–	–	–	–	–
(Lysine)	++	++	++	++	+

Table V

Extent of cross-reaction of various animal cytoplasmic phenylalanyl-tRNA synthetases with antibodies to the hen liver cytoplasmic enzyme expressed as percentage of the homologous reaction

Source of the Enzyme	extent of cross-reaction
Birds	
hen (Gallus gallus)	100
turkey (Meleagris galloparo)	82
pigeon (Columba livia)	67
duck (Anas platyrhynchos)	64
goose (Anser anser)	56
Reptiles	
turtle (Chelonia spp.)	19
Amphibia	
frog (Rana ridibunda)	14
Artiodactyls	
ox (Bos taurus)	12
pig (Sus scrofa)	10
Species of other mammalian orders	
cat (Felis silvestris)	10
horse (Equus caballus)	8
Fishes	
carp (Cyprinus carpio)	6
trout (Salmo fario)	5
Non-vertebrate species	
Drosophila melanogaster	–
Caenorhabditis elegans	–
Ascomycetes	
Saccharomyces cerevisiae	–
Enterobacteriaceae	
Escherichia coli	–

structure. They catalyse the same reaction and handle the same substrate. In spite of these facts they differ greatly in their enzymatic mechanism, and in the mode and strategy of rejection of incorrect substrates. We have recently included immunological methods in our comparative studies. The results of cross-reactivity of hen liver phenylalanyl-tRNA synthetase antibody with various phenylalanyl-tRNA synthetases are given in Table V (Gabius *et al.* 1983b). Apparently, the phenylalanyl-tRNA synthetase gene sequence is not well preserved. Thus, it becomes understandable, that mechanistic features also change in the course of evolution.

REFERENCES

Cramer, F., von der Haar, F. & Igloi G. L. (1979) *Transfer RNA: Structure, Properties and Recognition* (Ed. P. R. Schimmel, D. Söll, J. N. Abelson), pp. 267–279, Cold Spring Harbor Laboratory, USA.

Fasiolo, F., Remy, P. & Holler, E. (1981) *Biochemistry 20,* 3851–3856.

Fersht, A. R. (1977a) *Enzyme Structure and Mechanism,* Freeman, London.

Fersht, A. R. (1977b) *Biochemistry 16,* 1025–1030.

Fersht, A. R. & Dingwall, C. (1979a) *Biochemistry 18,* 1238–1245.

Fersht, A. R. & Dingwall, C. (1979b) *Biochemistry 18,* 1245–1249.

Fersht, A. R. & Kaethner, M. M. (1976) *Biochemistry 15,* 3342–3346.

Gabius, H. J. & Cramer, F. (1982) *Biochem. Biophys. Res. Commun. 106,* 325–330.

Gabius, H. J., von der Haar, F. & Cramer, F. (1983a) *Biochemistry 22,* 2331–2339.

Gabius, H. J., Engelhardt, R., Schröder, F. R. & Cramer, F. (1983b) *Biochemistry* (in press).

Gray, M. W. & Doolittle, W. F. (1982) *Microbiol. Rev. 46,* 1–42.

Hopfield, J. J. (1974) *Proc. Natl. Acad. Sci. USA 71,* 4135–4139.

Hopfield, J. J., Yamane, T., Yue, V. & Coutts, S. M. (1976) *Proc. Natl. Acad. Sci. USA 73,* 1164–1168.

Igloi, G. L. & Cramer, F. (1978) *Transfer RNA* (Ed. S. Altman), pp. 294–349, MIT Press, Cambridge, Mass., USA.

Igloi, G. L., von der Haar, F. & Cramer, F. (1978) *Biochemistry 17,* 3459–3468.

Jakubowski, H. & Fersht, A. R. (1981) *Nucleic Acids Res. 9,* 3105–3117.

Lea, P. J. & Norris, R. D. (1977) *Progr. Phytochem. 4,* 121–167.

Loftfield, R. B. & Vanderjagt, D. (1972) *Biochem. J. 128,* 1353–1356.

Santi, D. V., Danenberg, P. V. & Satterly, P. (1971) *Biochemistry 10,* 4804–4812.

Santi, D. V. & Webster, R. W. (1976) *J. Med. Chem. 19,* 1276–1279.

Savageau, M. A. & Freter, R. R. (1979) *Proc. Natl. Acad. Sci. USA 76,* 4507–4510.

Tsui, W. C. & Fersht, A. R. (1981) *Nucleic Acids Res. 9,* 4627–4637.

von der Haar, F. (1977) *FEBS Letters 79,* 225–228.

von der Haar, F. & Cramer, F. (1975) *FEBS Letters 56,* 215–217.

von der Haar, F. & Cramer, F. (1976) *Biochemistry 15,* 4131–4138.

von der Haar, F., Gabius, H. J. & Cramer, F. (1981) *Angew. Chem. 93,* 250–256.

von der Haar, F. & Gaertner, E. (1975) *Proc. Natl. Acad. Sci. USA 72,* 1378–1382.

DISCUSSION

ERDMANN: I was wondering about your antibody cross-reaction sites. Have you also compared the chargeability of the tRNAs with the different enzymes? Do you see a similar pattern in terms of chargeability from the different organisms?

CRAMER: We have not done quantitative studies with chargeability. We have tested these enzymes with yeast tRNA, but otherwise we have no quantitative data.

SCHIMMEL: I am interested in your antibody studies, particularly from the standpoint of the overall structure of the various synthetases. As you know, in *E. coli*, yeast, and rat liver this enzyme is an $\alpha_2\beta_2$ enzyme, and the question is, whether it has been possible for you to explore the arrangement of the quaternary structure in some of these other species as well? Is it $\alpha_2\beta_2$ throughout all of these organisms?

CRAMER: Yes, it is $\alpha_2\beta_2$. These enzymes were, however, not purified to 100% homogeneity but to around 50% homogeneity. In any case the main bands were the 2, the α and the β subunits. This pattern is undistinguishable between *E. coli* and higher animals.

SCHIMMEL: This is probably the best documented case of exploring this question in various organisms. It is clear from limited data on the various synthetases that this preservation of quaternary structure throughout different organisms does occur in general, even though there are a diversity of quaternary structures for different synthetases.

CRAMER: Yes, and as the amino acid sequence is not preserved in evolution, we would easily see 10% preservation of sequence. There must be a kind of convergent evolution to this $\alpha_2\beta_2$.

SCHIMMEL: That raises another interesting question: whether antibody cross-reactivity is an effective screen for sequence homology and if so how sensitive is it. We have cloned isozymes (for a metabolic enzyme) which should not cross-react immunologically, and yet we can detect them by cross-hybridization with cDNA probes. I think it remains an open issue as to whether Phe-tRNA synthetase from diverse organisms, which do not cross-react very well, might still have a considerable amount of sequence relationship.

CRAMER: Yes.

KURLAND: I wonder if we can take your correlation between proof-reading intensity and phylo-genetic development in a somewhat different way. I think it may be important to distinguish between for example yeast proof-reading systems and say host proof-reading systems, because one group of organisms is that which growth rate competition is important like *E. coli* or yeast, and other organisms may have other boundary conditions on their energetic economy. In general, you can show that for a growth rate competing organism it is an advantage to sacrifice a certain amount of accuracy in order to save a certain amount of energy. But you can imagine more complicated organisms like horses and people, etc., having boundary conditions on their energy consumption such that they will be willing to throw away a lot more GTP to have a more accurate polypeptide. Is it possible that it is really the life style that is given in the correlation and not phylo-genetic complexity?

CRAMER: Is this actually different? Of course, lifeform, our lifeform, mammalian, is higher in evolution and has this different kind of protein biosynthesis ...

KURLAND: No, what I am suggesting is, if we found a bacteria that had a different constraint.

KLUG: *E. coli* is a highly evolved organism.

KURLAND: Yes, and we might find a bacteria, which had a different energy constraint, different from simple growth competition, which then would look very much like a chicken with respect to its proof-reading.

CRAMER: Yes, it is an interesting idea.

KLUG: After all, is it not true that bacteria do not have any introns? Isn't the feeling more and more that these may have existed once and were lost? So they are highly evolved.

GALLANT: That is a perfect example of wasting energy!

KLUG: Saving energy!

CRAMER: Following up Dr. Kurland's remarks, what matters with higher animals is that they reach a certain age and preserve their structure for longer periods.

KLUG: On the question of the immunological cross-reactivity I don't find it too surprising that you get almost no correlation between the widely different kingdoms. The recognition regions may be on the inside of the proteins, and the outsides could just drift or be involved in other interactions. So it is, I think, a case where you may get a false impression from the immunolgical studies.

CRAMER: Our data are indicative to a certain extent, but must be taken with caution.

KLUG: No, I believe the data, it is the interpretation.

GRUNBERG-MANAGO: It is possible now to make antibodies with denatured protein, they are doing that at the Pasteur Institute. Such antibodies may be more suitable for studying sequence homologies.

KJELDGAARD: What is known about the accuracy throughout the animal and bacterial kingdom. Is that the same all over, or can some organisms permit higher inaccuracy in the protein synthesis than others?

CRAMER: It seems that the overall accuracy in, for example, the case of phenylalanine is almost the same within the limits of error throughout the evolutionary kingdom from *E. coli* to turkey liver, e.g., the distinction between phenylalanine and leucine is 750,000 for turkey liver and 650,000 for *E. coli*. So, the total accuracy seems to be at least in the same order, but the strategy to achieve it is slightly different.

ERDMANN: Do you see any correlation between the accuracy and the autohydrolysis of charged tRNA?

CRAMER: Do you mean the non-enzymatic?

ERDMANN: Yes.

CRAMER: No, we don't see a correlation.

KJELDGAARD: The mischarging, is that only within the 2′-group enzymes and the 3′-group, or would you get mischarging across the different groups?

CRAMER: Yes, there are also such mischargings. This depends more on the chemical nature of the amino acid, whether it is similar or not. But it is extremely difficult to mischarge, e.g., phenylalanine onto the 3′ position of its own tRNA. This is not actually mischarging. The barrier between 2′ and 3′ is very strict, and only under so-called mischarging conditions can one force the Phe-tRNA synthetase to attach the phenylalanine to the 3′ hydroxyl of its own tRNA.

SPRINZL: There are cases where such mischarging is possible for example the lysine tRNA from *E. coli* can be very easily phenylalanylated although the lysine $tRNA^{Lys}$ is a 3′ acceptor and the phenylalanine $tRNA^{Phe}$ a 2′ acceptor. So, in the case of amino acids belonging to different groups the mischarging seems to be possible.

CLARK: Something which happened at the recent tRNA workshop in Japan has intrigued me. If we go back to the formative days of protein synthesis, we recognized a facile migration of the amino acid between the hydroxyl groups. We, therefore, never worried about the position of charging, because we always assumed that the migration is 100 times or 1,000 times faster than the rate of peptide bond formation. Some Japanese workers have re-investigated this and found that the rate of migration is not so fast, so that the position of charging now does play a role in the rate of protein synthesis. Have you any comment on that?

CRAMER: I think the work is very good and reliable. The specificity with respect to 2′, 3′ originally was thought to be an academic kind of curiosity, because of this fast migration. Because the migration is not so fast, this will have some meaning. It could well be that EF-Tu moves the equilibrium to one side and that this is a necessary function of the factor or of some other part of the ribosomal machinery.

SCHATZ: Have any attempts been made to separate genetically the charging from the proof-reading function? If you selected *E. coli* mutants that were either more resistent or more susceptible to fluorophenylalanine, yet were not simply permeation mutants, you might already have your foot in the door. Has this been tried?

CRAMER: We have thought about it, but so far nothing seriously has been done. It would, as you say, be a foot in the door.

Translational Accuracy and Bacterial Growth

C. G. Kurland, D. I. Andersson, S. G. E. Andersson, K. Bohman, F. Bouadloun, M. Ehrenberg, P. C. Jelenc & T. Ruusala

INTRODUCTION

Physiological studies of streptomycin-dependent bacteria led to the idea that the site of action of this antibiotic is the ribosome (Spotts & Stanier 1961). Confirmation of this, as well as the demonstration that bacteria with altered response to streptomycin have mutated ribosomes (Flaks *et al.* 1962, Speyer *et al.* 1962) created a major cottage industry that thrives to this very day. Indeed, few molecular details have enjoyed the sustained interest that this antibiotic has commanded for so many years. One particularly good reason for this fascination with streptomycin was provided by Gorini & Kataja (1964a, b), who showed that the accuracy of translation is lowered by the antibiotic.

The latter effect is a partial theme of the present report. Here, we attempt to identify the mechanism of streptomycin's influence on translational fidelity as well as the nature of the kinetic adjustments made by mutant ribosomes with relevant phenotypes. In addition to its pertinence to the mechanisms of translation, the present analysis provides a way of viewing the optimization of the accuracy of translation. The important fact is that one can isolate mutants with either higher or lower error frequencies than those of wild type. This implies that the level of translational accuracy can be selected just like any other parameter of the bacterial phenotype. Here, we suggest that the optimum accuracy level, namely that which supports the maximum bacterial growth rate, is limited by the energetic cost of accuracy.

Department of Molecular Biology, The Biomedical Center, S-751 24, Uppsala, Sweden.

GENE EXPRESSION, Alfred Benzon Symposium 19.
Editors: Brian F. C. Clark & Hans Uffe Petersen, Munksgaard, Copenhagen 1984.

THE PARTICULARITIES OF THE PROBLEM

Although other components have been implicated, the 2 ribosomal proteins most often identified as the yin and yang of translational accuracy are S4 and S12. Both of these proteins contribute to ribosome functions related to the effects of streptomycin, and the history of mutations affecting them goes back to the very earliest applications of antibiotics.

Mutant bacteria with marked resistance (SmR) to, and even dependence (SmD) on, streptomycin were identified in the 1940's (Miller & Bohnhoff 1947, Newcomb & Hawirko 1949). It was recognized that the 2 distinct phenotypes were likely due to allelic mutations (Newcombe & Nyholm 1950, Hashimoto 1960), as was eventually confirmed by the direct functional identification of the altered component in SmD ribosomes (Birge & Kurland 1969), and SmR ribosomes (Ozaki *et al.* 1969) as S12. That streptomycin induced errors that can be restricted by SmR ribosomes, could have the trivial explanation that the binding of antibiotic is weaker than with wild type ribosomes. However, the demonstration that in the absence of streptomycin the activities of suppressor tRNA's are restricted in SmR bacteria (Couturier *et al.* 1964, Lederberg *et al.* 1964, Gartner & Orias 1966), as are spontaneous suppression events (Gorini *et al.* 1966), suggested that the structure of this protein influences the accuracy of tRNA selection. Indeed, recognition of the quantitative restriction of translational errors by mutant forms of S12 became a central preoccupation in the 1970's thanks to the efforts of L. Gorini (1971).

The discovery of the involvement of S4 in ribosome phenotypes associated with streptomycin and accuracy is a somewhat more complicated story. To begin, it is relatively easy to obtain streptomycin-independent "revertants" from cultures of SmD bacteria. Most of these retain the altered form of S12 but contain in addition a second mutation that suppresses the SmD phenotype (Hashimoto 1960, Brownstein & Lewandoski 1967, Turnock 1969). When one such double mutant was studied in detail, it could be shown by ribosome reconstitution with mutant and wild type proteins that the suppressor of SmD is S4 (Birge & Kurland 1970). The identification of electrophoretic variants of S4 in other suppressed-SmD strains confirmed the generality of this pattern of interaction between mutant forms of S12 and S4 (Deusser *et al.* 1970, Olsson *et al.* 1974).

At about the same time, a mutation affecting ribosomes and raising the translational error frequency in the absence of antibiotics was characterized by Rossett & Gorini (1969). This ribosomal ambiguity mutant (Ram) was then shown to carry an altered form of S4 (Zimmerman *et al.* 1971). A comparison of a number of independently selected mutants with altered forms of S4 has shown that, in each case the suppression of the SmD phenotype and the enhancement of translational suppression are correlated pleiotropic effects of the individual mutations in the gene coding for S4, *rpsD* (Olsson *et al.* 1979a). It might be tempting to use the parallel between the effects of Sm and the Ram alteration on the accuracy of ribosome function as a means of explaining the suppression of the SmD phenotype by S4 variants. Nevertheless, these phenotypes are far too complicated to be explained so neatly. There is good reason to believe that the requirement for streptomycin in SmD bacteria is, at least in part, to repair a defect in ribosome assembly (Spotts 1962, Hummel & Böck 1983). Likewise, an SmD suppressing mutant of S4 could be shown to have a defective ribosomal RNA binding site (Green & Kurland 1971), a property later found in other S4 mutants (Daya-Grosjean *et al.* 1972). This binding defect leads to the temperature sensitivity of 30S subunit assembly in S4 mutants (Olsson *et al.* 1974), as well as to disturbances in the regulation of ribosomal protein synthesis (Olsson & Isaksson 1979b, Olsson & Gausing, 1980, Yates *et al.* 1980). All of these pleiotropic effects of Ram mutations are simultaneously expressed to varying degrees in all of the S4 mutants studied so far (Olsson & Isaksson 1979a). Therefore, while some of the effects of S4 mutations on the SmD phenotype are expressed in the function of some doubly mutated ribosomes (Birge & Kurland, 1970), there are other, potentially more decisive, interactions between S4 and S12 variants at the level of ribosome formation.

It is, however, undeniable that the opposing effects of S4 and S12 variants on translational accuracy provides a fascinating problem of mechanism. Thus, the influences of these proteins are provocative because they are not what is expected from the simplest interpretation of the adapter and messenger hypotheses. Part of the fascination may also be aesthetic in that the symmetrical effects of these 2 proteins are very appealing as well as highly suggestive of a fine tuning mechanism within the translational machinery.

A WAY TO GET AT IT

A large array of *in vivo* experiments summarized by Gorini (1971) showed that the ambiguities of translation were influenced by the structure of the tRNA at sites separate from the anticodon, as well as by the phenotype of the ribosome.

These data could be interpreted in terms of a very simple kinetic model as shown by Ninio (1974). One version of Ninio's model had particularly attractive consequences: if the rate-limiting step of polypeptide elongation is the same as one of the kinetic steps that control the accuracy of tRNA-selection, the different ribosome phenotypes discussed above should have distinguishable kinetic characteristics easily measured *in vivo.* Thus, the more accurate streptomycin-resistant ribosomes should be slower than wild-type ribosomes, and the less accurate Ram ribosomes should be faster (Galas & Branscomb 1976).

The initial comparisons of β-galactosidase synthesis rates *in vivo* confirmed the expectation that SmR ribosomes elongate more slowly than wild type ribosomes (Galas & Branscomb 1976, Zengel *et al.* 1977). In contrast, it was later found that there are other streptomycin-resistant ribosomes which, like most Ram ribosomes, elongate polypeptides *in vivo* at rates that are indistinguishable from those of wild-type (Piepersberg *et al.* 1979, Andersson *et al.* 1982). The falsification of a simple reciprocal relationship between the speed and accuracy of elongation was unfortunate. In particular, that the kinetic changes responsible for characteristic changes in the accuracy with which mutant ribosomes function are not necessarily reflected in the *in vivo* rates of elongation, has one very unfortunate methodological consequence. It means that the search for the kinetic mechanisms underlying the accuracy of tRNA selection must be carried out largely with *in vitro* systems.

Now, the obvious problem with *in vitro* systems is that they are, literally, artifacts. The uncertainties of such systems are compounded when it is kinetic data that we are after. There is a vast amount of data concerning the mechanisms of translation that has been obtained with what we call "conventional" *in vitro* systems; these are systems which appear to be supporting polypeptide synthesis at rates below 0.05 peptide bonds per second per ribosome, and with missense error frequencies anywhere from 1 to 50%. In addition, there are many experiments performed under these conditions that are directed at the accuracy of translation, but for which there are no error measurements to help in determining how relevant the results are to the problem. In view of the present space limitations, we will not discuss this important literature, but refer the curious to references in the much more limited literature covered here.

We have set as our goal *in vitro* systems that have performance characteristics comparable to those of the living bacteria. This means elongation rates close to 17 peptide bonds per second per ribosome, and missense frequencies in the range of 10^{-4} to 10^{-3} (Parker *et al.* 1980, Ellis & Gallant 1982, Bouadloun &

Kurland 1983). We have not quite met these standards, but we are getting reasonably close.

The accuracy of translation seems now to be relatively easy to reproduce *in vitro* if 2 experimental conditions are met. One is that the substrates for protein biosynthesis, aminoacyl-tRNA's and nucleoside triphosphates, must be displaced very far from equilibrium, as in bacteria, and the other is that a complex mixture of polyvalent ions, resembling that found *in vivo,* must be used to buffer the system (Jelenc & Kurland 1979, Pettersson & Kurland 1980). Matching the speed of translation *in vivo* has been more problematic.

One discovery that we made in the course of these developmental studies is that most of the *in vitro* systems used until recently, are not rate-limited in elongation, but rather are limited at the initiation of polypeptide synthesis. This initiation limitation can be overcome for poly(U) translation by preinitiating the systems with N-Acetyl-Phe-tRNA (Lucas-Lenard & Lipmann 1967), and when this is combined with the accuracy-optimizing conditions, poly (Phe) synthesis at a rate up to 12, but usually closer to 10 peptide bonds per second per ribosome is obtained at a leucine missense frequency of 4×10^{-4} (Wagner *et al.* 1982, Andersson *et al.* 1982).

Although 10 is not 17, it certainly is a lot closer than 0.05. The point is that we now can be reasonably certain that the kinetic effects we detect in our systems concern elongation rather than initiation of translation. Furthermore, the observed performance characteristics are not specific for poly(U). Indeed the same *in vitro* system programmed with poly(UG) and pre-initiated with N-Acetyl-Val-tRNA elongates poly(Cys Val) at a rate between 8 and 12 peptide bonds per second per ribosome (Andersson & Kurland 1983). Here, the missense substitution of Trp for Cys is 10^{-4} and that of Met for Val is close to 10^{-3}. Indeed, our finding that poly(U) and poly(UG) elongate at comparable rates suggests that the GC content of codon and anticodon has no marked influence on the rate-limiting step of elongation. This conclusion may be relevant to current theories of codon-preference (Grosjean & Fiers 1982).

The reproducible difference between the rate of ribosome function in our *in vitro* system and in the bacterium has been a source of some concern, but we have failed to find any new factor or parameter to close the gap. It is just possible that there is in fact no discrepancy between these 2 figures. Thus, Talkad *et al.* (1976) have studied the distribution of the rates of β-galactosidase elongation *in vivo* and they conclude that different ribosomes translate at different rates that are between 8 and 15 peptide bonds per second per ribosome. The figure of 17 per second usually quoted for the elongation rate would correspond to the fastest ribosomes, while our figure from *in vitro* systems is an average, which fits well within the distribution observed by Talkad *et al.* (1976).

WHERE DOES IT COME FROM?

The *in vitro* system described above is indeed an artifact, but one designed with certain ends in mind. One of these is to determine how aminoacyl-tRNA is selected on a codon-programmed ribosome. This had become a contentious issue, because model-system studies rather clearly indicate that unassisted triplet-triplet interactions do not have the sequence specificity required to support the accuracy of translation (Eissinger *et al.* 1971, Grosjean *et al.* 1976). There have been 2 extreme opinions on how to solve this problem.

The more interesting suggestion of the 2 is that of Hopfield (1974) and Ninio (1975). This kind of model is called a proof-reading scheme: it works by the repetition of editing steps driven at the cost of an excess dissipation of a second substrate, which is GTP in protein synthesis. Indeed, it is the excess hydrolysis of GTP, or more precisely, the excess dissipation of ternary complexes containing EF-Tu, GTP, and non-cognate aminoacyl-tRNA, that signals that ribosomes are carrying out proof-reading. The compelling aspect of such schemes is that they link the accuracy of translation to an energetic cost, which is the excess GTP flux.

The alternative view is more mundane: it suggests that the sequence specificity is in fact to be found in certain microstates of triplet interactions and that these more accurate conformational states are selected on the ribosome (Kurland *et al.* 1975). According to this model, a large non-specific binding-free energy term can be used by the ribosome to maximize the occupancy of the more sequence-specific states of the tRNA during its interaction with the codon. The virtue of this model is that it makes explicit use of the sort of data discussed by Gorini (1971) which attributes functions to both the ribosome and to the non-anticodon portions of the tRNA in the selection process.

It is unfortunate from the point of view of maximum clarity that these alternative descriptions are not really mutually exclusive. Our experimental situation is as follows: we still do not know enough to make an intelligent decision on whether or not the accuracy of codon-anticodon interactions limits the accuracy attainable in a single selection step on the ribosome. A different decision on this issue has been reached by Thompson & Karim (1982). There is, nevertheless, mounting, but by no means definitive, evidence that there is proof-reading on ribosomes.

If there were proof-reading at any intermediate stage of peptide-bond formation, the probability with which a ternary complex, once bound to the ribosome forms a peptide bond must be less than one, even for a ternary complex containing cognate tRNA, and much less than one for those containing non-cognate tRNA. This state of affairs will be reflected in the efficiency with which EF-Tu-GTP supports polypeptide synthesis when the ratio of cognate to non-cognate tRNA species is changed. That is so, if the flow of aminoacyl-tRNA over the ribosome is strictly dependent on EF-Tu and GTP in the *in vitro* system. The demonstration that the rate of polypeptide synthesis can be stimulated more than one thousandfold by EF-Ts when the system is limited by low concentrations of EF-Tu, permits the conclusion that to an excellent first approximation, all peptide-bond formation is dependent on EF-Tu and GTP in the optimized system (Ruusala *et al.* 1982a). It was then possible to compare the stoichiometry with which ternary complex makes peptide bonds with Phe-tRNA (f_c) and with Leu-tRNA (f_w) in a poly(U)-programmed system.

The ratio (f_w/f_c) of these coefficients is the proof-reading factor, F. During steady-state poly(Phe) synthesis this factor could be shown to vary from greater than 100 for $tRNA_2^{Leu}$, to close to 70 for $tRNA_3^{Leu}$, and to as low as 50 for $tRNA_4^{Leu}$ (Ruusala *et al.* 1982b). Although these experiments unambiguously demonstrate that there is a non-productive dissipation of ternary complexes in the translation system, they suffer from at least one serious ambiguity. This is that only a fraction of the purified ribosomes, usually between 10 and 25%, are fully active in polypeptide synthesis. Therefore, we cannot be certain that the same ribosomes that are translating poly(U) are also responsible for the excess dissipation of ternary complexes (Ruusala *et al.* 1982b). Data to be discussed in the next section persuade us that the possibility is remote that the appearences of proof-reading in our system are only the artifactual activities of a "killer" ribosome; nevertheless, as remote as it is, the possibility still does exist.

It should also be stressed that the above experiments do not identify the proof-reading function with EF-Tu. Indeed, it is formally possible that the proof-reading of aminoacyl-tRNA suggested by our data, is a function of EF-G. It is for that matter quite attractive to consider the possible interactions of multiple steps of proof-reading associated with EF-Tu as well as with EF-G and including error coupling between missense events and reading frame errors (Kurland 1979), or polypeptidyl-tRNA drop-off (Menninger 1978). It seems most prudent at this point in the development of the problem to be open to all of the possibilities that at last can be distinguished experimentally.

THE MAIN POINT

We return now to the mutant ribosomes and antibiotic effects that form the main concern of this progress report. By far the most straightforward set of mutants that we have studied are the Ram mutants containing altered forms of S4. We have studied 3 such mutants that can be ordered according to increasing ability to suppress UGA codons *in vivo* (Andersson *et al.* 1982). When we analyse the competition between $tRNA^{Phe}$ and either $tRNA_2^{Leu}$ or $tRNA_4^{Leu}$ in a poly (U)-programmed system, we find that compared to wild type, the missense frequencies with $tRNA_2^{Leu}$ increase between 6- and 12-fold, while for $tRNA_4^{Leu}$ the increase is between 4- and 8-fold. When we compare the proof-reading factors for these different ribosomes, we observe a very tight correlation between this decreased F and the increased error. It seems that within our experimental uncertainty of 20%, all the increased error is due to a decreased proof-reading capacity in the Ram mutants (Andersson & Kurland 1983).

The nominally streptomycin-resistant mutants that we have studied are a far more varied lot (Bohman, unpublished data). Some of these are able to grow independently of streptomycin, while others are weakly stimulated by the antibiotic. Two of the independent ones seem to work at slightly higher error frequencies than do wild type ribosomes *in vitro* in the absence of streptomycin, but their proof-reading factors are unchanged, which implies that the initial selectivity of these mutants is perturbed. The more restrictive mutants do seem to have improved proof-reading functions that at least roughly account for much of their greater accuracy of translation.

The effect of streptomycin itself is somewhat complicated (Ruusala, unpublished data). The effects on the speed and accuracy of wild-type ribosomes saturate at concentrations of antibiotic comparable to those of the ribosomes, and include a modest inhibition of the rate of cognate aminoacid incorporation as well as a very large stimulation of non-cognate flows. In the poly(U)-programmed system, the antibiotic stimulates the initial selection error of $tRNA^{Leu}$ by 3- to 5-fold, and the error due to decreased proof-reading by 30- to 50-fold, depending upon which $tRNA^{Leu}$ species is involved. The effect of the antibiotic in Ram ribosomes is less pronounced, as expected, because they already have lost some of their proof-reading functions. The related antibiotic kanamycin affects the error frequencies in a different way that is somewhat dependent on the bacterial strain, (Jelenc, unpublished data). Roughly speaking kanamycin has about as much effect on the initial selection as on the proof-reading selection, which in both cases can be as high as a 10-fold stimulation of the error.

Now, we have described all of these detailed results for one and only one reason: if the dissipative flows of ternary complex that we have described are an artifact due to unproductive "killer" ribosomes, it should be impossible to raise or lower the errors of translation in our system without there being a reciprocal lowering or raising of the proof-reading factor. In fact, we can by mutation or antibiotic contradict that prediction, which suggests that our proof-reading data are reliable. Therefore we suggest that the principal mechanism through which S4 and S12 variants exert their opposing effects on the accuracy is by their opposite tuning effects on the proof-reading flows.

ACCURACY AND GROWTH

If the only effect of a lower accuracy level for translation was that a smaller fraction of the proteins produced are serviceable, we would expect mutant ribosomes to be less accurate than wild type ribosomes. That is to say, evolution should have maximized the accuracy of translation. On the other hand, if there is an extra energetic cost associated with high accuracy, as for a proof-reading ribosome, evolution should have optimized the trade off between accuracy and cost. The latter optimization would explain the existence of the highly restrictive streptomycin mutants. Indeed, a rather striking illustration of this idea is provided by one of the mutants we have studied.

Zengel *et al.* (1977) have described a nominally streptomycin-resistant mutant (SM3) that grows at about half the rate of wild type in the absence of streptomycin, where it has a comparably reduced rate of polypeptide elongation; both rates are modestly stimulated by the antibiotic. We have studied the ribosomes from a strain into which this mutation was transduced, 017–1204. The donor strain SM3 was the generous gift of M. Nomura.

Briefly, we have found that the addition of antibiotic does in fact influence these mutant ribosomes at concentrations comparable to those that affect wild type ribosomes *in vitro*, a characteristic that sets this mutant apart from the other resistant ones we have studied (Ruusala, unpublished data). In addition, the details of its response to the antibiotic *in vitro* are quite suggestive.

The ribosomes of 017–1204 translate poly(U) at about two-thirds the rate of wild type and the addition of streptomycin has no effect on this rate difference as long as there are saturating amounts of EF-Tu present, as *in vivo* (Gouy & Grantham 1980). Nevertheless, the antibiotic does have other effects that could explain the stimulation. Thus, the ribosomes of 017–1204 in the absence of antibiotic, require the dissipation of roughly twice as many ternary complexes per peptide bond formed than does the wild type, i.e., its f_c is almost twice that of wild type. When streptomycin is added to mutant ribosomes, the f_c decreases to wild type levels, i.e., to close to a value of one. In other words, part of this mutant's kinetic adjustment to streptomycin leads to a lower energetic cost of translation, which may account for the stimulation by antibiotic. Conversely, its slower rate of biosynthesis in the absence of antibiotic *in vivo*, may be an indirect effect of the unusually high consumption of GTP during translation.

We are optimistic that the experimental tools that we have developed to study polypeptide synthesis *in vitro* will provide the sort of data needed to understand the connections between the rates, energetic costs and accuracy of translation. The results that we have obtained for the detailed effects of antibiotics and mutants illustrate another use to which our systems can be put: they may help us to understand the parameters and strategies that regulate the growth of bacteria.

REFERENCES

Andersson, D. I., Bohman, K., Isaksson, L. A. & Kurland, C. G. (1982) *Mol. Gen. Genet. 187,* 467–472.

Andersson, D. I. & Kurland, C. G. (1983) *Mol. Gen. Genet.* (in press).

Andersson, S. G. E. & Kurland, C. G. (1983) (in preparation).

Birge, E. A. & Kurland, C. G. (1969) *Science 166*, 1282–1284.

Birge, E. A. & Kurland, C. G. (1970) *Mol. Gen. Genet. 109,* 356–359.

Bouadloun, F. & Kurland, C. G. (1983) *EMBO J. 2, 1351–1356.*

Brownstein, B. L. & Lewandowski, L. J. (1967) *J. Mol. Biol. 25,* 99–106.

Couturier, M., Desmet L. & Thomas, R. (1964) *Biochem. Biophys. Res. Commun. 16,* 244–248.

Daya-Grosjen, L., Garret, R. A., Pongs, O., Stöffler, G. & Wittmann, H. G. (1972) *Mol. Gen. Genet. 119,* 277–284.

Deusser, E., Stöffler, G., Wittmann, H. G. & Apirion, G. (1970) *Mol. Gen. Genet. 109,* 298–302.

Ellis, N. & Gallant, J. (1982) *Mol. Gen. Genet. 188,* 169–172.

Eisinger, J., Feur, B. & Yamane, T. (1971) *Nature New Biology 231*, 126–128.

Flaks, J. G., Cox, E. C., Witting, M. L. & White, J. R. (1962) *Biochem. Biophys. Res. Commun. 7,* 390–393.

Galas, J. P. & Branscomb, W. B. (1976) *Nature 262,* 617–619.

Gartner, T. K. & Orias, E. (1966) *J. Bact. 91,* 1021–1026.

Gorini, L. & Kataja, E. (1964a) *Proc. Natl. Acad. Sci. USA 51,* 487–493.

Gorini, L. & Kataja, E. (1964b) *Proc. Natl. Acad. Sci. USA 51,* 993–1001.

Gorini, L., Jacoby, A. G. & Breckenridge, L. (1966) *Cold Spring Harbor Symp. Quant. Biol. 31,* 657–664.

Gorini, L. (1971) *Nature New Biology 234,* 262–264.

Gouy, M. & Grantham, R. (1980) *FEBS Lett. 115,* 151–155.

Green, M. & Kurland, C. G. (1971) *Nature New Biol. 234,* 273–275.

Grosjean, D. H., De Henau, S. & Crothers, D. M. (1978) *Proc. Natl. Acad. Sci. USA 75,* 610–614.

Grosjean, H. & Fiers, W. (1982) *Gene 18,* 199–209.

Hashimoto, K. (1960) *Genetics 45,* 49–62.

Hopfield, J. J. (1974) *Proc. Natl. Acad. Sci. USA 71,* 4135–4139.

Hummel, H. & Böck, A. (1983) *Mol. Gene. Genet.* (in press).

Jelenc, P. C. & Kurland, C. G. (1979) *Proc. Natl. Acad. Sci. USA 76,* 3174–3178.

Kurland, C. G., Rigler, R., Ehrenberg, M. & Blomberg, C. (1975) *Proc. Natl. Acad. Sci. USA 72,* 4248–4251.

Kurland, C. G. (1978) *Biophys. 3. 22,* 373–388.

Lederberg, E. M., Cavalli-Sforza, L. & Lederberg, J. (1964) *Proc. Natl. Acad. Sci. USA 51,* 678–681.

Lucas-Lenard, J. & Lipmann, F. (1967) *Proc. Natl. Acad. Sci. USA 57,* 1050–1057.

Menninger, J. R. (1978) *J. Biol. Chem. 253,* 6808–6815.

Miller, C. P. & Bohnhoff, M. (1947) *J. Bact. 54,* 467–475.

Newcombe, H. B. & Hawirko, R. (1949) *J. Bact. 56,* 565–571.

Newcombe, H. B. & Nyholm, M. H. (1950) *Genetics 35,* 603–608.

Ninio, J. (1974) *J. Mol. Biol. 84,* 297–313.

Ninio, J. (1975) *Biochimie 57,* 587–595.

Olsson, M. O., Isaksson, L. A. & Kurland, C. G. (1974) *Mol. Gen. Genet. 135,* 191–202.

Olsson, M. O. & Isaksson, L. A. (1979a) *Mol. Gen. Genet. 169,* 251–257.

Olsson, M. O. & Isaksson, L. A. (1979b) *Mol. Gen. Genet. 169,* 271–278.

Olsson, M. O. & Gausing, K. (1980) *Nature 283,* 599–600.

Ozaki, M., Mizushima, S. & Nomura, M. (1969) *Nature 222,* 333–339.

Parker, J., Johnston, T. C. & Borgia, P. T. (1980) *Mol. Gen. Genet. 180,* 275–281.

Pettersson, I. & Kurland, C. G. (1980) *Proc. Natl. Acad. Sci. USA 77,* 4007–4010.

Piepersberg, W., Noseda, V. & Böck, A. (1979) *Mol. Gen. Genet. 171,* 23–24.

Rosset, R. & Gorini, L. (1969) *J. Mol. Biol. 39,* 95–103.

Ruusala, T., Ehrenberg, M. & Kurland, C. G. (1982a) *EMBO J., 1,* 741–745.

Ruusala, T., Ehrenberg, M. & Kurland, C. G. (1982b) *EMBO J., 1,* 75–78.

Speyer, J. F., Lengyel, P. & Basilio, C. (1962) *Proc. Natl. Acad. Sci. USA 48,* 684–686.

Spotts, C. R. & Stainer, R. Y. (1961) *Nature 192,* 633–637.

Spotts, C. R. (1962) *J. Gen. Microbiol. 28,* 347–358.

Talkad, V., Schneider, E. & Kennel, D. (1976) *J. Mol. Biol. 104,* 299–303.

Thompson, R. C. & Karim, A. M. (1982) *Proc. Natl. Acad. Sci. USA 79,* 4922–4926.

Turnock, G. (1969) *Mol. Gen. Genet. 104,* 295–301.

Wagner, E. G. H., Jelenc, P. C., Ehrenberg, M. & Kurland, C. G. (1982) *Eur. J. Biochem. 122,* 193–197.

Yates, L. J., Arfsten, E. A. & Nomura, M. (1980) *Proc. Natl. Acad. Sci. USA 77,* 1837–1841.

Zengel, J. M., Young, R., Dennis, P. P. & Nomura, M. (1977) *J. Bacteriol. 129,* 1320–1329.

Zimmermann, R. A., Rosset, R. & Gorini, L. (1971) *Proc. Natl. Acad. Sci. USA 68,* 2263–2267.

DISCUSSION

MAALØE: I assume that the effect you observe on the kinetics with the mutant implies that proof-reading occurs to similar degrees for all different tRNAs?

KURLAND: Yes.

MAALØE: To strengthen the argument, if you add up the energy costs of protein synthesis in growing bacteria, it comes to 80, 90, or even over 90% of ATP turnover.

KURLAND: Another calculation gives 50.

MAALØE: Our calculations give a different result, but that does not matter so much because either way it means that if you increase the total costs in making proteins, it would have a very dramatic effect on the energy balance and the growth conditions.

KURLAND: Yes. I consider a factor of 2 very significant, and what we are suggesting is a paradigm for bacteria which might have only an increase of 10%. That 10%, all other things being equal, is going to make such bacteria less adaptive than their competitors, but we have much greater difficulties, measuring the smaller increases with other mutants. I should also say that this cost is not a fixed number; its significance will vary with the medium. We are trying to develop a rigorous theory with which to predict this wonderful, phenomenological rule that Maaløe and his group have discovered in *E. coli* and *Salmonella.* We think we can predict a lot of these things from considerations of the sort that I talked about today.

SCHATZ: Expanding on the point Dr. Maaløe has made, I wonder whether it would be reasonable to follow up your observations with careful determinations of molar yields of your bacteria.

KURLAND: Oh yes, it would be very reasonable. We believe that this optimization, of energy vs. error, is but one pair of parameters among others relevant to the maximization of growth. The exponentially growing bacteria will be optimized for growing, but if you make human brain or human pancreas the constraint is very different. We believe that the error functions, the elongation rates, and the energy costs for protein synthesis are going to vary. They may even be different in a single cell for cytoplasma, chloroplast and mitochondria.

SCHATZ: Molar growth yields were very popular until about 10 or 15 years ago. When things became "molecular", they fell into disregard. Now, however, it is obvious that we are again asking: "What happens in the cell itself?" Measuring molar growth yields, which makes only relatively few assumptions, is a very powerful technique to answer this question.

KERR: Coming back to your first experiment, the mis-incorporation of the Cys-Arg, is that totally random?

KURLAND: It is only at one codon that we have measured the Cys-Arg substitution. We think we have accidentally sampled the high end of the range and we know that there are lower error frequencies. I am not saying that the error frequencies we measured are relevant for all codons in all contexts.

SAFER: Does the cost of your proof-reading vary with the rate of translation? In other words under more energy-rich circumstances can the cell afford to spend more energy?

KURLAND: Not *per se*. Perhaps there is some confusion about this. The error for a 2-step proof-reading function is determined by 2 parameters: one is the intrinsic discrimination of the system, the other is a measure of the efficiency with which a substrate passes forward versus the chance that it falls off. The more efficiently you use substrate, the more error you are going to have. So, you have a trade-off between that efficiency of substrate usage and the error; if you make the system very efficient it is going to kill your accuracy. If you are making everything very accurate, your cost is going to go up. You are trapped between these 2 effects. That is the basis of this argument about optimization.

HERSHEY: Is it possible that in the Tu-limiting system you have actually altered the rate of error occurrence?

KURLAND: Yes, there certainly is such an effect, but we can correct for it. It is normally small. But there is a bigger effect during starvation.

HERSHEY: Can you explain that in more detail.

KURLAND: The substrate that is going on to the ribosome is the Tu:GTP:aa-tRNA complex, and everything said and done what you are getting out is a tRNA+GDP+Tu. We pump this GDP very efficiently back to GTP, which means that the accumulation of Tu-GDP is going to be small. It will be there, and in fact we are very interested in it, but its effect is normally small.

INNIS: I would like to expand on the point that Dr. Gallant brought up earlier on the effects of starvation on errors. Years ago there were some very nice experiments showing that if you do amino acid starvation, you would see spots shifting all over on the 2-dimensional gels. This is also a very practical problem, because we have observed changes in specific activities of certain enzymes that are regulated from amino acid promotors by amino acid starvation. You can overcome this effect to some extent by adding back amino acids and by changing conditions for the growth of these organisms. Does magic spot have a role directly in the modulation of the error?

KURLAND: We suspect that it does, but we have been unable to prove it *in vivo*.

INNIS: But can you overcome this effect in a practical sense by supplementing with amino acids in the growth media?

KURLAND: Oh, yes.

KLUG: In your system you cannot measure the ATP splitting, can you?

KURLAND: We can. The problem is that it is not a meaningful number, because there are so many background reactions, e.g., an uncoupled G-reaction which we could minimize by using tricks, but we cannot get rid of it. So the pyrophosphate hydrolysis has an enormous background that makes it very difficult for us to measure ATP and GTP hydrolysis rates directly. That is why we are very fortunate that the Tu-GTP system had this EF-Ts dependence. That gave us an experimental edge to avoid looking at total GTP hydrolyses, we could just look at cycle times. If it wasn't for that we would have been in trouble.

KLUG: I see, so the key to your experiments is actually to avoid ...

KURLAND: ... measuring GTP hydrolysis directly, exactly, so we measure instead Tu-GTP cycle number.

EBEL: There are many cases where adaptation of the tRNA pattern to the codon composition in a messenger have been described. When this adaptation does not exist, has it been experimentally proven that the number of Ts cycles is changed?

KURLAND: I wish I could answer that question, because we think that the codon preference game is related to this tRNA isoacceptor strategy. I hope in some time to be able to give you a table with more of those numbers, but what I can do now is give data for 3 tRNAs. For these the answer is yes.

The Ribosome's Frame of Mind

Robert Weiss, James Murphy, Gerhart Wagner & Jonathan Gallant

INTRODUCTION

We are interested in the educational psychology of ribosomes, that is, how they count. They count off exactly 3 messenger residues in each cycle of chain elongation, so as to maintain the correct reading frame. How is this done?

In principle, one can envision 2 classes of mechanism for the maintenance of reading frame. One is a gating mechanism determined by the sterochemistry of translocation. On this view, ribosomes are *constructed* to count by 3s, irrespective of what is being translated. An alternative mechanism might couple the exactness of translocation (or of the registration of mRNA with the ribosome's A site) to the decoding process itself, i.e., the interaction of aminoacyl-tRNA molecules with messenger RNA codons. On this view, there will be strong interaction between what is being translated, or mistranslated, and the maintenance of reading frame. Our general conclusion is that the latter mechanism contributes very significantly to the ribosome's frame of mind.

READING FRAME AND AMINOACYL-tRNA POOL BIAS

We have shown elsewhere that aminoacyl-tRNA pool biases affect reading-frame maintenance (Gallant & Foley 1980), and we have recently published a detailed analysis of frame-shifting at a single position in the $r_{II}B$ message (Weiss & Gallant 1983). This is amino acid position 32, the sole inframe tryptophan (UGG) codon in the dispensable N-terminal third of the $r_{II}B$ message (Pribnow *et al.* 1981). Limitation for trp-tRNA elicits strong phenotypic suppression of (+) frameshift alleles just downstream of this position, with production of full-length $r_{II}B$ protein at 10–20% of wildtype efficiency. Mutational alteration of

Genetics Department, University of Washington, Seattle, Washington, U.S.A.

GENE EXPRESSION, Alfred Benzon Symposium 19.
Editors: Brian F. C. Clark & Hans Uffe Petersen, Munksgaard, Copenhagen 1984.

this UGG codon abolishes the effect, indicating that a phenotypic shift into the (−) reading frame at this position is responsible for the phenotypic suppression (Weiss & Gallant 1983).

Both theoretical considerations (Kurland 1979, Kurland & Gallant 1983) and our genetic results suggested that frame-shifting may be a consequence of binding of a non-cognate aminoacyl-tRNA species at the "hungry" position, favoured by limitation for the cognate species. To test this hypothesis, we sought to redress the pool bias by limiting simultaneously for trp-tRNA and for each of the potential non-cognate competitors. We found that trp-limited phenotypic suppression was reversed by simultaneous limitation for leucine or leu-tRNA, but was essentially unaffected by simultaneous limitation for each of 7 other aminoacyl-tRNA species (Table III Weiss & Gallant 1983). The specificity of this result supports our hypothesis, and strongly suggests that the culprit is a leu-tRNA, presumably the one which normally reads UUG.

Double amino acid limitation offers a simple and powerful stratagem for defining competitive relationships of this sort. However, the conventional wisdom in bacterial physiology is that it may not be possible to starve for 2 amino acids (or aa-tRNAs) simultaneously. Therefore, we have taken some pains to characterize the effect of double amino acid limitation. In one type of experiment, we used UGA leakiness as a metric of the activity of trp-tRNA in protein synthesis, based on the evidence that this species is responsible for UGA readthrough (Hirsh & Gold 1971, Grosjean *et al.* 1980). We showed that trp-tRNA limitation did indeed reduce UGA leakiness, and that simultaneous leucine limitation did not restore it to normal, indicating that trp-tRNA was not recharged (Weiss & Gallant 1983, Table III). We have now confirmed this conclusion by a direct assay of tryptophanyl-tRNA (Table I). One might also bear in mind that direct measurements of intracellular amino acid pools have shown that starvation for 2 amino acids does drain both pools (Broda 1971). It is, therefore, not surprising that aa-tRNA pools behave the same way, contrary to the conventional wisdom. We are currently using the double-starvation stratagem to characterize other frame-shifting events.

Context rules of shiftiness

We have tested 22 frame-shift alleles in $r_{II}B$ for suppression during limitation for a variety of aa-tRNAs. We will report this matrix of over 100 tests elsewhere (if a journal with sufficiently large pages can be found), and confine ourselves here to the general conclusions which emerge from the pattern of positive and negative responses. First, let us consider what one might expect to find.

The sequence established by Pribnow *et al.* (1981) reveals which codons are present in each frame-shift mutations's "suppression window", defined simply as the region in which a compensating phenotypic frame-shift *could* suppress that mutation. The limits of the suppression windows are defined by the positions of barriers or out-of-frame terminator codons: if a terminator codon is read between a given codon and a given frame-shift allele, then suppression cannot occur because termination intervenes before restoration of the correct reading frame.

If all codons are equally prone to frame-shifting during limitation for their cognate aa-tRNAs, then every frame-shift mutation should be suppressed by limitation for each aa-tRNA called for by codons within its suppression window. This is not so: phenotypic suppression was observed in only a few cases. Thus, there must be context rules governing the shiftiness of individual positions.

One context rule appears to be the identity of the 3′ base following a codon. FCO^+ shifts reading into the (+) frame, and, thereby, generates a downstream UGG codon before the barrier, and, thus, within the suppression window of

Table I
Levels of aminoacyl-tRNA

Condition	pmo/ml trp-tRNA	pmol/ml lys-tRNA
25° (trp-tRNA sufficient)	3.6	2.7
40° (trp-tRNA limited)	0.29	2.5
40° (trp and leu limited)	0.55	2.4

A $trpS^{ts}$ derivative of CP79 was grown at 25°. Limitation for trp-tRNA was done by transfer of the strain to 40° for 10 min before pulse labelling. Simultaneous limitation for leucine was done by filtering and resuspending the cells in the absence of leucine upon temperature upshift (the strain is a leucine auxotroph). The cells were pulse-labelled with ^{3}H-trp and ^{14}C-lys (as an internal control). After 40 and 60 sec, samples were precipitated in 5% TCA and chilled in ice with the addition of 0.01% SDS. The precipitate was filtered, washed in ice-cold 5% TCA and then with 95% ethanol. The dried filters were incubated with 20 mM $CuSO_4$ in 0.25 potassium acetate buffer, pH 5.5, for 90 min at 37°, in order to hydrolyze aminoacyl-tRNA on the filter. TCA was then added to 14% together with carrier BSA (100 μg/ml), and the chilled precipitate was centrifuged down. An aliquot of the supernatant was counted in a liquid scintillation spectrometer to detect labelled amino acid released by tRNA hydrolysis. The radioactivity released by $CuSO_4$ hydrolysis of aa-tRNA was corrected by a blank for each incubation, which was pulse-labelled in the same way, chased for several minutes with an excess of cold amino acids, then processed in the same way as the other samples.

Table II
Context effect on trp-limited phenotypic suppression

$r_{II}B$ genotype	%wildtype phage yield normal infection	trp-limited	fold-increase
638Δ	0.004	0.004	1.0
$FC38^+$	0.028	13.5	483
FCO^+	0.023	0.53	2.3
$N24^{UAA} \cdot FC38^+$	0.44	1.24	2.8

Phage yield relative to wildtype was determined as described by Weiss and Gallant (1983) in HfrH 3000 (lambda Cl_{857}) *relA*⁻ *sup*F66 at 30°. Tryptophan limitation was produced by 20 μg/ml indoleacrylic acid during the first 30 min of the lytic cycle, as described (Weiss & Gallant 1983). 638 Δ is a deletion of $r_{II}B$, tested as a control for physiological leakiness.

FCO^+. Yet trp-tRNA limitation produces only a 2–3 fold suppression of FCO^+, contrasted with the several 100-fold effect we observe at the other UGG at position 32 described above. The (+) frame UGG downstream of FCO^+ is followed by U, whereas the UGG at position 32 is followed by C. We tested the significance of this single context relationship by examining the double mutant $N24^{UAA} \cdot FC38^+$. The ochre mutant is in the codon immediately following UGG at position 32 (the UAA replaces the normal CAA). Thus, this double mutant differs from $FC38^+$ only in the presence of a U rather than a C following the UGG at position 32. In this case, trp-tRNA limitation elicits only a 3-fold suppression of $FC38^+$, similar to the weak suppression of FCO^+ (Table II). Therefore, it appears that shifting into the (−) frame at UGG codons occurs much more frequently when the next base is C than when it is U.

Context rules apparently determine the *direction* as well as the frequency of shiftiness. Limitation for trp-tRNA suppresses (+) frame-shift alleles downstream of position 32, but not one (−) frame-shift allele in the same region. Evidently, the UGG at position 32 is shifty in the (−) direction but not the (+) direction. On the other hand, we find that lys-tRNA limitation strongly suppresses two (−) frame-shift alleles at, and downstream of, an AAA (lys) codon at another position. Thus, phenotypic shifting in one direction or the other is favoured at different positions. More detailed analysis of these context rules will perhaps give us some insight into the complex relationship between decoding and reading frame maintenance.

THE ROLE OF ppGpp

All of the aforementioned studies were conducted with *rel*A$^-$ mutant strains, defective in ppGpp production. Their *rel*A$^+$ counterparts exhibit much lower frequencies of phenotypic frame-shifting. We assume this is due to the well-documented effect of ppGpp in restricting non-cognate aa-tRNA selection (Gallant & Foley 1980, Wagner & Kurland 1980, Wagner *et al.* 1982). In order to test this postulate directly, we have turned to translation of MS2 RNA *in vitro*. Atkins *et al.* (1979) identified elongated variants of MS2 coat and replicase proteins which are almost surely due to frame-shifting upstream of the normal terminators, with consequent readthrough and termination at downstream out-of-frame terminators. (We refer readers to the paper by Atkins *et al.* for the evidence. Recent work in Ray Gesteland's lab and in ours provides further confirmation, in the case of the coat variant).

We find that limitation of *in vitro* incubations for the amino acids ala, ser, or pro stimulates the production of the frame-shift variant of coat (Fig. 1). Codons

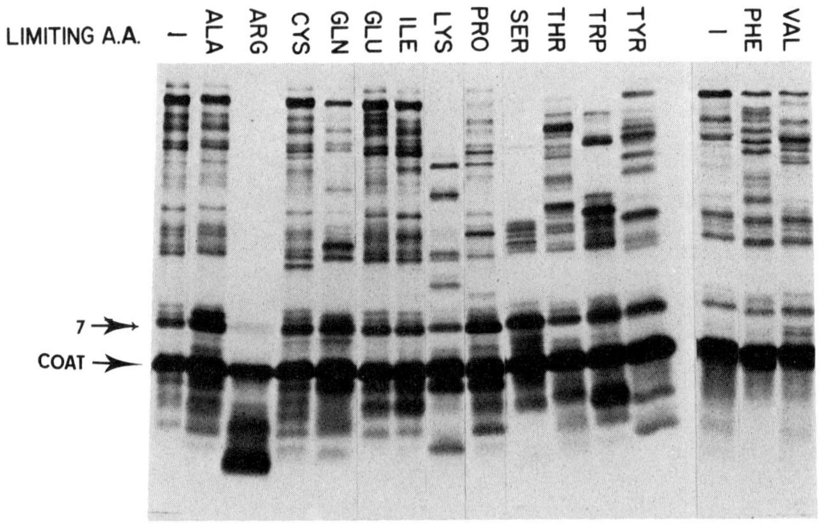

Fig. 1. Translation of MS2 RNA in amino-acid-limited incubations. Phage MS2 RNA was translated *in vitro* as described by Atkins *et al.* (1979), labelled with ^{35}S-methionine. Each incubation was supplemented with a complete amino acid mixture (control) or a mixture lacking a single amino acid, as indicated. The products of each incubation were resolved by 17.5% polyacrylamide SDS gel electrophoresis, according to Atkins *et al.* (1979), on the indicated lanes. Protein 7 is the frame-shifted variant of coat; note its high level in reactions starved for ala, ser, or pro.

for each of these amino acids are present within the suppression window of the normal coat terminator. Limitation for ala-tRNA, accomplished through the use of S_{30} extracts deficient in alanyl-tRNA synthetase activity, produces the same result as alanine limitation.

In both cases, the presence of a high concentration of ppGpp virtually eliminates the production of the frame-shift variant (Figs. 2, 3). Proline starvation also stimulates production of the putative frame-shift variant of replicase, and this too is eliminated by ppGpp (Fig. 2). This effect of ppGpp is consistent with our view that frame-shifting is a consequence of non-cognate aa-tRNA binding at hungry codon positions.

THE MECHANISM OF ERROR PREVENTION BY ppGpp

Wagner *et al.* (1982) have proposed a subtle kinetic mechanism for the error-preventing effect of ppGpp, based on the notion that ppGpp sequesters EF-T_u,

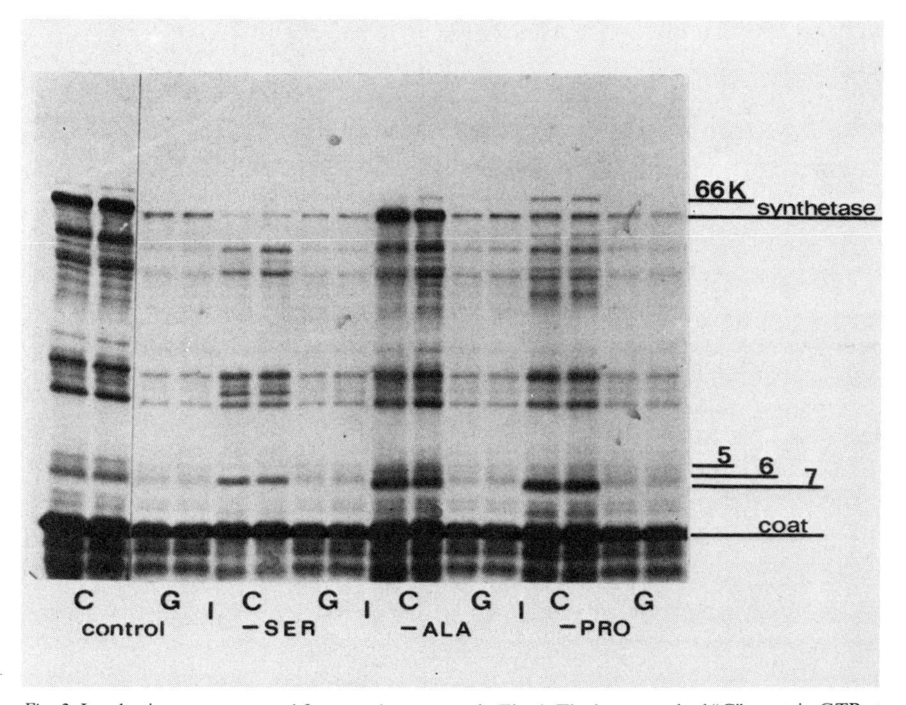

Fig. 2. Incubations were starved for ser, ala or pro as in Fig. 1. The lanes marked "C" contain GTP at 1 mM; the lanes marked "G" contain ppGpp in place of GTP at 1 mM.

thus lowering the effective level of binary complex. This will naturally reduce the formation of ternary complex for non-limiting, fully charged aa-tRNA species, including whichever ones can misread hungry codons which call for a limiting

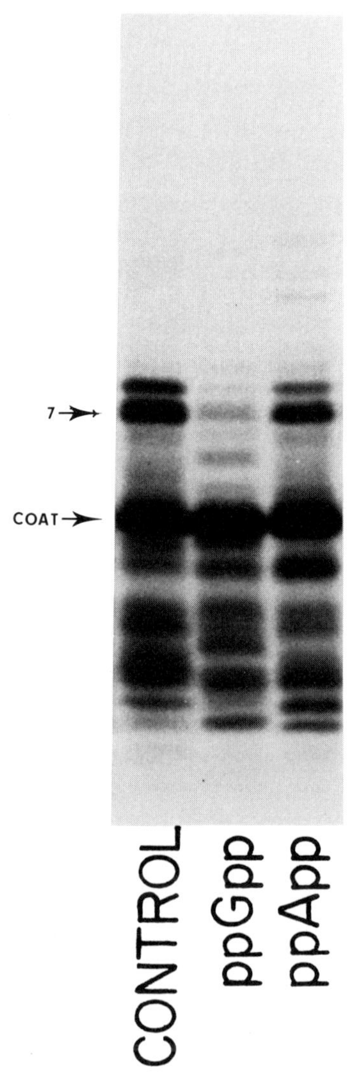

Fig. 3. The incubations were limited for ala-tRNA through the use of a 95:5 mixture of S_{30} protein derived from *ala*S^{ts} and *ala*S^{+}. The lane marked "ppGpp" had 1 mM ppGpp in place of GTP. The lane marked "ppApp" had 1 mM ppApp in place of ATP, and serves as a negative control for the effect of ppGpp on the production of frame-shifted coat protein.

aa-tRNA species. The reduction in ternary complex occasioned by ppGpp may be less severe for the limiting species, because in this case recharging of the tRNA can compensate for the reduction in binary complex. In this way, ppGpp would force non-limiting ternary complex levels down toward that of the limiting one, and, thus, even-out competition between them.

We have tested this model *in vitro,* by measuring total and ternary complex bound aa-tRNA corresponding to a limiting and a non-limiting species. The results (Table III) are in agreement with the model's primary postulate. Note that ppGpp reduces ternary complex by only 22% for the limiting species (alanine), but by 64% for a non-limiting species (serine).

Wagner *et al.* (1982) clouded their readers' minds by adding a curious secondary postulate to their model: they supposed that most of the total aa-

Table III

Aminoacyl-tRNA in free and ternary complex bound form.

		*ala*S in excess		*ala*S limiting	
		control	ppGpp	control	ppGpp
	Total	6.1	6.3	0.45	1.1
ala-tRNA	Complex	3.8	0.8	0.32	0.25
	Free	2.2	5.5	0.13	0.85
	Total	3.2	3.4	3.6	3.9
ser-tRNA	Complex	2.2	0.6	2.8	1.0
	Free	1.0	2.8	0.8	2.9

Effect of ppGpp on free and ternary complex-bound aa-tRNA. Phage MS2 RNA was translated *in vitro* by crude S_{30} preparations, exactly as described by Atkins *et al.* (1979). The left-hand columns (*ala*S in excess) were done with an S_{30} prepared from a *rel*A$^-$ *ala*S$^+$ strain. The right-hand columns (*ala*S limiting) were done with a mixture of 5% S_{30} from the latter strain, and 95% S_{30} from an isogenic *ala*S^{ts} strain. Both incubations yield steady-state translation for about 30 min, decreased about 6-fold in the *ala*S-limiting incubation. Incubations with ppGpp contained the nucleotide in place of GTP at 1 mM (omission of GTP scarcely affects the rate of translation in crude S_{30} extracts, which contain endogenous GTP sufficient to provide it at up to 40 μM in the incubations). The incubations were labelled with ^{3}H-ala and ^{14}C-ser, and one set of aliquots were processed as described in Table I to detect total aa-tRNA of each type. Duplicate aliquots were treated with ribonuclease for 10 sec at 0°, to hydrolyze free aa-tRNA, leaving ternary complex-bound aa-tRNA protected (Knowlton & Yarus 1980), and then processed as above in order to detect ternary complex-bound aa-tRNA. The table reports the average of 3 determinations. Free aa-tRNA is calculated from the difference between total and ternary complex-bound (RNase resistant) aa-tRNA. The specificity of the technique is validated by the reduction in ala-tRNA, but not ser-tRNA, observed in the *ala*S-limiting incubations; and by the reduction in ternary complex-bound forms of both aa-tRNAs produced by ppGpp.

tRNA corresponding to the limiting species remains in ternary complex form in the presence of ppGpp, despite the recharging of free aa-tRNA of this species which is the essence of the model, while most of the non-limiting aa-tRNA remains in free rather than ternary complex form. This stipulation demands that the rate constants of ternary complex formation be different for the limiting and non-limiting species, which is most unlikely. Our results (Table III) show that the ratio of free to ternary complex bound aa-tRNA is about the same for limiting and non-limiting species under all conditions, just as one would expect.

The logical and empirical defect in this secondary postulate of the model does not vitiate its main point, which can be justified algebraically for a steady state system. Nonetheless, the relevance of the model, and of our confirmation of its main point, to the *in vivo* situation is problematic. Our experiment was done at a ratio of ppGpp to GTP much higher than that which obtains *in vivo* during the stringent response of *rel*$^+$ cells. At physiological ratios, about 1:1, the effect of ppGpp on ternary complex formation is very slight, in our experience and that of other workers. Moreover, the model demands that ppGpp elicit significant recharging of the limiting aa-tRNA species, but measurements of the limiting aa-tRNA in isogenic *rel*$^+$ and *rel*$^-$ cells do not confirm this prediction (Kurland & Gallant 1983). We have recently confirmed that recharging is *not* observed in aminoacyl-tRNA-limited *rel*$^+$ cells, using a technique similar to that of Table I. This is extremely puzzling, given the clear evidence that ppGpp does restrict protein synthesis *in vivo* (Laffler & Gallant 1974, O'Farrell 1978).

We can only conclude that the mechanism through which ppGpp restricts translational error is still mysterious. The absence of recharging, in particular, implies that ppGpp effects on kinetic amplification or proof-reading ought to be seriously considered (Gallant & Foley 1980). Further work, co-ordinating *in vitro* and *in vivo* approaches to the problem is required. Our work does show that ppGpp restricts reading-frame errors as well as mistranslation errors, and reveals a coupling between them which remains to be elucidated in detail.

REFERENCES

Atkins, J. F., Gesteland, R. F., Reid, B. R. & Anderson, C. W. (1979) *Cell 18,* 1119–1131.

Broda, P. (1971) *Monatshefte für Chemie 102,* 811–823.

Gallant, J. & Foley, D. (1980) In: "*Ribosomes, Structure, Function, and Genetics,*" pp. 615–638 University Park Press, Baltimore.

Grosjean, H., deHenau, S., Houssier, C. & Buckingham, R. H. (1980) *Arch. Int. Physiol. Biochim. 88,* 168–179.

Hirsh, D. & Gold, L. (1971) *J. Mol. Biol. 58*, 459–468.

Knowlton, R. & Yarus, M. (1980) *J. Mol. Biol. 139,* 721–732.

Kurland, C. G. (1979) In: "*Nonsense Mutations and tRNA Suppressors,*" pp. 98–108 Academic Press, New York.

Kurland, C. G. & Gallant, J. (1983) In: "*Accuracy in Biology*" Marcel Dekker, New York.

Laffler, T. & Gallant, J. (1974) *Cell 3,* 47–49.

O'Farrell, P. H. (1978) *Cell 14,* 545–557.

Wagner, G. & Kurland, C. G. (1980) *Mol. Gen. Genet. 180,* 139–145.

Wagner, G., Ehrenberg, M. & Kurland, C. G. (1982) *Mol. Gen. Genet. 185,* 269–274.

Weiss, R. & Gallant, J. (1983) *Nature 302,* 389–393.

Pribnow, D., Sigurdson, D. C., Gold, L., Singer, B. S., Brosius, J., Dull, T. J. & Noller, H. F. (1981) *J. Mol. Biol. 149,* 337–376.

DISCUSSION

KURLAND: I should say that Dr. Gallant has been kind and tried to avoid saying outright that there is an error in our earlier kinetic theory. There is. It is a trivial, but lethal mistake. The model will not work the way we suggested it would. In addition, the assumption of our earlier model, that magic spot does not change the accuracy of translation unless the translation system is starved, is questionable. We have seen effects of magic spot on the translation accuracy even when you have a full complement of the aminoacyl tRNAs and this has re-stimulated us to look again for effects of magic spot on some editing function such as proof-reading.

GALLANT: I should like to add that I suspect that both things are correct, that one has both a recharging effect and an editing effect of magic spot. If you would allow me 2 minutes, I will explain what the puzzle about the charging levels is. In Table I we have a case of what might be, because of the high levels of magic spot, a sort of *in vitro* artifact. We get an effect compatible with the first assumption of your model, and of course we see recharging of the limiting tRNA. That's what the first postulate depends on: limiting tRNA is recharged because one is decreasing the drain. So, there is a good 2-fold increase in the level of limiting tRNA because of magic spot. If that is true in real life then amino acid starved rel^+ cells would have to maintain higher charging levels of limiting tRNA than rel^- cells which don't make magic spots. The trouble is there is a considerable literature showing that that is not the case. Three different laboratories have reported that, and we have rechecked it using our simple method of measuring aminoacyl tRNA, and it seems to be correct. We have limited Trp activation with different levels of indoacrylic acid in a $rel^{+/-}$ isogenic pair. The level of Trp-tRNA goes down in about the same way with increasing inhibition of the activating enzyme in the rel mutant and the rel^+ despite the big difference in level of magic spot. So, the great puzzle in this little field is the fact that there is no evidence of recharging *in vivo* in rel^+ cells. (Table I).

ROSENBERG: You indicated that when you change the nucleotide immediately beyond the UGG codon you reduce the amount of frameshifting because you have altered the context. Does that mean that anytime you have a Trp codon followed by that particular nucleotide, you can expect a frameshift, or is there more to the whole context problem then just the nucleotide beyond?

GALLANT: Certainly the latter. The following 3′ nucleotide is the particular case I described.

KJELDGAARD: You only had the plus frame mutants. You said if you had a minus frame mutant it did not work. Is that due to the context?

GALLANT: We are using the classical Cambridge collection which Dr. Gold has been kind enough to supply us with, and as it happens there are several plus frameshift mutations in this region; there is one minus, 176 minus, in the same region, and it does not respond to Trp starvation.

KJELDGAARD: But could that be due to the fact that it is in the wrong context?

GALLANT: I don't know. My point was that it is not due to a general rule against frameshifting in that direction, because elsewhere there are minus frameshifts that are suppressed through limitation for an amino acid in their suppression windows.

SPRINZL: What is the influence of magic spot on the level of aminoacyl tRNAs in ternary complexes. If we consider the known binding constants of ppGpp for Tu, is it theoretically possible that they would severely influence the concentration of ternary complexes? I recall that Pingoud published a paper, where he said that this is not to be expected.

GALLANT: Pingoud's calculation is that at normal concentrations of ppGpp and GTP one would expect only a very small decrease in ternary complex. The same thing was evident in David Miller's old experiments which are the first ones to show an effect of ppGpp on Tu, and the same thing is true, if you look at the numbers, in a recent paper from Bob Thompson's group. That is why I emphasize that here we are using very high levels of magic spot relative to GTP. Reasoning from *in vitro* cases to whole cells always has its ambiguities, so the question is what really could one expect in whole cells; that is, does ppGpp inhibit ternary complex formation significantly in whole cells? We are doing experiments to test that.

KURLAND: The calculations of Pingoud are based on the idea that you can measure the inactivation of Tu as a Tu:magic spot binary complex, if it is allowed to equilibrate. We know protein synthesis does not take place in equilibrium, and, in fact, we can show experimentally that the major way that magic spot inhibits ternary complex function is by inactivating Ts and preventing it from regenerating ternary complex from inactive Tu-GDP complex. The way that works is, a binary complex of Tu:magic spot will be attacked by Ts thinking that it is Tu:GDP and be held up for a longer time than the usual GDP complex would hold up the Ts. As a consequence, the Ts cannot keep up with the production of Tu:GDP and the system's activity sinks. It is a purely kinetic effect. It is not an equilibrium effect. So the Pingoud calculations are simply irrelevant.

Table I

Tryptophanyl-tRNA in a relA^{+}/$^{-}$ pair of strains.

Strain	Growth Condition	trp-tRNA pmol/ml	trp tRNA % Control
117	control	1.47	100
(relA^{+})	30 μg/ml Ia	1.04	71
	100 μg Ia	0.054	3.6
	150 μg/ml IA	0.030	2.0
	250 μg/ml IA +300 μg/ml chloramphenicol	0.20	14
119	control	1.0	100
(*rel*A^{-})	30 μg/ml IA	0.56	56
	100 μg/ml IA	0.10	10
	250 μg/ml IA	0.012	1.2
	250 μg/ml IA +300 μg/ml chloramphenicol	0.11	11

Tryptophanyl-tRNA synthetase was inhibited with varying levels of indoleacrylic acid (IA). After 10 min, the cells were pulse-labelled with ^{3}H-tryptophan for 30 sec, one aliquot was then precipitated in 80% ethanol: 5% TCA at ethanol-dry ice temperature another aliquot was chased for 10 min with cold tryptophan, then precipitated and treated identically to the first. In each case, the "chased" sample serves as a blank. The precipitated samples were filtered, and labelled tryptophan stripped from tRNA on the filter by treatment with 1.0 ml 5 mH NaOH for 10 min at 37°. The samples were then reprecipitated in 5% cold TCA with carrier bovine serum albumin, centrifuged, and an aliquot of the supernatant counted in a liquid scintillation counter. In each case, labelled tryptophan found in the supernatant is corrected for the corresponding value (generally much lower) found in the "chase" blank. The culture inhibited with IA and chloramphenicol shows that recharging of the limiting Trp-tRNA due to additional protein synthesis inhibition, by chloramphenicol, *can* be detected.

IV. Components of the Biosynthetic Apparatus.
I. tRNA, mRNA, Decoding

Two Approaches to Mapping Functional Domains of an Enzyme: Applications to an Aminoacyl tRNA Synthetase

Maria Jasin & Paul Schimmel

INTRODUCTION

The 20 aminoacyl tRNA synthetases, which correspond to each of the amino acids, have highly diversified subunit sizes and quaternary structures (Table I) (Ofengand 1977, Schimmel & Söll 1979). In *E.coli,* there are 4 different quaternary structures (α, α_2, α_4, and $\alpha_2\beta_2$), and subunit sizes span a 3-fold range of molecular weights. As proteins which establish rules of the genetic code by matching amino acids with trinucleotide sequences (contained within tRNA molecules), they undoubtedly arose early in evolution; thus, strong selective pressures have brought about or maintained their diversity.

One hypothesis is that size diversity is inherent to arranging a variety of active sites which are specific for the different amino acid side chains and tRNA species. This explanation seems unlikely because of the small length variation seen for immunoglobulin chains, in spite of the great diversity of molecules that they recognize (Porter 1973). Additionally, the tRNA substrates themselves are not highly diversified in size and they fold into the same basic three-dimensional structure (Rich & RajBhandary 1976). It is not obvious that a size variation in synthetases is required to interact with a relatively invariant tRNA surface.

A second hypothesis is that synthetases have a core enzyme, which is specific for a given amino acid and tRNA (Jasin *et al.* 1983). Additional polypeptide structure, beyond that required for the core enzyme, is acquired by fusion of

Department of Biology, Massachusetts Institute of Technology, Cambridge, Massahusetts 02139, USA.

GENE EXPRESSION, Alfred Benzon Symposium 19.
Editors: Brian F. C. Clark & Hans Uffe Petersen, Munksgaard, Copenhagen 1984.

Table I
Quaternary Structures and Subunit Sizes of E.coli *Aminoacyl tRNA Synthetases*

Quaternary Structure	Enzyme and Subunit Molecular Weight
α	Arg (63–74,000), Gln (60,000), Ile (112,000), Leu (105,000), Val (110,000)
α_2	His (42,000), Lys (52,000), Met (85,000), Pro (47,000), Ser (48,000), Thr (76,000), Trp (37,000), Tyr (48,000)
$\alpha_2\beta_2$	Gly (α=33,000; β=75,000), Phe (α=39,000; β=94,000)
α_4	Ala (95,000)

Data are from Schimmel and Söll (1979), Putney *et al.* (1981a), Yamao *et al.* (1982), and Webster *et al.* (1983).

additional segments to the core protein. These segments would impart additional functions to the enzymes.

Although far from established, 2 lines of evidence support the second hypothesis. First, one aminoacyl tRNA synthetase has been shown *in vitro* to bind to its promoter sequence and regulate gene transcription (Putney & Schimmel 1981). The cognate amino acid is a co-repressor in regulation. Presumably, additional polypeptide sequences have been joined to the enzyme in order to accomplish gene regulation.

The second piece of supporting evidence comes from studies of fragments of aminoacyl tRNA synthetases. Proteolytic digestion of the methionine enzyme results in removal of approximately 200 carboxyl terminal residues from the 800 amino acid subunit (Cassio & Waller 1971, Koch & Bruton 1974). Removal of these sequences converts the normally dimeric enzyme to a monomer. The monomeric species has aminoacylation activity *in vitro* and *in vivo* (Cassio & Waller 1971, Barker *et al.* 1982).

Another example is alanine tRNA synthetase. The native enzyme is a tetramer of identical, 875 amino acid polypeptide chains (Putney *et al.* 1981a, 1981b). Proteolytic digestion gives an amino terminal fragment of approximately 440 residues; the fragment has full activity for aminoacyl adenylate synthesis, but cannot aminoacylate tRNA. More recent work (see below) showed that only a few additional amino acids are required to give an enzyme which aminoacylates $tRNA^{Ala}$ *in vivo.* Thus, in this case, removal of over 40% of the protein still leaves a viable enzyme.

To explore further the structural arrangement of aminoacyl tRNA synthetases, and to identify the putative core segment required for catalysis, requires development of approaches which modify the gene coding for the enzyme rather than approaches which attempt to alter the native gene product. For example, few defined enzyme fragments can be produced by proteolytic digestion. The functional importance of specific amino acids can be approached through chemical modification of the native enzyme, but generally it is possible to modify only a few specific positions in a polypeptide chain.

Described below are 2 approaches for mapping functional regions along the sequence of an enzyme. These generate altered polypeptides through site-specific gene modification. In one approach, amino acid substitutions are introduced through nucleotide sequence alterations at defined sites in the gene. In the other, site-specific gene deletions are created and these give rise to defined enzyme fragments.

AMINO ACID SUBSTITUTIONS INTRODUCED BY D-LOOP MUTAGENESIS

With this approach, base changes are directed to defined sections of the gene-coding region. Locations of those mutations which do not interfere with catalytic function define those parts of the polypeptide not associated with catalysis. Locations of those which inactivate the enzyme operationally define regions that are important for catalysis. In addition to mutations in the active site region, those which fall outside of this region, but which create an instability in the protein, are also scored.

Limited proteolysis of Ala-tRNA synthetase showed that the N-terminal half of the chain contains the site for aminoacyl adenylate synthesis (Putney *et al.* 1981b). The location of the site for the tRNA-dependent step was not defined by these experiments. To determine whether sequences near the C-terminus were essential for the tRNA-dependent reaction, initial experiments directed clusters of mutations to the carboxyl terminal-coding region of the gene for Ala-tRNA synthetase (*alaS*). This was accomplished by a variation of the site-directed mutagenesis procedure of Shortle *et al.* (1980) and it is illustrated in Fig. 1. A single-stranded fragment of 100–300 nucleotides is isolated from a defined region of the gene and mixed with supercoiled plasmid in the presence of *recA* protein and ATP. The *recA* protein catalyzes single strand uptake by the supercoiled plasmid, through hybridization of the single strand to its complementary sequence in the plasmid (Shibata *et al.* 1979, McEntee *et al.* 1979). This creates a D-loop which in turn is treated with sodium bisulfite. Bisulfite catalyzes deamination of cytosine to give uracil at single-stranded regions (Hayatsu 1976). The treated DNA is transformed into an *ung⁻* *E.coli* strain (deficient in uracil N-glycosidase) (Duncan *et al.* 1978), which enables fixation of mutations. Because reaction of sodium bisulfite with cytosine is pseudo first-order, the extent of deamination (and therefore of mutagenesis) is adjusted by varying the time of bisulfite treatment or the concentration of bisulfite. Independently, another group has taken a similar approach (Peden & Nathans 1982).

In initial experiments, extensive clusters of mutations were introduced into the carboxyl terminal-coding region of *alaS*. The gene-coding region is numbered from +1 to +2625 and a *Pst*I fragment, corresponding to nucleotide numbers 2067 to 2583, was isolated and used for the D-looping procedure. The fragment was treated with exonuclease III (Richardson 1964) to generate single strands of approximately half-length. The mixture or single strands was then D-looped with supercoiled pMJ801 (a pBR322 plasmid with an *alaS* insert) in the presence of *recA* protein, treated with 3 M sodium bisulfite for 4 h (Shortle & Nathans 1978), and then transformed into an *ung⁻* strain. From the clones which contained mutant plasmids, two plasmids were selected for further study.

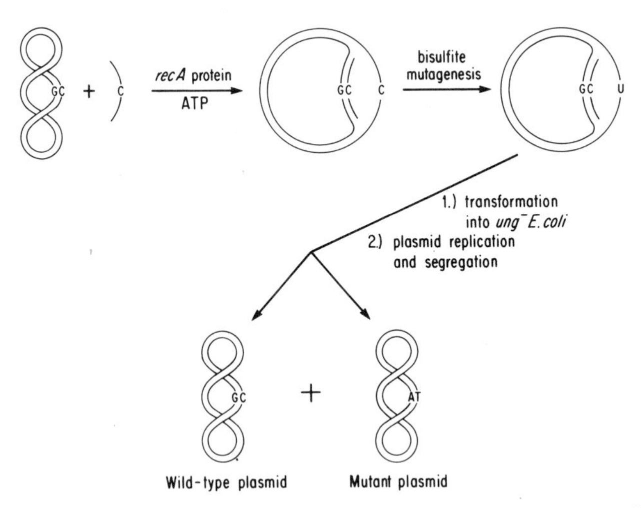

Fig. 1. D-Loop Mutagenesis Procedure. A single-stranded segment, derived from a specific region of the supercoiled plasmid, directs formation of a site-specific D-loop. The D-looped structure is treated with bisulfite which selectively deaminates C residues in single-stranded sections.

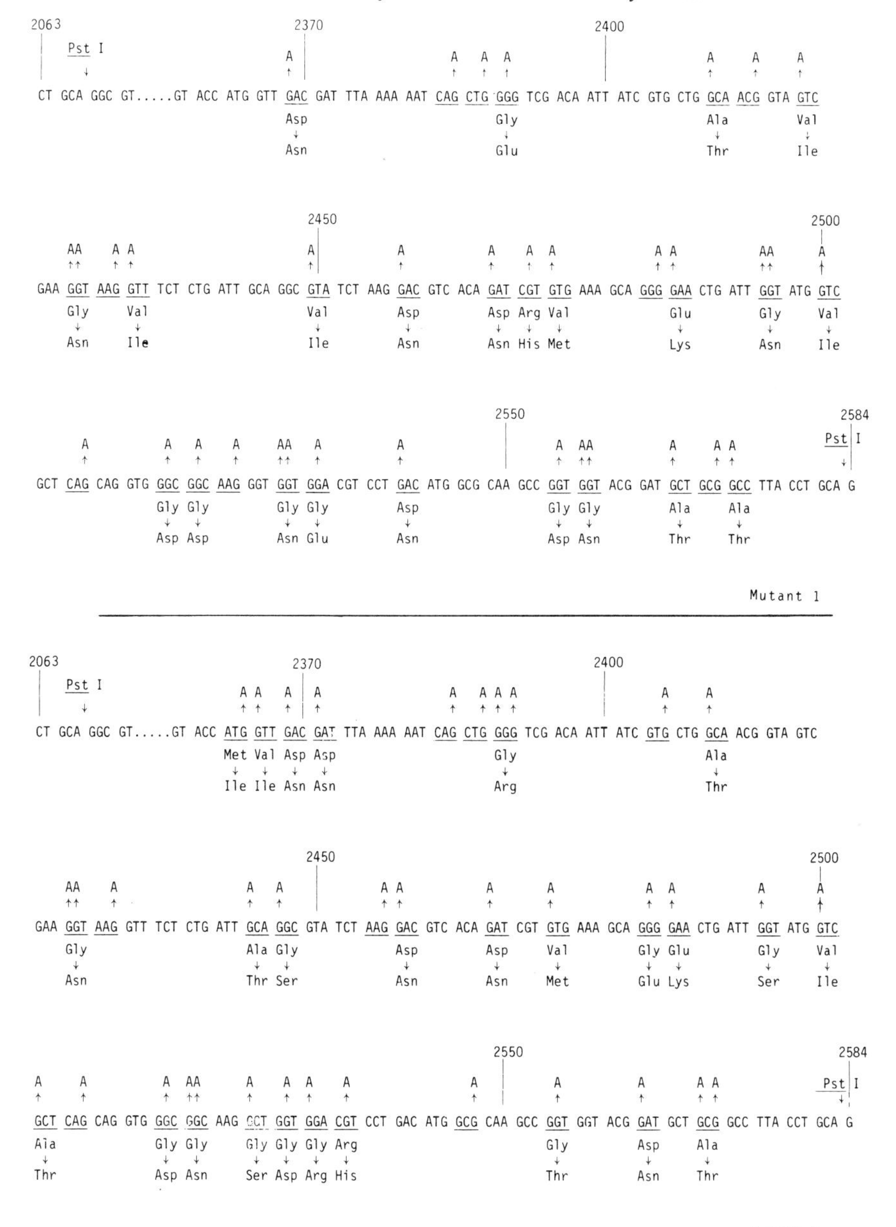

Fig. 2. DNA sequence of two mutants generated by the D-loop procedure shown in Figure 1. Wild-type sequence of the C-terminal *Pst*1 fragment of *alaS* is shown. Nucleotide alterations are shown above the sequence. Predicted amino acid substitutions are shown below the sequence.

The sequence analysis of these 2 mutants is shown in Fig. 2. For mutant 1, there are no base changes from nucleotides 2067 to 2367. All changes occur in the section from 2368 to 2572. Therefore, the D-loop used to produce these mutants was created from a single strand roughly half the size of the *Pst*I fragment. This is expected from exonuclease III treatment. In the altered part of the sequence, there are 35 G to A transitions, and these, in turn, give 23 amino acid substitutions.

The second mutant has base changes in the same section as those for mutant 1,

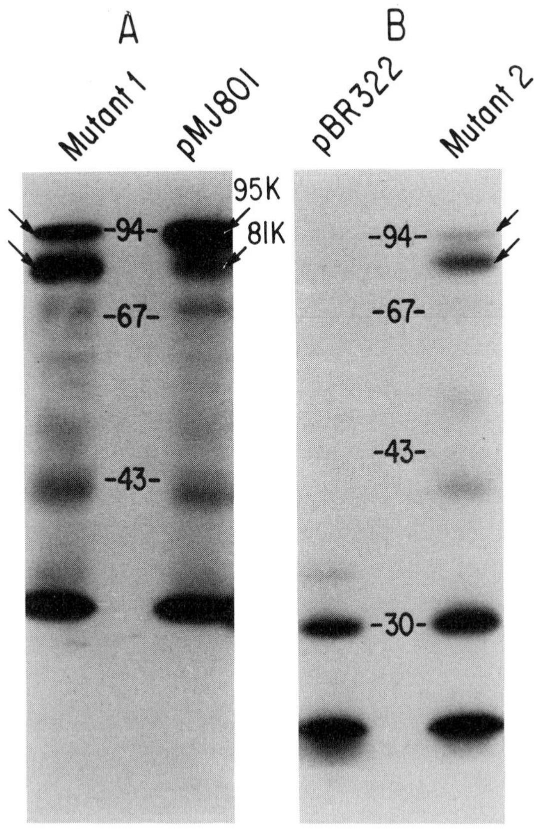

Fig. 3. Autoradiographs of SDS-polyacrylamide gels of maxicell extracts: A. An 8.75% gel of extracts of maxicells containing D-loop mutant 1 and of maxicells containing the parent plasmid pMJ801 (which contains wild-type *alaS*). B. A 10% gel of extracts of maxicells containing pBR322 and of maxicells containing D-loop mutant 2. The upper and lower arrows for the 2 mutants indicate full length mutant protein and a degradation product respectively. The small amount of 81 K degradation product of native enzyme is also indicated.

but these changes are distinct. The first base substitution occurs at nucleotide 2364. Altogether, there are 37 G to A transitions which give 26 amino acid changes.

Synthesis of mutant proteins was visualized by electrophoresis on SDS gels of extracts from maxicells (Sancar *et al.* 1979) which contained the mutant plasmids (Fig. 3). In maxicells containing a plasmid with wild-type *alaS,* the native 95 K protein is overproduced and a small amount is converted to 81 K by proteolysis. Overproduction of mutant proteins also occurs, but in both cases most of the 95 K product is converted to the 81 K protein. Presumably this occurs because multiple substitutions in the carboxyl terminal sequences give a disorganized polypeptide chain in that region.

The catalytic function of these mutant polypeptides was tested *in vivo* by introduction of the mutated plasmids into genetic backgrounds containing 2 different *alaS* mutant alleles. The alleles, *alaS4* and *alaS5,* are temperature-sensitive lethals (Theall *et al.* 1977). They encode enzyme molecules with low activity which cannot sustain cell growth at 42° (the restrictive temperature). In addition, the mutant proteins from *alaS4* and *alaS5* each are partially or fully dissociated, with a mutant subunit molecluar weight identical to that of the native subunit (Theall *et al.* 1977, Jasin *et al.* in preparation). A missense point mutation is believed responsible for each alteration.

Both of the *alaS* mutants produced by D-loop mutagenesis (Fig. 2) complement *alaS4* and *alaS5.* There are 2 possible mechanisms for this complementation. For one mechanism, the mutant plasmid-encoded protein itself has sufficient activity to sustain growth at the restrictive temperature. The other mechanism is *trans*-complementation through heterologous interactions between mutant proteins. For example, *alaS4* and *alaS5* mutant proteins might be partially reactivated through complex formation with an inactive plasmid-encoded mutant protein.

In other studies, *alaS* deletions have been created which preserve the oligomerization function of the enzyme, but remove catalytic sites (Jasin, *et al.* in preparation). These inactive fragments complement *alaS5,* presumably by interaction with, and partial reactivation of, the *alaS5* mutant protein. These same fragments with the native protein do not complement *alaS4,* possibly because the *alaS4* mutation prevents protein-protein interaction between synthetase molecules. For these reasons, complementation of both *alaS4* and *alaS5* by the D-loop mutants (Fig. 2) argues that the mutant proteins themselves have catalytic activity.

As shown in Fig. 3, both mutant proteins are significantly degraded, in crude extracts, to 81 K fragments. The complementation results suggest that these fragments possess sufficient activity to sustain protein synthesis *in vivo*. This has been confirmed by the creation of a series of deletion polypeptides.

GENE DELETIONS

The second approach is to delete sections of the gene-coding region to generate polypeptide fragments. Because the removal of sequences may generate unstable structures, products synthesized from deleted genes are first visualized in maxicell extracts. Those fragments which are stable are then subjected to various *in vitro* and *in vivo* assays.

Fragments of the 875 amino-acid native Ala-tRNA synthetase were made by creation of deletions within *alaS*. These deletions were made *in vitro* on recombinant plasmids which contained *alaS*. The resected plasmids were then introduced into *E.coli* via transformation.

Truncated proteins from 2 different types of deletions were investigated. The first was a nested set of carboxyl terminal deletions which produced polypeptide fragments that ranged in length from 257 to 852 amino acids. The second was a group of 8 internal deletions in which a specific segment of the coding region was removed and flanking gene sections were joined together in the same reading frame.

These protein fragments were tested in 3 different ways. Protein extracts were prepared from plasmid-containing cells and tested for ability of the specific fragments of Ala-tRNA synthetase to synthesize aminoacyl adenylate and to assemble into an oligomeric structure. The third test was to determine whether protein fragments aminoacylate $tRNA^{Ala}$ to a degree sufficient to sustain protein synthesis *in vivo*. For this, gene deletions on the plasmid were introduced via transformation into an *alaS* chromosomal deletion background. Complementation by the gene fragment of the chromosomal deletion mutant is evidence that the fragment sustains protein synthesis *in vivo*.

These studies established locations in the primary structure of residues essential for aminoacyl adenylate synthesis, interaction with tRNA, and oligomerization. A fragment of 385 amino acids is a monomer which synthesizes alanyl adenylate, but does not aminoacylate $tRNA^{Ala}$. Extension of the chain to 461 amino acids gives a polypeptide which not only synthesizes alanyl adenylate but aminoacylates $tRNA^{Ala}$ sufficiently to sustain *in vivo* protein synthesis. This is true in a genetic background in which the chromosomal allele has been deleted, so that cell viability depends entirely upon a fully functional fragment. These results indicate that amino acids 257 to 385 are required for aminoacyl adenylate synthesis and amino terminal residues up to 461 are required for aminoacylation of tRNA.

Amino terminal fragments up to 699 amino acids exist as monomers. In view of results cited above, oligomerization is not required for *in vivo* catalytic function. Extension of the chain to 808 amino acids restores the tetrameric structure. Therefore, residues essential for oligomerization are located between amino acids 699 and 808.

These experiments define the carboxyl terminal boundary of specific functional domains, but they do not define the corresponding amino terminal boundary. Most important is the observation that over 400 amino acids may be removed from the carboxyl terminus and still yield a protein which functions *in vivo.* This supports the concept of a core synthetase, located in the amino terminal half of the protein, to which additional sequences have been fused. One role for these additional sequences is to enable the protein to assemble into a tetramer. The tetramer structure is not required for activity, but other data suggest that the α_4 structure is required for the protein's regulatory role (Putney & Schimmel 1981).

The internal deletions removed variable amounts of sequences within the amino terminal half of the protein. In the largest deletion, 7 amino acids from the native protein's amino terminus were fused to a carboxyl terminal half which begins at amino acid 449. All of these deletions destroy catalytic activity, but retain the previously defined domain required for tetramer assembly. When synthesized *in vivo,* these fragments were shown to interact in *trans* with an *alaS* polypeptide produced from the *alaS5* mutant allele. This showed that the functional domain for oligomerization can act independently of that required for core enzyme activity.

CONCLUDING REMARKS

The 2 approaches outlined here are a general way to map functional domains in large proteins, and overcome the limitations of mapping procedures based on partial proteolysis of native enzyme molecules. The success of these approaches, as applied to Ala-tRNA synthetase is in part dependent upon the inherent design of the enzyme itself. In this instance, the enzyme is a modular arrangement of functional domains and removal of one domain does not prevent function of another. In a system where amino acids at one location interact with those 200 or 300 amino acids further along in the sequence, it would be difficult to obtain definitive answers through deletion mutagenesis, because internal deletions, located between the interacting segments, would change the relative positions of those interactions and most likely disrupt them. However, such examples may be relatively rare since functional domains in proteins may typically encompass 50–150 amino acids.

ACKNOWLEDGEMENT

This work was supported by Grant No. GM23562 from the National Institutes of Health. Helpful advice and assistance with this work was received from Lynne Regan.

REFERENCES

Barker, D. G., Ebel, J.-P., Jakes, R. & Bruton, C. J. (1982) *Eur. J. Biochem. 127,* 449–457.
Cassio, D. & Waller, J.-P. (1971) *Eur. J. Biochem. 20,* 283–300.
Duncan, B. K., Rockstroh, P. A. & Warner, H. R. (1978) *J. Bacteriol. 134,* 1039–1045.
Hayatsu, H. (1976) *Prog. Nuc. Acid. Res. Mol. Biol. 16,* 75–124.
Jasin, M., Regan, L. & Schimmel, P. (1983) *Nature* (in press).
Koch, G. L. E. & Bruton, C. J. (1974) *FEBS Letters 40,* 180–182.
McEntee, K., Weinstock, G. M. & Lehman, I. R. (1979) *Proc. Nat. Acad. Sci. USA 76,* 2615–2619.
Ofengand, J. (1977) In: *Molecular Mechanisms of Protein Biosynthesis* (Weissbach, H. & Pestka, S., eds.), pp. 7–79, Academic Press, New York.
Peden, K. W. C. & Nathans, D. (1982) *Proc. Nat. Acad. Sci. USA 79,* 7214–7217.
Porter, R. R. (1973) In: *MTP International Review of Science, Defense and Recognition, Biochemistry Series One 10* (Porter, R. R., ed.), pp. 159–197, Butterworth, London.
Putney, S. D. & Schimmel, P. (1981) *Nature 291,* 632–635.
Putney, S. D., Sauer, R. T. & Schimmel, P. R. (1981a) *J. Biol. Chem. 256,* 198–204.
Putney, S. D., Royal, N. J., Neuman de Vegvar, H., Herlihy, W. C., Biemann, K. & Schimmel, P. (1981b) *Science 213,* 1497–1501.
Rich, A. & RajBhandary, U. L. (1976) *Ann. Rev. Biochem. 45,* 805–860.
Richardson, C. C., Lehman, I. R. & Kornberg, A. (1964) *J. Biol. Chem. 239,* 251–258.
Sancar, A., Hack, A. M. & Rupp, W. D. (1979) *J. Bacteriol. 137,* 692–693.
Schimmel, P. R. & Söll, D. (1979) *Ann. Rev. Biochem. 48,* 601–648.
Shibata, T., DasGupta, C., Cunningham, R. P. & Radding, C. M. (1979) *Proc. Nat. Acad. Sci. USA* 76, 1638–1642.
Shortle, D., Koshland, D., Weinstock, G. M. & Botstein, D. (1980) *Proc. Nat. Acad. Sci. USA 77,* 5375–5379.
Shortle, D. & Nathans, D. (1978) *Proc. Nat. Acad. Sci. USA 75,* 2170–2174.
Theall, G., Low, K. B. & Söll, D. (1977) *Mol. Gen. Genet. 156,* 221–227.
Webster, T. A., Gibson, B. W., Keng, T., Biemann, K. & Schimmel, P. (1983) *J. Biol. Chem. 258,* 10637–10641.
Yamao, F., Inokuchi, H., Cheung, A., Ozeki, H. & Söll, D. (1982) *J. Biol. Chem. 257,* 11639–11643.

DISCUSSION

GRUNBERG-MANAGO: When you look at mutants which do not oligomerize, are there effects on the synthesis of the alanine tRNA synthetase?

SCHIMMEL: The problem is as follows: in order to repress *alaS* transcription, alanine is required as a corepressor. We believe that, under the conditions in which we grow *E. coli*, the gene is derepressed. *In vitro* experiments suggest that you have to go above the Km for alanine (which is about 2 mM) in order to achieve autogeneous repression. The intracellular concentration of alanine in *E. coli* is estimated at about 1 mM, which suggests that the native enzyme is probably at redepressed levels. The fragments (which do not oligomerize) presumably do not repress and would have no strong effect on observed levels. To test further these ideas may require strains which overproduce alanine so that substantial repression of native enzyme synthesis can be demonstrated *in vivo*. We are unaware of any strains with this characteristic.

KURLAND: You are stressing the fact that you have a tandem arrangement of domains. Is there any other way of arranging them except in tandem?

SCHIMMEL: The point is that the domains are organized from tandem sections of the amino acid sequence. There is no reason to believe that, *a priori*, this should be the case. For example, the carboxyl terminus of Ala-tRNA synthetase might combine with residues from the amino terminus to form a domain. Similarly, other domains might also be fabricated from segments far apart in the sequence. In the case of Ala-tRNA synthetase, we have a long (875 amino acids) polypeptide which has domains arranged linearly along the sequence. This arrangement in turn leads to a simple explanation for the size polymorphism of aminoacyl tRNA synthetases.

EBEL: What is the proof that the tetrameric structure is needed for the recognition of the gene by the amino acid-enzyme complex?

SCHIMMEL: The proof is incomplete. However, we have tested *in vitro* a fragment which is monomeric and it does not repress.

OFENGAND: There are residues at the carboxyl end that do not seem to have any function, yet. Do you want to speculate on whether they are just superfluous or whether they are important for something as yet to be determined?

SCHIMMEL: That is an interesting question. The *in vitro* aminoacylation activity drops considerably upon removal of some of the carboxyl terminal sequence. This effect occurs even with the removal of just 23 amino acids from the C-terminus to give an 952 amino acid N-terminal fragment. This fragment forms a stable tetramer, so that the drop in activity cannot be ascribed to effects on subunit association. Although the catalytic determinants have been mapped to the first 461 amino acids, there must be communication between these determinants and C-terminal sequences.

MAALØE: You said you might say something about the yields of the various fragments.

SCHIMMEL: There is significant variability in the amounts which are synthetized. For example, the N-terminal 699 amino acid fragment is abundantly synthesized but the 695-mer is barely detectable. Has the removal of 4 amino acids opened up the end or a domain, so that cellular proteases can attack more readily? The surprising observation is that the 695-mer is not degraded to a stable intermediate size, such as the 48 kd N-terminal fragment which can be produced *in vitro* by proteolysis. Does this mean that, *in vivo*, at least some proteolysis pathways are processive? Is there a uniquitine-like system in *E. coli* which marks protein for degradation?

ROSENBERG: We have had some similar results looking at the λ-phage C11 protein. We have introduced many point changes, some of which cause the protein to be turned over in *E. coli* with great rapidity.

PESTKA: I would like you to elaborate on the carboxyl terminal deletions with respect to their *in vivo* and *in vitro* activity.

SCHIMMEL: Fragments which aminoacylate tRNA *in vivo* generally show the level of *in vitro* aminoacylation activity. We have investigated extracts containing some of these proteins and compared their activities with those estimated from an "*in situ*" aminoacylation assay. The "*in situ*" assay is done by treating cells with toluene (to permeabilize them) and diffusing in the substrates (including tRNA). This assay gives higher activity than when we do a standard *in vitro* assay with extracts. This suggests that the fragments are unstable and start to lose activity upon preparation of extracts.

PESTKA: I would like to comment on that. We have done experiments with interferon in which we have removed some carboxyl terminal amino acids that are not necessary for activity. The truncated protein is degraded rapidly in *E. coli* even though it has (as far as we can tell) complete function.

ROSENBERG: With your internal deletions, you have constructions which still give tetramer formation. Do you have any evidence whether those tetramers still bind DNA?

SCHIMMEL: Not yet. From the standpoint of structural organization, there are some interesting possibilities. In the λ and *lac* repressors, the amino terminal parts bind to DNA and the carboxyl terminal segments assemble the tetrameric structures. You might speculate that, in the synthetase, there is a similar organization, except that we place the catalytic site for aminoacylation activity between the 2 parts. If this is true, then some of the amino terminal deletions may assemble into a tetramer, but they may not bind to DNA. We are investigating that.

Structural and Functional Studies on the Anticodon and T-loop of tRNA

M. Sprinzl, B. Helk & U. Baumann

THE PRIMARY STRUCTURE OF tRNA AND ITS INVARIANT FEATURES

The number of sequenced tRNAs increased very rapidly in the past few years. This is mainly due to the progress in RNA-sequencing methods, as well as to the available technology for the isolation of DNA fragments containing tRNA genes and their sequencing. Compared to recent reviews on tRNA (Rich & RajBhandary 1976, Clark 1978), the new sequence data, however, did not provide additional information to the accepted generalized cloverleaf structure of tRNA (Fig. 1). Two main objectives obviously influenced the selection of the tRNA species, which were sequenced (Gauss & Sprinzl 1983). First, since the tRNA sequences are suitable for phylogenetic studies, recently special attention has been turned to organisms providing tRNA sequences relevant to this problem. For instance, a number of tRNAs from different archaebacteria were recently sequenced and sequencing of tRNAs from cell organelles was also at the centre of interest. Second, the sequences of many mammalian tRNAs were determined in order to study the relation between the distribution of isoacceptors and tRNA modification and the cell differentiation, transformation or growth rate.

Although about 650 tRNAs and tRNA genes have been sequenced up to summer 1983, there is still no single cell type in which the sequences of all tRNA isoacceptors are known. The quantitative data on the concentrations of particular tRNAs, information which is important for understanding the regulatory mechanism of the translational apparatus (Konigsberg & Godson 1983), are still not available, even in the case of the most intensively studied organisms as *E. coli* or yeast (Ikemura 1981).

Universität Bayreuth, Laboratorium für Biochemie, Bayreuth, Germany

GENE EXPRESSION, Alfred Benzon Symposium 19.
Editors: Brian F. C. Clark & Hans Uffe Petersen, Munksgaard, Copenhagen 1984.

The possibility of folding the tRNA sequences into a generalized cloverleaf structure, allows their alignment according to functional and structural domains and a statistical evaluation of the occurrence of nucleosides at a particular position (Schimmel *et al.* 1979). The generalized structure is compatible with all tRNA sequences starting from bacterial tRNAs up to tRNAs originating from mammalian cells. The only exceptions to this rule are the tRNAs from mammalian mitochondria, some of which cannot be fitted into the generalized scheme shown in Fig. 1, indicating that the translational machinery in the mammalian mitochondrium is more divergent from that in its own cytoplasma than that of the mammalian cytoplasma from the one in bacteria. It is interesting

Fig. 1. Cloverleaf structure of tRNA with the numbering system adopted at the Cold Spring Harbor meeting on tRNA held in 1978 (Schimmel *et al.* 1979). The variable positions 47:1 to 47:16 were omitted in this drawing.

that the tRNAs isolated from mitochondria of lower eukaryotes and plants, although often having unusual features which are not found in elongator tRNAs

Table I

Occurrence of nucleotides in particular positions of tRNA sequences[1)]

Pos.[3)]	Occurrence[2)]	Number of mod. nucleosides[4)]
Aminoacylation stem (5')		
1	179 20 / 33 24	2
2	105 94 / 18 40	-
3	81 106 / 27 40	-
4	99 59 / 39 45	10
5	102 69 / 49 37	-
6	81 62 / 47 67	9
7	121 9 / 74 53	1
8	1 1 / 4 251	47
9	97 10 / 143 6	63
Dihydrouridine stem (5')		
10	227 8 / 12 9	67
11	10 149 / 11 86	-
12	71 56 / 21 100	14
13	34 120 / 26 73	38
D-Loop		
14	1 0 / 245 8	4
15	170 3 / 78 3	8
16	5 52 / 16 177	105
17	2 33 / 7 96	63
18	235 0 / 5 3	62
19	238 0 / 0 3	1
20	18 32 / 21 170	157
20:1	5 21 / 20 67	52
20:2	3 8 / 9 14	8
21	17 0 / 222 3	3
Dihydrouridine stem (3')		
22	136 4 / 82 31	3
23	54 71 / 108 21	0
24	158 9 / 77 10	0
25	7 188 / 5 56	4
26	125 12 / 102 17	81
Anticodon stem (5')		
27	34 96 / 34 92	61
28	46 88 / 40 82	23
29	80 41 / 74 51	-
30	180 53 / 14 9	-
31	54 75 / 95 32	13
Anticodon loop		
32	1 173 / 1 81	83
33	0 8 / 0 248	
34	——	
35	32 56 / 104 64	4
36	41 65 / 73 77	2
37	253 2 / 1	176
38	7 55 / 168 26	31
Anticodon stem (3')		
39	79 55 / 23 99	96
40	54 172 / 5 25	14
41	41 82 / 51 73	-
42	113 27 / 76 40	-
43	99 22 / 92 43	-
Extra arm		
44	45 25 / 123 63	7
45	167 11 / 39 39	1
46	145 14 / 64 30	-
47	15 28 / 7 141	73
48	3 177 / 7 69	65
T-stem (5')		
49	109 80 / 55 12	48
50	65 95 / 38 58	8
51	107 48 / 62 39	
52	188 48 / 49 13	1
53	246 1 / 8 1	-
T-loop		
54	0 3 / 16 237	209
55	6 7 / 4 239	228
56	0 242 / 6 8	16
57	175 0 / 76 5	11
58	1 2 / 250 3	83
59	52 17 / 109 76	1
60	0 41 / 25 184	-
T-stem (3')		
61	1 247 / 2 6	-
62	7 186 / 12 51	-
63	56 105 / 27 68	-
64	98 49 / 54 55	5
65	78 84 / 13 81	4
Aminoacylation stem (3')		
66	10 116 / 52 78	-
67	71 67 / 51 67	2
68	78 86 / 32 61	1
69	73 91 / 46 46	-
70	107 74 / 41 34	-
71	96 91 / 38 31	-
72	11 166 / 30 49	3
73	61 6 / 169 20	-
74	0 256 / 0 0	-
75	0 256 / 0 0	-
76	0 0 / 256 0	-

1) 256 tRNA sequences included into the 1983 compilation, were evaluated. The modified nucleosides were considered in the form of their precursors. In the cases, where the structure of modified nucleosides was not determined, it was assumed that the most abundant nucleoside is modified. Position 34 was not evaluated. In position 37 all modified nucleosides were assumed to be purines.

2) Key: G | C / A | U

3) The numbering of mammalian mitochondrial tRNAs is arbitrary. Other tRNAs are numbered according to $tRNA^{Phe}$ from yeast.

4) The nucleosides of unknown structure are considered to be modified.

isolated from cytoplasma, can be well fitted into the generalized cloverleaf structure. This fact implies a considerable divergence between the translational system in mammalian mitochondrium and the mitochondrium of lower eukaryotes and plants.

As the amount of sequenced tRNAs increased, also more exceptions from the originally postulated "invariant features" in the tRNA secondary structure appeared. According to the data shown in Table I, the only absolute invariant part in 256 tRNA sequences compiled by the end of 1982 (Gauss & Sprinzl 1983) is the 3′-terminal CCA end. A small level of variability occurs at positions 8, 10, 14, 15, 18, 19, 21, 32, 33, 37, 53, 54, 55, 56, 58 and 61.

The semiinvariant positions in tRNA sequences

Position 8 is occupied by uridine and it was suggested that this nucleobase residue is covalently attached to the aminoacyl-tRNA synthetase during the aminoacylation reaction (Starzyk *et al.* 1982). Some mammalian mitochondrial tRNAs and T5 phage-coded tRNAs, however, do not contain an U-8 residue. The nucleoside in position 14 is usually an adenosine. Phage T5-coded $tRNA^{Asp}$ and $tRNA^{Asn}$ have G or U, respectively, at this position. Some mitochondrial tRNAs from yeast, *Neurospora crassa,* or mammalian cells have here uridine or modified uridine. The positions 18 and 19 are occupied by guanosines, the position 18 being modified to a high extent. Some mammalian mitochondrial tRNAs have deletions in this region of the D-loop. The position 18 in some tRNAs from the mitochondria of lower eukaryotes may be occupied by adenosine. Exceptions with respect to invariant features at positions 18 and 19 are also the tRNAs participating in the synthesis of bacterial cell wall. The nucleoside 21 is usually an adenosine. There are, however, frequent exceptions from this rule. $tRNA^{Cys}$ from *E. coli* has a dihydrouridine, $tRNA^{Ile}$ from *H. volcanii* and uridine, $tRNAs^{Leu}$ from *E. coli* and $tRNA^{Met}$ from *T.acidophilum* guanosines at this position. Some tRNAs from organelles of eukaryotic cells have also guanosine at the position 21.

Positions 32, 33 and 37 represent the invariant features in the anticodon loop of the tRNA. Position 32 is always occupied by a pyrimidine, which is frequently modified. Only the $tRNA^{Thr}$ from spinach chloroplast and the $tRNA_f^{Met}$ from yeast mitochondrium are exceptions from this rule. The position 33 is an uridine, exceptions being only some mammalian mitochondrial tRNAs and the initiator tRNAs from the cytoplasma of plants and higher eukaryotic cells which have a cytidine at this position. The position 37 is in most cases a hypermodified purine. tRNAs from *S. epidermis* functioning in cell wall synthesis have a cytidine at this position. A very significant exception concerning the position 37 is the $tRNA^{Ala}$ from mitochondrium of *N. crassa,* in which the purine is replaced by uridine.

The highly conserved features in the T-region of tRNAs consist of a G 53-C 61 base pair and a uridine or modified uridine in position 54 followed by uridine or modified uridine-55, cytidine-56 and purine- (mostly guanosine)-57. Exceptions with respect to positions 53 and 61 are the phage T5-coded $tRNA^{His}$ and some tRNAs from mammalian mitochondrium. In the case of *bombyx mori* $tRNAs^{Ala}$ the position 54 is occupied by adenosine. This is the only exception where an elongator tRNA from cytoplasma has a nucleoside other than uridine or its derivative at this position. Initiator tRNAs isolated from the cytoplasma of eukaryotic cells have here an adenosine. The position 56 is invariably occupied by a cytidine, even in the case of the initiator tRNAs. The exceptions are again the very unusual mammalian mitochondrial species. The same situation applies for the position 57 which is normally occupied by a purine.

Analysing the statistical evaluation of the tRNA sequences as depicted in Table I, it becomes apparent that in addition to the highly conserved nucleosides discussed above many other positions possess an uneven distribution of nucleosides, considerably exceeding the statistical value of 25% for each of the 4 bases. Also positions, which are generally considered as variable are occupied to a high extent by the same base or the same type of base. If tRNAs of the same amino acid specificity, but originating from different organisms, or tRNAs with different specificity from one kingdom of organism (bacteria, plants, mammals) are compared, the amount of the positions in the sequence in which the distribution of nucleosides is far from statistical value is even larger. Several structural features, as G-U, a non-Watson-Crick base pair, composition of base pairs (G-C or A-U), or the polarity of base pairs, purine-pyrimidine (pyrimidine-purine), are not randomly distributed in the secondary structure of the tRNAs. Most probably these properties are important for the stabilization of the tRNA structure. The precise structural or functional meaning of these semi-invariances is not known but there are data in the literature, which indicate that in addition to the basic structural elements mostly determined by the Watson-Crick basepairing and stacking, the polarity and the context of the base pair are also important for the precise tuning of the structure of nucleic acids (Borer *et al.* 1974, Mizuno & Sundaralingam 1978). The comparative studies of tRNA sequences are a very useful tool for identification of these features.

The Modification of tRNAs

The structure of more than 50 modified nucleosides originating from tRNA was determined, and several more modified nucleosides in tRNA were detected, the structure of which remains to be elucidated (Nishimura 1979). The types of modification range from simple methylation to the introduction of hyper-modified bases with complicated side chains. The modified bases in tRNA are not distributed randomly (Table II):

a) Some positions in the sequence are almost always modified in a typical manner,
b) other positions are modified frequently, but with many different types of modified nucleosides and, finally,
c) some positions are very seldom or never modified.

The positions, which are frequently modified with a similar type of modified nucleosides, are especially interesting, as an occurrence of a certain type of modification may suggest a functional feature of the particular part of the tRNA. An example for such a consideration is the 2′-O-methylation. For instance, 2′-O-methylguanosine occurs at position 18 of tRNAs isolated from some eubacteria, archaebacteria or mammalian cells. In one case it could be shown that the extent of modification of G-18 to Gm-18 is directly related to the growth conditions of a thermophilic bacterium (Kumagai *et al.* 1980). 2′-O-Methylguanosine is also present in position 34 of some phenylalanine-specific tRNAs, and it was suggested that this modification increases the stacking interactions and stabilizes the anticodon loop structure in the $tRNA^{Phe}$ from yeast (Maelicke *et al.* 1975, Sprinzl & Cramer 1979). In analogy the 2′-O-methylation of the nucleosides in positions 4, 10, 18, 32, 39, 44, 54, 56 and 64 also most probably facilitates the stacking interactions.

The opposite case is the modification of the uridine to dihydrouridine, which can occur only at a position where stacking interactions are not present. This is related to the non-planar structure of the dihydrouridine ring (Bhanot *et al.* 1977). There are only 5 positions in tRNAs where dihydrouridines are usually placed: 16, 17, 20, 20:1, and 47. In one case, the position 20:1 is occupied by 3-(3-amino-3-carboxypropyl)uridine (acp^3U). The compatibility of the dihydrouridine and acp^3U is even more evident in the case of the position 47, which was found 55 times modified to dihydrouridine and 15 times to acp^3U (Table II). The x-ray structure of $tRNA^{Asp}$ from yeast (Ebel, J.-P., personal communication) as well as that of the $tRNA^{Phe}$ from yeast (Rich & RajBhandary 1976, Clark 1978), reveal that the residue 47 is not stacked but looped-out from the stack formed by the nucleosides 46, 21 and 48. Recently, it was suggested that single bases forming a similar bulge in stacked structures may serve as protein-

Table II

Occurrence of modified nucleosides in 256 sequences of tRNAs (Gauss. Sprinzl. 1983)

Pos.[1]	Modified Nucleosides[2] Aminoacylation stem (5′)
1	Ψ(2)
2	
3	
4	Um(6), Cm(4)
5	
6	m^2G(9)
7	m^2G(1)
8	Ψ(1), s^4U(46)
9	m^1G(37), m^1A(13), m^2G(1), s^4U(2), N(1)
Dihydrouridine stem (5')	
10	m^2_2G(11), m^2G(56), N(1)
11	
12	ac^4C(14)
13	Ψ(39), Cm(1)
Dihydrouridine loop	
14	D(1), m^1A(2), N(1)
15	N(8)
16	D(105)
17	D(63)
17:1	D(2)
18	Gm(61)
19	Gm(1)
20	D(157)
20:1	D(50), X(2)
20:2	D(6), Ψ(2)
21	
Dihydrouridine stem (3')	
22	m^1A(2), N(1)
23	
24	
25	Ψ(3), N(1)
26	Ψ(3), m^2_2G(58), m^2G(18), N(2)

Pos.[1]	Modified Nucleosides[2] Anticodon stem (5′)
27	Ψ(60), N(1)
28	D(1), (2), N(1)
29	
30	
31	Ψ(13)
Anticodon loop	
32	Um(7), s^2C(2), N(3), Cm(36), m^3C(5), m(1) F(29)
33	N(1)
34	o^5U(4), I(22), Gm(12), m^5C(1), mcm^5U(3), Q(8), N(43), mam^5s^2U(3), mcm^5s^2U(5), Cm(7), ac^4C(1), mo^5U(3), $cmnm^5U$(1), $cmnm^5s^2U$(1), s^2U(1), man Q(3)
35	Ψ(4)
36	m^7G(2)
37	m^1G(40), m^1I(4), m^2A(13), t^6A(42), ms^2i^6A(24), i^6A(15), N(19), mt^6A(4), m^6A(7), YW(3) oxyW(4), ms^2t^6A(2)
38	Ψ(19), m^5C(9), N(3)
Anticodon stem (5')	
39	Ψ(88), Gm(2), m^5C(1), N(1), Ψm(4)
40	Ψ(9), m^5C(4), N(1)
41	
42	
43	
Extra arm	
44	Um(7), N(1)
45	Ψ(1)
46	m^7G(105), Ψ(3)
47	D(55), X(15), N(3)
47:1	Ψ(1)
47:2	
47:3	m^3C(2)
48	m^5C(53), N(1)

Pos.[1]	Modified Nucleosides[2] T-stem (5′)
49	m^5C(46), N(2)
50	Ψ(1), m^5C(7)
51	
52	Ψ(1)
53	
T-loop	
54	T(179), Ψ(8), N(3), s^2T(2), Tm(4), $m^1Ψ$(13)
55	Ψ(228)
56	Cm(16)
57	m^2G(9), N(2)
58	m^1A(83), N(1)
59	m^1A(1)
60	
T-stem (3')	
61	
62	
63	
64	Gm(1), N(4)
65	Ψ(3), N(1)
Aminoacylation stem (3')	
66	
67	Ψ(2)
68	Ψ(1)
69	
70	
71	
72	Ψ(1), m^5C(2)
3'-terminal region	
73	N(1)
74	
75	
76	

[1] Numbering according to $tRNA^{Phe}$ from yeast

[2] The number in brackets give the occurence of the particular modified nucleoside in the compiled sequences.

nucleic acid interaction sites of the ribosomal RNAs (Peattie *et al.* 1981). The residue 47 in tRNA may function in a similar way. Indeed a modification of several tRNAs from *E. coli* with fluorescamine leads to facile attachment of the fluorescent label to the acp^3U residue-47 indicating a high accessibility of this modified nucleoside. The modification does not inactivate the tRNA with respect to the enzymatic aminoacylation reaction. The interaction of phenylalanyl-tRNA synthetase from *E. coli* with *E. coli* $tRNA^{Phe}$ carrying a fluorescamine label at position 47 leads to a strong increase in the fluorescence intensity indicating a structural change in the near proximity of this modified nucleoside upon tRNA :phenylalanyl- synthetase complex formation (Sprinzl & Faulhammer 1978).

EFFECT OF MODIFICATION AT POSITION 32 ON THE STRUCTURE AND FUNCTION OF $tRNA^{Arg}$ FROM *E. COLI*

The reactivity of the pyrimidine at position 32

The position 32 of tRNAs is occupied by a pyrimidine nucleoside which is very frequently modified. It was shown by chemical modification studies that the pyrimidine-32, although being a part of a loop structure, is not reactive against some types of reagents, e.g., bisulfite (Sprinzl & Cramer 1979), or methoxyamine (Rhodes 1975). As the product of the addition of bisulfite on the 5,6-double bond of the pyrimidine ring is non-planar and has a chair conformation, it cannot participate in stacking interactions (Bhanot & Chambers 1977). It is conceivable that, if the product of a reaction does not fit into the structure of a macromolecular system, in this case of the anticodon loop of tRNA, the reaction leading to such a product will be inhibited. This situation is observed in the case of the pyrimidine 32-residue. Due to stacking interactions, the uridine-32 in $tRNA^{Ala}$ from yeast is much less reactive towards bisulfite compared with the non-stacked pyrimidines in the same molecule (Bhanot & Chambers 1977). The modification of the cytidine-32 to 2′-O-methylcytidine enhances the stacking and leads to even higher resistance toward bilsulfite (Sprinzl & Cramer 1979).

In $tRNA^{Ser}$ and $tRNA^{Arg}$ from *E. coli* at position 32 2-thiocytidine (s^2C) was found. This modified nucleoside is reactive towards iodoacetamide giving a carbamoylmethyl-2-thiocytidine ($cams^2C$) (Kröger *et al.* 1976). The alkylated product can be converted to cytidine by treatment with 2-mecaptoethanol under mild conditions. As compared with the bisulfite reaction discussed above, the alkylation of the 2-thiocytidine, as well as the nucleophilic replacement of the 2-alkylthio-group do not change the planar character of the nucleobase and consequently these reactions proceed with high yield.

The modification of the pyrimidine-32 affects the structure and decoding properties of tRNA^{Arg}

The possibility of a facile chemical conversion of the modified nucleoside (2-thiocytidine) into its precursor (cytidine) in tRNA^{Arg} from *E. coli* prompted us to investigate the effect of this modification on the structure of the anticodon loop and on the codon-anticodon interaction (Sprinzl *et al.* 1978). The structural effects in the anticodon loops of different tRNAs^{Arg} were monitored by nuclease S1 mapping using sequencing gels. It was demonstrated by Wrede *et al.* (1979), that nuclease S 1 cleaves the initiator tRNAs preferentially at 2 sites 3′ to residues 34 and 35. The elongator tRNAs were cleaved at 4 sites, namely 3′ to the residues 33, 34, 35 and 36 (Fig. 2). This difference in the nuclease S1-cleavage pattern was ascribed to the presence of the 3 G-C base pairs (G29-C41, G30-C40, and G31-C39) adjacent to the anticodon loop of all initiator tRNAs. The tRNA^{Arg} from *E. coli* is an elongator tRNA, which has an anticodon stem with base pairs C29-G41, G30-C40, and G31-C39 very similar to that of tRNA$_f^{Met}$ from *E. coli* (Fig. 2). The S1 cleavage pattern of tRNA^{Arg} shows, as in the case of initiator tRNA, 2 main hits corresponding to cleavage after inosine-34 and cytidine-35, but weak cleavage sites after U-33 and G-36 -typical for elongator tRNAs- are also present. The anticodon loop of the tRNA^{Arg} has, therefore, with respect to S1 nuclease cleavage a structure, which is intermediate between the tRNA$_f^{Met}$ and tRNA$_m^{Met}$ types. This result is in favour of the suggestion that the structure of the anticodon stem has an influence on the conformation of the anticodon loop (Wrede *et al.* 1979).

We extended these studies and investigated the effect of the modification of the residue 32 on the conformation of the anticodon loop. The results summarized in Fig. 2 show that the modification of the C-32 residue strongly affects the accessibility of the anticodon loop for the S1 nuclease cleavage, reflecting a different conformation of the loop in dependence on the modification of this residue.

The different modified tRNA^{Arg} species were also tested in an *in vitro* translation system using the S 30 supernatant from *E. coli* and MS2 RNA as messenger. In this system a production of a coat protein-related polypeptide was described and characterized, which is produced by ribosomal frame-shifting in a well-defined part of the MS 2 message (Atkins *et al.* 1979). The result of a typical

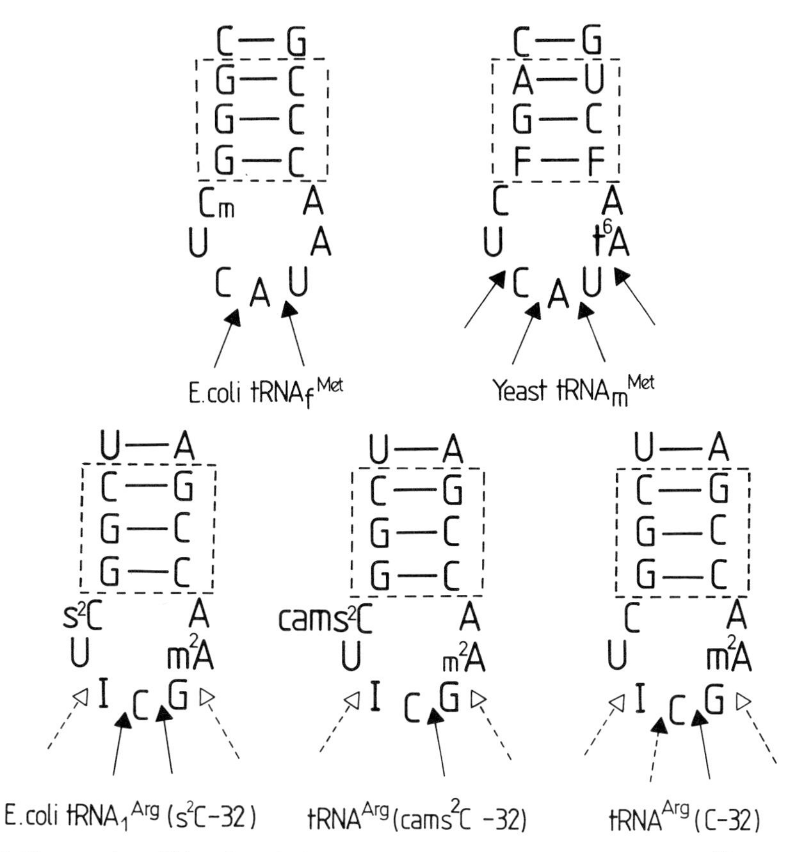

Fig. 2. Cleavage sites of S1 nuclease in the anticodon loops of different tRNAs. The 5′-[^{32}P]-labelled tRNA-species were partially digested with S1 nuclease in a buffer containing 25 mM sodium acetate pH 4.5, 5 mM magnesium chloride, 50 mM potassium chloride, 1 mM zinc acetate, for 1–20 min at 37 °C, with 1 unit enzyme/μg of tRNA. The digestion products were analyzed by slab gel electrophoresis and autoradiography. The different arrows indicate: ——▶ heavy band, -----▶ weak band, - - -▷ very weak band.

experiment from our studies is shown in Fig. 3. The differences in the polypeptide composition in the lines 1, 2, 3 and 4 of Fig. 3 represent the effect of different modification of the C-32 residue of the $tRNA^{Arg}$, which was the only variable in the system. It is clear, that the modification of C-32 to s^2C-32 considerably suppresses the frame shifting and the appearance of the band (arrow) with the molecular weight of 18000, which is the frame-shift product 7 characterized by Atkins *et al.* (1979).

THE INVARIANT REGION IN THE T-LOOP OF ELONGATOR tRNAs

Function of the T-loop of tRNA during translation

The GpTpψpCp region of tRNA occurs in all elongator tRNAs with the exception of the mammalian mitochondrial species and $tRNA^{Ala}$ from *bombyx mori.* The sequence d(GpTpTpCp) was found to be important as an intragenic promotor in eukaryotic tRNA genes (Sharp *et al.* 1981). Furthermore, the role of the TpψpCp-sequence in tRNA for the stabilization of its tertiary structure is well established (Rich & RajBhandary 1976, Goddard 1977, Clark 1978). Does this invariant sequence fulfill some additional function?

It was suggested by several authors that the conserved GpTpψpCp-sequence of elongator tRNAs may function as a binding region, which interacts with some

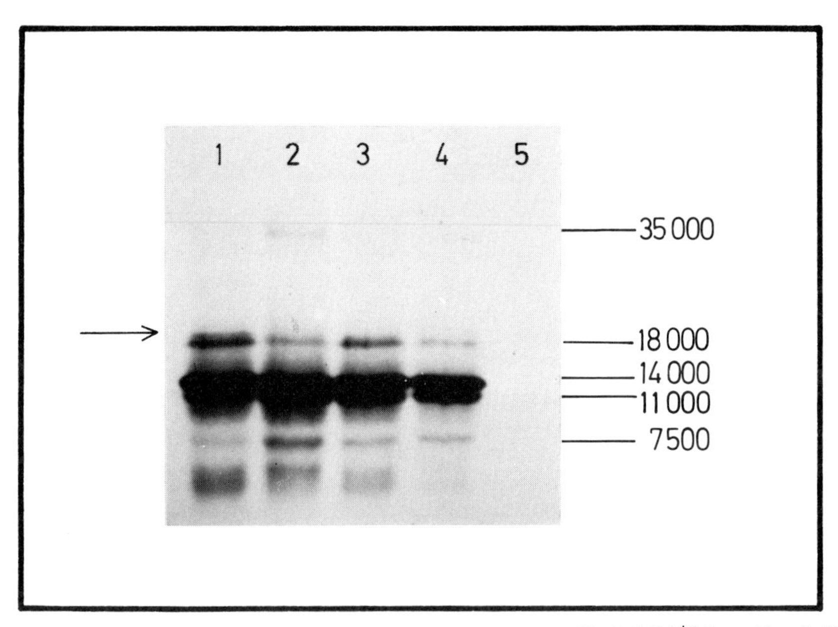

Fig. 3. Biosynthesis of MS2-related proteins with native and modified $tRNA^{Arg}$ from *E. coli.* The translation assay was carried out according to Atkins *et al.* (1979) except that the Mg^{2+}-concentration was lowered to 7.3 mM. The S 30 extract was prepared from *E. coli* K 12 cells. A 25 μl assay contains in addition to the salt and energy mix and the S30 extract: 0.3 A_{260} units MS 2 RNA, 0.5 A_{260} units $tRNA^{bulk}$ from *E. coli,* 11 μCi of [^{35}S] Methionine (1000 Ci/mmole) and lane 1: no tRNA addition, lane 2: 0.4 A units native $tRNA^{Arg}$ (s^2C-32), lane 3: 0.4 A_{260} units $tRNA^{Arg}$ (C-32), lane 4: 0.4 A_{260} units $tRNA^{Arg}$ ($cams^2$ C-32). The translation products were analyzed by SDS-polyacrylamide gelelectrophoresis according to Laemmli (1970). The lane 5 is a control without MS 2 messenger RNA. The arrow indicates the band 7 according to Atkins (1979) corresponding to a frame-shift product.

ribosomal RNA segment (Forget & Weissman 1967, Ofengand & Henes 1969, Shimizu *et al.* 1970). Indirect evidence was in support of an interpretation that a CpApApGp sequence occurring in all ribosomal 5S RNA species is the complementary binding site for the GpTpψpC sequence of the tRNA (Erdmann *et al.* 1973). However, it was not possible to provide conclusive evidence for this suggestion. In contrary, the recent progress made in the elucidation of the secondary structure of 5S RNA (Delihas & Andersen 1982) as well as the work with the ribosomes modified at the putative tRNA binding site of the 5S RNA (Pace *et al.* 1982), or experiments with tRNAs modified in the T-loop (Yarus & Breeden 1981), are not in support of the above hypothesis. Nevertheless, the potential of the TpψpCpGp oligonucleotide excised from tRNA to interact with ribosomes is well established (Sprinzl *et al.* 1976). It is, however, possible that the interaction does not take place between the T-loop of tRNA and 5S RNA, but with other ribosomal RNA or ribosomal proteins. Recently it was shown that the modification of the residue 54 in tRNA is related with the efficiency and fidelity of translation, and it is likely that the high extent of modification of the residues 54 and 55 as well as the modification of the invariant cytidine 56 facilitate the tRNA-ribosomal RNA interaction (Roe & Tsen 1977, Kersten *et al.* 1981). The types of modification at positions 54 (T, s^2T, Tm, $m^1\Psi$) and 56 (Cm) are in favour of this suggestion, as they are expected to promote stacking interactions and to facilitate the formation of RNA-RNA hybrids. On the other hand, if the tRNA retains its tertiary structure during the protein elongation steps, then the T-loop modification can serve mostly to stabilize the tertiary interactions. Indeed it could be shown with NMR studies, that the tRNA modified in the T-loop is more stable than the unmodified species and that the modification at the residue 54 may influence the structure of the anticodon region of tRNA (Davanloo *et al.* 1979). On the other hand there is supportive evidence from experiments performed with isolated tRNA in the absence of ribosomes indicating that upon codon-anticodon interaction the tertiary interactions in the tRNA can be replaced by the intermolecular interactions of its T-loop with complementary oligonucleotides (Möller *et al.* 1979, Labuda & Pörschke 1980).

Basepairing potential of the T-loop of tRNA

The ability of the conserved GpTpψpCp sequence in the T-loop of tRNA to participate in RNA-RNA interactions was not directly demonstrated. The attempts to bind complementary oligonucleotides to this region were not successful since intramolecular interactions in the tRNA render the T-loop inaccessible to intermolecular interactions (Pongs *et al.* 1976). Recently, we performed some experiments which demonstrate that the structure of the seven-membered T-loop of tRNA is different from that of the seven-membered anticodon loop. Whereas, in the anticodon loop, the bases 34, 35 and 36 located in the middle of the loop represent the anticodon and participate in codon-

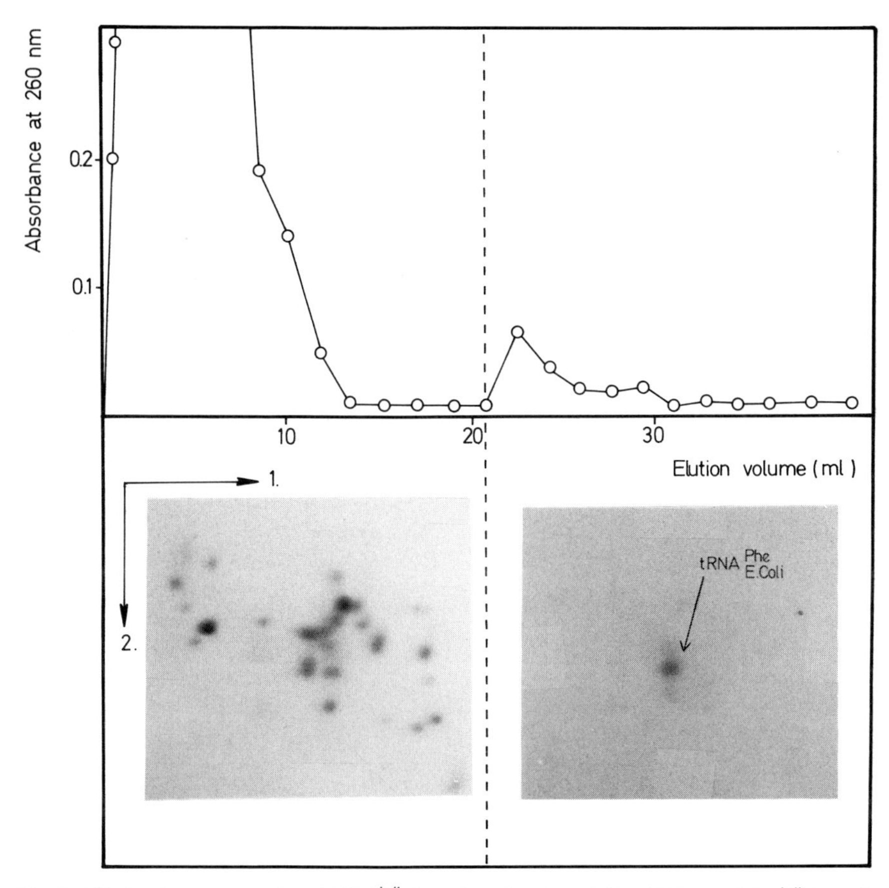

Fig. 4. Affinity chromatography of $tRNA^{bulk}$ from *E. coli* on immobilized T-loop. $tRNA^{bulk}$ from *E. coli* (1.0 mM, 150 μl) was applied onto the column of immobilized $tRNA^{Phe}$ (47–76) fragment. The column was eluted first at 4 °C with a buffer containing 10 mM sodium acetate pH 4.8, 1.0 M sodium chloride and 10 mM magnesium chloride, then at 30°C (indicated by a dashed line) with the same buffer, in which the magnesium chloride was replaced by 10 mM EDTA. The fractions were analyzed by two-dimensional electrophoresis according to Fradin *et al.* (1975), left: the break-through fraction; right: the retarded fraction. The $tRNA^{Phe}$ was identified by sequencing according to Donis-Keller *et al.* (1977).

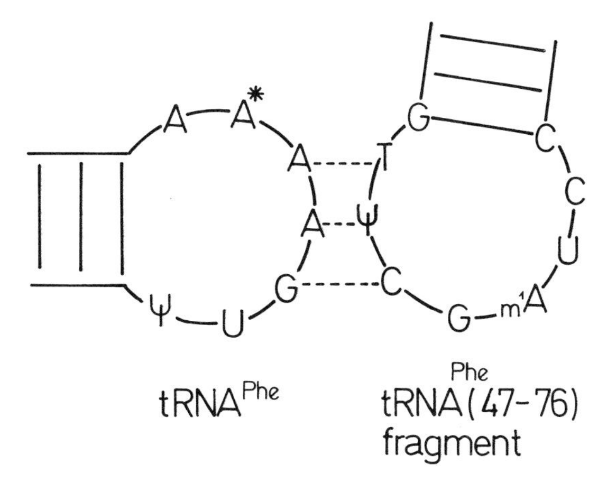

Fig. 5. Proposed interaction of the anticodon loop of $tRNA^{Phe}$ from *E. coli* with the T-loop of the $tRNA^{Phe}$ (47–76) fragment from yeast. A* = 2-methylthio-N6-isopentenyladenosine.

anticodon interactions, in the case of the seven-membered T-loop the bases located on its 5′-side, 54, 55 and 56, are most efficient in RNA-RNA interactions. This could be demonstrated by affinity chromatography experiments, in which an immobilized (47–76) fragment of $tRNA^{Phe}$ from yeast served as a model of a tRNA with a free T-loop. The hairpin structure of this fragment was proved by S1 nuclease mapping. If a mixture of tRNAs from *E. coli* was passed through this affinity column only the $tRNA^{Phe}$ with the anticodon GpApAp, complementary to the TpψpCp-sequence of the immobilized T-loop, was retained (Fig. 4). This experiment demonstrates the potential of the invariant nucleosides at positions 54, 55 and 56 of the tRNA to participate more efficiently in intermolecular RNA-RNA interactions than the other nucleosides in the T-loop (Fig. 5). It is interesting to note that there is a certain homology in the invariant features of the anticodon and T-loop of tRNA concerning the pyrimidine nucleotides located at the 5′-side of the seven-membered loop. Nevertheless, when the interaction of the anticodon loops of intact tRNAs was investigated an involvement of these pyrimidine residues in RNA-RNA interactions was not noticed (Grosjean *et al.* 1976). Therefore, it can be concluded that the T-loop of tRNA possesses a conformation different from the conformation of the anticodon loop, which favours participation of the invariant pyrimidines in intermolecular interactions with complementary RNA sequences.

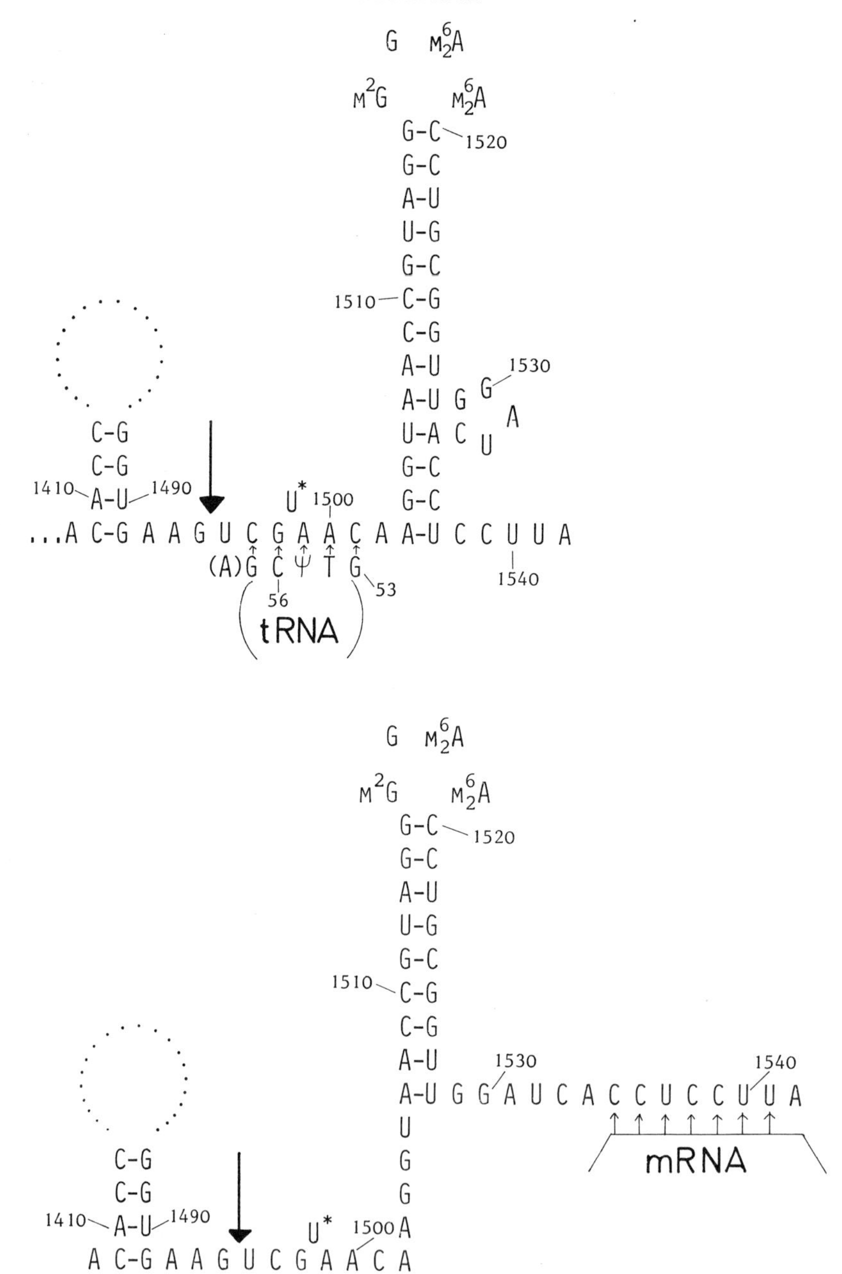

Fig. 6. Alternative structures of the 3′-end of the 16S ribosomal RNA from *E. coli* according to Wickstrom (1983). Top: the structure with 2 adjacent double-stranded regions. The potential complementary binding site of the 53-57 region of tRNA to 1496-1501 region of 16S RNA is indicated by small arrows. The large arrow depicts the Colicin E 3 cleavage site. Looping-out the

The RNA of the small ribosomal subunit is a possible binding site for the interactions with the T-loop of tRNA

There is a very good evidence showing that during the protein elongation the anticodon loop of the tRNA is placed in the vicinity of the 3′-end of the 16S rRNA in *E. coli* ribosomes (Prince *et al.* 1982). Furthermore, there are several reports indicating that the binding of the aminoacyl tRNA to ribosomes has an influence on the accessibility of the 3′-end of 16S RNA towards enzymatic cleavage by nucleases (Kaufmann & Zamir 1973).

In search of a sequence in ribosomal RNAs which is complementary to the GpTpψpCpG(A)pA(m^1A)p sequence of tRNA we found that the sequence in the region 1495-1501 of the ribosomal RNA from the small subunit has such complementarity. This part of 16S RNA forms a single stranded region containing a sequence UpCpGpUpApApCp which is connecting 2 helical regions of the molecule. These 3 structural features, namely the 2 helical parts connected with a single stranded stretch of RNA, are typical for the 3′-end of all ribosomal RNAs from the small ribosomal subunit (Stiegler *et al.* 1981). Furthermore, a structure of this type is well suited to participate in intermolecular interactions since the new base pairs will extend and finally connect the 2 pre-existing stacked helical regions. In the case of bacterial ribosomes it will, however, depend on the alternative secondary structures (Fig. 6) of the 16S RNA, if such stabilization can take place. It was shown that the 16S RNA has during initiation a structure in which the stem region near the 3′-end is partially unfolded (Wickstrom 1983). Following the above argument the interaction of the T-loop of tRNA with the putative region of the 16S RNA would be less probable during the involvement of the 30S ribosomal subunit in the initiation of protein synthesis (Fig. 6).

The complementarity between the GpTpψpCpG(A)p region of tRNA and UpCpGpUp ApApCp region of 16S RNA of *E. coli* ribosomes is not perfect due to the modified uridine residue 1498, which is most probably a derivative of 3-alkyluridine (Johnson & Horowitz 1971). In analogy with a similar modification of tRNA at position 47, where a 3-alkyluridine (acp^3 U) is often present, it can be assumed that the U* residue in 16S RNA is like the acp^3U residue in tRNA

---

modified uridine (U*) 1498 and the adenosine 1502 will allow a stacking of the tRNA: 16S RNA hybrid to the 3′-stem of the 16S RNA. Bottom: This is not possible with the structure of the 16S RNA, which is probably present during the initiation (Wickstrom 1983), where the m-RNA interacts with the 3′-terminal part of the 16 S RNA.

looped-out, if a stacked structure is formed. It is obvious that the 3-alkyluridine cannot be involved in base pairing due to its modification, but on the other hand, as suggested recently, a single nucleoside forming a bulge in a stacked structure may serve as an attachment site for proteins (Peattie *et al.* 1981).

The presented hypothesis is based mainly on sequence comparison studies performed with the EMBO Database sequences. The experimental work, which is at present in progress in our laboratory favours morc thc hypothcsis involving the interaction between the 16S RNA and the T-loop of the tRNA as depicted in Fig. 6 than the previous model involving 5S-RNA: tRNA interaction (Erdmann *et al.* 1973).

REFERENCES

Atkins, J. F. Gesteland, R. F., Reid, B. R. & Anderson, C. W. (1979) *Cell 18,* 1119–1131.

Bhanot, O. S. & Chambers, R. W. (1977) *J. Biol. Chem. 252,* 2551–2559.

Borer, P. N., Dengler, B., Tinoco, I. & Uhlenbeck, O. C. (1974) *J. Mol. Biol. 86,* 843–853.

Clark, B. F. C. (1978) In: *Transfer RNA,* S. Altman, Ed., MIT Press Cambridge, Massachussetts and London, England, pp. 14–47.

Davanloo, P., Sprinzl, M. & Cramer, F. (1979) *Biochem. 18,* 3189–3199.

Delihas, N. & Andersen, J. (1982) *Nucl. Acids Res. 10,* 7323–7344.

Donis-Keller, H., Maxam, A. M. & Gilbert, W. (1977) *Nucl. Acids Res. 4,* 2527–2538.

Erdmann, V. A., Sprinzl, M. & Pongs, O. (1973) *Biochem. Biophys. Res. Commun. 54,* 942–948.

Fradin, A., Gruhl, H. & Feldmann, H. (1975) *FEBS-Lett. 50,* 185–189.

Forget, B. G. & Weissman, S. M. (1967) *Science 158,* 1695–1699.

Gauss, D. & Sprinzl, M. (1983) *Nucl. Acids Res. 11,* r1-r103.

Goddard, J. P. (1977) *Progr. Biophys. Molec. Biol. 32,* 233–308.

Grosjean, H., Söll, D. G. & Crothers, D. M. (1976) *J. Mol. Biol. 103,* 499–519.

Ikemura, T. (1981) *J. Mol. Biol. 146,* 1–21.

Johnson, J. D. & Horowitz, J. (1971) *Biochim. Biophys. Acta 247,* 262–279.

Kaufmann, Y. & Zamir, A. (1973) *FEBS-Lett. 36,* 277–280.

Kersten, H., Albani, M., Mannlein, E., Praisler, R., Wurmbach, P. & Nierhaus, K.-H. (1981) *Eur. J. Biochem. 114,* 451–456.

Konigsberg, W. & Nigel Godson, G. (1983) *Proc. Natl. Acad. Sci. USA 80,* 687–691.

Kröger, M., Sprinzl, M. & Cramer, F. (1976) *Liebigs Ann. Chem.,* 1395–1405.

Kumagai, I., Watanabe, K. & Oshima, T. (1980) *Proc. Natl. Acad. Sci. USA 77,* 1922–1926.

Labuda, D. & Pörschke, D. (1980) *Biochem. 19,* 3799–3805.

Laemmli, U. K. (1970) *Nature 227,* 680–685.

Maelicke, A., von der Haar, F., Sprinzl, M. & Cramer, F. (1975) *Biopolymers 14,* 155–171.

Mizuno, H. & Sundaralingam, M. (1978) *Nucl. Acids Res. 5,* 4451–4461.

Möller, A., Wild, U., Riesner, D. & Gassen, H. G. (1979) *Proc. Natl. Acad. Sci. USA 76,* 3266–3270.

Nishimura, S. (1979) In: *Transfer RNA: Structure, Properties and Recognition.* Schimmel, P. R., Söll, D. & Abelson, J. N. (Eds.), Cold Spring Harbour Laboratory, pp 59–79.

Ofengand, J. & Henes, C. (1969) *J. Biol. Chem. 244,* 6241–6253.

Pace, B., Matthews, E. A., Johnson, K. D., Cantor, Ch. R. & Pace, N. R. (1982) *Proc. Natl. Acad. Sci. USA 79,* 36–40.
Peattie, D. A., Douthwaite, S., Garret, R. A. & Noller, H. F. (1981) *Proc. Natl. Acad. Sci. USA 78,* 7331–7335.
Pongs, O., Wrede, P., Erdmann, V. A. & Sprinzl, M. (1976) *Biochem. Biophys. Res. Commun. 71,* 1025–1033.
Prince, J. B., Taylor, B. H., Thurlow, D. L., Ofengand, J. & Zimmermann, R. A. (1982) *Proc. Natl. Acad. Sci. USA 79,* 5450–5454.
Rhodes, D. (1975) *J. Mol. Biol. 94,* 449–460.
Rich, A. & RajBhandary, U. L. (1976) *Annual Review of Biochem. 45,* 805–860.
Roe, B. A. & Tsen, H.-Y. (1977) *Proc. Natl. Acad. Sci. USA 74,* 3696–3700.
Schimmel, P. R., Söll, D. & Abelson, J. N. (1979) In: *Transfer RNA: Structure, Properties and Recognition,* Cold Spring Harbor Laboratory, pp. 518–519.
Sharp, S., de Franco, D., Dingermann, T., Farrell, P. & Söll, D. (1981) *Proc. Natl. Acad. Sci. USA 78,* 6657–6661.
Shimizu, N., Hayashi, H. & Miura, K. (1970) *J. Biochem. 67,* 373–387.
Sprinzl, M., Wagner, T., Lorenz, S. & Erdmann, V. A. (1976) *Biochem. 15,* 3031–3039.
Sprinzl, M., Siboska, G. E. & Pedersen, J. A. (1978) *Nucl. Acids Res. 5,* 861–877.
Sprinzl, M. & Faulhammer, H. F. (1978) *Nucl. Acids Res. 5,* 4837–4853.
Sprinzl, M. & Cramer, F. (1979) *Progr. Nucl. Acid Res. Mol. Biol. 22,* 1–69.
Starzyk, R. M., Koontz, S. W. & Schimmel, P. (1982) *Nature 298,* 136–140.
Stiegler, P., Carbon, Ph., Ebel, J.-P. & Ehresmann, Ch. (1981) *Eur. J. Biochem. 120,* 487–495.
Wickstrom, E. (1983) *Nucl. Acids Res. 7,* 2035–2052.
Wrede, P., Wood, N. H. & Rich, A. (1979) *Proc. Natl. Acad. Sci. USA 76,* 3289–3293.
Yarus, M. & Breeden, L. (1981) *Cell 25,* 815–823.

DISCUSSION

GARRETT: Do you really have any evidence for the interaction that you show in Fig. 5? You are assuming that the tRNA structure is conserved in the tRNA fragment?

SPRINZL: Yes, by partial digestion with nuclease Sl we can demonstrate that the secondary structure of the $tRNA^{Phe}$ (47–76) fragment contains a hairpin composed of the T-stem and T-loop. This is very clear and was also previously demonstrated by others.

GARRETT: But you are still speculating about that interaction?

SPRINZL: Yes, I am still speculating about that interaction. Surely, the affinity chromatography does not provide a direct evidence about the nucleobases, which are involved in base-pairing. But the high specificity of the interaction, when only one tRNA with the anticodon G_mpApA, which is complementary to TpψpCp remains attached to the column is a strong indication that the interaction (Fig. 5) takes place.

GARRETT: I think the possibility of an interaction between the T-loop of tRNA and 5S RNA has been completely eliminated by the experiments of Pace *et al.* (1982), where the CpGpApA-sequence of the 5S RNA was deleted and the ribosomes were still active.

SPRINZL: No, I don't agree. This experiment only shows that for the synthesis of a polypeptide on the *E. coli* ribosome the CpGpApA-sequence of the 5S RNA is not an absolute requirement.

GARRETT: Finally, phylogenetic evidence is also against the hypothesis of a tRNA – 5S RNA interaction, because this sequence is not conserved in the eucaryotic tRNAs.

SPRINZL: The mammalian mitochondrial translation apparatus may be a special case. At the present time we simply do not know if the tRNA interacts with 5S RNA or not. Our recent experiments indicate that the T-loop of tRNA interacts most probably with 16S RNA of *E. coli* ribosomes as it is depicted in Fig. 6. This interaction involved the conserved single-stranded region of the 16S RNA near its 3′-end. In this region the Colicin cleavage site is located. The binding of tRNA or tRNA fragments to 70S ribosomes inhibits the Colicin-dependent cleavage of the 16S RNA. Ribonuclease H, which is specific for RNA-DNA hybrids cleaves the 16S RNA of the *E. coli* ribosomes near its 3′-end in the presence of d(GpTpTpCpGpA) oligonucleotide, which is homologous to the part of the T-loop of tRNA.

OFENGAND: You showed that modifications at one part of the tRNA can influence Sl nuclease cleavage in the anticodon. Would you like to comment on what that means for footprinting experiments in terms of the direct versus long-range kind of interaction?

SPRINZL: It means that if a protein binds to one region of tRNA it can potentially change the structure in regions distant from the place where the binding occurs.

KLUG: I want to comment on 2 points. The first is on the use of nuclease to probe structure. I am not very impressed by those differences, when you use nuclease Sl. Nuclease Sl cleaves single-stranded regions of both DNA and RNA and it depends upon how much of the RNA or DNA it has to hold before it cuts. That will depend upon the strength of the base pair as well as on the length of the double helical region. My colleague, Horace Drew, has been doing studies on loops of DNA, using a variety of nucleases, and he finds similar variations, which I don't think reflect *large* differences in the structure. They may represent differences in stability of the double helical parts, which the enzyme holds onto. There is also base specificity, as I keep stressing. No nuclease is known, which is unspecific, they all have some, maybe not very high, degree of specificity, or base preferences. So, I do think one has to be careful about relying on some change in structure based on nuclease cleavage.

SPRINZL: I agree with you, it is certainly not very good to compare tRNAs with different sequences. In our experiments, however, we compare one tRNA with a change in only one modification, the sequence remains otherwise the same.

KLUG: The second point I want to make, is about the role of the Tψ C sequence. Everybody seems to be looking at the ribosome and if we were at a meeting where people were concerned with DNA and gene expression, they would probably give a different argument, namely that the T-loop of tRNA functions as an intragenic promotor and may have nothing to do with protein synthesis.

SPRINZL: I know that the TψCG sequence, as also shown by Söll, is an intragenic promotor in the eukaryotic tRNA genes. The second function would be to stabilize the tertiary interactions in the tRNA. This does not really exclude that this sequence has a third function, too. This would be to bind to the ribosomes. Maybe this multiple function of invariant sequences of tRNA is reasonable.

ERDMANN: Concerning the TψCG interaction, of course, what we have seen is only indirect evidence, but I think we cannot avoid the evidence that the TψCG fragment is giving us a specific inhibition of the A-site binding and also that it can trigger magic spot synthesis. But I agree this is a problem we have discussed a lot over the last 10 years. We need good experiments to show what is going on. Wagner, in Wittmann's department, has shown that if you bind tRNA to the ribosomal P-site then the GAAC-sequence of 5S RNA becomes exposed for chemical modification.

CLARK: We have studied nucleases for a long time very carefully, and we did not use Sl nuclease for the kind of reasons we have been discussing. The first problem, of course, is the lack of pH optimum at the physiological conditions. The other is, the results are irreproducible, and if we go back to the structural work of Paul Sigler and Rich's group, where originally the $tRNA_f^{Met}$ was supposed to have a different anticodon-arm structure than $tRNA_m^{Met}$, and some of that, of course, is based on the Wrede-Rich experiment. The crystallography shows that the structures of the anticodon arms of both the initiator and elongator tRNAs are very similar. So, it looks like the original interpretation was rather glib and too early.

SPRINZL: Sure, as I said, it is dangerous to compare tRNA's with different sequences, but the $tRNA_f^{Met}$ and the $tRNA_m^{Met}$ in the Wrede-Rich experiment have very similar anticodon loops. The differences they see in the Sl patterns of both tRNAs are, therefore, correctly interpreted.

SCHATZ: The discussions we have had yesterday and today on the probing of nucleic acids with digestive enzymes remind me of the discussions at the symposia, which focus on interactions between proteins. There are 2 technical points which might be stressed here: the first point is illustrated by cytocrome *c*. It is one of the best studied proteins, and has the same crystal structure in the reduced and the oxidized state, even if analyzed at high resolution. Yet, every biochemist knows that reduced and the oxidized cytochrome *c* differ in their resistance to trypsin. The protease may detect a very subtle change which might have little relevance to the overall structure and the function of the protein. The second technical point is the following: in many of your probing experiments, the incubation with the nuclease or a modifying chemical, takes many mintues or even several hours. How can you interpret such an experiment, when the modifying reaction is an irreversible, covalent change, which you allow to take place at time periods many orders of magnitude longer than the off rates of some of the biologically relevant complexes.

SPRINZL: Yes, it is very clear that all these enzymatic digestions have to be made under kinetically controlled conditions, where one analyzes the first hit. That is how our experiments were done.

EBEL: I would like to make a comment about the Tψ C loop interaction with ribosomal components. Nobody could find a photo-cross-reaction between tRNA and ribosomal RNA except that which will be reported here by Dr. Ofengand.

SPRINZL: We tried to irradiate the tRNA and to make a cross-link. We did not get any positive results. The question is, whether we can expect to get some, because the case, which Dr. Ofengand has, is a special case. The 5-alkoxy uridine of *E. coli* $tRNA^{Val}$ is a photo-reactive residue and only due to this special structural feature, he gets a cross-linking. But on the other hand, it does not mean that if you don't get the cross-link the interaction between ribosomal RNA and the Tψ C-region does not exist.

OFENGAND: I would like to make 2 comments. First, although it is true that the cross-link I will discuss tomorrow is a special case of RNA-RNA cross-linking, it is clearly possible to cross-link unmodified tRNAs with UV light as Roger Brimacombe, among others, has shown. However, I agree that the absence of cross-linking does not mean the absence of interaction. Second, with regard to the Tψ CG story, we have been talking about base pairing potentials her, yet the question really is, does Tψ CG behave any differently than UUCG. Put another way, why has the cell bothered to make 2 uridine modifications if U is as good as T or ψ?

SPRINZL: In our experiments with the oligonucleotides TψCG and UUCG we cannot detect any differences due to modification of the uridine residues. However, in experiments using intact tRNA the modification of U_{54} to T_{54} was found to influence the structure of tRNA as well as its function during the translation.

GARRETT: I would just like to make the point that there are 3 studies of tRNA structure on the ribosome, 3 of them on the P-site and one study also on the A-site, and in none of those studies is there evidence for a substantial change in the tRNA comfirmation. I think that argues strongly against your proposed interactions.

SPRINZL: I don't think that it is arguing strongly against. The tRNA has a tertiary structure in the solution. When it interacts with the ribosome the residues involved in tertiary structure, probably get involved in some new interactions on the ribosome. So, it does not mean that the tRNA has to be accessible when it enters the ribosome and its tertiary structure changes.

A Model for the Structure of a Small Mitochondrial tRNA

M. H. L. de Bruijn & A. Klug

Comprehensive analysis of the human, bovine and murine mitochondrial (mt) genomes (Anderson *et al.* 1981, 1982a, Bibb *et al.* 1981) has shown each to contain 22 tRNA coding sequences. The 3 sets of tRNAs are homologous to one another and the bovine ones have been confirmed by direct RNA sequence analysis (de Bruijn *et al.* 1980, Arcari & Brownlee, 1980, Roe *et al.* 1981, 1982a, B. A. Roe, personal communication). The limited number of tRNAs encoded is thought to be sufficient to translate the complete mammalian mt genetic code, with the possible exception of methionine codons, using a unique pattern of codon recognition (Barrell *et al.* 1980). No methionine elongator tRNA has as yet been identified.

All but one of the identified tRNAs can form the familiar cloverleaf secondary structure (Anderson *et al.* 1982b). However, they lack many of the "universal" features of prokaryotic and non-mitochondrial eukaryotic tRNAs (Gauss & Sprinzl 1983a, 1983b, Dirheimer *et al.* 1979, Clark & Klug 1975), and their primary sequences are generally shorter. The semi-invariant sequence T-Ψ-C-R-A is absent, as well as a number of invariant bases in the D-loop and elsewhere. Their TΨC loops do not have the constant 7-nucleotide size, but are generally shorter and can vary considerably in size and sequence between homologous tRNA species (c.f. human and bovine mt tRNA^{Thr}: Anderson *et al.* 1982b). This also applies to the "extra" (or "variable") loop and D-loop, and reflects the shorter but variable lengths of the primary tRNA sequences as a whole.

The stems of mammalian mt tRNAs contain more than the usual number of non-standard base pairs. The predominant non-standard pairs, G-U and A-C, are non-randomly distributed both within the individual tRNAs and within the

---

MRC Laboratory of Molecular Biology, Hills Road, Cambridge CB2 2QH, England.

GENE EXPRESSION, Alfred Benzon Symposium 19.
Editors: Brian F. C. Clark & Hans Uffe Petersen, Munksgaard, Copenhagen 1984.

total mt tRNA population (Anderson *et al.* 1982a). A-C pairs, which normally occur primarily in the aminoacyl and D-stems, are found mainly in the anticodon stems and TΨC stems of the mitochondrial tRNAs. Furthermore, while A-C pairs are found exclusively in tRNAs derived from L-strand sense transcripts, most of the G-U pairs are found in the H-strand sense tRNAs.

A striking feature of mitochondrial tRNAs in general and of animal mitochondrial tRNAs in particular is their high A+U content, which is in marked contrast to the base composition of all non-mitochondrial tRNAs sequenced to date. Thus, the G+C/A+C ratio for the latter usually varies between 2 and one (Eigen & Winkler-Oswatitsch 1981). But for human and bovine mt tRNAs it ranges from 0.94 to 0.25 (Anderson *et al.* 1982a), and in a recently analysed mt $tRNA^{Asp}$ from *Drosophila melanogaster* it even drops to 0.11 (de Bruijn 1983). Consequently, the stem regions have a high proportion of A.U base pairs. This together with the absence of otherwise universally conserved features and a higher incidence of non-standard base pairs must result in a considerable weakening of overall tertiary structure. The smaller size of many of the TΨC and D-loops as well as their variability are almost certainly additional factors influencing tertiary structure when compared to non-mitochondrial tRNAs.

The most extreme example, in this respect, is mammalian mt $tRNA^{Ser}_{AGY}$ which lacks the entire D-arm (de Bruijn *et al.* 1980, Arcari & Brownlee 1980) and whose secondary structure is consequently reduced to a "truncated cloverleaf" (c.f. Fig. 1). Nevertheless, this tRNA is presumed to function in mt protein synthesis. It can be specifically aminoacylated with serine and the corresponding AGY codons, which occur in almost all identified mammalian mt protein-coding genes, have been shown to specify serine (de Bruijn *et al.* 1980). Two important questions then remain. How can it fold into a tertiary structure enabling serine to be placed in the correct position on the ribosome to allow chain elongation? And, how can its structure be compatible with that of other tRNAs, that interact with the same ribosome and display a complete cloverleaf secondary structure?

Experimental evidence on the structure has been obtained by monitoring the conformation of both the bovine and human $tRNA^{Ser}_{AGY}$ using 3 structural probes. We have constructed a tertiary structure model of bovine mt $tRNA^{Ser}_{AGY}$ based on the results of this probing, on a comparison of $tRNA^{Ser}_{AGY}$ secondary structures from 9 different mammalian species, and on analogies with the crystal structure of yeast $tRNA^{Phe}$. The structural model has implications for other mammalian mt tRNAs.

PROBING OF BOVINE AND HUMAN mt $tRNA^{Ser}_{AGY}$ CONFORMATION

Conformation of the 2 tRNAs was explored using the chemical probing method of Peattie and Gilbert (1980). Dimethyl sulphate is used to monitor the availability of N-7 in guanosines and of N-3 in cytidines to methylation and diethyl pyrocarbonate (DEP) the availability of N-7 in adenosines to carbethoxylation. In a folded RNA chain the bases are available for chemical modification if the specified atoms are not involved in hydrogen bonding. DEP has also been found to monitor stacking of adenosines in yeast $tRNA^{Phe}$ (Peattie & Gilbert 1980). The advantage of the chemical probes over those employed in earlier chemical accessibility studies on tRNA (cf. Robertus *et al.* 1974, Rhodes 1977) is that they lead, after treatment with aniline to strand scission at the corresponding ribose, so that the sites of modification can be read from a polyacrylamide gel on which the resulting fragments are fractionated.

Chemical probing of both mt tRNAs was carried out at a series of 5° intervals between 5 and 90°C and under conditions of ionic strength and pH that promote a stable tertiary structure. This was used as an alternative to optical determination of a melting curve for each of the tRNAs, for which there was, in any case, insufficient material. In effect, the chemical approach is much more extensive and precise, as it allows one to follow not only the melting of the tertiary and secondary structures as a whole, but also that of individual base pairs relative to each other. The results for bovine $tRNA^{Ser}_{AGY}$ are summarized in Fig. 1A-B (de Bruijn & Klug 1983). Three general classes of bases are evident: (i) those that are not protected from chemical modification at any temperature; (ii) those that are protected from modification up to 40°C or higher; and (iii) those that become exposed to modification at temperatures below 40°C. The last 2 classes, therefore, represent bases involved in interactions that are lost above and below 40°C, respectively.

Unfortunately, there is as yet no chemical probe available for monitoring the hydrogen bonding of uridines by this method. Yet it is clear from Fig. 1A-B that the single stranded regions that are likely to play a role in the formation of tertiary structure do contain several uridines. A number of these however, are replaced by cytidines in the corresponding human tRNA. As cytidines can be structurally monitored, it was useful to duplicate the bovine $tRNA^{Ser}$ data by probing the human variant. These results are summarised in Fig. 1C-D using the same classification as described for the bovine tRNA. It is clear that the bovine and human tRNA species show differences in melting behaviour. Some bases in the human tRNA become chemically modified at a higher temperature than the

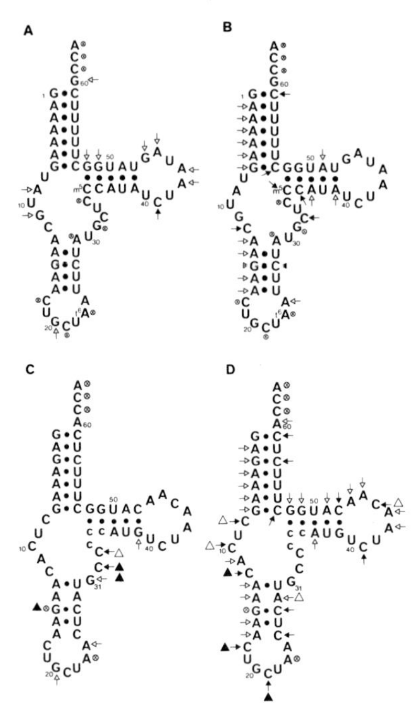

Fig. 1. Points of chemical modification in bovine and human mt tRNA$^{Ser}_{AGY}$. The diagrams are a summary of more detailed results obtained with chemical probing of the 2 RNAs (de Bruijn & Klug 1983). Nucleotides in the "truncated cloverleaf" secondary structure are numbered in the 5' to 3' direction. Black dots signify Watson-Crick base pairs. (A) Chemical modifications in the bovine tRNA occurring between 5 and 40°C. Arrows with open heads indicate that N-7 of the corresponding purines became modified at a temperature within the given temperature range but was protected below that temperature. Arrows with solid heads signify the same for N-3 of the corresponding cytidines. (B) Chemical modifications of the bovine tRNA occurring between 40 and 90°C. The arrows in (A) are not repeated. (C) Chemical modifications in the human tRNA occurring between 5 and 40°C. (D) Further chemical modifications in the human tRNA between 40 and 90°C. The symbol $\otimes$ indicates that the corresponding base is modified at all temperatures and is thus never protected from the modifying chemical. Nucleotide numbering in the human tRNA is according to the bovine tRNA sequence. An open triangle signifies that the corresponding position in the bovine tRNA contains uridine. A solid triangle signifies that the probing results at the corresponding positions in the two tRNAs are substantially different. For G_1 and the uridines no probing data are available.

corresponding bases in the bovine tRNA, or show a markedly different pattern of reactivity towards the chemicals. This is not unexpected, since the human anticodon stem shows poor Watson-Crick base pairing and the nucleotide sequences of TΨC, variable and "D-arm replacement" loop differ substantially.

MODEL BUILDING

Chemical probing by itself is insufficient to establish a tertiary structure. It allows a detailed insight into which bases are involved in certain interactions. Thus, N-3 positions in cytidines may be involved in Watson-Crick or reverse Watson-Crick hydrogen bonding (Jack *et al.* 1976), whereas N-7 positions in guanosines are involved in weaker tertiary interactions and in co-ordination reactions with ions or water molecules (Jack *et al.* 1976, 1977, Holbrook *et al.* 1977, Hingerty *et al.* 1978). N-7 positions in adenosines are hydrogen bonded in reverse-Hoogsteen base pairs and in certain basetriples as defined in the crystal structure of yeast $tRNA^{Phe}$ (Jack *et al.* 1976). But other interactions are only partly monitored or not monitored at all because the chemical probes have no reactivity towards the atomic positions involved in hydrogen bonding. These include interactions between a base and the ribose-phosphate backbone, or between 2 positions in the backbone. In addition, no information is obtained as to the identity of the second member of an interacting base pair.

Nevertheless, results of chemical probing have generally been in good agreement with tRNA crystal structure data (Robertus *et al.* 1974b, Rich & Rajbhandary 1976). They are, therefore, useful to test an existing tRNA structure or a preconceived structural model. We have built such a model of the bovine mt $tRNA^{Ser}_{AGY}$ structure using Labquip parts (Fig. 2). The proposed base-pair interactions are shown in Fig. 3A, and in Fig. 3B the structure is schematically represented by a tracing of the ribose-phosphate backbone. In the following sections the assumptions made in building the model are described and justified in the light of the probing data.

Fig. 2. Comparison of the proposed tertiary structure of bovine mt $tRNA^{Ser}_{AGY}$ with the established crystal structure of yeast $tRNA^{Phe}$ (Ladner *et al.* 1975). The bovine tRNA is to the right. Both tRNA structures are represented by "Labquip" molecular models (1 cm per Å).

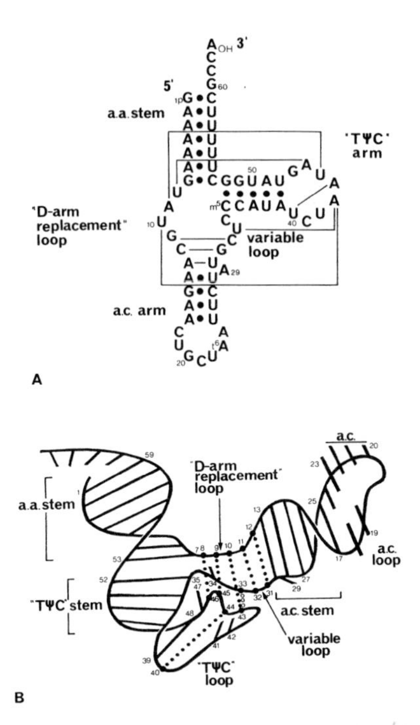

Fig. 3. Schematic diagram showing (A) the proposed "truncated cloverleaf" formula and proposed tertiary connections and, (B), the chain-folding and proposed tertiary interactions between bases in bovine mt $tRNA^{Ser}_{AGY}$. Long straight lines indicate base pairs in double-helical stems. Shorter lines represent unpaired bases. Dotted lines represent proposed base pairs in addition to those in the double helical stems. The schematic drawing of the structure is based on the "Labquip" model (Fig. 2, to the right) which is viewed in a direction perpendicular to the plane through the 3 double-helical stems. Note that in (A), as compared to Fig. 1, the base A_{29} has been looped out.

THE "TRUNCATED CLOVERLEAF" SECONDARY STRUCTURE

From Fig. 1B it is clear that the most stable interactions (i.e., those lost only above 40°C) mostly correspond to interactions composing the "truncated cloverleaf" secondary structure. Protection of adenosines, and probably also guanosine, in the 3 stem regions are interpreted as due to stacking and of cytidines due to Watson-Crick base pairing with guanosines. As mentioned earlier, interpretation of the human data is less straight forward. The majority of interactions (Fig. 1C-D) appear to melt in a one-step cooperative process. Most of the discrepancies between the human and bovine probing data (indicated by solid triangles in Fig. 1C-D) occur in the lower half of the "truncated cloverleaf" and are probably related to the occurrence of the non-standard pairs A_{17}-C_{25} and A_{14}-A_{28} in the human tRNA. Nevertheless, the human probing data are consistent with the secondary interactions defined in the "truncated cloverleaf". We, therefore, conclude that this secondary structure is real and can serve as a basis for spatial folding.

The proposed non-standard pairs A_{13}-A_{29} in the bovine tRNA and A_{14}-A_{28} in the human tRNA unexpectedly show distinctly different behaviour towards DEP. In the human tRNA both A_{14} and A_{28} appear to be stacked, but only A_{13} in the bovine tRNA shows this effect with A_{29} always exposed to the modifying chemical. We conclude from this that A_{13}-A_{29} is not a pair, but that instead the standard pair $A_{13}.U_{30}$ is formed. A_{29} is then looped-out of the double-helical stem (Fig. 3A) by rotation about its phosphodiester bonds without disturbing the helical continuity of its nearest neighbour base pairs (Lomant & Fresco 1975). Both tRNAs now have 4 nucleotides in their variable loop.

A-C AND A-A PAIRS IN tRNA DOUBLE HELICAL STEMS

The anticodon stem of human $tRNA^{Ser}_{AGY}$ contains the non-complementary pairs A_{17}-C_{25} and A_{14}-A_{28}. Analysis of published tRNA sequences (Gauss & Sprinzl 1983a) shows that such pairs, in addition to G-U, are not uncommon and occur with significant frequency. It is, therefore, reasonable to assume that A-C and A-A pairs in tRNA stems use their intrinsic capacity for hydrogen bonding even though they are likely to be less stable than Watson-Crick pairs.

As mentioned earlier, A-C pairs are found exclusively in mammalian mt tRNAs derived from L-strand transcripts and G-U pairs mostly in H-strand sense tRNAs. The occurrence of one pair or the other in a tRNA, therefore, seems to be directly related to the base composition of the coding strand. This, in turn, suggests that structurally the 2 pairs must have a measure of equivalence, at least in mammalian mt tRNAs.

How are A-C pairs made? There is as yet no crystallographic evidence, as there is for G-U, so one can only list the possibilities. There are 2 general ways of doing this. The first, illustrated in the left-hand column of Fig. 4, is by analogy to a G-U wobble base pair (Fig. 4a). In the case of A-C only one hydrogen bond can be formally made, but there is a possibility that in this situation the N-1 of the A base may be protonated to give 2 hydrogen bonds (Fig. 4c). More likely, however, a cation could be picked up to mediate an interaction between N-1 of the A and 0-2 of the C (Fig. 4b).

4

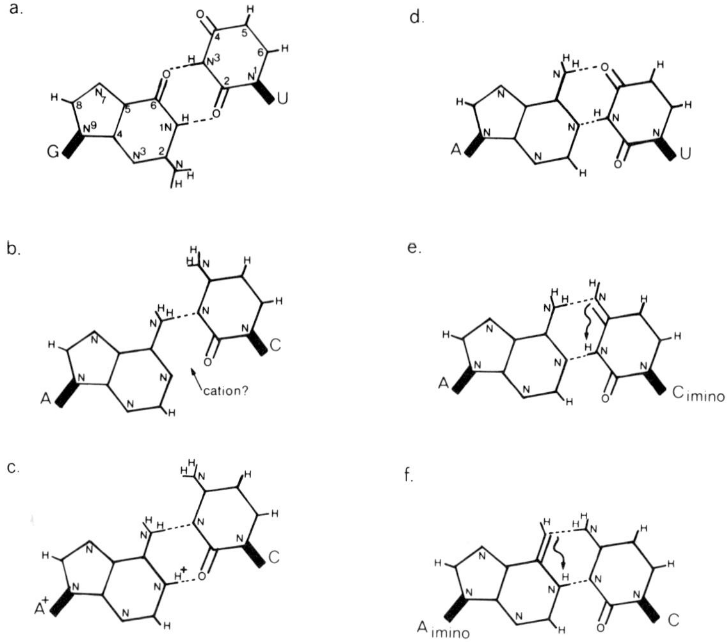

a. G U

d. A U

b. A C cation?

e. A C imino

c. A+ C

f. A imino C

5

a. A A

c. A imino A syn

b. A A

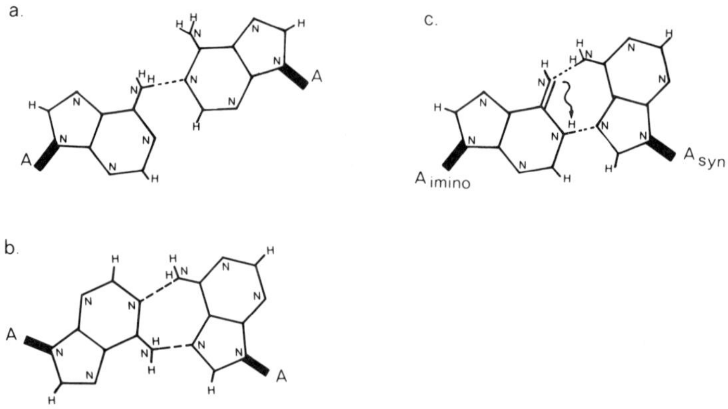

The second way of making an A-C base pair is to make it like the standard A-U base pair (Fig. 4d), by a tautomeric shift which changes either of the 2 bases from the amino form to the less favoured imino form (Figs. 4e, f). Indirect, but good, evidence for such hydrogen-bonded A-C pairs has come from melting studies on long I-C double helices containing very small proportions of opposing A and C residues (Lomant & Fresco 1975, Fresco *et al.* 1980). The occurrence of these A-C base pairs is explained if the "cost" in free energy of forming them is compensated, in the structure as a whole, by, for example, improved stacking. The double helical stem regions of tRNAs are much shorter than the double helices used in Lomant & Fresco's experiments, so there is less "pressure" to make the tautomeric shift. For this reason, we think that the A-C wobble type pairs described above are more likely. The chemical probing data for A_{17}-C_{25} in human $tRNA^{Ser}_{AGY}$ are consistent with all 4 schemes and all 4 have anti-parallel glycosyl bonds and bond angles which are compatible with a double-helical configuration.

At least 3 bonding schemes for an A-A pair are possible (Fig. 5). With one hydrogen bond the distance between the glycosyl bonds is probably too great to fit into a double helix (Fig. 5a). The second scheme, in which one member of the pair may take up the *syn* conformation, has approximately parallel glycosyl bonds but could, nevertheless, fit with some distortion of the backbone (Fig. 5b). The third scheme (Fig. 5c), in which there is no such distortion, requires the second member of the pair to undergo a tautomeric shift to its unfavoured imino form (Topal & Fresco 1976a). The chemical probing data (details not shown) for the human pair, A_{14}-A_{28}, are consistent with the last 2 schemes.

Fig. 4. Possible hydrogen bonding schemes for A-C pairs in tRNA double-helical regions (Klug *et al.* 1974, Topal & Fresco 1976a, b, de Bruuijn & Klug 1983). (a) The wobble-pair G-U which is included for reference. (b) An A-C wobble-pair with one hydrogen bond; this pair has one unsatisfied hydrogen bond acceptor site (0–2 on C) unless 0–2 of C is shielded by a cation. (c) An A-C wobble-pair made possible by protonation of N-1 on the A member. (d) The standard pair A-U for reference. (e, f) A-C base pairs with 2 hydrogen bonds requiring either of the 2 bases to change from the favoured amino to the less favoured imino tautomeric form involving the transfer of one proton.

Fig. 5. (a) An A-A wobble-pair with one hydrogen bond in which the 2 glycosyl bonds are far apart. (b) An A-A pair with parallel glycosyl bonds also requiring some distortion of the double-helical backbone. (c) An A-A pair requiring one of the members to take up its unfavoured imino tautomeric form and the second member to be in the *syn* conformation (Topal & Fresco 1976a, b).

COMPARISON OF HOMOLOGOUS "TRUNCATED CLOVERLEAF" SECONDARY STRUCTURES

In addition to the nucleotide sequences of bovine and human mt $tRNA^{Ser}_{AGY}$, RNA or DNA sequences for this tRNA have been reported from 7 other mammalian species, and all can be folded in the truncated cloverleaf secondary structure (Figure 6). However, they vary considerably in sequence and also vary in the size of their TΨC and variable loops. If their spatial folding is to follow a general plan, then equivalent tertiary base-base interactions should occur in the 9 tRNAs, some of which should be apparent through the presence of co-ordinated base changes (Levitt 1969, Klug *et al.* 1974). Many such co-ordinated changes are evident in the stem regions, where purines and pyrimidines remain opposite each other and maintain base pairing. At least 2 additional sets of co-ordinated base changes can be detected. If A-C base pairing is permitted, then 7 of the 9 tRNAs can extend the anticodon stem upwards by 2 base pairs. The 2 apparent exceptions are the bovine and gibbon tRNAs. As discussed earlier, there are indications that A_{29} in the bovine tRNA is extra-helical, in which case similar extension of the anticodon stem is possible. If this argument is extended to the gibbon tRNA and A_{28} or C_{29} is looped-out, then not only can the anticodon stem be extended by 2 base pairs, but overall base pairing in that stem would also improve considerably. For the bovine tRNA the probing data clearly support the additional base pairs $G_{11}.C_{32}$ and $C_{12}.G_{31}$, while for the human tRNA they are at least compatible with such an extension.

COMMON ELEMENTS IN THE TERTIARY STRUCTURES OF YEAST $tRNA^{Phe}$ AND BOVINE mt $tRNA^{Ser}_{AGY}$.

The tertiary structure of yeast $tRNA^{Phe}$ (Ladner *et al.* 1975, Jack *et al.* 1976) is made up of 2 arms which are almost at right angles to each other (Fig. 7). The left arm is a long double helix containing the aminoacyl stem stacked end-to-end on the TΨC stem and on top of which the single-stranded sequence CCA is stacked. The right arm is composed of the anticodon loop which stacks on top of the double-helical anticodon stem. The tips of the 2 arms are ~80 Å apart. The thorax region connecting the 2 arms consists of the D-stem, nucleotides 8 and 9, the extra or variable loop and part of the D-loop. The tertiary interactions between these 4 structural elements are arranged in such a way that the continuous stack of bases in the D-helix is extended to form the so-called "augmented D-helix". The interactions between the (semi) invariant bases of

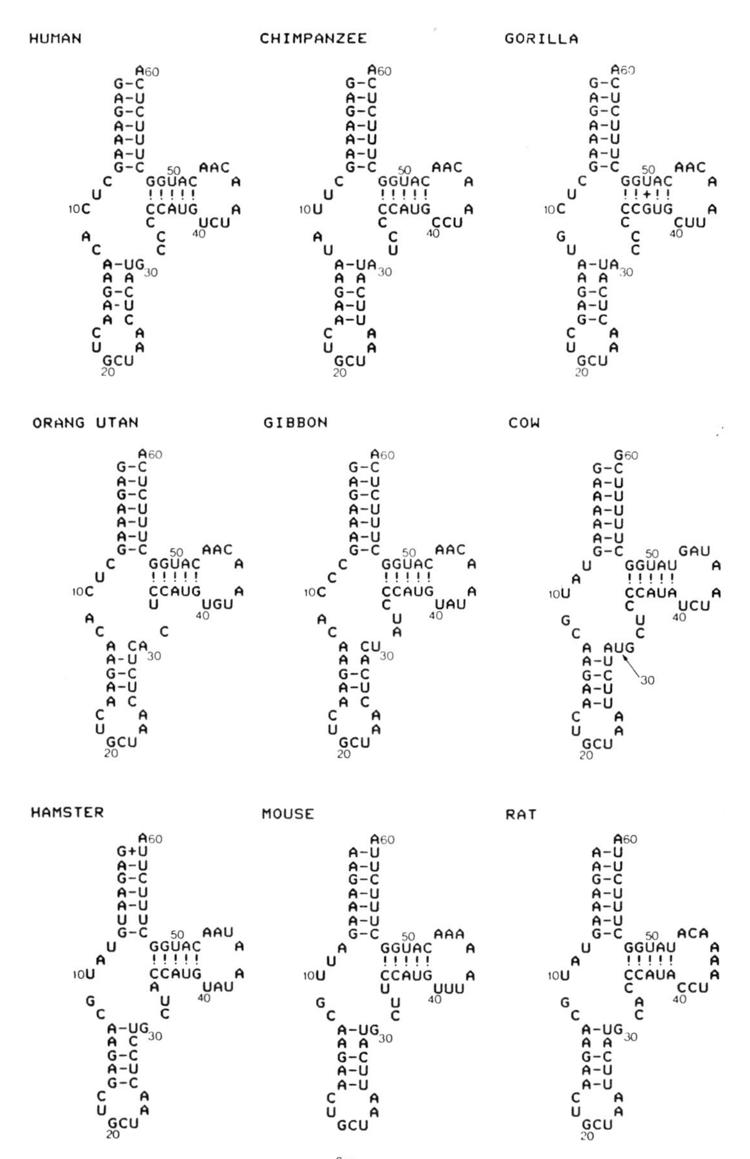

Fig. 6. The known mammalian mt tRNA$^{Ser}_{AGY}$ sequences arranged in their predicted "truncated cloverleaf" secondary structures. The species from man, cow and hamster (de Bruijn *et al.* 1980, Arcari & Brownlee 1980, Baer & Dubin 1980, de Bruijn & Klug 1983) have been determined or verified by direct RNA sequence analysis. The RNA sequences for the remaining 6 tRNA species have been inferred from the sequence of their genes (Bibb *et al.* 1981, Grosskopf & Feldmann 1981, Brown *et al.* 1982). Post-transcriptional modifications have been omitted. Nucleotide numbering is according to that in the bovine tRNA sequence and disregards the differing lengths of TUC and variable loops.

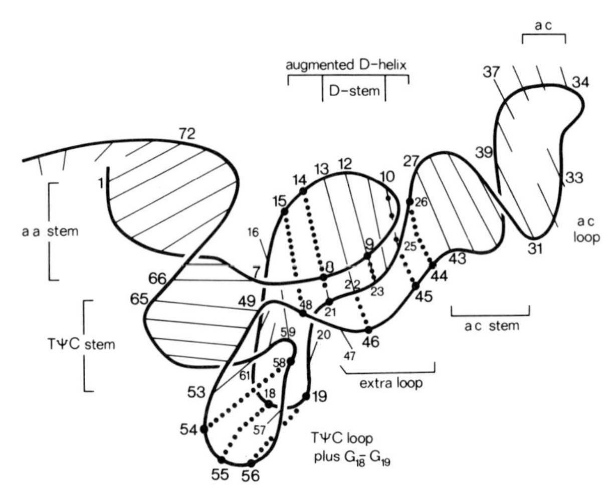

Fig. 7. Schematic diagram showing the chain folding and tertiary interactions between bases in yeast phenylalanine tRNA (from Ladner *et al.* 1975). Conventions are as described in the legend to Fig. 3.

TΨC and D-loops are crucial in maintaining the angle between the 2 arms. The TΨC loop is anchored by T_{54}-m^1A_{58} which is stacked on the end of the left arm. Specificity of the hydrogen bonding system gives the tRNA its characteristic shape, but its stability is probably mainly due to extensive stacking.

The 3 stem regions in bovine mt $tRNA^{Ser}_{AGY}$ (Fig. 3) appear quite normal and it is reasonable to assume that they are arranged as in yeast $tRNA^{Phe}$ so as to give a similar relationship between the anticodon and CCA ends. It is clear however, that the thorax region must be different since the D-arm is absent. It is important to note that the 9 known $tRNA^{Ser}_{AGY}$ sequences (Fig. 6) have 8 or 9 (in rat) rather than 7 nucleotides in the TΨC loop. One nucleotide extra in this loop, if the backbone is locally stretched, can increase the size of the loop by ~7 Å; this is equivalent to increasing the length of a stack by 2 base pairs (c.f. the distance between residues 8 and 9 in Fig. 7). The TΨC loop therefore has the potential to interact with both the variable loop and the "D-arm replacement" loop.

The "TΨC" loop of yeast $tRNA^{Phe}$ contains one intra-loop base pair, T_{54}-m^1A_{58}, which is of the "reverse-Hoogsteen" type (Hoogsteen 1963, Ladner *et al.* 1975, Fig. 8a). Six of the 9 mt $tRNA^{Ser}_{AGY}$ structures could make the equivalent reverse-Hoogsteen pair U_{40}-A_{44}, thus dividing the enlarged loop into 2 domains and so restricting its interactive capabilities. However, in gorilla, chimpanzee and rat the corresponding pair is C_{40}-A_{44}, which cannot be of the "reverse-Hoogsteen" type because N-3 of C_{40} is not protonated. Interestingly, an A-C pair

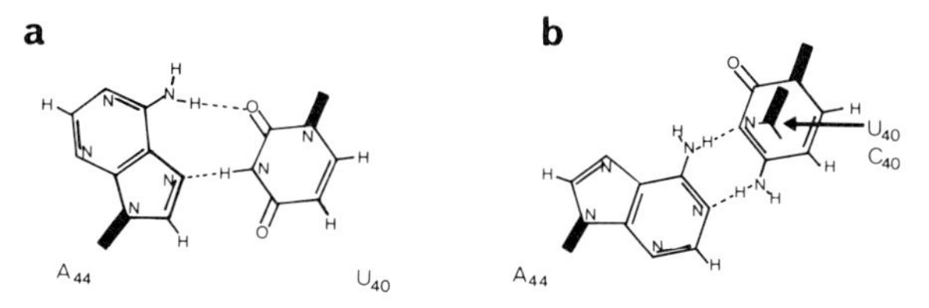

Fig. 8. Proposed intra-loop base pairing in the TψU loop of mammalian mt serine (AGY) tRNAs. (a) Reverse-Hoogsteen-type base pairing (Hoogsteen 1963) between U_{40} and A_{44} in the tRNAs from man, orang utan, gibbon, cow, hamster and mouse. The analogous base pair T_{54}-m^1A_{58} is made in the crystal structure of yeast $tRNA^{Phe}$ (Ladner *et al.* 1975). (b) Alternative but equivalent base pairing between C_{40} and A_{44} in the tRNAs from chimpanzee, gorilla and rat. For reference, the relative position of the glycosyl bond for U_{40} (see b) is shown when the glycosyl bond for A_{44} is held fixed.

with parallel glycosyl bonds is an excellent alternative for a reverse-Hoogsteen pair. In Fig. 8, the A-C pair is shown overlapped with U_{40}-A_{44} and aligned with respect to the glycosyl bond of A_{44}. The glycosyl bonds of U_{40} and C_{40} have almost the same angle and the shift between the 2 bonds is of the order produced by a wobble base; this can easily be accommodated. The probing data for both bovine and human $tRNA^{Ser}_{AGY}$ (de Bruijn & Klug 1983) clearly support involvement of N-7 of A_{44} in hydrogen bonding and are, therefore, consistent with the proposed reverse-Hoogsteen pair.

THE TERTIARY STRUCTURE MODEL OF BOVINE mt $tRNA^{Ser}_{AGY}$.

Each of the arguments put forward regarding the spatial folding of bovine $tRNA^{Ser}_{AGY}$ and the base pairs defining its structure, is plausible by itself and supported by the available chemical probing data. The molecular model (Figs. 2, 3) demonstrates that they are also compatible with one another. The 2 arms of the structure have about the same angle between them as in yeast $tRNA^{Phe}$ and are connected by the anti-parallel chains of the variable and "D-arm replacement" loops. The anticodon stem has been extended upwards by the Watson-Crick pairs $G_{11}.C_{32}$ and $C_{12}.G_{31}$ as predicted from the "truncated cloverleaf" comparisons. The reverse-Hoogsteen pair U_{40}-A_{44} stacks on the end of the TΨC stem, by analogy to T_{54}-m^1A_{58} in yeast $tRNA^{Phe}$, thus dividing the TΨC loop into 2 domains.

The model itself suggested 3 additional tertiary interactions. Two of these, U_8-A_{46} and A_9-U_{45}, are standard Watson-Crick pairs. The third interaction is the base triple U_{10}-A_{43}-U_{33} (de Bruijn & Klug 1983, not shown), in which the U_{10}-A_{43}

connection is of the reverse-Watson-Crick type and hydrogen bonding between A_{43} and U_{33} is according to Hoogsteen (1963).

It is not clear whether the 3 additional tertiary interactions suggested by model building are equally applicable to the other mt $tRNA^{Ser}_{AGY}$ structures. With the exception of A_{46} and A_{43}, none of the involved bases are universally conserved and no additional co-ordinate base changes are apparent. As the gibbon and orang utan tRNAs effectively have a variable loop of only 3 nucleotides, it is almost impossible to introduce an equivalent base triple without "overstretching" the variable loop locally. The other serine tRNAs probably use alternative base pairing arrangements in this part of the thorax region.

DIMENSIONS OF BOVINE mt $tRNA^{Ser}_{AGY}$ AND OF OTHER BOVINE mt tRNAs.

There is one additional major difference between our model of bovine mt $tRNA^{Ser}_{AGY}$ and yeast $tRNA^{Phe}$ (Figs. 3, 7). While the thorax of yeast $tRNA^{Phe}$ is composed of an array of 7 stacked base pairs and triples (the "augmented D-helix"), the bovine thorax region only has 5. The 5-nucleotide size of the "D-arm replacement" loop and the backbone topology of the model could possibly allow for extention of the number of stacks to 6, through stacking of unpaired bases, but a stack of 7 would be virtually impossible. As all bovine mt tRNAs must be able to interact with the same mt ribosome, they should have the same dimensions. Our model, therefore, implies that all other bovine mt tRNAs have smaller dimensions as well.

There are 2 additional observations that support this contention. First, many mammalian mt tRNAs have TΨC and D-loops of only 3–4 nucleotides (Anderson *et al.* 1982b) against at least 7 in non-mitochondrial tRNAs. It is, therefore, doubtful whether these tRNAs can fold in the image of yeast $tRNA^{Phe}$. If mammalian mt tRNA structures were slightly smaller this would again be possible. Second, at least 3 mammalian mt tRNAs, $tRNA^{Ser}_{AGY}$ from orang utan and gibbon (the latter in its revised form, as described earlier), and bovine $tRNA^{Leu}_{CUN}$ (Anderson *et al.* 1972b), have only 3 nucleotides in the extra (or variable) loop. In a structure like that of yeast $tRNA^{Phe}$ a 3-nucleotide loop is physically impossible requiring the ribose-phosphate backbone to stretch beyond its capacity (Clark & Klug 1975). If the thorax of these tRNAs were only one stacked base pair shorter (~3.5 Å), a 3-nucleotide variable loop could be accommodated.

CONCLUDING REMARKS

The bovine mt $tRNA^{Ser}_{AGY}$ structural model is no more than a reasonable prediction. Its value lies not in accuracy of detail but in satisfactorily combining a variety of data in a structure that makes sense in terms of design and stability. The structure is probably weaker than those of more conventional tRNAs by having more unusual and energetically less favourable base pairs. It should, however, derive a considerable degree of stability from extensive stacking throughout the molecule which can be as great as in yeast $tRNA^{Phe}$. Moreover, it has only a limited use since it is needed for recognition of less than a fifth of all serine codons used in mammalian mt genomes (Anderson *et al.* 1982a).

It is remarkable that bovine mt $tRNA^{Ser}_{AGY}$, despite having an anomalous structure and a unique set of interactions, is capable of folding into a structure resembling yeast $tRNA^{Phe}$ (Fig. 2). The absent D-loop is partly compensated for by an extra nucleotide in the TΨC loop which comes to rest in the thorax region, where it interacts with 3 residues in the "D-arm replacement" loop. The remaining 2 residues in the latter form base pairs with the variable loop. Thus, all 5 bases of the "D-arm replacement" loop are used and stack on the anticodon stem replacing the augmented D-helix in the thorax of the molecule.

The bovine mt $tRNA^{Ser}_{AGY}$ structure indicates that all bovine mt tRNAs are at least 3.5 Å shorter than non-mitochondrial tRNAs. However, one of them, mt $tRNA^{Leu}_{UUR}$, does not necessarily follow this general scheme. It is unique in having retained most of the "universal" features of non-mitochondrial tRNAs (Anderson *et al.* 1982b) and, therefore, could fold with the same dimensions as yeast $tRNA^{Phe}$. What is interesting, is that the mt $tRNA^{Leu}_{UUR}$ gene lies in a unique position in the genome, immediately behind the highly expressed ribosomal RNA genes and immediately in front of the moderately expressed protein-coding region (Anderson *et al.* 1981, 1982a, Gelfand & Attardi 1981). Because these 2 domains of the genome show such different levels of expression, the region immediately preceding the $tRNA^{Leu}_{UUR}$ gene has been suggested to contain a site for termination or attenuation of transcription. The 3' end of the large ribosomal RNA gene in HeLa cell mitochondria does contain a short hairpin, which could serve in this capacity (Dubin *et al.* 1982), but it is only poorly conserved in other mammalian mtDNAs. The mt $tRNA^{Leu}_{UUR}$ gene however, is highly conserved between mammals (Anderson *et al.* 1982b) and could play a similar role. Because the $tRNA^{Leu}_{UUR}$ transcript, when folded, contains most classical tRNA features, it might function as a very efficient recognition signal for the RNAse P-type activity which, in mammalian mitochondria, has been proposed to be responsible for the processing of large multicistronic primary transcripts at various stages of their synthesis (Anderson *et al.* 1981). Alternatively, the mt $tRNA^{Leu}_{UUR}$ gene could contain a moderately active promoter sequence acting on the downstream coding sequences. The (semi-) invariant sequences present in the $T\Psi C$ and D-loop of mt $tRNA^{Leu}_{UUR}$ have been identified as promoter elements for transcription of non-mitochondrial eukaryotic tRNA genes (Sharp *et al.* 1981, Galli *et al.* 1981). It is worth noting, that in *Drosophila melanogaster* mt $tRNA^{Leu}_{UUR}$ has retained none of the "universal" tRNA characteristics (de Bruijn 1983). At the same time, its gene has a different position in the mt genome. The highly expressed ribosomal RNA genes in *D. melanogaster* mtDNA are on a different strand to most of the moderately expressed protein genes. It is, therefore, clear that in this case differential expression is mediated by different promoters. Thus, the "universal" features of mt $tRNA^{Leu}_{UUR}$ may have a more specific role at the level of transcription rather than at the level of translation and do not, *per se,* specify a tRNA tertiary structure that would be incompatible with other mammalian mt tRNAs.

REFERENCES

Anderson, S., Bankier, A. T., Barrell, B. G., de Bruijn, M. H. L., Coulson, A. R., Drouin, J., Eperon, I. C., Nierlich, D. P., Roe, B. A., Sanger, F., Schreier, P. H., Smith, A. J. H., Staden, R. & Young, I. G. (1981) *Nature 290,* 457–465.

Anderson, S., de Bruijn, M. H. L., Coulson, A. R., Eperon, I. C., Sanger, F. & Young, I. G. (1982a) *J. Mol. Biol. 156,* 683–717.

Anderson, S., Bankier, Barrell, B. G., de Bruijn, M. H. L., Coulson, A. R., Drouin, J., Eperon, I. C., Nierlich, D. P., Roe, B. A., Sanger, F., Schreier, P. H., Smith, A. J. H., Staden, R. & Young, I. G. (1982b) In: *Mitochondrial Genes* (eds. P. Slonimski, P. Borst & G. Attardi) pp 5–43. Cold Spring Harbor Laboratory, Cold Spring Harbor, New York.

Arcari, P. & Brownlee, G. G. (1980) *Nucl. Acids Res. 8,* 5207–5212.

Baer, R. J. & Dubin, D. T. (1980) *Nucl. Acids Res. 8,* 3603–3610.

Barrell, B. G., Anderson, S., Bankier, A. T., de Bruijn, M. H. L., Chen, E. Coulson, A. R., Drouin, J., Eperon, I. C., Nierlich, D. P., Roe, B. A., Sanger, F., Schreier, P. H., Smith, A. J. H., Staden, R. & Young, I. G. (1980). *Proc. Natl. Acad. Sci. USA 77,* 3164–3166.

Bibb, M. J., van Etten, R. A., Wright, C. T., Walberg, M. W. & Clayton, D. A. (1981) *Cell 26,* 167–180.

Brown, W. M., Prager, E. M., Wang, A. & Wilson, A. C. (1982) *J. Mol. Evol. 18,* 225–239.

de Bruijn, M. H. L. (1983) *Nature 304,* 234–241.

de Bruijn, M. H. L. & Klug, A. (1983) *EMBO J. 2,* 1309–1321.

de Bruijn, M. H. L., Schreier, P. H., Eperon, I. C., Barrell, B. G., Chen, E. Y., Armstrong, P. W., Wong, J. F. H. & Roe, B. A. (1980) *Nucl. Acids Res. 8,* 5213–5222.

Clark, B. F. C. & Klug, A. (1975) In: *Proceedings of the Tenth FEBS Meeting,* vol. 39 (eds. F. Chapeville & M. Grunberg-Manago) pp. 183–205. North Holland, Amsterdam.

Dirheimer, G., Keith, G., Sibler, A. P. & Martin, R. P. (1979) In: *Transfer RNA: Structure, Properties and Recognition* (eds. P. R. Schimmel, D. Söll & J. N. Abelson) pp. 19–41. Cold Spring Harbor Laboratory, Cold Spring Harbor, New York.

Dubin, D. T., Montoya, J., Timko, K. D. & Attardi, G. (1982) *J. Mol. Biol. 157,* 1–19.

Eigen, M. & Winkler-Oswatitsch, R. (1981) *Die Naturwissenschaften 68,* 217–228.

Fresco, J. R., Broitman, S. & Lane, A. E. (1980) In: *Mechanistic studies of DNA replication and genetic recombination* ICN-UCLA Symp. Mol. Cell Biol. (eds. B. Alberts & C. F. Fox) Vol. XIX, pp. 753–768. Academic Press, New York.

Galli, G., Hofstetter, H. & Birnstiel, M. L. (1981) *Nature 294,* 626–631.

Gauss, D. H. & Sprinzl, M. (1983a) *Nucl. Acids Res. 11,* r1–r53.

Gauss, D. H. & Sprinzl, M. (1983b) *Nucl. Acids Res. 11,* r55–r103.

Gelfand, R. & Attardi, G. (1981) *Mol. Cell Biol. 1,* 497–511.

Grosskopf, R. & Feldmann, H. (1981) *Curr. Genet. 4,* 191–196.

Hingerty, B., Brown, R. S. & Jack, A. (1978) *J. Mol. Biol. 124,* 523–534.

Holbrook, S. R., Sussman, J. L., Warrant, R. W., Church, G. M. & Kim, S.-H. (1977) *Nucl. Acids Res. 4,* 2811–2820.

Hoogsteen, K. (1963) *Acta Crystallogr. 16,* 907–916.

Jack, A., Ladner, J. E. & Klug, A. (1976) *J. Mol. Biol. 108,* 619–649.

Jack, A., Ladner, J. E., Rhodes, D., Brown, R. S. & Klug, A. (1977) *J. Mol. Biol. 111,* 315–328.

Klug, A., Ladner, J. & Robertus, J. D. (1974) *J. Mol. Biol. 89,* 511–516.

Ladner, J. E., Jack, A., Robertus, J. D., Brown, R. S., Rhodes, D., Clark, B. F. C. & Klug, A. (1975) *Proc. Natl. Acad. Sci. USA 72,* 4414–4418.

Levitt, M. (1969) *Nature 224,* 759–763.

Lomant, A. J. & Fresco, J. R. (1975) *Prog. Nucl. Acids Res. Molec. Biol. 15,* 185–218.

Peattie, D. A. & Gilbert, W. (1980) *Proc. Natl. Acad. Sci. USA 77,* 4679–4682.

Rhodes, D. (1977) *Eur. J. Biochem. 81,* 91–101.

Rich, A. & RajBhandary, U. L. (1976) *Ann. Rev. Biochem. 45,* 805–860.

Robertus, J. D., Ladner, J. E., Finch, J. T., Rhodes, D., Brown, R. S., Clark, B. F. C. & Klug, A. (1974a) *Nature 250,* 546–551.

Robertus, J. D., Ladner, J. E., Finch, J. T., Rhodes, D., Brown, R. S., Clark, B. F. C. & Klug, A. (1974b) *Nucl. Acids Res. 1,* 927–932.

Roe, B. A., Wong, J. F. H., Chen, E. Y. & Armstrong, P. A. (1981) In: *Recombinant DNA, Proc. 3rd Cleveland Symp. on Macromolecules* (ed. A. G. Walton) pp. 167–176. Elsevier Scientific Publishing Company, Amsterdam.

Roe, B. A., Wong, J. F. H., Chen, E. Y., Armstrong, P. W., Stankiewicz, A., Ma, D.-P. & McDonough, J. (1982) In: *Mitochondrial Genes* (eds. P. Slonimski, P. Borst & G. Attardi) pp. 45–49. Cold Spring Harbor Laboratory, Cold Spring Harbor, New York.

Sharp, S., DeFranco, D., Dingermann, T., Farrell, P. & Söll, D. (1981) *Proc. Natl. Acad. Sci. USA 78,* 6657–6661.

Topal, M. D. & Fresco, J. R. (1976a) *Nature 263,* 285–289.

Topal, M. D. & Fresco, J. R. (1976b) *Nature 263,* 289–293.

DISCUSSION

UHLENBECK: At a meeting 2 weeks ago I heard of NMR evidence for an A-C pair similar to the middle one you showed. The data clearly indicated an exchangeable proton in the DNA helix. Patel also has good evidence for A-G pairs.

KLUG: There are A-G pairs in yeast $tRNA^{Phe}$. These, like A-C pairs, are not usually allowed in the rules for secondary structure prediction of nucleic acids. However, there is not enough flexibility in the rules. Double helical stems should be allowed to contain A-Cs, but it is difficult to estimate the "cost". Also A-A, which is required in the mitochondrial tRNA's. Even the long A-A pair I showed (Fig. 6a) could be put in at the end of a helix. In think these things are going to turn up, but we must wait for them to be confirmed by crystallography.

KURLAND: Your comment that one must be able to make G-U and A-C pairs and so on, has interesting consequences. The beauty of the DNA double helix was that it made us think that we understood physically what put one nucleotide opposite another one in DNA. We are now beginning to understand that that choice in nucleic acids is not simply being made by invariant structural rules for the nucletides themselves, but could be influenced by the action of protein, because of the flexibility in the base pairing interactions.

KLUG: Yes, it is a problem of energy, and also of the kinetics. The strange base pairs are much more transient during molecular breathing. The chemical probing methods measure the average accessibility.

ERDMANN: Could you make a comment on the nucleotide A_{29}, whether or not it is looped out in the model building or is simply turned into the double helix of the stem?

KLUG: We have looped it out, to produce continuous stacking on the rest. Fresco and Alberts showed more than 20 years ago that this is possible, though there is a little strain in the structure. In popular language, A_{29} would be squeezed out by the pressure to continue the double helix for 3 more base pairs.

ERDMANN: Do you have evidence from the chemical probing that this is a weak point in the structure?

KLUG: Yes. The helical stem is less stable, and the same happens at an A-C pair. Our experiments through a range of temperatures give a base by base picture of the melting of the helices.

OFENGAND: One thing that was striking was the distinct 2-base pair difference between this tRNA and the yeast tRNA. The other mitochondrial tRNAs all have similar curtailed secondary structures. Would they all have the shorter length, when you look at their structures and make models?

KLUG: We haven't of course tried to build models for all. But the fact is that only the mitochondrial $tRNA^{Leu}_{UUR}$ could fold exactly like yeast Phe. All the others are abnormal in one way or another. They have a smaller or larger number of bases and various loops and many of the invariant interactions are lost. Our feeling is that all of them would be shorter, except that for leucine (UUR). But this might have a special role, as we said, unconnected with protein synthesis.

OFENGAND: My point only was that if you have a defined slot size on the mitochondrial ribosome and it has to be shorter or smaller to accommodate serine tRNA, then it also has to accomodate the other tRNAs.

KLUG: Yes, that's right.

EBEL: Two remarks: the first is that these tRNAs are recognized by mitochondrial aminoacyl-tRNA synthetases which are of nuclear origin. They are specific because generally there is no cross-reaction with the cytoplasmic enzymes. So, there was a parallel pressure in evolution to make enzymes different from the cytoplasmic ones and adapted to recognize these special tRNA conformations. My second remark is that there is another class of odd tRNA-like structures located at the 3′ ends of the RNA of some plant viruses, particularly TYMV RNA. Some very artificial models were postulated, but using similar techniques to those you described, it was possile to detect the accessible sites in the structures and also to build a model of secondary structure (Florehtz *et al.* 1982). There are 3 loops, corresponding to the D-loop, the anticodon loop and T-loop, but there is no real amino acid acceptor stem. There is a complicated structure at the 3′-end which gives a kind of double-helical stem, maintaining the distance between the CCA and the anticodon similar to that existing in tRNAs (Rietveld *et al.* 1983).

MAALØE: I shudder a little at this, because of the efforts to make computer predictions for 23S RNA messenger RNAs and so on. If we add some of your looser rules, what would the effect be?

KLUG: We began "shuddering" some years ago, when we found in $tRNA^{Phe}$ that the invariant base A_{21} fixes a ribose in a very critical position. Computer models do not take such base-ribose or base-phosphate interactions into account. So I think it is in general impossible to predict a detailed 3-dimensional RNA structure. Here it is possible to do so, because we have yeast Phe structure as a basis. It would not have been possible otherwise, with the chemical data alone.

Florentz, C., Briand, J. P., Romby, P., Hirth, L., Ebel, J. P. & Giege, R. (1982) *EMBO J 1,* 269–276.
Rietveld, K., Pleis, C. W. A. & Bosch, L. (1983) *EMBO J. 2,* 1079–1085.

Role of Post-Transcriptional mRNA Modification in the Maintenance of Eucaryotic mRNA Levels

Fritz M. Rottman, Sally A. Camper & Richard P. Woychik

INTRODUCTION

Regulation of eucaryotic gene expression can occur at a variety of levels, several of which involve steps subsequent to the synthesis of nuclear mRNA precursors and prior to cytoplasmic mRNA translation. Collectively these events result in the establishment of the steady-state level of cytoplasmic mRNAs. Although the bulk of gene regulation most likely occurs at the transcriptional level, post-transcriptional control may be important in "fine tuning" the amounts of individual mRNAs in response to a variety of physiological signals. In an attempt to further define the role of posttranscriptional regulatory events, we are focusing on 3 mRNA processing reactions: the addition of 3′-terminal poly(A) tracts to mRNA, 5′-terminal capping, and methylation of mRNA molecules including internal N^6-methyladenosine (m^6A) residues. We have employed both cDNA and genomic clones for bovine pituitary growth hormone and prolactin in an attempt to define the regulated expression of specific transcripts. The major focus of these studies involves methylation of mRNA and its possible role in mRNA metabolism.

Earlier work from several laboratories established the presence of 5′-terminal cap structures and internal m^6A residues in eucaryotic mRNAs, as shown in Fig. 1 (Rottman 1978, Banerjee 1980). Capping of mRNA appears to be an early post-transcriptional event, and the enzymatic details concerning the addition of guanosine and the 7-methyl group in the formation of cap structures, has been

Department of Molecular Biology and Microbiology, Case Western Reserve University, Cleveland, USA.

GENE EXPRESSION, Alfred Benzon Symposium 19.
Editors: Brian F. C. Clark & Hans Uffe Petersen, Munksgaard, Copenhagen 1984.

$^{5'}m^7GpppN'$ (m)pN″ (m) pN‴p.....Npm^{6}ApNp.....Poly(A)$^{3'}$

cap 0	[m^{7}GpppN′]pN″ ....
cap 1	[m^{7}GpppN′m]pN″p ....
cap 2	[m^{7}GpppN′mpN″m]pN‴p ....

Fig. 1. Post-transcriptional modifications of eucaryotic mRNA. Shown are the 5′-terminal cap structures, 2′ *O*-methylnucleosides, internal N^6-methyladenosine and 3′-terminal poly(A).

described in detail for several biological systems. Although mRNA from higher eucaryotic organisms appear to always contain an intact “cap 1” structure (Fig. 1) which is characterized by the presence of a 2′*O*-methylnucleoside (N′m) in the penultimate position, the second 2′*O* methylnucleoside (N″m), found only in “cap 2” structures, appears to be the result of an optional cytoplasmic methylation reaction. Addition of internal m^{6}A is a nuclear event and a good correlation can be drawn between the presence of m^{6}A in the RNA of viruses which replicate in the nucleus as opposed to viruses which replicate in the cytoplasm and contain no m^{6}A. The cap structure has been shown in numerous studies to function in translation initiation reactions and has also been implicated to play a role in mRNA stability from experiments involving the injection of capped and uncapped mRNA into Xenopus oocytes (Green *et al.* 1983). The possible function of 2′ *O*-methylnucleoside residues and internal m^{6}A, on the other hand, is less clear. It has been hypothesized that m^{6}A residues may function in nuclear mRNA processing, possibly in splicing reactions involving the removal of intervening sequences, but little data supporting such a role has been forthcoming.

Stoltzfus *et al.* (1982) examined the potential role of methylation in the maturation of viral RNA and obtained evidence that methylation may be required for RNA processing. These model systems offer certain advantages that mainly relate to the ease of isolation of large amounts of specific transcripts. Because only some of the viral transcripts are spliced during processing, and the possibility that mRNA processing may be altered in infected cells, it is important to examine the role of methylation in the processing of normal cellular mRNAs. Our approach to studying the role of methylation in cellular mRNA processing is to utilize an inhibitor which blocks mRNA methylation, but is, nonetheless, non-toxic to cells. Such studies are best performed with specific mRNA transcripts, however, rather than with pooled mRNA populations, and require cloned sequences for isolation of individual mRNAs.

RESULTS AND DISCUSSION

Effect of mRNA methylation on mRNA metabolism

One approach to establishing the function of mRNA methylation involves the use of inhibitors which result in the formation of undermethylated mRNA sequences. In earlier studies we showed that the inhibitor S-tubercidinyl-homocysteine (STH), the 7-deazo analog of S-adenosylhomocysteine, blocked certain mRNA methylation reactions in Novikoff hepatoma cells (Kaehler *et al.* 1977). As a result of treatment with STH, mRNA in the polysomal fraction of Novikoff cells was found to contain "cap 0" structures, or caps which lacked methylation of the 2′ *O*-ribose in the N′ position (Fig. 1). Such undermethylated cap structures were not observed in untreated Novikoff cells. In addition to the decrease in ribose methylation in the N′ and N″ positions of the cap, the levels of internal m^6A were also dramatically decreased, whereas, no effect on the levels of m^7G in cytoplasmic mRNA was observed (Kaehler *et al.* 1979). From these studies we concluded that a complete complement of mRNA methyl groups was not an absolute requirement for the processing and presumed translation of at least some Novikoff mRNA species. Therefore, we decided to examine the possible role of mRNA methylation on the kinetic properties of mRNA, including its cytoplasmic stability and the rate of entry of newly synthesized mRNA sequences into the cytoplasm.

Although our earlier studies suggested that undermethylated mRNAs containing fewer than normal levels of 2′ *O*-methylnucleosides and internal m^6A residues appear to be processed and translated, there was also evidence to suggest that the composition of individual mRNA species in the total population of cytoplasmic mRNA was altered. Analysis of cap structures from mRNA synthesized in the presence of the inhibitor STH, indicated an increase in pyrimidine-containing cap structures (Kaehler *et al.* 1977). One possible explanation for this result is that methylation is involved in cytoplasmic mRNA stability and/or processing and that undermethylation of mRNA could, thereby, alter the relative population of mRNA species. As the major effect of STH on mRNA methylation appears to involve 2′ *O*-methylation and internal m^6A methylation, we believe that any observed modification of mRNA metabolism is likely due to alteration of methylation levels at these sites rather than undermethylation at the m^7G of the cap. To test the possibility that undermethylated cytoplasmic mRNAs may decay with altered kinetics, we recently measured the half-life of total HeLa cell cytoplasmic poly(A) containing RNA which was synthesized in the presence and absence of STH. Before conducting these experiments, however, we examined the effect of this inhibitor on other cellular functions. Vital staining of HeLa cells over a 24 h course of treatment indicated no detectable toxicity. Protein synthesis, as measured by (^{35}S)-methionine incorporation, remained unchanged during a 10 h course of treatment with STH. Autoradiographs of proteins labelled in the presence of the inhibitor and resolved by polyacrylamide gel electrophoresis, were indistinguishable from untreated cells. The effective half-life of total HeLa cell

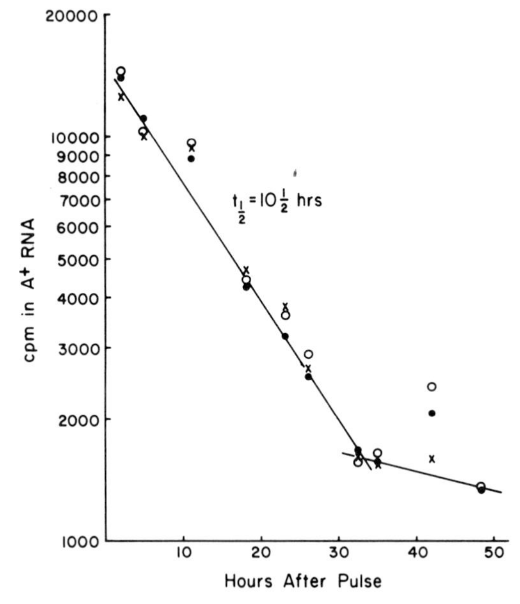

Fig. 2. Turnover of poly(A) containing RNA. HeLa cells were concentrated to 2×10^6/ml, and treated for one h with 0 (●—●), 50 μM (x—x) or 500 μM STH (O—O). Media was made 0.125 μM in (5,6-^{3}H)-uridine and cells were labelled for 2.5 h. Radioactive media was removed by washing with phosphate buffered saline. Cells were resuspended at 3.5×10^5/ml in media containing 10 mM uridine and 5 mM cytidine. 2×10^6 cells were removed at the times indicated. Cells were washed in phosphate buffered saline and incubated 10 min on ice in buffer containing 0.01 M NaCl, 10 mM Tris-Cl, pH 7.5 and 1 mM magnesium acetate. Brij-58 and sodium deoxycholate were each added to a final concentration of 0.24% to lyse the cells. Nuclei were removed by centrifugation. Poly(A) containing RNA was prepared by 2 passes over oligo(dT)-cellulose (Collaborative Research) in 0.5 M LiCl, 10 mM Tris Cl, pH 7.5, 0.5% SDS and eluted in the same buffer without LiCl. The RNA was heated at 95° C for 5 min and quenched on ice before the second binding to oligo (dT).

cytoplasmic mRNA was measured in a pulse-chase experiment in which cells were pretreated with 0, 50 or 500 uM STH for 1.5 h, then labelled with (5, 6, ^{3}H)-uridine for 2.5 h, and subsequently chased with 10 mM uridine and 5 mM cytidine. Cytoplasmic poly(A) containing RNA was isolated at various times after starting the chase. At a concentration of STH where m^{6}A was reduced by 83% and the 2′ *O*-methylribose of N′m by approximately 50%, there was no measurable change in cytoplasmic mRNA half-life (Fig. 2).

Earlier studies from our laboratory indicated that mRNA methylation is not absolutely required for nuclear processing, transport into the cytoplasm and translation on ribosomes. However, the possibility remains that methylation may influence the efficiency of nuclear mRNA splicing, polyadenylation and transport. To address this question, we measured the rate of cytoplasmic mRNA appearance in STH-inhibited and control Hela cells. Cells were prelabelled for 12 h with 220 μM (^{14}C)-uridine as an internal standard for extraction efficiency, then incubated with 0 or 500 μM STH for 1.5 h and labelled with (5,6-^{3}H)-uridine. Cytoplasmic poly(A) containing RNA was isolated from aliquots of the

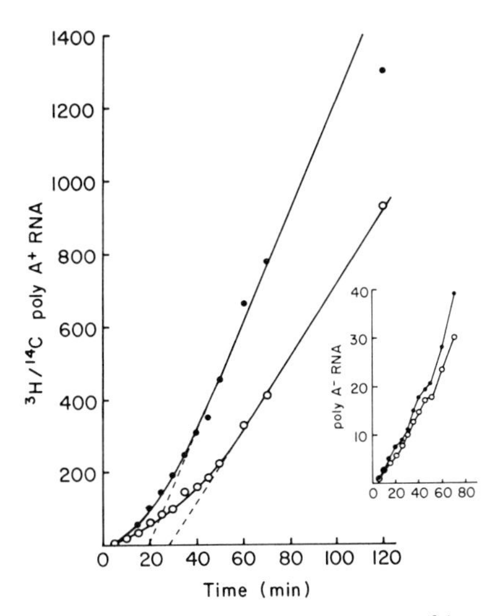

Fig. 3. Time of Cytoplasmic Appearance of RNA. HeLa cells at 4×10^5/ml were prelabelled with 220 μM (^{14}C)-uridine for 12 hours and adjusted to 2.5×10^6/ml, treated with 0 or 500 μM STH for 1.5 h, and labelled with 4 μm (5,6-^{3}H)-uridine. At various times after addition of (^{3}H)-uridine, 3.75×10^6 cells were removed and placed directly into ice-cold phosphate buffered saline. Cytoplasmic RNA was prepared as described in the legend to Fig. 2. The ratio of (^{3}H)-uridine to (^{14}C)-uridine in cytoplasmic poly(A) containing RNA and RNA lacking poly(A) (figure inset) is presented. Control (●—●), STH treated (○—○).

cells at various times (Fig. 3). These results indicate a delay in the entry of mRNA into the cytoplasm of HeLa cells from 20 min in normal cells to 28 min in STH-treated cells. This difference of nearly 50%, is likely to represent a dramatic effect on some nuclear mRNA-processing events. The delay in cytoplasmic mRNA appearance is not a general property of all classes of newly synthesized RNA, however, because a comparable effect on non-poly(A)-containing RNA is not observed in STH-treated cells (Fig. 3 inset). These data suggest that perturbation of methylation interferes with nuclear mRNA processing and/or transport and, thereby, inhibits the appearance of mRNA in the cytoplasm. Whether or not this is a generalized phenomenon affecting all mRNA species or rather, that specific mRNA sequences are differentially sensitive to undermethylation remains to be established. In any event, these results strongly indicate a correlation between mRNA methylation and processing events.

Utilization of bovine growth hormone and bovine prolactin cDNA and genomic clones

The availability of a cloned gene sequence facilitates the analysis of its corresponding mRNA. In this regard, we have generated full-length cDNA and genomic clones for the bovine pituitary hormones, prolactin and growth hormone (Woychik *et al.* 1982, Sasavage *et al.* 1982a). The steady-state concentrations of the mRNAs coding for both these hormones, aside from being highly abundant in the anterior pituitary, are developmentally regulated and change in response to thyroid hormone and glucocorticoids. Therefore, we reasoned that analysis of the expression of these 2 genes would offer the advantages associated with studying a relatively abundant mRNA, and afford the opportunity to study the factors that may be involved in the developmental, tissue-specific or hormonal regulation of specific mRNA levels. The transcriptional activity of these 2 genes, as determined by rates of accumulation of their respective mRNAs in primary bovine pituitary cell cultures and *in vitro* nuclear run-off experiments, however, was found to be very low (unpublished results). Thus, the high level of prolactin and growth hormone mRNA in the anterior pituitary gland, which accounts for at least 50% of the total poly(A)-containing mRNA in this tissue, must be a reflection of low turnover rates and not high rates of transcription. Such low endogenous transcription rates preclude significant incorporation of (^{3}H)-methyl groups and the subsequent characterization of mRNA methylation patterns. As an alternative, we are using DNA-mediated gene-transfer techniques to express a growth-hormone genomic clone. One example of this approach involved the subcloning of a 4.3 Kb Eco RI genomic fragment containing the entire growth-hormone gene into an expression plasmid containing the SV40 "enhancer" sequence (pSVB3, Fig. 4). Removal of BamHI fragment from this plasmid juxtaposes the transcription start site of the growth-hormone gene with the late SV40 promoter (pSVB3/Ba, Fig. 4). Transfection of the pSVB3/Ba plasmid into COS-1 cells using the calcium-phosphate technique resulted in a high level of expression of growth-hormone gene (Fig. 5). The large amounts of growth hormone mRNA made under these circumstances contains a normal 3′-terminal poly(A) region and codes for immunologically reactive growth-hormone protein (unpublished results). This expression system, utilizing the late SV40 promoter to drive the growth-hormone gene, has already proven to be an adequate model for several analyses. The pSVB3 plasmid, which contains the growth-hormone promoter and 5′ flanking sequence, however, does not appear to express the growth-hormone gene with high efficiency.

The full-length bovine growth hormone and prolactin cDNA clones we isolated and characterized, were used to study the polyadenylation sites of these

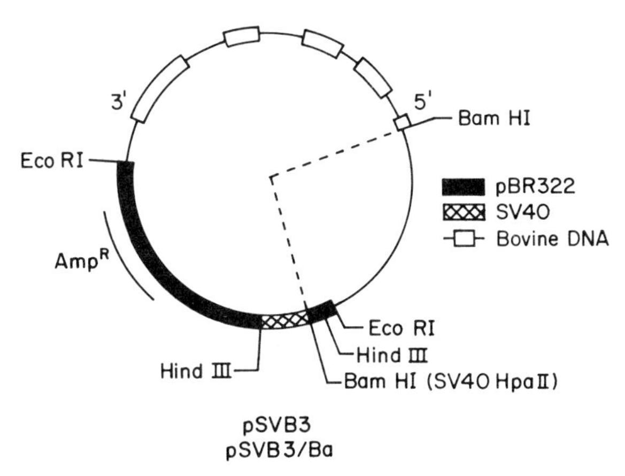

Fig. 4. Structure of the Expression Plasmids. The 4.3 kb Eco RI bovine genomic fragment was introduced into a plasmid expression vector containing the SV40 enhancer sequence (pSVB3). The SV40 sequence within this molecule corresponds to the approximately 400 bp Hind III-Hpa II restriction fragment containing the SV40 replication origin. Deletion of the BamHI fragment (designated with a dashed line) juxtaposes the transcription start site of the growth hormone gene with the late SV40 promoter to produce pSVB3/Ba. Therefore, using pSVB3, the growth-hormone gene is driven off its own promoter, while with the pSVB3/Ba the growth-hormone gene is driven off the SV40 late promoter.

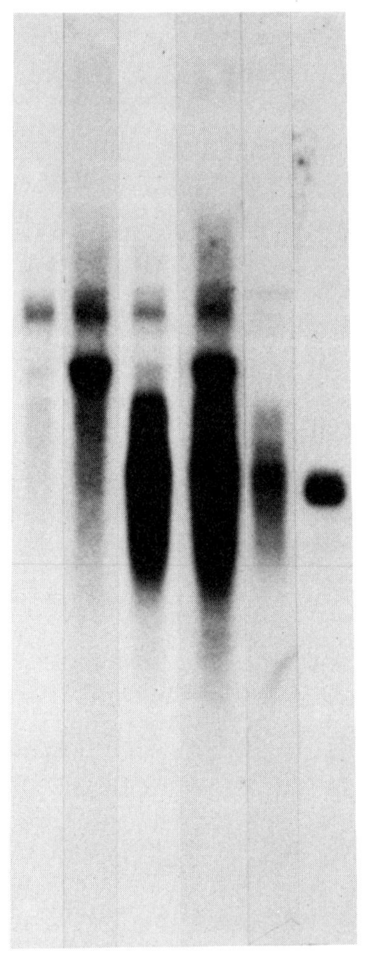

Fig. 5. Analysis of pSVB3 and PSVB3/Ba transcripts. Autoradiogram showing a Northern Blot of poly(A) containing RNA extracted from COS-l cells transfected with plasmid constructs containing the growth-hormone gene (lanes A-E) or poly(A) containing RNA extracted from a bovine pituitary (lane F). (Lanes A-F, from *l-r*). Lanes A-E: 1/4 of the RNA extracted from a confluent monolayer of COS-l cells on a 175 cm^2 flask. Lane F: 25 ng. of poly(A) containing RNA extracted from a bovine pituitary. Lanes A and B contained RNA from COS-l cells transfected with pSVB3 while lanes C and D were from cells transfected with pSVB3/Ba. Lane E is a shorter exposure of lane C. Lanes A through F were hybridized with a bovine growth hormone cDNA probe while lanes B and D were additionally hybridized with an α-tubulin probe (middle band) to permit relative comparison of mRNA levels with an endogenous transcript.

specific mRNAs. Prior to these studies polyadenylation of a given mRNA sequence was assumed to always occur at the same nucleotide. We devised a method to determine the sequence of nucleotides adjacent to the 3′-terminal poly(A) tail in a given mRNA. A series of oligonucleotides of the sequence d(pT_8-N-N′) were used as specific primers to span the poly(A) junction in single mRNA species and, thus, initiate synthesis of cDNA by reverse transcriptase. Only the primers containing N and N′ nucleotides complementary to those in mRNA were effectively used by the transcriptase. The nucleotide sequence of the region immediately upstream to the poly(A) tail was then determined using the dideoxy sequencing method. Bovine growth hormone mRNA showed a single site for polyadenylation, corresponding to the activity of only a single oligonucleotide primer, d(pT8-G-C) in the reverse transcriptase reaction (Sasavage *et al.* 1980). By contrast, 3 different primers were utilized by reverse transcriptase when bovine prolactin mRNA was used as template. Sequence analysis suggested 3 distinct polyadenylation sites within a 12 nucleotide region spanning the 3′ end of the prolactin message (Sasavage *et al.* 1982b). The presence of multiple polyadenylation sites on the prolactin mRNA was confirmed by sequencing individual bovine prolactin cDNA clones. Several other mRNAs have been shown to exhibit multiple polyadenylation sites over longer stretches of nucleotides (Early *et al.* 1980, Tosi *et al.* 1981). In some cases the genes coding for these transcripts contain multiple AAUAAA sequences, a conserved nucleotide sequence that has been shown to function in polyadenylation (Fitzgerald & Shenk 1981). The bovine prolactin gene contains only a single AAUAAA and has been determined to be a single-copy gene (unpublished results). Therefore, there must be some signal in addition to the AAUAAA which directs the polyadenylation reaction. Since RNA polymerase II is known to transcribe genes beyond the polyadenylation site (Hofer & Darnell 1981), this additional signal may be found within the 3′ flanking sequence.

SUMMARY

The methylation inhibitor, STH, is a useful tool for studying the role of mRNA methylation. STH has no cytotoxicity, as determined by vital staining procedures, and shows no significant inhibition in the synthesis of cellular proteins or polyamines (Hibasami *et al.* 1980). Treatment with STH significantly decreases the extent of 2′ *O*-methylation within the cap structure itself, and also efficiently reduces the extent of m^6A methylation within the mRNA molecule. Thus, the metabolism of undermethylated mRNA sequences can be studied in an environment in which other macromolecular synthesis and general cellular metabolism appears to be relatively normal.

STH had no effect on the apparent half-life of cytoplasmic mRNA. The significant delay in cytoplasmic appearance of mRNA in the presence of STH, however, suggests some interference with nuclear mRNA metabolism. We plan to determine whether the delay in appearance results from an inhibition of splicing, polyadenylation and/or transport by examining the nuclear processing of individual mRNAs. The use of currently available techniques for DNA-mediated gene transfer make it possible to study these processes with prolactin and growth hormone mRNA by virtue of their enhanced transcription. Furthermore, the availability of transfection systems in which growth hormone and prolactin genes are efficiently transcribed to produce mRNA sequences with 3′-termini corresponding to wild-type transcripts provides an opportunity to examine the details of post-transcriptional polyadenylation signals.

REFERENCES

Banerjee, A. M. (1980) *Microbiol. Rev. 44,* 175–205.

Early, P., Rogers, J. Davis, M., Calame, K., Bond, M. Wall, R. & Hood, L. (1980) *Cell 20,* 313–319.

Fitzgerald, M. & Shenk, T. (1981) *Cell 24,* 251–260.

Green, M. R., Maniatis, T. & Melton, D. A. (1983) *Cell 32,* 681–694.

Hibasami, H., Borchardt, R. T., Chen, S. Y., Coward, J. K. & Pegg, A. E. (1980) *Biochem. J. 187,* 419–428.

Hofer, E. & Darnell, Jr., J. E. (1981) *Cell 23,* 585–593.

Kaehler, M., Coward, J. & Rottman, F. (1977) *Biochemistry 16,* 5770–5775.

Kaehler, M., Coward, J. & Rottman, F. (1979) *Nucleic Acids Res.* 6, 1161–1175.

Rottman, F. M. (1978). In: B. F. C. Clark (ed.) *Biochem. of Nucleic Acids 17,* 45–73. University Park Press, Baltimore.

Sasavage, N. L., Smith, M., Gillam, S., Astell, C., Nilson, J. H. & Rottman, F. M. (1980) *Biochemistry 19,* 1737–1743.

Sasavage, N. L., Nilson, J. H., Horowitz, S. & Rottman, F. M. (1982a) *J. Biol. Chem. 257,* 678–681.

Sasavage, N. L., Smith, M., Gilam, S., Woychik, R. P. & Rottman, F. M. (1982b). *Proc. Natl. Acad. Sci. USA 79,* 223–227.

Stoltzfus, C. M. & Dane, R. W. (1982) *J. Virol. 42,* 918–931.

Tosi, M., Young, R. A., Hagenbuchle, O. & Schibler, U. (1981) *Nucleic Acids Research 9,* 2313–2323.

Woychik, R. P., Camper, S. A., Lyons, R. H., Horowitz, S., Goodwin, E. C. & Rottman, F. M. (1982) *Nucleic Acids Res. 10,* 7197–7210.

DISCUSSION

SCHATZ: There is now increasing evidence, both from prokaryotes and from eukaryotes that rapid carboxymethylation of proteins and methylation of phospholipids play some regulatory role in differentiation, chemotaxis and possibly a host of other phenomena. When you add a methylation inhibitor, which would be expected to interfere with all these other methylations as well. I worry that whatever effect you see in terms of cytoplasmic mRNA entry might reflect events other than methylation of mRNA.

ROTTMAN: Indeed, the question of specificity is a problem one must always contend with in using an inhibitor to study these reactions *in vivo.* However, the time span involved in studies on cytoplasmic mRNA is very short. I don't know about the time span of these protein modifications, but the effects we observe are within minutes after the addition of labelled uridine.

SCHATZ: The other methylation reactions are very fast, too!

ROTTMAN: We certainly cannot exclude the possibility of STH inhibiting other methylation events within the cell.

KERR: I thought you said that the methylation inhibitor resulted in a substantial change in the ratio of pyrimidines and purines in N_1 and N_2 position of the cap, implying a substantial change in the mRNA population, and yet you see no overall marked change in the pattern of protein synthesis. Those seem to be paradoxical observations.

ROTTMAN: There was no shift in the ratio of pyrimidine and purine containing caps in HeLa cells, where we looked at the inhibition of protein synthesis. However, even if there was a shift in the ratio of labelled purine and pyrimidine containing caps in the presence of STH, I'm not sure we would have detected this at the level of translation because the bulk of the mRNA present would have been methylated prior to the addition of the inhibitor. With the translational assay you are mainly looking at abundant mRNA species. I believe that the only way we will be able to sort this out is to look at specific mRNA sequences, which we are now doing. There could well be changes in mRNA population which are rapidly turning over that would be observed in mRNA labelling studies yet not influence the overall translation products of steady state mRNA.

KERR: Am I right in thinking that some of the published studies on virus-associated methylating enzymes indicate inhibition at the m^7-position of the cap, and if so, have you looked at the effect of this inhibitor on viral RNA methylation?

ROTTMAN: Yes, there are reports that STH does inhibit specific viral methylases, such as the inhibition of the $m^7{\sim}G$ methylase of Newcastle Disease Virus and also Vaccinia, but from our earlier work it apparently does not inhibit the Novikoff hepatoma cellular enzyme. There may be a definite order in the sensitivity of different methylation reactions in response to restricted levels of methionine or in the presence of methylation inhibitors and the least sensitive of these methylation reactions to inhibition might be the m^7-position of the cap. Whether this is a protective strategy on the part of the cell to protect a critical mRNA methylation event remains speculative.

V. Components of the Biosynthetic Apparatus.
II. Ribosomes

On the Structural Organization of the tRNA-ribosome Complex

James Ofengand, Piotr Gornicki, Kelvin Nurse & Miloslav Boublik

INTRODUCTION

Although much is known about the detailed architecture of the ribosome, especially that of *E. coli* (Liljas 1982), and the three-dimensional structure of several tRNAs have been determined (Moras *et al.* 1980 and references therein), the structural organization of the ribosome-tRNA complex is still only poorly understood. Earlier attempts to analyze this complex used a variety of chemical and enzymatic techniques. More recent work has focussed on affinity-labelling methods (reviewed in Ofengand 1980) because this approach, when properly controlled, provides unequivocal information about tRNA neighborhoods. Coupling of the identification of crosslinked ribosomal components with immunoelectron microscopic localization of the same components on the ribosomal surface then defines the tRNA binding sites. In a few cases, the crosslinked tRNA (Keren-Zur *et al.* 1979, Gornicki *et al.* 1983), or antibiotic marker (Olson *et al.* 1980, 1982, Lührmann *et al.* 1981a, Stöffler & Stöffler-Meilicke 1983) has been directly visualized by immunoelectron microscopy.

Up to now, affinity-labelling has mostly identified ribosomal proteins. Implicit in this work has been the assumption that the crosslink site in a given protein will correspond to its antigenic site. Although this may not be true for elongated proteins possessing multiple antigenic sites, as the number of such proteins continues to decrease (Stöffler & Stöffler-Meilicke 1983), this assumption may acquire a more general validity.

Two functional domains of the tRNA binding site have been localized by this approach. Affinity-labelling from the 3′-end of AA-tRNA or with antibiotics whose mechanism of action is believed to be at the peptidyl transferase center has identified a set of 8 50S proteins, L2, 11, 14, 15, 16, 18, 23, and 27, and 2 30S proteins, S14 and S18, as being in the vicinity of this center (Table I). No

Roche Institute of Molecular Biology, Roche Research Center, Nutley, New Jersey 07110, U.S.A.

GENE EXPRESSION, Alfred Benzon Symposium 19.
Editors: Brian F. C. Clark & Hans Uffe Petersen, Munksgaard, Copenhagen 1984.

Table I

Affinity-labelling identification of ribosomal proteins at tRNA binding sites

Site	Ligand	Labelled proteins: Major	Labelled proteins: Minor	References
Peptidyl transferase center	N-substituted AA tRNA	L2, L11, L15, L16, L18, L27	L5, L14, L33	Ofengand (1980), Cooperman (1980), Johnson & Cantor (1980)
	Puromycin and analogs	L18, L23, S14		Cooperman (1980), Nicholson *et al.* (1982), Olson *et al.* (1980, 1982)
	Chloramphenicol and analogs	L2, L11, L16, L27		Cooperman (1980), Le Goffic *et al.* (1980)
	Linomycin analog	L14		Cooperman (1980)
	Pleuromutilin analog	L2, L27, S18		Hogenauer *et al.* (1981)
Decoding site	mRNA analogs	S1, S3, S4, S5, S12, S18, S21		Ofengand (1980), Cooperman (1980)
	MS2 RNA	S3, S4, S9, S18	S5, S7	Broude *et al.* (1983)
Unspecified	Tetracycline (A site)	S7		Goldman *et al.* (1980)
	tRNA: random C (P site)	S10		Riehl *et al.* (1982)
	tRNA: random G			
	A site:	S9, S15–19, L8/9, L13, L15, L27		Babkina *et al.* (1983)
	P site:	S5, S9, S11–13, S19–S21, L14, L24, L27, L31, L33		
	tRNA: random by UV			
	A site:	S5, S9, S10, L2, L6, L16		Abdurashidova *et al.* (1979)
	P site:	S9, S11, L2, L4, L7/12, L27		

distinction between A and P sites has been made since the aminoacyl end of both P and A site bound tRNAs should be close together in the functioning ribosome (Ofengand 1980). Location of these proteins on the 50S subunit defines the peptidyl transferase center (Fig. 1A). Three of the proteins mapped (L15, L18, L27) are clustered, but 2 (L11, L23) are separate. Although L11 is distant from L15, L18, and L27, it is one of the best characterized antigenic sites (Stöffler & Stöffler-Meilicke 1983). L23 and L18 in about equal amount are the major puromycin-labelled proteins, yet direct affinity immunoelectron microscopy (AIM) places the main puromycin site in the L15-L18 region (Olson *et al.* 1982, Lührmann, *et al.* 1981a). Possibly the major site (78%) between the left and central lobes is L18-specific, and the minor amount of puromycin crosslinking observed to the base of the 50S subunit (Olson *et al.* 1982) represents adventitious crosslinking to L23. In any case, direct AIM of crosslinked chloramphenicol (Stöffler & Stöffler-Meilicke 1983) confirms the region shown

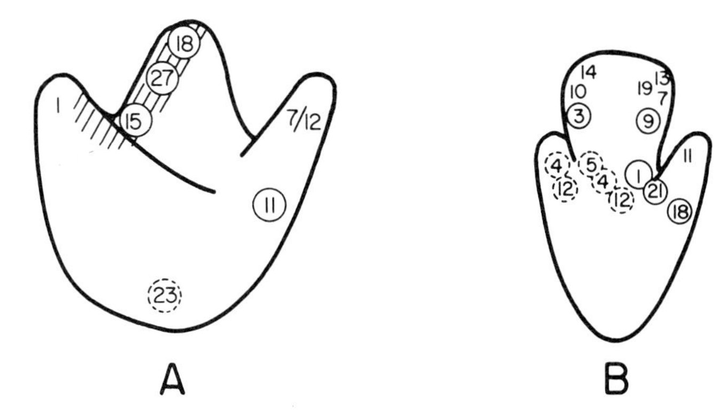

Fig. 1. Proteins on the surface on the *E. coli* ribosome which have been crosslinked with tRNA or analogs. (A) 50S subunit. Shape of the subunits and location of the proteins by immunoelectron microscopy is a composite of Stöffler and Stöffler-Meilicke (1983) and Lake and Strycharz (1981). The shape according to Boublik *et al.* (1977) is similar. (B) 30S subunit. Shape and protein location is a composite of Stöffler and Stöffler-Meilicke (1983), Stöffler-Meilicke and Stöffler (1982), Winkelmann *et al.* (1982), Sillers and Moore (1981), and Boublik *et al.* (1982). Two sites are shown for S4 and S12 as Lake and Stöffler disagree about their location. There is reasonable agreement for the other proteins shown, except that Winkelmann *et al.* (1982) place S3, S10, and S14 on the surface away from the viewer and include a second S19 site next to S10 and S14, while Stöffler and Stöffler-Meilicke (1983) locate S3, S10, and S14 at the periphery of the head, and, thus, detectable from both surfaces. They do not report a second S19 site. Circled numbers in (A) are proteins near to the peptidyl transferase center, and in (B) near to the decoding site according to the data of Table I. Dashed circles denote a protein on the surface away from the viewer, and plain numbers the sites for other proteins. The shaded area in (A) marks the putative peptidyl transferase center.

in Fig. 1A as the peptidyl transferase center. This region should correspond to the location of the 3′-end of both A and P site bound tRNA.

Although both S14 and S18 have been identified at the peptidyl transferase region of the 30S subunit, inspection of Fig. 1B shows that both identifications cannot be correct. As the S14 region has been directly identified by AIM with puromycin (Olson *et al.* 1980), see also Fig. 3, it is more likely to be the correct site. Correspondence of S14 with the peptidyl transferase region is in accord with other studies which place the head of the small subunit near to L1 and far from L7/12 in the 70S ribosome (Kastner *et al.* 1981, Lake 1982, Stöffler & Stöffler-Meilicke 1983).

The decoding site, corresponding to the tRNA anticodon, has been delineated with the use of mRNA analogs (Table I). A set of 8 proteins has been identified. Again, no distinction between A and P site is made since the 2 tRNA anticodons must be adjacent to each other during elongation. The proteins form a reasonably compact cluster (Fig. 1B) in the vicinity of the partition between the head and body of the subunit. Although the crosslinking of S3 has been clearly shown to be tRNA-dependent (Ofengand 1980), it is located outside the main area. This discrepancy is not understood. S3 has been shown to be important for ribosomal binding of tRNA by chemical modification, reconstitution, and protection studies (Ohsawa & Gualerzi 1983), Gimautdinova *et al.* (1981) have reported extensive protein labelling with mRNA analogs. From their results, additional 30S proteins are S11 and S13. These proteins occur in the same general vicinity as the other mRNA proteins. Numerous 50S proteins which span the subunit from L1 to L7/12 were also labelled.

The tRNA A site also includes S7, which is crosslinked to tetracycline. The P site includes S10, which was crosslinked from an unspecified C residue in the tRNA. The proximity of S10 to S14 suggests that the C residue may be near the 3′-end of the tRNA, although it is not C_{74} or C_{75}. A number of other proteins of both subunits have been labelled with randomly modified or reacted tRNAs in either the P or A sites. However, the number and distribution of the proteins labelled, coupled with a lack of information about which labelled protein makes contact with which segment of tRNA, make these latter experiments difficult to interpret.

THE DECODING SITE OF THE RIBOSOME

Crosslinking of the anticodon of tRNA to 16S rRNA at the ribosomal P site

The lack of precision in the location of the decoding site evident in Fig. 1B

Table II

Crosslinking of E. coli *AcVal-tRNA$_1^{Val}$ to ribosomes from prokaryotes, eukaryotes, and eukaryotic organelles*

Ribosomes	Percent Crosslinking	Reference
E. coli	48.	Ofengand *et al.* 1979
Yeast cytoplasm	31.	Ofengand *et al.* 1982
A. salina	46.	Nurse & Ofengand ,unpublished results
Spinach chloroplasts	31.	Ofengand *et al.* 1982
Yeast mitochondria	0.7	Nurse & Ofengand, unpublished results

Crosslink formation and assay conditions were as described in Ofengand, *et al.* (1982) with poly (U_2,G) as a source of codons, and irradiation at 300 nm without the Mylar filter (290–325 nm). Binding reactions for *A. salina* and yeast mitochondria were essentially the same as for yeast cytoplasm (Ofengand *et al.* 1982). Crosslinking to yeast mitochondrial rRNA was assayed by Sephacryl S-300 chromatography in 0.5% SDS, 1 mM EDTA after disruption of irradiated ribosome-tRNA complexes in 1% SDS. This procedure was considerably more sensitive than the usual Millipore filtration assay.

prompted our continued interest in potential higher resolution methods for defining this important functional site on the ribosome. The serendipitous discovery that the anticodon of certain tRNAs could be crosslinked to the ribosome by irradiation with light >310 nm (Schwartz & Ofengand 1974, 1978) made it possible to explore the architecture of the decoding site in some detail. The crosslink was uniquely to the 16S rRNA component of the 70S ribosome, occurred at high (>50%) efficiency, was specifically from P site-bound tRNA, and only occurred with tRNAs containing either cmo^{5}U[1] or mo^{5}U at the 5′-anticodon position, which was also deduced to be the crosslinking site (Ofengand *et al.* 1979). Furthermore, crosslinking was shown to be codon-specific (Ofengand & Liou 1981).

In contrast to this high specificity for tRNA, ribosomal site, ribosomal component, and codon, ribosomes from a variety of sources could be crosslinked, albeit at somewhat different efficiencies (Table II). It is noteworthy, however, that yeast mitochondrial ribosomes could not be crosslinked. For yeast and *A. salina,* crosslinking was entirely to 18S rRNA and occurred at the P site, just as with *E. coli* ribosomes. For the yeast mitochondrial ribosomes which did not crosslink, the bound AcVal-tRNA was shown to be puromycin-sensitive, and, thus, at the P site.

[1] cmo^{5}U, 5-carboxymethoxyuridine; mo^{5}U, 5-methoxyuridine.

In *E. coli* ribosomes, the structure of the crosslink was found to be that of a pyrimidine-pyrimidine cyclobutane dimer (Ofengand & Liou 1980). The main diagnostic characteristic was *photoreversal* by 254 nm light, a property shared by the yeast ribosome crosslink (Ofengand *et al.* 1982). In the case of *A. salina,* 254 nm photolysis with the same kinetics as for *E. coli* was observed but reversal was not tested. It, thus, seems likely that in all cases, cyclobutane dimers were formed. This requires that the 2 bases stack on each other so that their C_5-C_6 π-π bonds can approach to within *ca.* 4Å of each other. As the rRNA crosslink site is not near an end (see below), such stacking requires that a loop of rRNA be able to approach the 6 base-pair stack made from the P and A site bound tRNA anticodon-codon complex. This is possible even though the 5′-anticodon base of the P site tRNA is at the center of the stack because in order for 2 tRNAs in their known crystallographic configuration to basepair with adjacent codons, the mRNA must kink between codons in order to accommodate the bulk of the tRNA (Sundaralingam *et al.* 1975). A possible model for the stacking of the rRNA pyrimidine on the 5′-anticodon tRNA pyrimidine without interfering with codon-anticodon recognition is given in Prince *et al.* (1982).

The site of crosslinking in both *E. coli* and yeast rRNA has been identified as C_{1400} in *E. coli* (Prince *et al.* 1982, Ehresmann *et al.* 1983), or C_{1626} in yeast (Ehresmann *et al.* 1983). In addition, the same C_{1400} was identified when *B. subtilis* $tRNA^{Val}$ with mo^5U instead of cmo^5U, i.e., with the carboxyl group absent, was crosslinked (Ehresmann *et al.* 1983). The crosslinked residue in tRNA was confirmed to be the 5′-anticodon base in all 3 experimental situations.

The C_{1400} (or C_{1626}) residue occurs near the 3′-end of the rRNA in the center of a sequence of 16 bases which has been conserved virtually throughout all species studied (Table III). The only variations are found in fungal mitochondria. Possibly the sequence variation around the C_{1400} residue in yeast mitochondria is sufficient to block crosslinking even though the C_{1400} residue itself is retained. This aspect needs further study.

The conserved region of rRNA shown in Table III is believed to be single stranded (Noller & Woese 1981, Stiegler *et al.* 1981a, Zwieb *et al.* 1981), and to be located near the surface of the 30S (Herr *et al.* 1979, Stiegler *et al.* 1981b, Vassilenko *et al.* 1981), at the interface between the 30S and 50S subunits (Herr *et al.* 1979, Vassilenko *et al.* 1981). It is one of only 3 such long sequence-conserved regions in 16-18S rRNA, the other 2 being residues 1492-1505 and 517-533. All are thought to be single-stranded (Noller & Woese 1981). No function has been assigned to any of these regions except for the anticodon

Table III

Conserved sequence in rRNAs from small subunit

Prokaryotes																
E. coli	G	U	A	C	A	C	A	C	C	G	m^4Cm	C	C	G	U	m^5C
P. vulgaris	G	U	A	C	A	C	A	C	C	G	m^4Cm	C	C	G	U	C
B. brevis	G	U	A	C	A	C	A	C	C	G	C	C	C	G	U	C
Mycoplasma sp. (6)[a]	G	U	A	C	A	C	A	C	C	G						
Methanogen sp. (15)[a]	G	C̲	A	C	A	C	A	C	C	G						
H. halobium, H. volcanii	G	C̲	A	C	A	C	A	C	C	G						
T. acidophilum	G	C̲	A	C	A	C	A	C	C	G						
Prochloron	G	U	A	C	A	C	A	C	C	G						
Eukaryote Cytoplasm																
S. cerevisiae	G	U	A	C	A	C	A	C	C	G	C	C	C	G	U	C
B. mori	G	U	A	C	A	C	A	C	C	G	C	C	C	G	U	C
			A̲	C̲												
D. melanogaster	G	U	A	C	A	C	A	C	C	G	C	C	C	G	U	C
X. laevus	G	U	A	C	A	C	A	C	C	G	Cm	C	C	G	U	C
Rat, mouse, rabbit	G	U	A	C	A	C	A	C	C	G	C	C	C	G	U	C
Eukaryote Organelle																
Z. mays chloroplast[b]	G	U	A	C	A	C	A	C	C	G	C	C	C	G	U	C
Hamster mitochondria	G	U	A	C	A	C	A	C	C	G	m^4C	C	m^5C	G	U	C
Mouse mitochondria	G	C̲	A	C	A	C	A	C	C	G	C	C	C	G	U	C
Human mitochondria	G	U	A	C	A	C	A	C	C	G	C	C	C	G	U	C
S. cerevisiae mitochondria	G	C̲	A	C	U̲	A̲	A	U̲	C	A̲	C	U̲	C	A̲	U	C
A. nidulans mitochondria	G	U	A	C	U̲	A̲	A	C	C	A̲	C	U̲	C	G	U	C
Residue No. for *E. coli*[c]	1392								1400							1407

[a] Nos. in brackets refer to number of species examined.

[b] Also *E. gracilis, C. reinhardii* chloroplasts.

[c] Bases numbered according to *E. coli* 16S rRNA (Noller & Woese 1981).

References for the sequences listed above are cited in Ofengand *et al.* (1982), and Ehresmann *et al.* (1983). *H. volcanii* is from Gupta *et al.* (1983). Underlined bases are those that differ from the *E. coli* sequence.

crosslink described above. Nevertheless, it seems highly unlikely that such long stretches of conserved sequences would have survived if they did not play some important role in ribosome assembly or function.

The rRNA crosslink site is highly specific. Despite the presence of an adjacent C_{1399} or C_{1625}, no crosslinking (<1%) could be detected to this residue (Ehresmann & Ofengand 1983), although it should have been equally photochemically reactive. This fact indicates that a very specific structure is formed between this region of the rRNA and the tRNA anticodon. The conservation of the sequence surrounding the crosslink site, its location in the rRNA chain, and its crosslinking ability all serve to reinforce this view, implicating this region of rRNA as well as the 5′-anticodon end of P site-bound tRNA in some function(s) essential to both prokaryotic and eukaryotic protein synthesis.

Our results also show that a tertiary folding of the rRNA in the 30S particle must be superimposed on the secondary structure. First, the C_{1400} residue in *E. coli*, 130 residues from the 5′-terminus of the Shine-Dalgarno region, A_{1531}, should be equivalent in location to the 3′-G residue of the initiator AUG codon since it is crosslinked to the 5′-anticodon base of P site bound tRNA. Yet in *Φ*XH mRNA, this G residue is only 5 nucleotides away from the U residue complementary to A_{1531} (Sanger *et al.* 1978). Second, Wagner *et al.* (1976) have shown that residues 462–474 are in the vicinity of the decoding site, yet this region is also distant from C_{1400} in the secondary structure models. A third crosslink reported by Thompson and Hearst (1983) brings still another loop of 16S RNA (619–625) into the vicinity of the decoding site (1420–1427).

Involvement of this region of 16S rRNA in the decoding process is reinforced by the mapping of paromomycin-resistance to a nearby site (Table IV), and by the crosslinking of S12, a protein well-known to be involved with the decoding process (Liljas 1982), to G_{1322} (Chiaruttini *et al.* 1982). Several other functional sites on the ribosome which intimately involve rRNA are also summarized in this Table. It is conceivable that most ribosomal functions will eventually be shown to be controlled by rRNA, with ribosomal protein serving mainly as structural elements to maintain correct rRNA conformation.

Location of the decoding site on the ribosome by immunoelectron microscopy

Topographic localization of the decoding site on the 30S ribosome would not only delineate that site, but would also localize C_{1400}, the 5′-end of the Shine-Dalgarno sequence, and rRNA residues 462–474. The technique of affinity immunoelectron microscopy, in which an antigenic group is crosslinked to the ribosome, was used. In a previous attempt (Keren-Zur *et al.* 1979), the added antigen was some 80Å from the crosslink site, necessitating a "triangulation" analysis to determine the approximate location of the site. Here, we have taken advantage of the presence of a carboxyl group in cmo^5U, the tRNA residue which is crosslinked, in order to attach a DNP group by a variable length spacer directly to the residue which is crosslinked (Fig. 2, Left). For this study we used the 21Å long adduct (n=1). Nuclease digestion and denaturing gel electrophoresis of [^{14}C]DNP-modified tRNA showed that the correct product was

Table IV

Functional sites in ribosomal RNA

RNA	Activity	Site*	Species	References
16S	Decoding site			
	P site anticodon	C_{1400}	Prokaryote, eukaryote	Ofengand *et al.* (1982)
	Paromomycin resistance (Miscoding)	$C_{1409} \rightarrow G$	Yeast mitochondria	Martin *et al.* (1982)
	mRNA codon binding	$G_{462} \ldots G_{474}$	Prokaryote	Wagner *et al.* (1976)
	mRNA base-pairing	$A_{1531}UCACCUCCUUA_{1542}$	Prokaryote	Steitz (1980)
	Kasugamycin resistance (Initiation)	$m_2^6Am_2^6A_{1519} \rightarrow AA$	Prokaryote, eukaryote?	Helser *et al.* (1972)
23S	Peptidyl transferase	N^+AAGC	Prokaryote	Leitner *et al.* (1982)
		UU_{2585}	Prokaryote	Barta *et al.* (1984)
	Thiostrepton resistance (GTPase center)	$A_{1067} \rightarrow Am$	Prokaryote	Thompson *et al.* (1982)
	α-sarcin nuclease (t-RNA A site)	G_{2661}	Prokaryote, eukaryote	Endo & Wool (1982)
	Erythromycin resistance	$A_{2058} \rightarrow m_2^6A$	Prokaryote	Skinner *et al.* (1983)
		$A_{2058} \rightarrow G$	Yeast mitochondria	Sor & Fukuhara (1982)
	Chloramphenicol resistance	$A_{2447} \rightarrow A$	Yeast mitochondria	Dujon (1980)
		$A_{2451} \rightarrow U$	Mouse mitochondria	Kearsey & Craig (1981)
		$C_{2452} \rightarrow A$	Human mitochondria	Blanc *et al.* (1981a)
		$A_{2503} \rightarrow C$	Yeast mitochondria	Dujon (1980)
		$U_{2504} \rightarrow C$	Mouse mitochondria	Kearsey & Craig (1981)
				Blanc *et al.* (1981b)
	EFG contact	G_{1055}–C_{1076}	Prokaryote	Sköld (1983)

* Numbering system according to *E. coli* rRNAs (Noller & Woese 1981, Brosius *et al.* 1980).

[+] Crosslinked N: either 1468, 1543, 1578, 1875, 1888 or 2756.

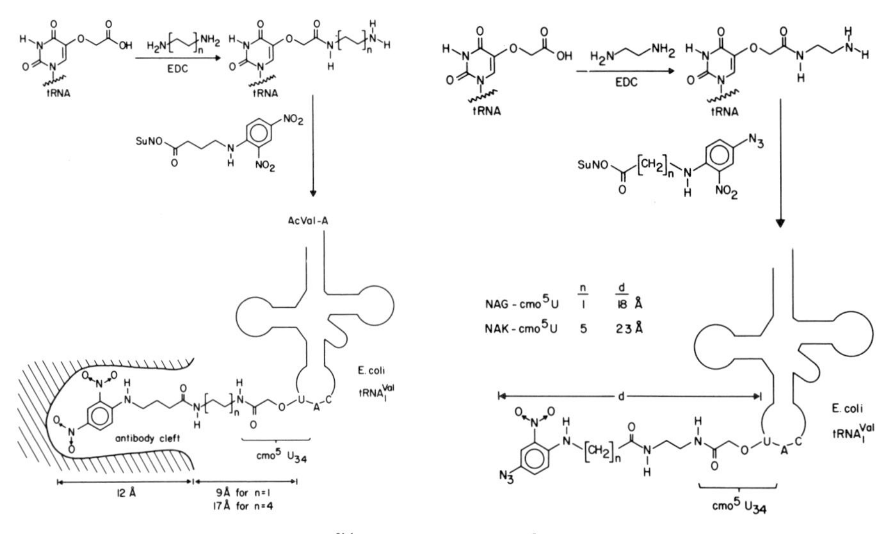

Fig. 2. Scheme for modifying $tRNA^{Val}$ of *E. coli* at its cmo^5U_{34} residue. *Left panel:* addition of the dinitrophenyl group; *Right panel:* addition of the 2-nitrophenyl azide group. SuNO, N-hydroxysuccinimide; EDC, 1-(3-dimethylaminopropyl)-3-ethylcarbodiimide. Condensation with ethylenediamine was done at pH 4.0 (Krzyzosiak *et al.* 1979) and reaction with the SuNO ester was at pH 8.1. Yields ranged from 90 to 95% based on the fraction retained on BD-cellulose when modified tRNAS were purified.

obtained. As the depth of the DNP-binding site is 12Å (Willan *et al.* 1977) the actual distance from the surface of the IgG to the crosslinked nucleotide is only 9Å. This is probably close to the minimum needed for accessibility to the DNP probe.

Both P site-binding and crosslinking of the modified tRNA was 60–70% that of the unmodified tRNA, and was codon and irradiation dependent. After crosslinking, ribosomes were dissociated into subunits by exposure to 2 mM Mg^{++}, and treated with a 7-fold excess of anti-DNP. Excess IgG was removed by gel filtration on Sepharose 6B, and the amount of antibody in the ribosome peak determined (Keren-Zur *et al.* 1979). Only when DNP-modified tRNA was irradiated and 70S ribosomes were dissociated could antibodies be detected in the ribosome region of the column. Neither crosslinked unmodified tRNA, uncrosslinked modified tRNA, nor undissociated ribosomes crosslinked to DNP-containing tRNA were able to bind antibody. The failure of DNP-modified tRNA-70S ribosome covalent complexes to react with antibody shows that the DNP group must be at or near the subunit interface. Such a location for

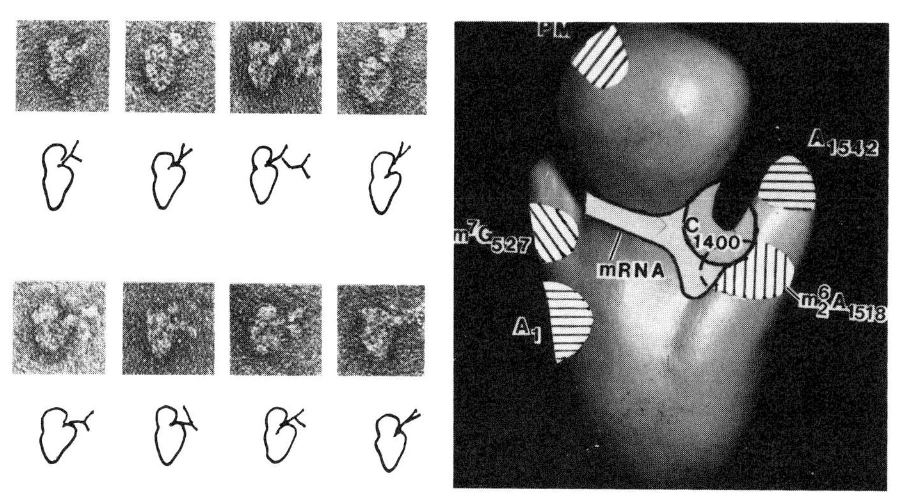

Fig. 3. Immunoelectron microscopic visualization of the decoding site on the *E. coli* ribosome. Panel A: Gallery of selected images of complexes between anti-DNP antibody and covalent complexes of AcVal-$tRNA^{DNP}$ and 30S subunits. Fractions from the Sepharose 6B column (see text) were examined by eletron microscopy (Keren-Zur *et al.* 1979). One-third of the measured antibody complexes were detected in the microscope. Representative monomers are shown with interpretive drawings. Of 80 images examined, 76 (95%) were like those in the gallery. Panel B: Model of 30S subunit with topographically identified rRNA sites. The 30S model of Boublik *et al.* (1982) is shown with all known rRNA sites indicated. C_{1400}, the RNA residue crosslinked to cmo^5U_{34} of $tRNA^{Val}$, denotes the decoding site identified in panel A (Gornicki *et al.* 1983). The locations of the 3′-end, A_{1542}, (Olson & Glitz 1979, Shatsky *et al.* 1979, Lührmann *et al.* 1981b, Stöffler-Meilicke *et al.* 1981), $m_2^6A_{1517-18}$ (Politz & Glitz 1977, Stöffler & Stöffler-Meilicke 1981), m^7G_{527} (Trempe *et al.* 1982), the 5′-end, A_1 (Mochalova *et al.* 1982), the puromycin crosslinking site (Olson *et al.* 1980), and the mRNA binding domain (Stöffler & Stöffler-Meilicke 1981) are also shown.

C_{1400} agrees with an earlier kethoxal-accessibility study of G_{1405} (Chapman & Noller 1977).

Electron microscopic examination of the complexes gave a consistent set of results (Fig. 3A). The gallery represents all of the various forms observed. All of the antibodies were bound to the region identified as C_{1400} in Fig. 3B, namely the deepest part of the cleft separating the head and the neck from the large protrusion. This region should correspond to the decoding site. The mRNA domain identified by Stöffler (Fig. 3B) and by Evstafieva *et al.* (1983) includes this area. The site is also close to the $m_2^6Am_2^6A$ domain which is 12 residues from the 5′-end of the Shine-Dalgarno region, noted above to be not more than 5 residues from C_{1400}. Figure 3B also shows the sites of all other topographically defined points on the 30S subunit, except for the ribosomal proteins. The conserved loop of 17 bases, 517–533, marked by m^7G_{527}, is on the opposite side of the subunit from C_{1400}, itself at the center of the conserved sequence 1392–1407 connecting 2 stems. The IF3 binding site (not shown) overlaps the cleft and C_{1400} areas (Stöffler & Stöffler-Meilicke 1981, 1983).

Crosslinking of the anticodon of tRNA to ribosomal A and P sites via an aromatic azide photo-affinity probe

In order to further explore the topography of the decoding site, an aromatic azide photo-affinity probe was attached to the cmo^5U_{34} residue in a manner similar to that used for addition of the DNP group (Fig. 2 Right). Two different chain-length products were used, $tRNA^{NAK}$ and $tRNA^{NAG}$, with the reactive azide group 23 and 18 Å, respectively, from U_{34}. When $tRNA^{NAK}$ was bound to the P and A sites and irradiated, 40–50% crosslinking was obtained. This is much higher than the usual 5–15% yield with aromatic azides (Ofengand *et al.* 1980, Kao *et al.* 1983). Although there was not much difference in yield between the P and A sites, the 5Å shorter chain of NAG depressed crosslinking 4 to 5-fold at both sites. Crosslinking was codon-, irradiation-,and probe-dependent. A site crosslinking was EFTu-dependent and P site cross-linked amino acid was released by puromycin treatment (Ofengand *et al.* 1979). Mercaptoethanol quenching at the A and P sites differed, the A site reaction being *ca.* 4-fold more resistant (50% quenching at 2.0 mM vs. 0.5 mM), but even the P site reaction was 5-fold more resistant than a similar azide crosslink to ribosomal protein S19 (Hsu *et al.* 1983). These results suggest that both the P and A subsites of the decoding site are relatively shielded from solvent, consistent with interaction deep in the cleft of the 30S subunit.

Preliminary analysis of the site of crosslinking revealed that both P and A site crosslinks are only to the 30S subunit, and that the A site crosslinks mainly (>70%) to 16S rRNA. The P site crosslink appears to be unstable and further work is needed to clarify its site of attachment. The stable A site crosslink to rRNA holds out the promise of determining additional areas of 16S rRNA that make up the decoding site. Preliminary analysis indicates that the rRNA fragment is from 10–16 residues long. Sequence determination should allow its exact placement in the 16S molecule.

TRANSLOCATION OF tRNA CROSSLINKED TO THE RIBOSOMAL A SITE

We have previously shown that tRNA, modified at the s^4U_8 position with the photo-affinity probe, *p*-azidophenacyl (APA), can be crosslinked to the

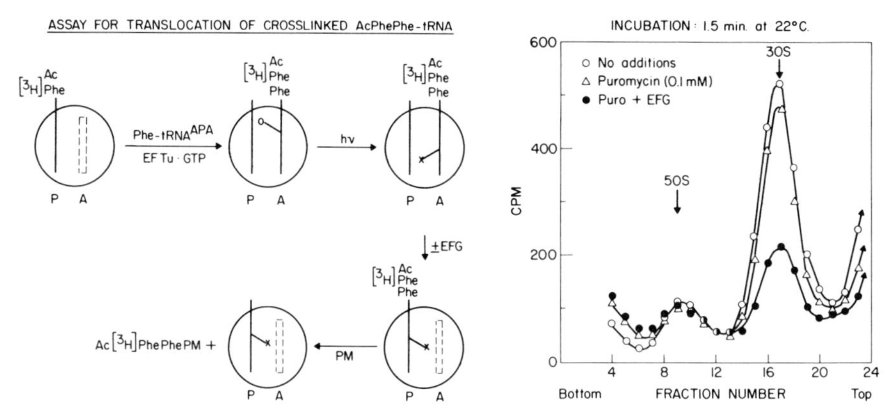

Fig. 4. Movement from A to P site of AcPhePhe-tRNA which had been covalently attached to the A site prior to translocation. *Left panel:* Schematic representation of the assay system devised to detect translocation of the aminoacyl end of tRNA following prior A site fixation. $tRNA^{APA}$, tRNA whose 4-thiouridine residue (position 8) was derivatized with *p*-azidophenacyl bromide (Ofengand *et al.* 1980, Hsu *et al.* 1983) PM, puromycin. *Right panel:* Sucrose gradient analysis of the experiment diagrammed in panel A. AcPhe-tRNA was bound by incubation at 30°C for 15 min. An excess of AcPhe-tRNA was used. $Phe\text{-}tRNA^{APA}$ (Hsu *et al.* 1983) was complexed with EFTu-GTP by incubation at 30°C for 5 min. The complex, in limiting amount, was mixed with the AcPhe-tRNA-ribosome complex. Binding to the A site and dipeptide formation occurred at 0°C for 40 min. Irradiation at >310 nm for 10 min. was also at 0°C (Hsu *et al.* 1983). Incubation with puromycin±EFG was as indicated. Sucrose gradient analysis was at 0.5 mM Mg^{++}. 70% of the $Ac[^3H]PhePhe$ was released from the 30S peak by addition of puromycin and EFG, but only 3% by puromycin alone.

ribosomal A site (Ofengand *et al.* 1980, Hsu *et al.* 1983) at protein S19 (Lin *et al.* 1983a). We wondered if such a crosslinked tRNA could be translocated by EFG to the P site, and designed an experiment to test this (Fig. 4, Left). Since only the P site-bound AcPhe-tRNA carried a radioactive label, and only the A site Phe-tRNA had the photo-affinity probe, the only way for radioactivity to become attached to the 30S subunit was by dipeptide formation before or after crosslinking because unmodified AcPhe-tRNA in the P site does not crosslink (Hsu *et al.* 1983). In these experiments, dipeptide formation preceded cross-linking. Dipeptidyl-tRNA also crosslinks only to S19 (Lin *et al.* 1983a).

The purpose of this experimental design was to reduce the background of crosslinking from A site-bound $Phe\text{-}tRNA^{APA}$ which had failed to form a dipeptide, since in our hands, Phe-tRNA does not translocate. The test for translocation was the ability of puromycin (PM) to react with the AcPhePhe moiety and remove it from the ribosome. The results (Fig. 4, Right) show that

EFG *was* able to catalyze the removal of AcPhePhe from a 30S subunit-bound state. This means that the 3′-end of A site-bound tRNA, even though crosslinked via the central fold of the tRNA, can still be moved into the P site part of the peptidyl transferase center. This result was confirmed by additional kinetic experiments at 25°C and 30° C which showed an almost complete dependence on added EFG. The maximum extent of translocation ranged between 70–80%. This result implies that the s^4U_8 region of tRNA maintains a rather constant distance from S19 whether the tRNA occupies either the A or P site. Thus, S19 is likely to be centrally located between the 2 sites. Strictly interpreted, this result shows only that the *aminoacyl end* of tRNA can be translocated. It may be possible to show that the tRNA anticodon can also be moved by making use of the P site-specific crosslinking to the anticodon of $tRNA^{Val}$ since APA-modified $tRNA^{Val}$ also crosslinks to the A site (Hsu *et al.* 1983).

HYPOTHETICAL MODEL FOR THE STRUCTURE OF THE tRNA-RIBOSOME COMPLEX

Any proposed arrangement of tRNA on the ribosome must comply with certain experimental facts. (1) The structure of tRNA does not deviate in any major way from its known crystal structure when it is on the ribosome at either the A or P site (Ofengand 1980, Farber & Cantor 1980, Peattie & Herr 1981, Douthwaite *et al.* 1983, Bertram *et al.* 1983). (2) There is strong evidence for 2 tRNA binding sites on the ribosome, the A and P sites. As evidence for an additional R (Lake 1977) and/or E (Rheinberger *et al.* 1981, Rheinberger & Nierhaus 1983, Grajevskaja *et al.* 1982, Kirillov *et al.* 1983) site is still controversial (Schmitt *et al.* 1982, Johnson & Cantor, 1980), we will focus on only the well-established A and P sites. (3) The anticodons of both tRNAs must be close together in order to translate adjacent codons, and are positioned deep in the cleft of the 30S subunit (see Fig. 3 above). (4) The aminoacyl ends of both A and P site tRNAs are located at the peptidyl transferase center on the 50S subunit (Fig. 1). (5) The 2 tRNAs in the A and P sites are positioned at an angle of 30°-90° with each other (Johnson *et al.* 1982, Paulsen *et al.* 1983), ruling out diamond-shaped arrangements (Ofengand 1980) for the tRNAs. (6) The relative arrangement of the 2 ribosomal subunits is known (Kastner *et al.* 1981, Lake 1982, Stöffler & Stöffler-Meilicke 1983). (7) Affinity labelling experiments using probes on either side of the tRNA "L" near the central fold, i.e., on s^4U_8 and acp^3U_{47} (Ofengand *et al.* 1980, 1981) (Fig. 5) show that, at the A site, the short (8Å) s^4U_8 probe (APA)

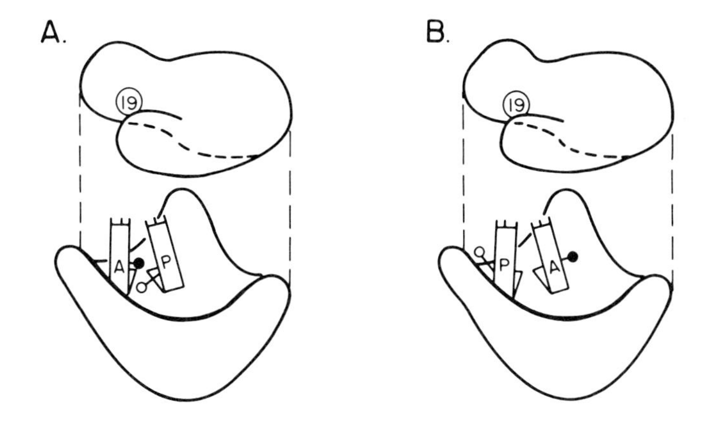

Fig. 5. Hypothetical model of the tRNA-ribosome complex. Two different orientations of A and P site tRNAs are shown on an exploded form of one version (Boublik *et al.* 1977) of the 70S particle. The tRNAs are indicated schematically, but in correct scale to the ribosome. The APA (solid ball) and NAK (open ball) photo-affinity probes are shown in their approximate location on the tRNA and only in the ribosomal sites in which crosslinking with the given probe occurred (Ofengand *et al.* 1980, 1981). See text for further discussion.

is close to the S19 site on the head of the 30S subunit near to the large projection (Lin *et al.* 1983b). At the P site, the longer (20Å) acp^3U_{47} probe (NAK) is equally close to both subunits (Ofengand *et al.* 1980).

These facts suggest placement of the plane of the tRNAs perpendicular to the subunit junction plane (Fig. 5). Models A and B differ only in that the A and P sites have been exchanged. The large projection of the 30S subunit is shown toward the viewer according to the models of Boublik *et al.* (1977), Lake (1982), Vasiliev *et al.* (1983), and the newest version of the Stöffler model (Stöffler & Stöffler-Meilicke 1983). There are 3 arguments against the A-P site arrangement of model B. First, EFTu crosslinked to tRNA at acp^3U_{47} via NAK can bind to the A site of ribosomes with a filled P site, and form dipeptide (Kao *et al.* 1983). In model B, this would place the crosslinked EFTu between the 2 tRNAs, which appears sterically difficult. Second, the structural requirements for pairing 2 tRNAs with adjacent codons of an mRNA leads to an A-P site orientation consistent with model, A, but not B (Sundaralingam 1975). Accommodation to model B would require an approximately 90° twist of both tRNAs around their anticodon stem axes, while keeping their anticodon loops fixed in place. This seems unlikely. Third, preliminary analysis of the major 30S proteins crosslinked by the P site tRNA have identified S11, S13, S18, and S19 (L. Kahan and J. Ofengand, unpublished results), all of which are located in the cleft region (Fig. 1B). This pattern is more consistent with model A.

We think it likely that the principle illustrated in model A, namely that the 2 tRNA "L" shapes wrap around the head and/or neck of the 30S subunit, will prove correct, although many of the details will no doubt need revision. There are already inconsistencies. The A site bound tRNA is not near the L7/12 stalk, a region thought to be intimately involved in EFTu-dependent A site binding (Liljas 1982). The TψCG loop of tRNA is well away from the 5S RNA region of the 50S, which is thought to be at the central protuberance (Fig. 1A) (Stöffler & Stöffler-Meilicke 1983). At one time this loop was thought to interact with 5S RNA (Ofengand 1980, Erdmann this volume).

Other tRNA-ribosome models have been proposed at various times. The Lake (1977) model is not compatible with facts 5 and 7 listed above. The Glitz-Cooperman model (Olson *et al.* 1982, Goldman *et al.* 1983) is not in agreement with the affinity labelling results (Fact 7), and the recent proposal by Spirin (1983) does not account for either the S19 site crosslinked (Lin *et al.* 1983b), or the A-P site orientation argument presented above.

Although we have used the Boublik model for illustration in Fig. 5, the latest models of Stöffler and Stöffler-Meilicke (1983), Lake (1982), and Vasiliev *et al.* (1983), which are similar, could also be accommodated with little additional difficulty. In actuality none of the extant models allow one to readily place the tRNAs in concordance with *all* the known facts. One is tempted to suggest that the static dimensions of the ribosome as determined by electron microscopy of subunits may not accurately reflect the dynamic structure of a functioning 70S ribosome. Additional affinity-labelling information, especially from other segments of tRNA will be helpful in this regard, as will newer methods (Wall 1979, Frank *et al.* 1982) for visualization of tRNA on the ribosome.

REFERENCES

Abdurashidova, G. G., Urchinsky, M. F., Aslanov, Kh. & Budowsky, E. I. (1979) *Nucleic Acids Res. 6,* 3891–09.

Babkina, G. T., Graifer, D., Karpova, G. G. & Matasova, N. B. (1983) *FEBS Lett. 153,* 03–306.

Barta, A., Steiner, G., Brasius, J., Noller, M. F. & Kuechler, E. (1984) *Proc. Natl. Acad. Sci. USA 81,* (in press).

Bertram, S., Göringer, U. & Wagner, R. (1983) *Nucleic Acids Res. 11,* 575–589.

Blanc, M., Adams, C. W. & Wallace, D. C. (1981a) *Nucleic Acids Res. 9,* 5785–5795.

Blanc, H:, Wright, C. T., Bibb, M. J., Wallace, D. C. & Clayton, D. A. (1981b) *Proc. Natl. Acad. Sci. USA 78,* 3789–3793.

Boublik, M., Hellmann, W. & Kleinschmidt, A. K., (1977) *Cytobiologie 14,* 293–300.

Boublik, M., Robakis, N., Hellmann, W. & Wall, J. S. (1982) *Eur. J. Cell Biol. 27,* 177–184.

Brosius, J., Dull, T. J. & Noller, H. F. (1980) *Proc. Natl. Proc. Sci. USA., 77* 201–204.

Broude, N. E., Kussova, K. S., Medvedeva, N. I. & Budowsky, E. I. (1983) *Eur. J. Biochem. 132,* 139–145.

Chapman, N. M. & Noller, H. F. (1977) *J. Mol. Biol. 109,* 131–149.

Chiaruttini, C., Expert-Bezancon, A., Hayes, D. & Ehresmann, B. (1982) *Nucleic Acids Res. 10,* 7657–7676.

Cooperman, B. S. (1980) In: *Ribosomes: Structure, Function, and Genetics* (Chambliss, G., Craven, G., Davies, J., Davis, K., Kahan, L. & Nomura, M., eds.), 531–554. University Park Press, Baltimore, MD.

Douthwaite, S., Garrett, R. A. & Wagner, R. (1983) *Eur. J. Biochem. 131,* 261–269.

Dujon, B. (1980) *Cell 20,* 185–197.

Ehresmann, C., Ehresmann, B., Millon, R., Ebel, J.-P., Nurse, K. & Ofengand, J. (1984) *Biochemistry* (in press).

Ehresmann, C. & Ofengand, J. (1984) *Biochemistry* (in press).

Endo, Y. & Wool, Ira, G. (1982) *J. Biol. Chem. 257,* 9054–9060.

Evstafieva, A. G., Shatsky, I. N., Bogdanov, A. A., Semenkov, Y. P. & Vasiliev, V. D. (1983) *EMBO J. 2,* 799–804.

Farber, N. & Cantor, C. R. (1980) *Proc. Natl. Acad. Sci. USA. 77,* 5135–5139.

Frank J., Verschoor, A. & Boublik, M. (1982) *J. Mol. Biol. 161,* 107–137.

Gimautdinova, O. I., Karpova, G. G., Knorre, D. G. & Kobetz, N. D. (1981) *Nucleic Acids Res. 9,* 3465–3481.

Goldman, R. A., Hasan, T., Hall, C. C., Strycharz, W. A. & Cooperman, B. S. (1983) *Biochemistry 22,* 359–368.

Gornicki, P., Nurse, K. & Ofengand, J. (1983) *Fed. Proc. 42,* 2186.

Grajevskaja, R. A., Ivanov, Y. V. & Saminsky, E. M. (1982) *Eur. J. Biochem. 128,* 47–52.

Gupta, R., Lanter, J. & Woese, C. (1983) *Science, 221,* 656–659.

Helser, T. L., Davies, J. E. & Dahlberg, J. E. (1982) *Nature New Biol. 235,* 6–9.

Herr, W., Chapman, N. M. & Noller, H. F. (1979) *J. Mol. Biol. 130,* 433–449.

Högenauer, G., Egger, H., Ruf, C. & Stumper, B. (1981) *Biochemistry 20,* 546–552.

Hsu, L., Lin, F.-L. & Ofengand, J. (1983) *J. Mol. Biol. 171,* (in press).

Johnson, A. E., Adkins, H. J., Matthews, E. A. & Cantor, C. (1982) *J. Mol. Biol. 156,* 113–140.

Johnson, A. E. & Cantor, C. R. (1980) *J. Mol. Biol. 138,* 273–297.

Kastner, B., Stöffler-Meilicke, M. & Stöffler, G. (1981) *Proc. Natl. Acad. Sci. USA 78,* 6652–6656.

Kao, T.-H., Miller, D. L., Abo, M. & Ofengand, J. (1983) *J. Mol. Biol. 166,* 383–405.

Kearsey, S. E. & Craig, I. W. (1981) *Nature 290,* 607–608.

Keren-Zur, M., Boublik, M. & Ofengand, J. (1979) *Proc. Natl. Acad. Sci. USA 76,* 1054–1058.

Kirillov, S. V., Makarov, E. M. & Semenkov, Y. P. (1983) *FEBS Lett. 157,* 91–94.

Krzyzosiak, W. J., Biernat, J., Ciesiolka, J., Gornicki, P. & Wiewiorowski, M. (1979) *Nucleic Acids Res. 7,* 1663–1674.

Lake, J. A. (1977) *Proc. Natl. Acad. Sci. USA. 74,* 1903–1907.

Lake, J. A. (1982) *J. Mol. Biol. 161,* 89–106.

Lake, J. A. & Strycharz, W. A. (1981) *J. Mol. Biol. 153,* 979–992.

Le Goffic, F. Capmau, M.-L., Chausson, L. & Bonnet, D. (1980) *Eur. J. Biochem. 106,* 667–674.

Leitner, M., Wilchek, M. & Zamir, A. (1982) *Eur. J. Biochem. 125,* 49–55.

Liljas, A. (1982) *Progr. Biophys. Molec. Biol. 40,* 161–228.

Lin, F.-L., Boublik, M. & Ofengand, J. (1983b) *J. Mol. Biol. 171*(in press).

Lin, F.-L., Kahan, L. & Ofengand, J. (1983a) *J. Mol. Biol. 171*(in press).

Lührmann, R., Bald, R., Stöffler-Meilicke, M. & Stöffler, G. (1981a) *Proc. Natl. Acad. Sci. USA. 78,* 7276–7280.

Lührmann, R., Stöffler-Meilicke, M. & Stöffler, G. (1981b) *Mol. Gen. Genet. 182,* 369–376.

Martin, R. P., Bordonné, R. & Dirheimer, G. (1982) in: *Cell function and differentiation,* Part B, Biogenesis of energy transducing membranes and membrane and protein energetics, FEBS vol. 65, pp. 355–365, Alan R. Liss, New York.

Mochalova, L. V., Shatsky, I. N., Bogdanov, A. A. & Vasiliev, V. D., (1982) *J. Mol. Biol. 159,* 637–650.

Moras, D., Comarmond, M. B., Fischer, J., Weiss, R., Thierry, J. C., Ebel, J. P. & Geige, R. (1980) *Nature 288,* 669–674.

Nicholson, A. W., Hall, C. C., Strycharz, W. A. & Cooperman, B. S. (1982) *Biochemistry 21,* 3797–3808.

Noller, H. F. & Woese, C. R. (1981) *Science 212,* 403–411.

Ofengand, J. (1980) In: *Ribosomes: Structure, Function, and Genetics* (Chambliss, G., Craven, G., Davies, J., Davis, K., Kahan, L. & Nomura, M., eds.), 497–529. University Park Press, Baltimore, MD.

Ofengand, J., Gornicki, P., Chakraburtty, K. & Nurse, K. (1982) *Proc. Natl. Acad. Sci. USA 79,* 2817–2821.

Ofengand, J., Lin, F.-L., Hsu, L. & Boublik, M. (1981) in: *Molecular Approaches to Gene Expression and Protein Structure,* (M. A. Siddiqui, M. Krauskopf & H. Weissbach, eds.), pp 1–31. Academic Press, Inc. New York.

Ofengand, J., Lin, F.-L., Hsu, L., Keren-Zur, M & Boublik, M. (1980), *Ann. N. Y. Acad. Sci. 346,* 324–354.

Ofengand, J. & Liou, R. (1980) *Biochemistry 19,* 4814–4822.

Ofengand, J. & Liou, R. (1981) *Biochemistry 20,* 552–559.

Ofengand, J., Liou, R., Kohut, III, J., Schwartz, I. & Zimmermann, A. (1979) *Biochemistry 18,* 4322–4332.

Ohsawa, H. & Gualerzi, C. (1983) *J. Biol. Chem. 258,* 150–156.

Olson, H. M., & Glitz, D. G. (1979) *Proc. Natl. Acad. Sci. USA. 76,* 3769–3773.

Olson, H. M., Grant, P. G., Cooperman, B. S. & Glitz, D. G. (1982) *J. Biol. Chem. 257,* 2649–2656.

Olson, H. M., Grant, P. G., Glitz, D. G. & Cooperman, B. S. (1980) *Proc. Natl. Acad. Sci. USA 77,* 890–894.

Paulsen, H., Robertson, J. M. & Wintermeyer, W. (1983) *J. Mol. Biol. 167,* 411–426.

Peattie, D. A. & Herr, W. (1981) *Proc. Natl. Acad. Sci. USA 78,* 2273–2277.

Politz, S. M. & Glitz, D. G. (1977) *Proc. Natl. Acad. Sci. USA 74,* 1468–1472.

Prince, J. B., Taylor, B. H., Thurlow, D. L., Ofengand, J. & Zimmermann, R. A. (1982) *Proc. Natl. Acad. Sci. USA 79,* 5450–5454.

Rheinberger, H.-J. & Nierhaus, K. H. (1983) *Proc. Natl. Acad. Sci. USA. 80,* 4213–4217.

Rheinberger, H.-J., Sternbach, H. & Nierhaus, K. H. (1981) *Proc. Natl. Acad. Sci. USA. 78,* 5310–5314.

Riehl, N., Remy, P., Ebel, J.-P. & Ehresmann, B. (1982) *Eur. J. Biochem. 128,* 427–433.

Sanger, F., Coulson, A. R., Friedmann, T., Air, G. M., Barrell, B. G., Brown, N. L., Fiddes, J. C., Hutchison, C. A. III, Slocombe, P. M. & Smith, M. (1978) *J. Mol. Biol. 125,* 225–246.

Schmitt, M., Neugebauer, U., Bergmann, C., Gassen, H. G. & Reisner, D. (1982) *Eur. J. Biochem. 127,* 525–529.

Schwartz, I. & Ofengand, J. (1974) *Proc. Natl. Acad. Sci. USA 71,* 3951–3955.

Schwartz, I. & Ofengand, J. (1978) *Biochemistry 17,* 2524–2530.

Shatsky, I. N., Mochalova, L. V., Kojouharova, M. S., Bogdanov, A. A. & Vasiliev, V. D. (1979) *J. Mol. Biol. 133,* 501–515.

Sillers, I.-Y. & Moore, P. B. (1981) *J. Mol. Biol. 153,* 761–780.

Skinner, R. H., Cundliffe, E. & Schmidt, F. J. (1983) *J. Biol. Chem. 258*(in press).

Sköld, S.-E., (1983) *Nucleic Acids Res. 11,* 4923–4932.

Sor, F. & Fukuhara, H. (1982) *Nucleic Acids Res. 10,* 6571–6577.

Spirin, A. S., (1983) *FEBS Lett. 156,* 217–221.

Steitz, J. A. (1980) In: *Ribosomes: Structure, Function, and Genetics,* (G. Chambliss, G. Craven, J. Davies, K. Davis, L. Kahan & M. Nomura, eds.), pp. 479–495. University Park Press, Baltimore, MD.

Stiegler, P., Carbon, P., Ebel, J.-P. & Ehresmann, C. (1981a) *Eur. J. Biochem. 120,* 487–495.

Stiegler, P., Carbon, P., Zuker, M., Ebel, J.-P. & Ehresmann, C. (1981b) *Nucleic Acids Res. 9,* 2153–2172.

Stöffler, G. & Stöffler-Meilicke, M. (1981) *International Cell Biology 1980/81* (ed) H. G. Schweiger, pp. 93–102. Springer Verlag, Berlin, Heidelberg, New York.

Stöffler, G. & Stöffler-Meilicke, M. (1983) In: *Modern Methods in Protein Chemistry* (ed. H. Tschesche) Walter de Gruyter Verlag, Berlin-New York, (in press).

Stöffler-Meilicke, M. & Stöffler, G. (1982) *Abstr. 10th International Congress Electron Microscopy, Hamburg 3,* 99–100.

Stöffler-Meilicke, M. Stöffler, G., Odom, O. W. Zinn, A., Kramer, G. & Hardesty, B. (1981) *Proc. Natl. Acad. Sci. USA 78,* 5538–5542.

Sundaralingam, M., Brennan, T., Yathindra, N. & Ichikawa, T. (1975) In: *Structure and Conformation of Nucleic Acids and Protein-Nucleic Acid Interactions,* (M. Sundaralingam & S. T. Rao, Eds.), pp. 101–115. University Park Press, Baltimore, MD.

Thompson, J. F. & Hearst, J. E. (1983) *Cell 32,* 1355–1365.

Thompson, J., Schmidt, F. & Cundliffe, E. (1982) *J. Biol. Chem. 257,* 7915–7917.

Trempe, M. R., Ohgi, K. & Glitz, D. G. (1982) *J. Biol. Chem. 257,* 9822–9829.

Vasiliev, V. D., Selivanova, O. M., Baranov, V. I. & Spirin, A. S. (1983) *FEBS Lett. 155,* 167–172.

Vassilenko, S. K., Carbon, P., Ebel, J.-P. & Ehresmann, C. (1981) *J. Mol. Biol. 152,* 699–721.

Wagner, R., Gassen, H. G., Ehresmann, C., Stiegler, P. & Ebel, J.-P. (1976) *FEBS Lett. 67,* 312–315.

Wall, J. S. (1979) In: *Introduction to Analytical Electron Microscopy* (J. J. Hren, J. J. Goldstein & D. C. Joy (Eds). pp. 333–342. Plenum Press, New York.

Willan, K. J., Marsh, D., Sunderland, C. A., Sutton, B. J., Wain-Hobson, S., Dwek, R. A. & Givol, D. (1977) *Biochem. J. 165,* 199–206.

Winkelmann, D. A., Kahan, L. & Lake, J. A. (1982) *Proc. Natl. Acad. Sci. USA 79,* 5184–5188.

Zwieb, C., Glotz, C. & Brimacombe, R. A. (1981) *Nucleic Acids Res. 9,* 3621–3640.

DISCUSSION

GARRETT: We have looked at the conserved sequence where you have a tRNA crosslink, and although there is no obvious base pairing, there is chemical and enzymatic evidence for higher order structure in that region (Douthwaite *et al.* 1983). G-1405 is accessible but most of the neighbouring residues are not. Nevertheless, it is clear that C_{1400} is available for crosslinking to tRNA.

OFENGAND: These studies are certainly very interesting, but as they were done on the isolated 16S rRNA, it is conceivable that the presence of ribosomal proteins and condensation into a 30S particle might unfold this particular sequence.

GRUNBERG-MANAGO: I think the IF3 binding site is very near to your tRNA crosslink and this would fit well, because it has been reported that one cannot have both IF3 and initiator tRNA on the ribosome at the same time.

OFENGAND: Yes, you are correct that the ribosome binding site for IF3 overlaps the anticodon cross-linking site, but as the IF3 site covers a large region, the correspondence could be partially fortuitous. Actually, as Fig. 3 shows, several functional sites tend to cluster in the same central region of the head and neck of the subunit. One would need higher resolution probes than we currently have to dissect out the fine topographical detail of these functional sites.

KURLAND: Could you not explain your translocation results by a certain flexibility in the A site bound tRNA? If the 3′ end of the tRNA shifts a few Å into the P-site, allowing puromycin to get into the A-site and generate some peptidyl transferase activity, then the rest of the tRNA, after being cross-linked, need not move in order to explain your results.

OFENGAND: That is a good point. For simplicity, I have described the reaction as the movement of the entire tRNA molecule, but I recognize that the assay actually measures only the movement of the 3′ end of the tRNA. We think, we have a way of measuring whether the anticodon also moves. We propose to do this by taking advantage of the specific P-site cross-linking of the anticodon of *E. coli* $tRNA^{Val}$. That is, after A-site cross-linking and treatment with EFG we will look for movement of the anticodon into the P-site, that is, into a cross-linkable position. A positive result would answer your objection at least in that the 2 ends would have moved. However, we have not yet done that experiment, and it is true that we may have moved only the 3′ end of the tRNA. Let me add, however, that before we did the experiment, it was not at all obvious that a cross-linked tRNA could be even partially translocated from one site to another.

KURLAND: As one of your arguments about the relative arrangement of A- and P-site tRNAs on the ribosome depends on there being 2 simultaneous codon-anticodon interactions, would you or anyone else like to comment on how strong you think the evidence is for this?

OFENGAND: It is probably impossible to really determine this, because what you are asking is to be able to somehow measure simultaneous events. For example, although one could show codon-dependent binding into the P-site, and then show codon-dependent binding into the A-site of the same ribosome, there is no way to ensure that codon *interaction* at the P-site was retained during A-site codon recognition.

KURLAND: That's my point. There may be requirement for 2 tRNAs to interact with mRNA simultaneously.

OFENGAND: I agree with you. It might be appropriate to mention the following experiment, as it deals peripherally with the question discussed here. We wanted to know if codon-dependent binding of a tRNA to the A-site would influence the ability of a tRNA already bound to the P-site to be cross-linked. We tested this by comparing the ability of $tRNA^{Val}$ to be bound and cross-linked in the presence of Phe-tRNA and GU_n where n varied from 2–6. While no Phe-tRNA binding occurred until n reached 5, both $tRNA^{Val}$ binding and cross-linking were unchanged over the whole range of n. Thus, cross-linking occurred as readily with an empty A-site as with a filled A-site. However, this need not indicate that codon-anticodon interaction was retained at the P-site. It may have been broken as the tRNA entered the A-site.

KURLAND: There seems to be no need for simultaneous recognition either. If you are worrying about the mechanisms of translation it is not obvious that one would require a second codon-anticodon interaction.

SPRINZL: You showed that irradiation was done at 290–320 nm. It that optimal? Also, did you try to cross-link some other tRNAs which do not have this 5-alkoxy-modified base?

OFENGAND: The irradiation band was chosen for technical and practical reasons. We do not know the optimal wave length, nor have we determined a true action spectrum. We do know that wave lengths above 320 nm are essentially inactive, that the band 310–320 nm allows cross-linking, and that the band 290–320 nm increases the rate of cross-liking 6–7 fold. We have studied the effect of other nucleotides in place of the cmo^{5}U or mo^{5}U residue by attempting to cross-link the appropriate tRNAs. Purines, cytidine, and 5-fluorouridine were all inactive in the 5′ anticodon position. The only tRNAs that we have found to be cross-linkable are those which contain either of the 2 alkoxy derivates of uridine. Thus, the valine, alanine, threonine, and serine-1 species are active, but they do not all cross-link with the same efficiency. There appear to be some subtleties of orientation in a given tRNA. These results can be found in more detail in Ofengand *et al.* (1979).

CASKEY: After hearing of this system I was wondering about the possibility of using the system as a means of measuring the position of translational fidelity on the ribosome. I have 2 questions with regard to that. What level of crosslinking do you get in the absence of codons or in the presence of inappropriate codons? Do you get crosslinking to *other* positions on the ribosome in the absence of codons, and what are those position?

OFENGAND: First, to answer the question directly, missense codons or the absence of codons, do not yield detectable amounts ($<1\%$) of cross-linking. This corresponds to a reduction of 50–100-fold from that in the presence of the correct codons. Your question brings up an interesting point however, which I might elaborate on. We have looked at the cross-linking reaction using several different "correct" codons. Because the anticodon is cmo^5 UAC, the codon set is GUA, GUG, and GUU. This latter codon which involves a modified U-U "base pair" is known to support binding to the ribosome although less efficiently than GUA or GUG. The dinucleotide pGpU will also stimulate binding almost as well as GUU. The surprising feature of the cross-link reaction is that when the cross-linking residue, cmo^{5}U, is paired with either A or G, cross-linking is blocked. The presence of U, or the absence of any nucleotide, allows efficient cross-linking. Our interpretation of this result is that, in the absence of a base-pairing directive influence, the 5′-anticodon base, cmo^5U, can move into and out of a position which overlaps C_{1400} when the tRNA is bound at the P-site. With continuous irradiation the base can then be trapped by cyclobutane dimer formation as it moves into the overlap position with C_{1400}. Eventually, 90% or more of the bound tRNA can be accumulated in cross-linked form. If, however, the position of the base is fixed by base-pairing with A or G, then this mobility property is lost. The experimental results can be accounted for if the position of cmo^5-U when base-paired with A or G is not conducive to crosslink formation, which has a fairly stringent stereochemical requirement of C_5-C_6 π orbital overlap. An expanded version of this interpretation can be found in our paper on this subject (Ofengand & Liou 1981).

Douthwaite, S., Christensen, A. & Garrett, R. A. (1983). *J. Mol. Biol. 169*, 249–279.
Ofengand, J., Liou, R., Kohut III, J., Schwartz, I. & Zimmermann, R. (1979) *Biochemistry 18,* 4322–4332.
Ofengand, J. & Liou, R. (1981). *Biochemistry 20,* 552–558.

RNA Surface Topography of the *E. coli* Ribosome

J. P. Ebel, C. Branlant, P. Carbon, B. Ehresmann, C. Ehresmann, A. Krol & P. Stiegler

INTRODUCTION

Protein biosynthesis is based on a co-ordinated interaction between the ribosomes and the macromolecules involved: mRNAs, tRNAs and translational factors. To achieve this complex function, the ribosomal particle possesses specific binding sites for the various ligands. The understanding of the functional role of the ribosome requires the identification and localization of ribosomal components located on the surface of the particle.

The determination of the three-dimensional location of the ribosomal proteins on the surface of ribosomal subunits and ribosomes has become possible mainly due to the use of the immune electron microscopy technique (Stöffler *et al.* 1980, Lake 1980). As far as ribosomal RNAs are concerned, because they represent the backbone around which the ribosomal proteins are assembled in the course of ribosome assembly, attention was mainly focused on the interaction sites between these RNAs and ribosomal proteins (Ehresmann *et al.* 1980, Zimmermann *et al.* 1980, Ebel *et al.* 1983a). But, more recently, it appeared that ribosomal RNAs may also play a functional role, as several regions within the RNAs have been found to interact directly with the macromolecules involved in protein synthesis. It was, therefore, necessary to determine the regions of ribosomal RNAs which are accessible at the surface of ribosomal subunits and of ribosomes.

Laboratoire de Biochimie, Institut de Biologie Moléculaire et Cellulaire, 15 rue René Descartes, 67000 Strasbourg, France

GENE EXPRESSION, Alfred Benzon Symposium 19.
Editors: Brian F. C. Clark & Hans Uffe Petersen, Munksgaard, Copenhagen 1984.

CHARACTERIZATION OF ACCESSIBLE REGIONS OF RIBOSOMAL RNAs AT THE SURFACE OF THE RIBOSOME

Extensive work has been performed on the structural organization of the *E. coli* ribosomal RNAs. The nucleotide sequences of the *E. coli* rRNAs have been determined either directly on the rRNAs or on the corresponding genes, and additional sequences of rRNAs from various species have been published afterwards (Ebel *et al.* 1983a, b). Completion of the nucleotide sequences of these rRNAs has allowed investigation of the secondary structure of the nucleotide chains. Using different approaches secondary structure models have been proposed for the *E. coli* 5SrRNA (Fox & Woese 1975), as well as for the *E. coli* large 16S (Stiegler *et al.* 1980, 1981a, Woese *et al.* 1980, Glotz & Brimacombe 1980, Brimacombe 1980), and 23SrRNAs (Branlant *et al.* 1981, Glotz *et al.* 1981, Noller *et al.* 1981). All proposed secondary structures were established independently and are in good agreement, a strong argument in favour of their validity. They are supported by i) enzymatic and chemical accessibility studies, ii) crosslinking experiments, iii) phylogenetic evidence (Ebel *et al.* 1983a, b). Phylogenetic studies revealed remarkable conservation, during the course of evolution, of the primary structure of some long single-stranded stretches, as well as of numerous secondarystructure motifs (Ebel *et al.* 1983b). These conserved structural features must obviously be important, either for the internal architecture of the ribosome (RNA tertiary structure, ribosomal proteinbinding sites), or for the interaction on the ribosome surface with the ligands involved in protein synthesis.

RNA surface topography data are essential clues for the assignment of RNA sites of functional importance. The most exposed RNA regions have been probed by specific chemical modifications (Noller 1974, Chapman & Noller 1977, Winship & Noller 1978, Herr *et al.* 1979, Brow & Noller 1983), or nuclease digestion (for the 30S subunit: Ehresmann *et al.* 1972, 1975, Santer & Santer 1973, Santer & Shane 1977, Ross & Brimacombe 1979, Vassilenko *et al.* 1981, Stiegler *et al.* 1981a, for the 50S subunit: Branlant *et al.* 1977, 1981, 1983). The guanine-specific reagent kethoxal (2-keto,3-ethoxy-n-butyraldehyde) reacts only with unpaired guanine residues. T_1 ribonuclease cuts the phosphodiester bond at the 3'end of a guanylic residue. Under the conditions of partial hydrolysis, only some of these bonds are split, indicating single-stranded regions. On the contrary, the cobra venom ribonuclease has the unusual property of preferentially hydrolyzing double-stranded regions without any base specificity. In our laboratory, both T_1 and cobra venom ribonucleases have been extensively used for an accurate mapping of residues located in either single-stranded or helical regions of both 16S and 23S rRNAs on the surface of the 30S and 50S subunits and of the 70S couple (Ehresmann *et al.* 1972, 1975, Branlant *et al.* 1977, 1981, 1983, Stiegler *et al.* 1981a, Vassilenko *et al.* 1981).

I. 16SrRNA SURFACE TOPOGRAPHY

16SrRNA surface topography in the 30S subunit

A survey of accessible sites at the surface of the 30S subunit is given in Fig. 1. The T_1 RNase cleavage sites on the native 30S subunits are observed at a higher density in the 3'-half of the 16S rRNA than in the 5' one. Three areas of the molecule are particularly sensitive to the T_1 RNase attack: nucleotides 684 to 843 in domain II, 1073 to 1165 in domain III and 1493 close to the 3'-end of 16S rRNA in domain IV. In these 3 areas clusters of T_1 cuts are observed. In addition to these highly susceptible regions, numerous guanine residues are always cleaved by T_1 RNase (guanines 6, 86, 296, 301, 361, 504, 505, 1267, 1278, 1315, 1360, 1400). These preferential cleavage sites must, therefore, be located in highly exposed RNA regions (Stiegler *et al.* 1981a).

There is a fine correlation between the sensitivity of guanine residues to RNase T_1 hydrolysis and kethoxal modification in active subunits, with both these structural probes having single-stranded guanine residues as targets. Among the 23 positions of kethoxal attack in 30S ribosomal subunits (Brow & Noller 1983), many correspond to preferential T_1 RNase cleavage sites, or sometimes they are located very near to them, thus providing further evidence for their location in highly exposed single-stranded RNA regions. On the other hand, cobra venom RNase has been used for investigation of nucleotides located in exposed double-stranded RNA regions on the surface of the 30S subunit (Vassilenko *et al.* 1981). A total of 21 cleavage sites have been precisely located. Their occurrence in a precise order and their distribution reveal a non-uniform sensitivity of 16S rRNA towards the nuclease digestion. The number of primary cuts is limited. Two of them occur in the 5'-area of the molecule, at positions 78 and 88, and 4 others in the 3' one-third of the RNA chain, at residues 1019/1020, 1161, 1407/1408 and 1488 ($\pm$3). Five of these primary sites are located in the vicinity of kethoxal reactive sites or of T_1 ribonuclease major cleavage sites. The good correlation between 3 different approaches, tends to confirm the hypothesis that the relevant RNA regions are highly exposed on the surface of the subunit. The secondary cuts are mostly produced in the 5′ two-thirds of the molecule, at

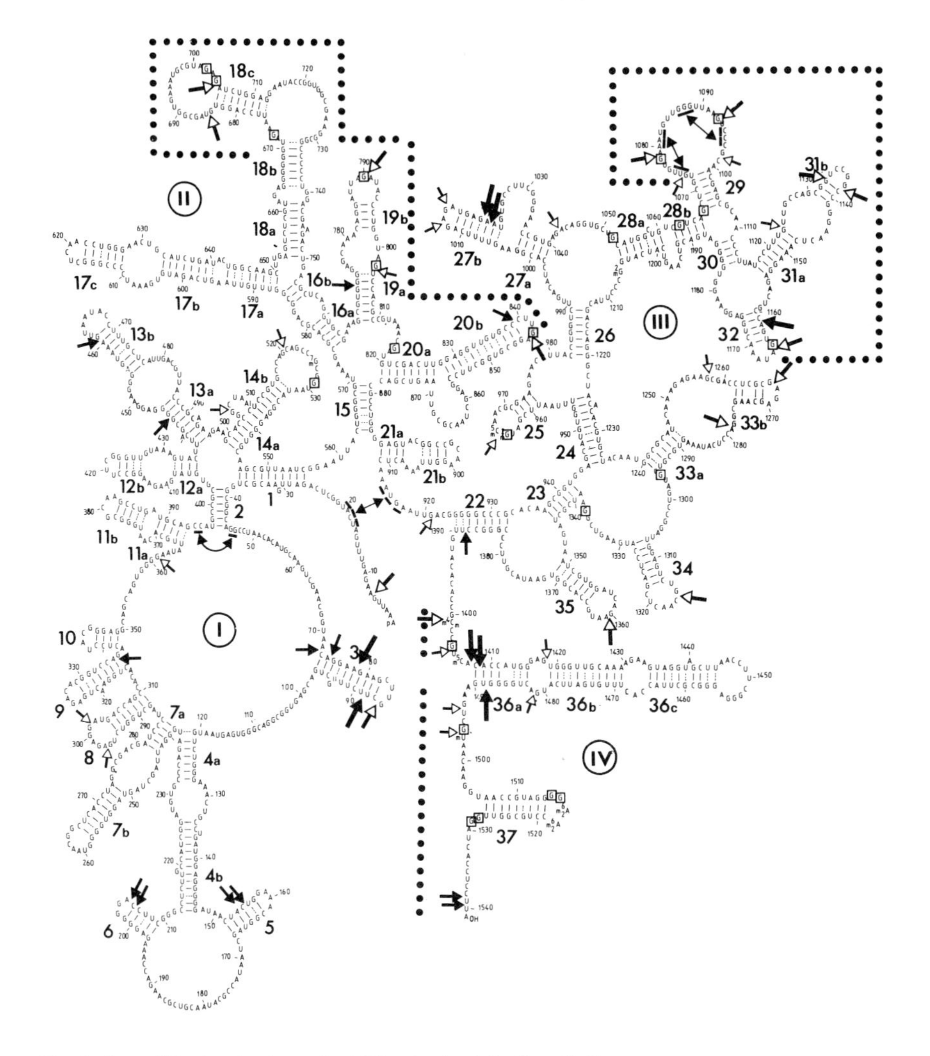

Fig. 1. Accessible sites in the native 30S subunit of the E. coli ribosome
Secondary structure model proposed by Stiegler *et al.*, 1980, 1981. Points of nuclease attack are indicated by arrows:

➡	major	RNase T₁ cuts (Stiegler *et al.* 1981a)
→	medium	
➡	primary	cobra venom RNase cuts (Vassilenko *et al.* 1981)
→	secondary	
G (boxed)	kethoxal modified G residues (Herr *et al.* 1979)	
• • • • •	RNA regions which are highly sensitive to RNase T₁ and to kethoxal modifications.	

positions 72, 74, 154/155, 204/208, 336, 443, 460, 772 and 839, whereas only 2 cuts occur in the 3'-area at positions 1388 and 1538/1539. The existence of the latter site suggests that the 3'-terminal region of 16SrRNA is base-paired within the 30S subunit with some non-identified region of 16SrRNA. Many of the cobra

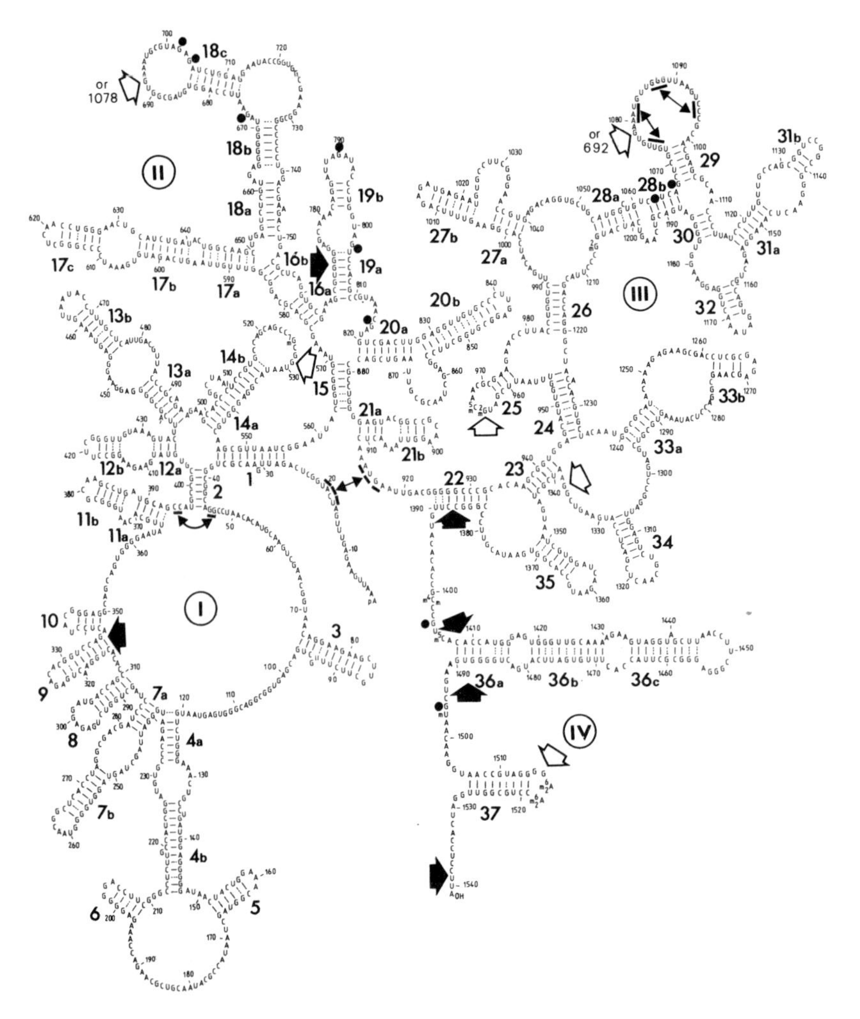

Fig. 2. Accessible sites in E. coli 70S couples

- ● guanine residues essential for subunit association (Herr *et al.* 1979).
- ➡ cobra venom cleavage sites protected after subunit association (Vassilenko *et al.* 1981)
- ⇨ guanine residues protected in polysomes relative to runoff ribosomes (Brow & Noller 1983).

venom RNase secondary cuts are located in areas where no kethoxal modifications or only a few RNase T_1 cuts are observed. This could suggest that these areas become accessible due to slight conformational changes occurring after the primary cleavages.

16SrRNA surface topography in the 70S couple

The effect of the association of 30S and 50S subunits on the accessibility of 16SrRNA has been investigated using single-stranded specific reagents such as kethoxal (Chapman & Noller 1977, Herr *et al.* 1979, Brow & Noller 1983), or RNase T_1 (Santer & Shane 1977). In our laboratory, the effect of binding of the 50S subunit to the 30S subunit on the accessibility of 16S RNA to cobra venom ribonuclease has been investigated (Vassilenko *et al.* 1981). We have found that 16S RNA is cleaved much more slowly in the 70S ribosome than in the isolated 30S subunit and that several cleavage sites are specifically protected due to the binding of the 50S subunit (Fig. 2). Four protected sites are clustered in the 3'-terminal 200 nucleotides of the molecule (at positions 1388, 1407/1408, 1488 ($\pm$3), 1538/1539), one in the middle (at position 772), and one in the 5'-domain (at position 336). These sites might be located at the interface of the 2 subunits and, therefore, directly protected by 50S subunit. The lack of reactivity might also result from a partial unwinding of the relevant helical regions or from some local reorganization between 16S rRNA and ribosomal proteins. On the other hand, the binding of the 50S subunit does not unmask any new cleavage sites. Except for position 336, the results on cobra venom ribonuclease accessibility are consistent with the data derived from kethoxal modification studies (Fig. 2). Probing RNA surface topography in 70S couples with kethoxal has shown that the central region (600–850) and the 3'-area (1350–1541) of 16SrRNA contain sequences which make contact with the 50S subunit. Some of the cobra venom RNase cleavage sites, which become unreactive after binding of the 50S subunit, are located in the vicinity of those G-residues essential for subunit association.

These studies undoubtedly show that the 3'-terminal region and the central region are in close proximity in the ribosome structure and lie at the interface of the two subunits. In a recent study (Brow & Noller 1983), polysomes, taken as functionally engaged ribosomes, have been probed using kethoxal. In 16SrRNA, 5 guanine residues (at positions 530, 693 or 1079, 966, 1338 and 1517) are protected in polysomes as compared to 16SrRNA reactivity in 70S couples. Polysome specific protection has been interpreted as the result of direct contact of bound ligands (tRNA, mRNA) or conformational changes in 70S ribosomes upon tRNA and/or mRNA binding.

Relationship between RNA surface topography and RNA function

Investigation of the extent of both nucleotide sequence and secondary structure conservation which might exist between the *E. coli* 16S rRNA and numerous

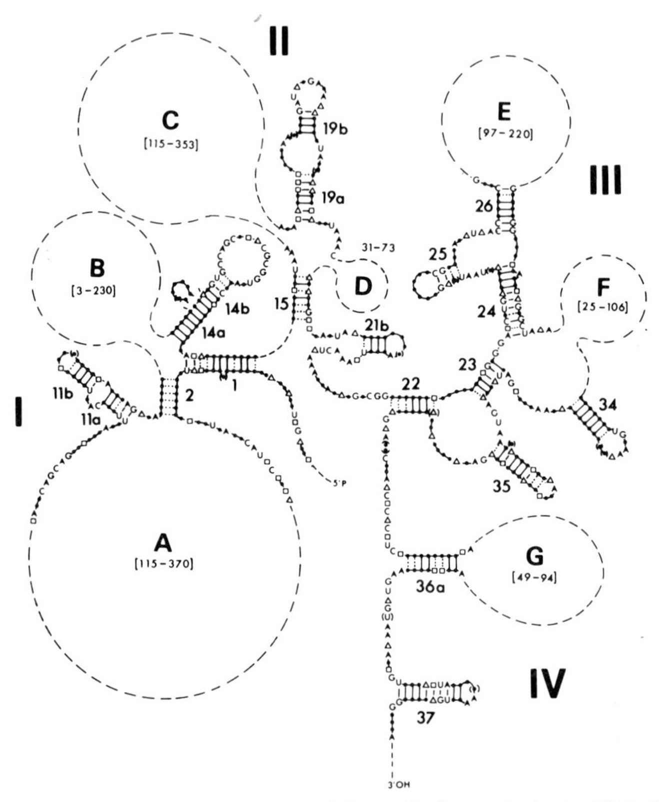

Fig. 3. Common basic structural organization of the small ribosomal subunits RNAs (Stiegler et al. 1982b).

Only secondary structure motifs which are common to all the RNAs studied are shown. The helices are numbered according to Stiegler *et al.* 1982b. Nucleotides are symbolized by filled circles and base pairs by bars. Dashed bars indicate that base-pairing between the relevant nucleotides is not possible in every case. The 4 structural domains (I–IV) are shown. Variable domains (A–G) are symbolized by a dashed line, their size in nucleotide residues varying between the 2 extreme values indicated in parentheses.Invariant nucleotides, found in homologous positions in all RNA sequences, are indicated. Semi-invariant residues are symbolized (□) for purines and (△) for pyrimidines. In some cases, a single residue was inserted or deleted and these are indicated in parentheses.

small and large ribosomal subunit rRNAs from various sources, covering diverse types of prokaryotes, eukaryotes and organelles, brought to light strong sequence conservations as well as a remarkable preservation of many secondary structure motifs (Stiegler *et al.* 1981b, Ebel *et al.* 1983a, b), (Fig. 3). It was interesting to check whether the 3 highly accessible regions in the 30S subunit which are also involved in 50S subunit association (Fig. 2), have been conserved during the course of evolution. Comparison of Figs. 1 and 3 shows that among the 3 major accessible regions, only part of domain II (motifs 19a and b), and the 3'-end region in domain IV (except the Shine & Dalgarno region, Shine & Dalgarno 1974), are strongly conserved. The other parts within the 3 major accessible regions are variable ones.

There is more and more evidence that the 3'-end region of 16S *E. coli* rRNA, which is a highly accessible one, and which also contains highly conserved sequences, plays a direct function in protein synthesis. The directive role of this region in initiation of protein synthesis is now well known (Steitz 1980). In prokaryotic ribosomal RNAs this region contains close to the 3'-end, the pyrimidine stretch which is complementary to the purine stretch located close to the initiator codon in prokaryotic messenger RNAs (Shine & Dalgarno interaction).

The 3'-end region of the small subunit ribosomal RNA has also been shown to be involved in the ribosomal decoding site. The anti-codon of *E. coli* and *B. subtilis* $tRNA^{Val}$ could be crosslinked, when bound at the ribosomal P site, to an identical site in *E. coli* 16S rRNA and yeast 18SrRNA, located in a highly conserved single-stranded region corresponding in *E. coli* 16S rRNA to the sequence 1392–1401 close to the 3′-end (Fig. 1), (Ofengand *et al.* 1979, Ofengand 1983, Ehresmann *et al.* 1983). Recent findings (Martin *et al.* 1983) reinforce the view that this area is intimately involved in the decoding process of protein synthesis. The authors have mapped the locus for the resistance to paromomycin, an antibiotic which promotes translational errors in the yeast mitochondrial gene coding for 15SrRNA, and found it at a position which is only 9 nucleotide residues to the 3'-site of the preceding crosslinking site. Finally, affinity labelling studies with an mRNA analogue have implicated residues 462–474 (located in domain IV) of *E. coli* 16SrRNA (Wagner *et al.* 1976). This site does not belong to the 3 highly exposed 16S rRNA regions, but it is close to the cobra venom cleavage site 336 which is protected after 50S subunit association.

II. 23SrRNA SURFACE TOPOGRAPHY

23SrRNA surface topography in the 50S subunit

The 23S rRNA phosphodiester bonds which are susceptible to T_1 and cobra venom ribonuclease attack within the 50S subunits (Branlant *et al.* 1977), as well as the guanines modified with kethoxal (Winship & Noller 1978), are indicated in Fig. 4. Our previous study on the 50S subunits (Branlant *et al.* 1977) showed that within the subunit 2 areas of 23S rRNA are highly susceptible to RNase T_1. These 2 areas encompass respectively domain II plus domain III (531–1195), and domain V plus domain VI (1651–2629) (Branlant *et al.* 1981). Large parts of

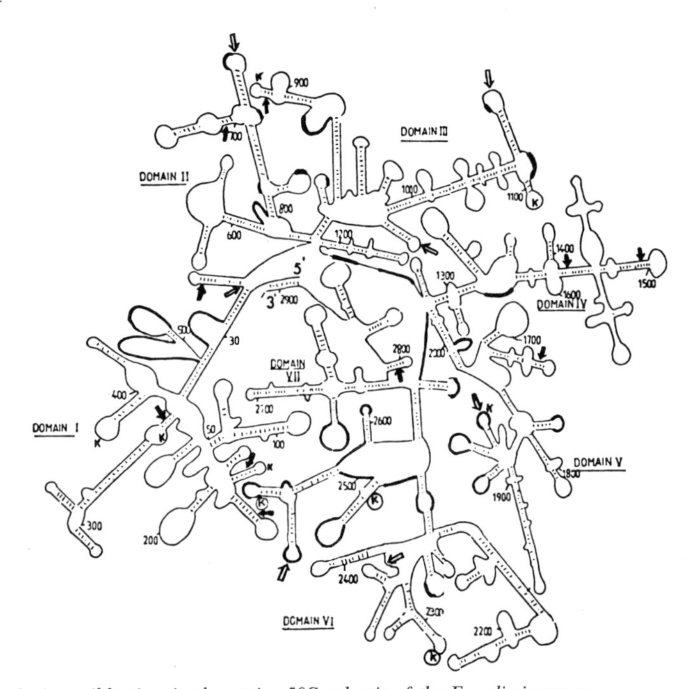

Fig. 4. Accessible sites in the native 50S subunit of the E. coli risosome.
A schematic representation of the secondary structure model proposed by Branlant *et al.* 1981. The thick lines in single-stranded regions correspond to sequences highly conserved throughout evolution. The 7 domains I to VII are indicated.

⇨ RNase T_1 cuts (Branlant *et al.* 1981)

➡ Cobra venom RNase cuts (Branlant *et al.* 1983)

K Kethoxal modification sites (Winship & Noller 1978)

Ⓚ Kethoxal modification sites protected in the presence of 30S subunit (Winship & Noller 1978, 1979).

these domains can be easily stripped off the subunit and, after such cleavage, a resistant core particle still exists which mainly consists of fragments from domain I, IV and VII and of a certain number of proteins. It can, therefore, be concluded that domains I, IV and VII probably have an important structural role.

Domains II and III each contain a guanine very sensitive to RNase T_1, G 748 and G 1068 respectively. Interestingly, G 748 is located in a hairpin loop close to a cluster of modified nucleotides $m^1G\psi T$ (Branlant *et al.* 1981). G 1068 is also located in a hairpin loop, and is adjacent to the adenine which is methylated in bacteria strains resistant to thiostrepton (Thompson *et al.* 1982). The primary structure of both hairpin loops is highly conserved throughout evolution. Altogether these observations suggest that these 2 loops play an important role. As to the loop containing G 1068, it might be involved in the subunit association, as thiostrepton impairs this process.

Domains V and VI (Branlant *et al.* 1981), contain three highly sensitive guanines G 1954, G 2391 and G 2532 (Branlant *et al.* 1977). As was the case for the sensitive guanines in domains II and III, these 3 guanines are located in single-stranded regions whose primary structure is highly conserved throughout evolution. A good agreement is observed with kethoxal results, since G 1954 and G 2391 are also attacked by kethoxal within 50S subunits. No highly sensitive guanines were detected in domains I, IV and VII. On the contrary, these 3 domains contain several phosphodiester bonds sensitive to the cobra venom ribonuclease: 3 cuts occur in domain I (500 nucleotides) as compared to one cut in domains V plus VI (1000 nucleotides). This means that the compact domains I, IV and VII display double-stranded regions accessible at the surface of the subunit, whereas domains II, III, V and VI contain accessible single-stranded regions, susceptible to play a functional role.

23S RNA surface topography in the 70S ribosome

We compared the accessibility of 23SrRNA to cobra venom RNase in the 50S subunit and in 70S ribosomes. Whereas 16S rRNA becomes protected against cobra venom enzyme digestion upon subunit association (Vassilenko *et al.* 1981), 23SrRNA becomes, on the contrary, more sensitive (Branlant *et al.* 1983). This means that during the course of the association process a conformational change occurs in the 50S subunit. The more likely explanation is that some of the 50S proteins, which interact with 23S RNA within the subunit, prevent this RNA from interacting with the 30S subunit, and possibly with 16S rRNA. This observation is in good agreement with the results obtained by Winship & Noller (1979) using kethoxal modification. It is noticeable that only the 2 guanines highly protected against kethoxal modification in the presence of the 30S subunit lie within domain VI.

Relationship between RNA surface topography and RNA function

The structural results show that domains I, IV and VII belong to a compact core within the subunit, and, therefore, play an important architectural role. On the contrary domains II, III, V and VI are loose regions easily removed from the particles and containing accessible single-stranded regions susceptible to play a functional role. This idea is reinforced by the fact that domains II, III, V and VI are highly conserved throughout evolution both in primary and secondary structure. In addition several experimental data favour the location of domain VI in the peptidyl transferase centre. Two punctual mutations in domain VI (2447 and 2504) lead to chloramphenicol resistance (Dujon 1980), another one in the same domain (2058) leads to erythromycin resistance (Sor & Fukuhara 1982), and, finally, a covalent linkage obtained between 23S rRNA and a puromycin derivative is probably located in domain VI (Greenwell *et al.* 1974).

CONCLUSION

This study shows that limited regions of 16S and 23S rRNAs from *E. coli* are exposed at the surface of ribosomal subunits. Both single- and double-stranded structures are present in these accessible regions. It is likely that some of these regions are involved in ribosome functions. Precise sites have been characterized which are located at the interface between the 2 subunits, as they are protected against chemical or enzymatic reagents in the 70S couple. They might be involved in subunit association.

There is much evidence that several of the exposed regions are involved in ribosome functions. The 3'-end region in the small subunit RNAs, which is a highly accessible one, and which also contains highly conserved sequences throughout evolution, seems to play an important role. In prokaryotes, it contains the Shine and Dalgarno region, which plays a directive role in initiation of protein synthesis by interacting with a messenger RNA region close to the initiator codon. In prokaryotes as well as in eukaryotes, a highly conserved sequence located in the 3'-end region is in close contact with the anti-codon of rRNAs when bound to the ribosomal P site, suggesting that this region is intimately involved in the ribosome decoding site.

In the large subunit, domain VI of 23S rRNA seems to play a functional role. It contains single-stranded regions, accessible in the isolated subunit, but protected in the 70S ribosome, highly conserved throughout evolution, and for which experimental evidence exists in favour of their location in the peptidyl-transferase centre.

Additional evidence must be brought to demonstrate the involvement of accessible regions at the surface of the ribosome in the different steps of protein synthesis. Several experimental approaches can be envisaged to prove such a direct role. One of them consists in producing specific cuts in these accessible regions, or to introduce changes in their nucleotide composition either directly on the RNA, or on the corresponding DNA, by site specific mutagenesis. Another approach is to characterize oligonucleotides arising from the ribosomal or messenger RNAs which strongly interact with the 30S or 50S subunits, or with the 70S ribosomes at their surface, and to study their possible inhibitory effect on specific ribosome functions. This approach is at present being explored in our laboratory.

REFERENCES

Brimacombe, R. (1980) *Biochem. Int. 1,* 162–171.

Branlant, C., Krol, A., Sri Widada, J. & Ebel, J. P. (1977) *J. Mol. Biol. 116,* 443–467.

Branlant, C., Krol, A., Machatt, M. A., Pouyet, J., Ebel, J. P., Edwards, K. & Kössel, H. (1981) *Nucl. Acids. Res. 9,* 4303–4324.

Branlant, C., Vassilenko, S., Krol, A. & Ebel, J. P. (1983) Unpublished results.

Brow, D. A. & Noller, H. F. (1983) *J. Mol. Biol. 163,* 27–46.

Chapman, N. M. & Noller, H. F. (1977) *J. Mol. Biol. 109,* 131–149.

Dujon, B. (1980) *Cell, 20,* 185–197.

Ebel, J. P., Branlant, C., Carbon, P., Ehresmann, B., Ehresmann, C., Krol, A. & Stiegler, P. (1983a) In: *Structure, Dynamics, Interactions and Evolution of Biological Macromolecules,* C. Hélène ed., pp. 177–193, D. Reidel Publishing Company, Dordrecht.

Ebel, J. P., Branlant, C., Carbon, P., Ehresmann, B., Ehresmann, C, Krol, A. & Stiegler, P. (1983b)

In: *Nucleic Acids: The Vectors of Life,* Pullman, B. & Jortner, J. Eds, pp. 387–401. D. Reidel Publishing Company, Dordrecht.

Ehresmann, B., Millon, R., Backendorf, C., Golinska, B., Olomucki, M., Ehresmann, C. & Ebel, J. P. (1980) In: *Biological implications of protein-nucleic acid interactions,* (Augustyniak, J., ed.). pp. 63–75. Adam Mickiewicz University Press, Poznan, Poland.

Ehresmann, C., Stiegler, P., Fellner, P. & Ebel, J. P. (1972) *Biochimie 54,* 901–967.

Ehresmann, C., Stiegler, P., Fellner, P. & Ebel, J. P. (1975) *Biochimie 57,* 711–748.

Ehresmann, C., Ehresmann, B., Millon, R., Ebel, J. P., Nurse, K. & Ofengand, J. (1983) *Biochemistry,* (in press).

Fox, G. E. & Woese, C. R. (1975) *Nature (London), 256,* 505–507.

Glotz, C. & Brimacombe, R. (1980) *Nucleic Acids Res., 8.* 2377–2395.

Glotz, C., Zwieb, C., Brimacombe, R., Edwards, K. & Kossel, H. (1981) *Nucl. Acids. Res. 9.* 3287–3306.

Greenwell, P., Harris, R. & Symons, R. (1974) *Eur. J. Biochem., 49.* 539–554.

Herr, W., Chapman, N. R. & Noller, H. F. (1979) *J. Mol. Biol. 130.* 433–449.

Lake, J. A. (1980) *in Ribosomes: Structure, Function and Genetics* (Chambliss, G.,Craven, G. R., Davies, J., Kahan, L. & Nomura, M., Eds. pp. 207–236, University Park Press, Baltimore.

Martin, R., Bordonne, R. & Dirheimer, G. (1982) In: *Cell function and differentiation, Part B, Biogenesis of energy transducing membranes and membranes and protein energetics, vol. 65,* pp. 355–365, Alan R. Liss Pub, New-York.

Noller, H. F. (1974) *Biochemistry, 13,* 4694–4703.

Noller, H. F., Kop, J., Wheaton, V., Brosius, J., Guttel, R., Kopylov, A., Dohme, F., Herr, W., Stahl, D., Gupta, R. & Woese, C. (1981) *Nucl. Acids Res. 9.* 6167–6189.

Ofengand, J. (1983) this volume.

Ofengand, J., Liou, R., Kohut, J., Schwartz, R. & Zimmermann, R. A. (1979) *Biochemistry, 18,* 4322–4332.

Ross, A. & Brimacombe, R. (1979) *Nature (London) 281,* 271–276.

Santer, M. & Santer, U. (1973) *J. Bacteriol. 116.* 1304–1313.

Santer, M. & Shane, S. (1977) *J. Bacteriol. 130.* 900–910.

Shine, J. & Dalgarno, L. (1974) *Proc. Natl. Acad. Sci. USA 71.* 1342–1346.

Sor, F. & Fukuhara, H. (1982) *Nucl. Acids. Res., 10.* 6571–6577.

Steitz, J. A. (1980) in: *Ribosomes: Structure, Function and Genetics.* (Chambliss, G., Craven, G. R., Davies, J., Davis, K., Kahan, L. & Nomura, M., eds). pp. 479–495. University Park Press, Baltimore.

Stiegler, P., Carbon, P., Zucker, M., Ebel, J. P. & Ehresmann, C. (1980) *C. R. Acad. Sci – Paris 291,* 937–940.

Stiegler, P., Carbon, P., Zucker, M., Ebel, J. P. & Ehresmann, C. (1981a) *Nucl. Acids. Res. 9.* 2153–2172.

Stiegler, P., Carbon, P., Ebel, J. P. & Ehresmann, C. (1981b) *Eur. J. Biochem. 120,* 487–495.

Stöffler, G., Bald, R., Kastner, B., Löhrmann, R., Stöffler-Mellicke, M. & Tischendorf, G. (1980) In: *Ribosomes: Structure, Function and Genetics* (Chambliss, G., Craven, G. R., Davies, J., Davis, K., Kahan, L. & Nomura, M. eds). pp. 171–205. University Park Press, Baltimore.

Thompson, J., Schmidt, F. & Cundliffe, E. (1982) *J. Biol. Chem. 257,* 7915–7917.

Vassilenko, S. K., Carbon, P., Ebel, J. P. & Ehresmann, C. (1981) *J. Mol. Biol. 152,* 699–721.

Wagner, R., Gassen, H. G., Ehresmann, C., Stiegler, P. & Ebel, J. P. (1976) *FEBS Lett. 67,* 312–315.

Winship, H. & Noller, H. F. (1978) *Biochemistry 17,* 307–315.

Winship, H. & Noller, H. F. (1979) *J. Mol. Biol. 130,* 421–432.
Woese, C. R., Magrum, L. J., Gupta, R., Siegel, R. B., Stahl, D. A., Kop, J., Crawford, N., Brosius, J., Guttel, R., Hegan, J. J. & Noller, H. F. (1980) *Nucl. Acids. Res. 8,* 2275–2293.
Zimmermann, R. A. (1980) in: *Ribosomes: Structure, Function and Genetics* (Chambliss, G., Craven, G. R., Davies, K., Kahan, L. & Nomura, M. eds) pp. 135–169. University Park Press, Baltimore.

DISCUSSION

UHLENBECK: I have a comment with regard to both of the last talks and probably also the next one. Although it may be obvious it is important to point out that both the chemical modification and the phylogenetic comparison methods for getting RNA secondary structures give essentially an average view of what the RNA structure in the ribosome is like. It is possible that different secondary structures exist at different steps in the translation mechanism and we are working at the sum of all of them. This is particularly clear now in that some of Hearst's long-range cross-links don't make much sense with respect to the secondary structures that are proposed by other methods.

EBEL: I completely agree with this point of view.

ERDMANN: The 2 G's you modify in the dimethyl A loop and which are sensitive to kethoxal in the 30S subunit, are those accessible in the 70S ribosome?

EBEL: No, according to Noller's results (Brown & Noller 1983), they are particularly protected in 70S ribosomes and completely protected in polysomes.

GARRETT: I would just like to reinforce the comment that you do get changes in going from the RNA to the subunit. One good example in the 16S rRNA is the "Shine & Dalgarno" at the 3′-terminus sequence. It is base paired in the free RNA, and opens up when the proteins, and in particular protein S21, assemble.

EBEL: It has actually been reported that the nucleotides close to the 3′ end of 16S rRNA are base paired in the 16S rRNA, but that this region opens when the proteins and especially S21 assemble. However in our experiments (Vassilenko *et al.* 1981) this region is cut by cobra venom RNase within the 30S subunit, as shown in Fig. 1, and this suggests that this base pairing may exist in the 30S subunit. But it must be pointed out that the hydrolysis was performed at 0°C and we cannot exclude that at that temperature additional base pairing may exist.

Brown, D. A. & Noller, H. F. (1983) *J. Mol. Biol. 163,* 27–46.
Vassilenko, S. K., Carbon, P., Ebel, J. P. & Ehresmann, C. (1981) *J. Mol. Biol. 152,* 699–721.

Mechanisms of Protein-RNA Recognition and Assembly in Ribosomes

R. A. Garrett, B. Vester, H. Leffers, P. M. Sørensen, J. Kjems, S. O. Olesen, A. Christensen, J. Christiansen & S. Douthwaite

INTRODUCTION

This article is concerned primarily with the questions: how do proteins recognize RNA in ribosomes, and how do they influence the RNA structure during the assembly and functioning of the ribosome?

About one third of the eubacterial proteins have unique binding sites on the ribosomal RNAs (Zimmermann 1980, Garrett 1979, 1983). These primary binding proteins appear to have a special, and as yet unknown, role in stabilizing, organizing and/or reversibly altering, the structure of the RNAs so as to produce and maintain a functionally active ribosome (Noller & Woese 1981). Evidence for this derives, primarily, from both protein-RNA binding experiments and *in vitro* assembly studies on ribosomal subunits. While we cannot exclude the possibility that additional proteins, or protein-protein complexes, interact directly with the RNAs, we can be fairly sure that the primary RNA binding proteins are especially important; this view is reinforced by the observation that some of them also interact with their own mRNAs and modulate the synthesis of groups of ribosomal proteins (Nomura *et al.* 1982).

Although the interactions of proteins and RNAs are important for many biological processes, including viral RNA replication and expression, DNA transcription, protein biosynthesis, protein transport through membranes and, possibly, in mRNA splicing, we currently know very little about the general mechanisms of recognition and interaction. The ribosomal binding proteins yield a rich supply of complexes for investigating any rules governing mechanisms of protein-RNA interactions. Here we summarize current results and views on this object.

Biostructural Chemistry, Department of Chemistry, Aarhus University, 8000 Aarhus C, Denmark

GENE EXPRESSION, Alfred Benzon Symposium 19.
Editors: Brian F. C. Clark & Hans Uffe Petersen, Munksgaard, Copenhagen 1984.

RNA BINDING PROTEINS AND THEIR SITES

The following proteins have been characterized that bind to the ribosomal RNAs of *E. coli:* S4, S7, S8, S15 and S20 associate with 16S RNA; L1, L2, L3, L4, $(L12)_4$–L10, L11, L15, L20, L23 and L24 interact with 23S RNA; and L5, L18 and L25 bind to 5S RNA. Various criteria have been employed to establish that site-specific complexes are formed (Zimmermann *et al.* 1980), nevertheless, uncertainty still surrounds some of the results. Proteins L15 and L20, for example, both have a tendency to bind unspecifically, i.e., they do not saturate completely at 1:1 molar protein:RNA ratios (Littlechild *et al.* 1977). Also, protein L5, and to a lesser extent L4 and L10, exhibit low solubilities in aqueous buffers, such that proof of specificity is difficult; in our hands the solubility of L5 is only increased, and its stoichiometry in a complex with 5S RNA enhanced, when L18 is present. Similarly, L10 only interacts strongly with 23S RNA when complexed with L12 dimers; both of these increased solubilities may reflect the formation of protein-protein complexes (Bear *et al.* 1977, Pettersson *et al.* 1976, Dijk *et al.* 1977). The contention that several additional proteins, including S3, S5, S9, S11, S12 and S18 bind specifically to 16S RNA that has been treated at low pH (Hochkeppel & Craven 1977) has proved irreproducible (Ungewickell *et al.* 1977), and, at present, should be considered unreliable. Moreover, preliminary evidence for the binding of proteins S2 and S5 to 16S RNA has not been further substantiated (Littlechild *et al.* 1977). Finally, the evidence for the binding to 23S RNA of additional proteins including L9, $(L12)_2$, L16, L18, L22 and L29 (Marquardt *et al.* 1979) has not been supported by rigorous criteria for the specificity of the interactions. We are left, therefore, with the aforementioned group of RNA-binding proteins.

The binding proteins exhibit a range of affinity constants. These have been estimated from nitrocellulose filter assays, quenching of ethidium bromide fluorescence and low angle X-ray scattering (Spicer *et al.* 1977, Feunteun *et al.* 1975, Österberg & Garrett 1977, Spierer *et al.* 1978). K_a values fall in the range 10^6 to 10^8, estimates which are generally too low to explain the high stability of some of the complexes under non-equilibrium conditions. The estimates are likely to be low, however, owing to the failure to correct for the often high fraction of denatured protein in the equilibrium mixture.

Some of the primary binding proteins protect their RNA sites against ribonuclease digestion. These protected regions fall into the following size ranges: 400–500 nucleotides (S4, L24), 200–300 nucleotides (S20), 100–200 nucleotides (L1, L23) and 50–100 nucleotides (S8, S15, L11, L25) (Ehresmann *et al.* 1980, Noller & Woese 1981, Noller *et al.* 1981, Garrett *et al.* 1981, Zimmermann 1980). The large sites for proteins S4 and L24 constitute most of the 5'-domains of the 16S and 23S RNAs, respectively. While both domains, and especially that of the 16S RNA, exhibit stable tertiary structures, proteins S4 and L24 both confer some extra stability. The remainder of the protected RNA fragments are smaller and probably reflect, more closely, the protein attachment site. These RNA regions are, in general, stable compared with the rest of the RNA, and this may be important for the early stages of ribosomal assembly.

Various approaches have been used to localize nucleotides within an RNA site that are interacting with, or masked by, a protein. Chemical modification methods have been employed including kethoxal modification of unpaired guanosines (Noller 1980), and the footprinting method of Peattie & Gilbert (1980) for monitoring the involvement of guanosines, adenosines and cytidines in higher order structure. Enzymic footprinting methods have also been commonly employed for probing both "unstructured" regions and double helices.

Cross-links induced by either ultraviolet light or chemical reagents have yielded additional information on the RNA environment of the proteins (Brimacombe *et al.* 1983). While the former cross-links are likely to involve the binding site of the protein, the latter could reflect any neighbouring RNA region. Cross-links induced by ultraviolet light have been especially useful for localizing contact sites of S4, S7, S8 and S20 in the RNA structure (Wower & Brimacombe 1983, Brimacombe *et al.* 1983).

HOW DO THE PROTEINS INTERACT?

Four sites have been investigated in some detail; 2 on the 16S RNA and 2 on the 5S RNA. Each contains a double helix with certain irregularities. Tentatively, they can be classified into 2 groups. Type I: those with a single nucleotide bulged-out from an otherwise regular helix, and type II: those constituting an irregular helix with multiple G·U pairings, and/or purine-purine pairings. They are considered separately, below.

Type I interactions

These are the most commonly observed RNA sites. Their existence is strongly supported by phylogenetic evidence (Noller & Woese 1981, Peattie *et al.* 1981).

S15–16S RNA. The putative secondary structure of the main protected double helix is depicted in Fig. 1. The upper part constitutes a regular and stable structure, whereas the lower part contains a bulged adenosine. Between these helices are 2 putative G·A juxtapositions with a bulged guanosine. The guanosines have been probed with kethoxal and the 2 in the centre, including the putatively bulged guanosine, are strongly protected by protein S15 (Müller *et al.* 1979); so far no attempt has been made to investigate the involvement of the bulged adenosine – 746 in the protein interaction.

Partial protection of the adjacent helix, which constitutes the S8 binding site (Fig. 1), has also been observed in the presence of S15 (Ungewickell *et al.* 1975, Zimmermann & Singh-Bergmann 1979); it is still unclear whether this result reflects a genuine protection, or just results from the known high ribonuclease resistance of the adjacent helix.

L18–5S RNA. This is the most extensively characterized of the protein-binding sites in the *E. coli* ribosome. It is centred on helix II of the 5S RNA; this is a regular helix with a highly conserved bulged nucleotide, adenosine – 66 (Fig. 2). Kethoxal sites at the extremities of the helix G_{23}, G_{24}, G_{69}, G_{107} are protected by L18 (Noller & Garrett 1979, Garrett & Noller 1979), and most guanosines within the helix exhibit diminished reactivity to dimethyl sulphate (although G_{61} shows enhanced reactivity) when L18 is bound (Peattie *et al.* 1981). Finally, all

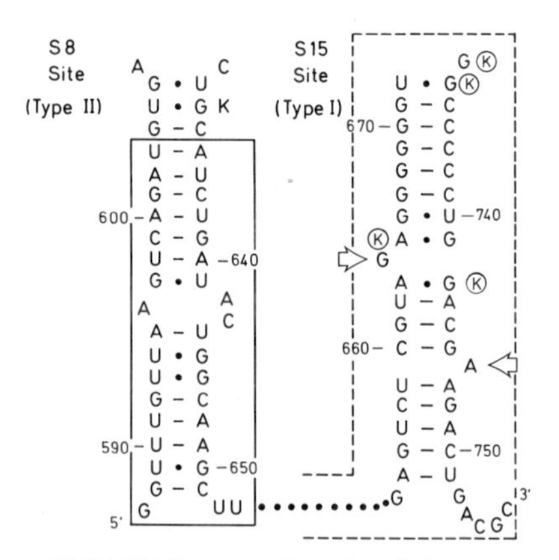

Fig. 1. Possible structures of 16S RNA fragments from E. coli *that are protected by proteins S8 and S15.* The main protected sequences are boxed. The kethoxal-reactive guanosines that are modified in the free 16S RNA, but protected by S15, are represented by encircled K's. Arrows denote the putative bulged nucleotides.

ribonuclease cuts within this helix are inhibited by protein L18, except for the cut at G_{20} which is enhanced (Douthwaite & Garrett 1981, Douthwaite *et al.* 1982). Direct evidence for the involvement of the bulged adenosine derives from 2 sources. First, the RNAse A cut directly preceding A_{66} is completely inhibited by L18. Second, rebinding experiments show that when 5S RNA molecules, each containing one or 2 randomly carbethoxylated adenosines, are complexed with L18 there is a strong selection against those molecules containing a modified A_{66} (Peattie *et al.* 1981).

Putative type I ribosomal RNA sites. The current secondary structural models of the 16S and 23S RNAs each contain other helices of the type described above. The 2 models exhibit approximately 55 and 100 short helical regions, respectively; of these, 10 contain bulged nucleotides in the 16S RNA, and 12 in

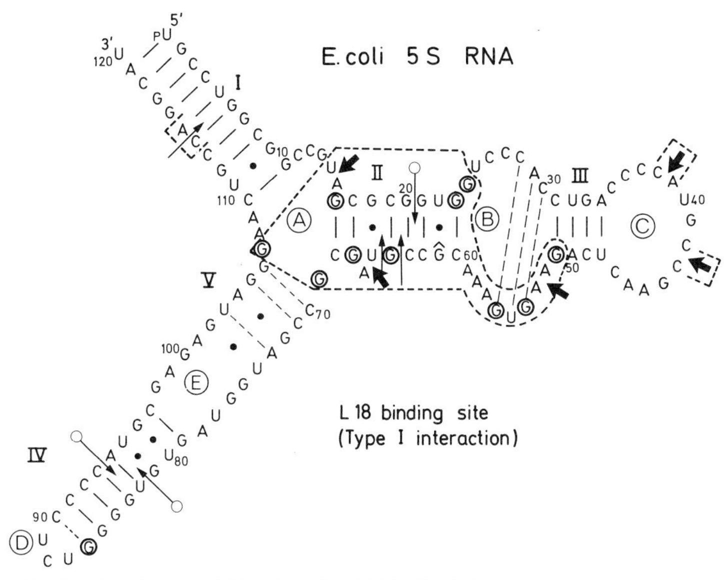

Fig 2. The binding site of protein L18 on E. coli *5S RNA.* Encircled nucleotides represent guanosines that are reactive to kethoxal and/or dimethyl sulphate in the native RNA structure but are protected in the L18-5S RNA complex (Garrett & Noller 1979, Peattie *et al.* 1981, Douthwaite *et al.* 1982). Two classes of enzyme cuts are shown. Those deriving from cobra venom ribonuclease that are specific for double helices are drawn across the polynucleotide chain and into the helical region. The ones from the single strand-specific ribonucleases A, T_1 and T_2 are drawn outside of the polynucleotide chain. All of the cuts shown are protected when L18 is bound, except for those with a circle at the extremity of the arrow, which are enhanced by L18 (Douthwaite *et al.* 1982). The boxed areas represent the putative L18 binding region.

the 23S RNA (Noller & Woese 1981, Noller & Van Knippenberg 1983, Brimacombe *et al.* 1983). While probing the 3'-minor domains of the small subunit RNAs, from eubacteria and eukaryotes, we found evidence for the helix depicted in Fig. 3 (Douthwaite *et al.* 1983). In the 2 eubacterial RNAs, the putative bulged adenosine is hyperreactive to carbethoxylation and is located next to a base-paired guanosine that is strongly methylated. Neighbouring double strand-specific cuts with the cobra venom ribonuclease confirm the presence of the double helix. A similar helix was also identified towards the 3'-end of the eubacterial 23S RNA (Fig. 4) which exhibited the same combination of bulged adenosine and adjacent base paired guanosine, both of which were hyperreactive with diethyl pyrocarbonate and dimethyl sulphate, respectively (A. Christensen, S. Douthwaite & R. A. Garrett, unpublished). The existence of this double helix was also strongly supported by the resistance of the adenosines and cytidines to carbethoxylation and methylation, respectively, and by phylogenetic evidence (Noller *et al.* 1981).

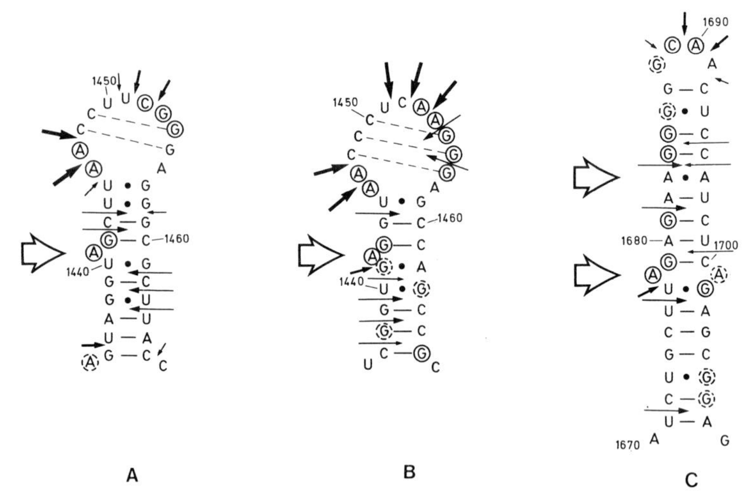

Fig. 3. Putative Type I protein-binding sites. "Hairpin" structures near the 3'-end of 16S RNA from (A) *E. coli* and (B) *B. stearothermophilus* and the corresponding helix (C) in the 18S RNA from *S. cerevisiae* (Douthwaite *et al.* 1983). The chemically reactive nucleotides (carbethoxylation of adenosines, methylation of guanosines and cytidines, Peattie & Gilbert 1980) are encircled as described for Fig. 2; the broken circles indicate weakly reactive nucleotides. The cuts specific for double helices and single-strand regions are also represented as described for Fig. 2. Dashed lines between nucleotides indicate possible base pairs.

```
              U   G
            G       A
            C — G
            U — A
            G — C ⁄ 2630
                   (A) ⇦
    2620— C — (G)
            A — U
            A — U
            G — C
            A • G
       U ← C — G
            U       U
```

Fig. 4. Putative Type I protein-binding site in the 3'-terminal domain of 23S RNA from E. coli *and* B. stearothermophilus. The sequence is conserved except for a change of C_{2615} to a uridine in the *E. coli* RNA. The bulged A_{2631} which is hyperreactive to carbethoxylation is shown; the adjacent guanosine is also very reactive (A. Christensen, S. Douthwaite & R. A. Garrett, unpublished).

Non-ribosomal nucleic acid sites. Several protein binding sites have been sequenced and drawn in energetically favourable secondary structures, some of which are supported by ribonuclease digestion data. A selection of these are shown in Fig. 5. All contain one or more nucleotides bulged from a regular helix. The coat protein binding site on the R17 (and $Q\beta$) phage RNA (Fig. 5A) exhibits a bulged adenosine in the polypurine sequence that is involved in the Shine and Dalgarno interaction between mRNA and 16S RNA (Gralla *et al.* 1974). Interestingly, this region of the R17 RNA is also protected against ribonuclease digestion by the ribosome. This raises the possibility that ribosomal protein S1, the mRNA binding protein, might also recognize this bulged nucleotide.

Protein L10 and its natural complex $L10.(L12)_4$ can bind to its own mRNA and prevent the translation of a group of ribosomal protein genes (Fiil *et al.* 1980). A region of the leader sequence from the L10 operon is shown in Fig. 5B that is important for the efficient translation of the mRNA *in vivo.* The corresponding region of a mRNA fragment, protected by $L10.(L12)_4$, lies adjacent to this structure (Johnsen *et al.* 1982). The structure contains a regular helix with a bulged cytidine. A putative secondary structure for part of the 3'-end of the tobacco mosaic viral RNA is presented in Fig. 5C. This constitutes part of the structure which initiates viral assembly by interacting with the protein discs; it contains 2 purines bulged from a helix (Zimmern 1977). The site of origin of second strand synthesis of Moloney murine leukemia pro viral DNA (Fig. 5D) also exhibits a possible bulged thymidine close to the point of initiation (Sutcliffe *et al.* 1980).

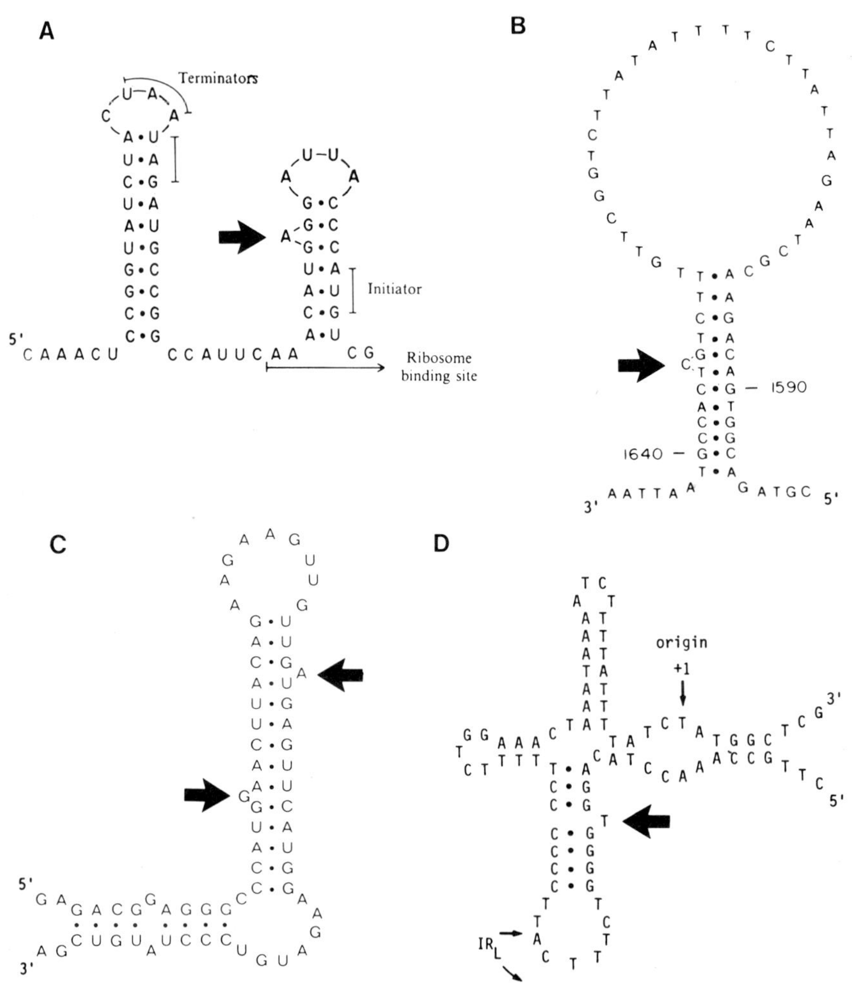

Fig. 5. Type I protein-nucleic acid-binding sites. (A) The coat protein binding site on the R17 phage RNA. The bulged adenosine occurs in the Shine & Dalgarno polypurine sequence (Gralla *et al.* 1974). (B) Part of the leader sequence of the L10 operon neighbouring the site where protein L10, or the L10.(L12)$_4$ complex, binds and inhibits translation of the mRNA (Johnson *et al.* 1982). (C) Part of the 3'-end of the tobacco mosaic viral RNA that initiates assembly to the protein discs. The 2 bulged nucleotides will lie on the same side of the helix in the three-dimensional structure (Zimmern 1977). (D) Site of origin of second-strand synthesis of the Moloney leukemia virus; the origin of replication is denoted by IR$_L$ (Sutcliffe *et al.* 1980).

Type II interactions

These are characterized by irregular double helices containing a number of non-Watson-Crick pairings including G·U and purine-purine pairings. There is strong phylogenetic evidence for the existence of these irregular structures; there is also a precedent for a G·A pairing in the structure of the yeast $tRNA^{Phe}$. These structures are highly resistant to ribonuclease digestion and chemical modification and, generally, they are stable at lower magnesium concentrations and exhibit a lower temperature resistance, than the type I helices. Two RNA sites have been investigated in detail. They are considered further below.

S8–16S RNA. The protected region is drawn in Fig. 1. It contains 5 G·U pairings concentrated in the lower part of the structure. Although it is very resistant to ribonucleases, and is stable at very low magnesium concentrations (Muto & Zimmermann 1978), it is relatively unstable on heating; S8 only interacts weakly with the RNA at 35–45°C (Schulte *et al.* 1974). There is evidence that the protected region does not constitute the whole protein binding site, however,

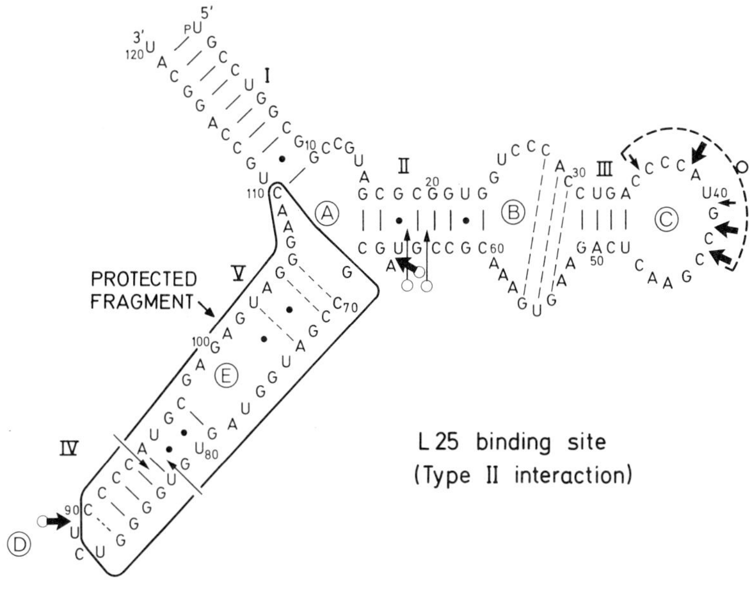

Fig. 6. L25-binding site on E. coli *5S RNA.* The RNA fragment protected by the protein is boxed (Douthwaite *et al.* 1979). Enzyme cuts are denoted by arrows as described for Fig. 2. The 2 cobra venom ribonuclease cuts contained therein are inhibited by the bound protein. All of the other cuts, which are encircled at their extremities, are enhanced in the presence of L25 (Douthwaite *et al.* 1982).

since strong rebinding of S8 to the RNA fragment only occurs when a part of the sequence above the helix (nucleotides 608–630) is present and in particular the sequence C-U-C_{620} A-A-C-C-U-G (Müller *et al.* 1979, Zimmermann & Singh-Bergmann 1979, Wower & Brimacombe 1983); a part of this sequence is also conserved in the putative mRNA binding site (Nomura *et al.* 1982).

L25–5S RNA. L25 binds to, and stabilizes, the most resistant part of the 5S RNA which is boxed in Fig. 6 (Douthwaite *et al.* 1979). "Footprinting" experiments with cobra venom ribonuclease have localized one attachment point within the regular double helix in the lower part of the fragment (Douthwaite *et al.* 1982). The upper part of the site has shown pronounced divergence during evolution, and these changes are illustrated in Fig. 7. In eubacteria it can form a very irregular helix; in some archaebacteria, for example *Sulfolobus acidocaldarius* (Stahl *et al.* 1981) and *Thermoplasma acidophilum* (Luehrsen *et al.* 1981), that grow under extreme conditions of pH and/or temperature, it can form a stable

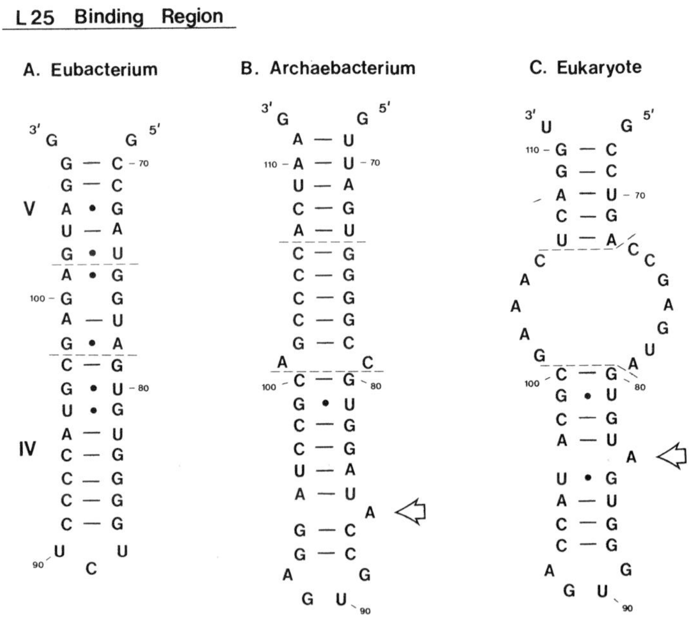

L25 Binding Region

A. Eubacterium

B. Archaebacterium

C. Eukaryote

Fig. 7. Evolution of the L25-binding region. (A) Represents the *E. coli* 5S RNA fragment drawn as an irregular helix (Type II site). For the archaebacterium *Sulfolobus acidocaldarius* a more regular helix is found (Stahl *et al.* 1981) and a possible bulged nucleotide occurs in the lower part of the helix. In the eukaryote, *S. cerevisiae,* regular helices occur in the upper and lower parts of the fragment with a partly accessible internal loop structure; the bulged nucleotide is very accessible in the lower helix (Garrett & Olesen 1982).

regular helix; in eukaryotes a regular helix and an internal loop structure are probably formed. The high resistance of this region in eubacteria to both enzymes and chemical reagents has reinforced speculation that an irregular helix with non-Watson-Crick base pairs does indeed form (Stahl *et al.* 1981, Garrett *et al.* 1981, De Wachter *et al.* 1982). This would also resemble the properties of the irregular helix in the S8 site (Fig. 1), in that it is stable at very low magnesium concentrations and unstable at moderate temperatures (Spierer & Zimmermann 1978, Douthwaite *et al.* 1979).

Recently, Kime & Moore (1982, 1983a, b) have investigated this fragment, in the absence and presence of protein L25, by high resolution NMR spectroscopy using nuclear Overhauser methods. They provide indirect evidence for non-Watson-Crick hydrogen bonding in the RNA. In addition, they conclude that a general perturbation of the structure occurs, on protein binding, which may be due to direct H-bonding between the protein side chains and the nucleotides.

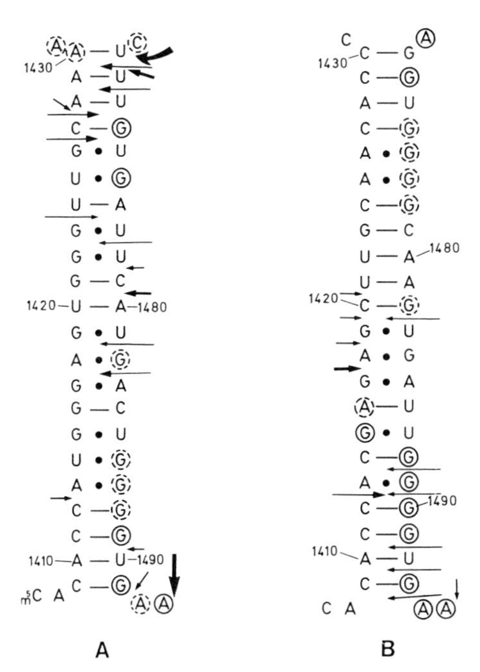

Fig. 8. Putative Type II protein-binding site. These highly irregular helices occur towards the 3'-ends of 16S RNA from (A) *E. coli* and (B) *B. stearothermophilus.* Chemically reactive nucleotides (Peattie & Gilbert 1980) and bonds cut by ribonucleases specific for double helices and single-stranded regions, in the free 16S RNAs, are represented as described for Figs. 2 and 3 (from Douthwaite *et al.* 1983).

Putative type II ribosomal RNA sites. In the current secondary structural models about 7 helices in 16S RNA and 16 in 23S RNA fall into this category (Noller & Woese 1981, Noller & Van Knippenberg 1983, Brimacombe *et al.* 1983). Some of them are strongly supported by phylogenetic evidence. One, in particular, that lies close to the 3'-end of 16S RNA has been probed by chemical reagents and ribonucleases in the free 16S RNA and there is good evidence for an irregular double helical structure. The results, which are summarized in Fig. 8, show the low level of reactivity of the adenosines and cytidines to diethyl pyrocarbonate and dimethyl sulphate, respectively, which is compatible with the presence of double helices. The positions of the cobra venom ribonuclease cuts, which are exclusive to double helices, are also included; they occur adjacent to and, in one case, between G·A juxtapositions (Douthwaite *et al.* 1983).

A further possible site neighbours a fragment which is protected by protein

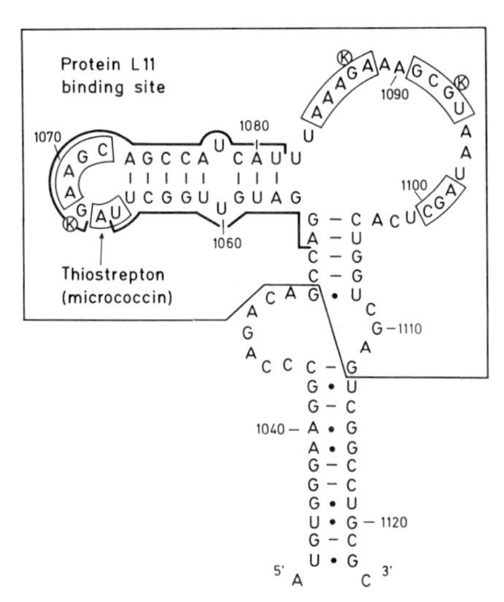

Fig. 9. Protein L11-binding site. The boxed fragment is protected by protein L11 (Thompson *et al.* 1981). This represents the EF-G-dependent GTP-ase centre on the ribosome. Boxed sequences are highly conserved during evolution. A_{1067} represents a binding site for the peptide antibiotic thiostrepton (and the structurally related micrococcin) which inhibit EF-G binding to the ribosome. The underlined fragment 1054-81 has been cross-linked to EF-G by diepoxybutane (Skjöld 1983). The general features of this secondary structure, proposed mainly on the basis of phylogenetic sequence comparisons (Noller & Van Knippenberg 1983), are supported both by the locations of kethoxal-reactive guanosines (marked with a K) and our own studies using chemical (Peattie & Gilbert 1980) and ribonuclease probes; in particular the irregular helix (1033–1043 and 1112–1122), a potential protein-binding site, is supported by our data (S.D., A.C. & R.G., unpublished work).

L11. The region, which is drawn in Fig. 9, is the centre of the EF-G-dependent GTPase reaction of the 50S subunit; it also contains the site of action of the related antibiotics thiostrepton and micrococcin (Thompson *et al.* 1981). The irregular helix, containing multiple G·U and G·A pairings, is drawn immediately below the boxed RNA fragment in Fig. 9. The presence of this helix is supported both by its resistance to chemical modification in the free 23S RNA and by its accessibility to the cobra venom ribonuclease (S.D., A.C. & R.G., unpublished work). It is a potential attachment site for protein L10 or another ribosomal protein.

PROTEIN INDUCED CONFORMATIONAL CHANGES

The capacity of RNA molecules to assume different and discrete local conformations may constitute a structural basis for ribosomal assembly and function. One specific functional hypothesis was the ratchet mechanism for tRNA binding to the mRNA which required the formation of alternating conformers in the anti-codon loop of tRNA (Woese 1980). Other conformational changes of importance for assembly or function have also been proposed and ribosomal proteins have been considered as likely effectors of such changes. However, while several lines of evidence support the occurrence of protein-induced changes in RNA conformation it is difficult, at present, to establish any biological significance. Most of the evidence has been accrued with isolated components using chemical modification and enzymic digestion approaches, or with such physical chemical methods as circular dichroism or temperature jump kinetics. The possibility remains that some, or all of the effects, are artifactual: ribosomal RNAs are always partially denatured during extraction such that, if renaturation is incomplete prior to protein binding, conformational changes in the RNA may just reflect protein-induced renaturation of the RNA structure.

The best documented effects occur in 5S RNA. L25 produces a large increase in the accessibility of nucleotides at the extremities of its binding site (Fig. 6), and a small decrease in the circular dichroism band of 5S RNA at 267 nm. L18 produces a minor increase in the accessibility of 2 nucleotides within its own binding site (Fig. 2), and a large decrease in the circular dichroism band at 267 nm; the latter was attributed either to an increase in higher order structure (Bear *et al.* 1977), or a perturbation of the helical binding site (Spierer *et al.* 1978). A change in the 5S RNA tertiary structure has also been detected by temperature jump kinetics, near neutral pH, that is base catalyzed and could also be mediated by the basic proteins (Kao & Crothers 1980). This change, which is also dependent on the magnesium concentration and temperature, has been characterized further by proton NMR studies (Kime & Moore 1982).

These data all suggest that relatively minor, and probably localized, changes occur in the RNA on protein binding. This view receives support from determination of the gross shape and structure of the RNA by low angle X-ray scattering (Österberg & Garrett 1977, Österberg *et al.* 1978, Österberg 1979), measurement of thermodynamic parameters (Spierer *et al.* 1978), and proton NMR studies (Kime & Moore 1982c).

More striking differences have been detected in the large 16S RNA by laser-light scattering (Bogdanov *et al.* 1978). At low Mg^{++} (1 mM) and 20°C the RNA was rendered more compact by the binding proteins, whereas, at high Mg^{++} (20 mM), and the same temperature, the RNA was already compact and binding of the proteins produced little additional change. Bogdanov (1982) correlates this result with the observation that primary binding proteins assemble on the large RNAs at 37°C when the RNA has an open conformation even at the higher magnesium concentrations. He concludes that the proteins must, therefore, be able to effect a more efficient packing of the RNAs that is important for assembly. More specific mechanisms of protein-RNA assembly are considered below.

ASSEMBLY MECHANISMS

Various mechanisms have been proposed for the co-operative assembly of proteins on the RNAs. The include direct protein-protein interactions, short range protein-induced effects in, for example, 5S RNA and long-range effects transmitted through the large RNA domains; they are summarized below.

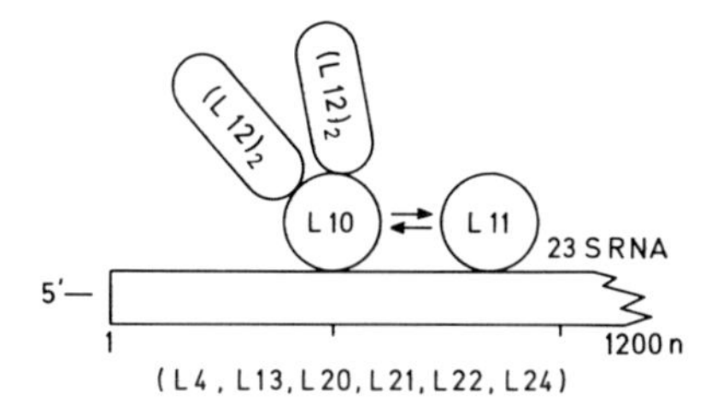

Fig. 10. An example of co-operative assembly of RNA-binding proteins. The binding of both protein L11 and the L10.$(L12)_4$ complex are enhanced when both are present (Dijk *et al.* 1979). They assemble to the 5'-one third of the 23S RNA of *E. coli* together with a group of proteins L4, L13, L20, L21, L22 and L24 which are important for the early assembly of the 50S subunit (Garrett 1982).

Mutual stimulation of binding proteins

Protein L11, and the complex L10.$(L12)_4$, stimulate one anothers binding within the 5'-one third of 23S RNA (Dijk *et al.* 1979). This effect, which may be important for 50S subunit assembly, is illustrated in Fig. 10. Since L11 and L10.$(L12)_4$ have been chemically cross-linked they are neighbours on the ribosome. They may also have RNA-binding sites in the same vicinity within the 23S RNA (Thompson *et al.* 1981, Johnson *et al.* 1982). The stimulatory effect may arise, therefore, from a direct interaction of the proteins when assembled on the RNA. Other pairs of RNA-binding proteins which co-operatively interact on the RNA include S8 and S15 (on 16S RNA), and L18 and L25 (on 5S RNA). Both pairs bind to adjacent RNA sites, but have not been chemically cross-linked (Figs. 1, 2, 6). For the L18 and L25 pair there is strong evidence that they influence the structure of their respective RNA-binding sites (ribonuclease digestion evidence is summarized in Figs. 2, 6), and this could be the basis for the co-operation.

Linkage proteins

Some of the binding proteins, including S4 and L18, contain very basic N-terminal sequences which do not appear to be involved in RNA binding (Garrett 1979, 1982). Although one line of evidence suggests that these basic regions may constitute functional domains and be involved, for example, in tRNA binding (Changchien *et al.* 1978), other experiments suggest that they are required for the co-operative assembly of the ribosome. In particular, the assembly of proteins L18 and L5 with 5S RNA was examined. The basic N-terminal sequence of L18 was essential for the co-operative assembly of protein L5. Moreover, L5 and the N-terminal region of L18 were necessary for assembly of 5S RNA to 23S RNA. Two hypothetical models, outlined in Fig. 11, require either that the N-terminal sequence directly links L5 with 23S RNA, or that it binds to L5 (possibly to the

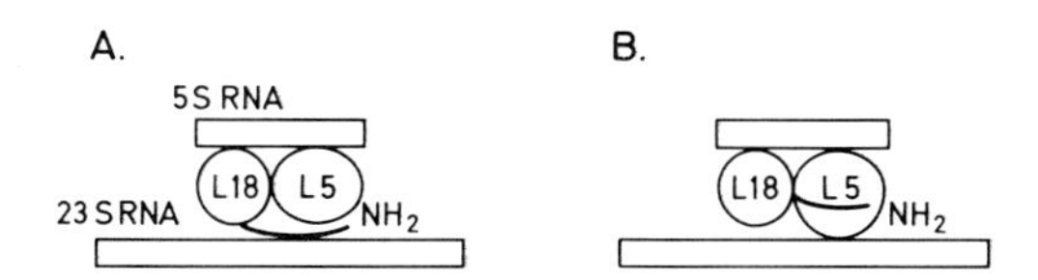

Fig. 11. Alternative hypothetical models for the role of the basic N-terminal region of L18 in the 5S RNA - 23S RNA assembly. In (A) a link is formed between L5 and the 23S RNA, and in (B) the basic terminal region binds to, and effects a conformational change in, L5 such that it can bind to 23S RNA.

central and acidic part of the sequence) and effects a conformational change such that L5 can interact with the 23S RNA (Newberry & Garrett 1980).

RNA cores

Longer-range effects have been suggested for the large ribosomal RNAs. The evidence derives mainly from studies on domain I of the 16S RNA which can be isolated, virtually intact, by mild digestion of 16S RNA with carrier-bound RNase A. It constitutes a compact and highly structured RNA region which may act as a nucleus for 30S subunit assembly (Garrett *et al.* 1977). When protein S4, which plays a critical role in 30S subunit assembly, is bound to this domain its melting behaviour changes from non-co-operative to highly co-operative (Bear *et al.* 1979). This effect, which is illustrated in Fig. 12, was assumed to reflect a "tuning" of the RNA structure. The effect could provide a mechanism for the transmission of allosteric effects throughout the RNA domain and possibly the whole RNA. More general evidence for the individual RNA domains acting as nuclei for the assembly of groups of ribosomal proteins has been accrued, and summarized, by Zimmermann (1980).

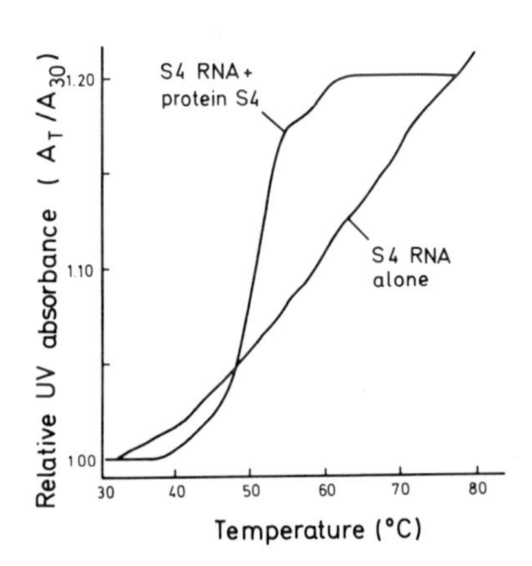

Fig. 12. Co-operative melting of domain I of the E. coli *16S RNA (the S4 RNA) in the free state and complexed with protein S4.* The complex was formed, and measurements made, in 30 mM Hepes-KOH, pH 7.6, 20 ml $MgCl_2$, 300 mM KCl. The protein-RNA complex was prepared at 20°C in the presence of a 3-fold molar excess of protein and exhibits, approximately, a 1:1 stoichiometry with the RNA (from Bear *et al.* 1979).

CONCLUSIONS

We have summarized current knowledge on the ribosomal RNA sites and possible mechanisms of protein-RNA recognition and assembly. The summary serves to emphasize what a rich and useful source of protein-RNA complexes the ribosome is, and also its advantages for deducing any general rules for mechanisms of interaction and assembly that may exist.

Clearly many of the concepts and hypotheses considered are tentative, and sometimes even speculative; nevertheless, they provide a focus for designing future experiments. The recent progress in developing secondary structural models for all of the RNAs, many parts of which receive strong experimental support, has facilitated this process, and more detailed characterization of other ribosomal protein binding sites, such as those of S20, L1, L2, L3, L4, L20 and L23, will help further.

There is a limitation, however, with all of the approaches described for investigating protein-RNA interactions further: it is difficult to locate individual nucleotides that interact with the protein, especially when they occur in double helices. As a possible improvement, X-ray diffraction analysis of crystals of protein-RNA fragment complexes does not look very promising in the medium term, although some progress has been made with the L25-fragment complex shown in Fig. 6. (P. Moore, personal communication). The recombinant DNA technology, however, is likely to prove very useful for further and more detailed analyses of RNA sites. The approaches used to produce deletions in the large ribosomal RNAs (Stark *et al.* 1982), and the point mutagenesis of the RNAs by chemical methods (Gourse *et al.* 1982) have a high potential, especially when developed to include oligonucleotide-directed mutagenesis, whereby a defined point mutation in a protein-binding site can be introduced.

ACKNOWLEDGEMENTS

R. G. particularly thanks Professor Harry Noller and members of his laboratory in Santa Cruz for many stimulating discussions and fruitful collaborations; provisions of travel grants by NATO for these mutual projects was also appreciated. The research from our laboratory, and three of the authors (S.O.O., J.C. and A.C.), were supported by the Danish Science Research Council. Additional research grants were provided by the Carlsberg Foundation and the Aarhus University Fund. A. Lindahl, O. Jensen and J. Christensen are thanked for help in preparing the manuscript.

REFERENCES

Bear, D. G., Schleich, T., Noller, H. F., Douthwaite, S. & Garrett, R. A. (1979) *FEBS Lett. 100,* 99–102.

Bear, D. G., Schleich, T., Noller, H. F. & Garrett, R. A. (1977) *Nucleic Acids Res. 4,* 2511–2526.

Bogdanov, A. A. (1981) In: *Soviet Science Rev. Biology,* Vol. 2. (Ed. V. P. Skulachev), pp. 1–45, Harwood Internat. Publ., New York.

Bogdanov, A., Zimmermann, R. A., Wang, C. C. & Ford, N. C. (1978) *Science 202,* 999–1001.

Brimacombe, R., Maly, P. & Zwieb, C. (1983) *Progr. Nucleic Acids Res. & Mol. Biol. 28,* 1–48.

Changchien, L. M., Schwarzbauer, J., Cantrell, M. & Craven, G. (1978) *Nucleic Acids Res. 5,* 2789–2803.

De Wachter, R., Chen, M.-W. & Vandenberghe, A. (1982) *Biochimie 64,* 311–329.

Dijk, J., Garrett, R. A. & Müller, R. (1979) *Nucleic Acids Res. 6,* 2717–2730.

Dijk, J., Littlechild, J. & Garrett, R. A. (1977) *FEBS Lett. 77,* 295–300.

Douthwaite, S., Christensen, A. & Garrett, R. A. (1982) *Biochem. 21,* 2313–2320.

Douthwaite, S., Christensen, A. & Garrett, R. A. (1983) *J. Mol. Biol. 169,* 249–279.

Douthwaite, S. & Garrett, R. A. (1981) *Biochem. 20,* 7301–7307.

Douthwaite, S., Garrett, R. A., Wagner, R. & Feunteun, J. (1979) *Nucleic Acids Res. 6,* 2453–2470.

Ehresmann, C., Stiegler, P., Carbon, P., Ungewickell, E. & Garrett, R. A. (1980) *Eur. J. Biochem. 103,* 439–466.

Feunteun, J., Monier, R., Garrett, R., Le Bret, M. & Le Pecq, J. B. (1975) *J. Mol. Biol. 93,* 535–541.

Fiil, N. P., Friesen, J. D., Downing, W. L. & Dennis, P. P. (1980) *Cell 19,* 837–844.

Garrett, R. A. (1979) *Int. Rev. Biochem., Vol 25,* pp. 121–177 (Ed. R. E. Offord), University Park Press, Baltimore.

Garrett, R. A. (1983) *Horizons in Biochemistry and Biophysics* (Ed. F. Palmieri), Vol. 7, 101–138. J. Wiley.

Garrett, R. A., Douthwaite, S. & Noller, H. F. (1981) *Trends Biochem. Sci. 6,* 137–139.

Garrett, R. A. & Noller, H. F. (1979) *J. Mol. Biol. 132,* 637–648.

Garrett, R. A. & Olesen, S. O. (1982) *Biochem. 21,* 4823–4830.

Garrett, R. A., Ungewickell, E., Newberry, V. N., Hunter, J. & Wagner, R. (1977) *Cell Biol. Int. Reps. 1,* 487–502.

Gourse, R. L., Stark, M. J. R. & Dahlberg, A. E. (1982) *J. Mol. Biol. 159,* 397–416.

Gralla, J., Steitz, J. A. & Crothers, D. M. (1974) *Nature 248,* 204–208.

Hochkeppel, H.-K. & Craven, G. R. (1977) *Mol. Gen. Genet. 153,* 325–329.

Johnsen, M., Christensen, T., Dennis, P. P. & Fiil, N. P. (1982) *EMBO Jnl. 1,* 999–1004.

Kao, T. H. & Crothers, D. M. (1980) *Proc. Natl. Acad. Sci. USA 77,* 3360–3364.

Kime, M. J. & Moore, P. B. (1982) *Nucleic Acids Res. 10,* 4973–4986.

Kime, M. J. & Moore, P. B. (1983a) *Biochem. 22,* 2615–2622.

Kime, M. J. & Moore, P. B. (1983b) *Biochem. 22,* 2622–2629.

Kime, M. J. & Moore, P. B. (1983c) *FEBS Lett. 155,* 199–203.

Littlechild, J., Dijk, J. & Garrett, R. A. (1977) *FEBS Lett. 74,* 292–294.

Luehrsen, K. R., Fox, G. E., Kilpatrick, M. W., Walker, R. T., Domdey, J., Krupp, G. & Gross, H. J. (1981) *Nucleic Acids Res. 9,* 965–970.

Marquardt, O., Roth, H. E., Wystrup, G. & Nierhaus, K. H. (1979) *Nucleic Acids Res. 6,* 3641–3649.

Müller, R., Garrett, R. A. & Noller, H. F. (1979) *J. Biol. Chem. 254,* 3873–3878.

Muto, A. & Zimmermann, R. A. (1978) *J. Mol. Biol. 121,* 1–15.
Newberry, V. N. & Garrett, R. A. (1980) *Nucleic Acids Res. 8,* 4131–4142.
Noller, H. F. (1980) In: *Ribosomes* (Ed. G. Chambliss *et al.*), pp. 3–22. University Park Press, Baltimore.
Noller, H. F. & Garrett, R. A. (1979) *J. Mol. Biol. 132,* 621–636.
Noller, H. F., Kop, J. A., Wheaton, V., Brosius, J., Gutell, R. G., Kopylov, A. M., Dohme, F., Herr, W., Stahl, D. A., Gupta, R. & Woese, C. R. (1981) *Nucleic Acids Res. 9,* 6167–6189.
Noller, H. F. & Van Knippenberg, P. (1983) *Horizons in Biochem. and Biophys,* (Ed. F. Palmieri), Vol. 7, 71–100, J. Wiley.
Noller, H. F. & Woese, C. R. (1981) *Science 212,* 403–411.
Nomura, M., Dean, D. & Yates, J. L. (1982) *Trends Biochem. Sci. 7,* 92–95.
Österberg, R. (1979) *Eur. J. Biochem. 97,* 463–469.
Österberg, R. & Garrett, R. A. (1977) *Eur. J. Biochem. 79,* 67–72.
Österberg, R., Sjöberg, B., Garrett, R. A. & Littlechild, J. (1978) *Nucleic Acids Res. 5,* 3579–3588.
Peattie, D., Douthwaite, S., Garrett, R. A. & Noller, H. F. (1981) *Proc. Natl. Acad. Sci. USA 78,* 7331–7335.
Peattie, D. A. & Gilbert, W. (1980) *Proc. Natl. Acad. Sci. USA, 77,* 4679–4682.
Pettersson, I., Hardy, S. J. S. & Liljas, A. (1976) *FEBS Lett. 64,* 135–138.
Schulte, C., Morrison, C. A. & Garrett, R. A. (1974) *Biochem. 13,* 1032–1037.
Skjöld, S. E. (1983) *Nucleic Acids Res. 11,* 4923–4932.
Spicer, E., Schwarzbauer, J. & Craven, G. R. (1977) *Nucleic Acids Res. 4,* 491–499.
Spierer, P., Bogdanov, A. A. & Zimmermann, R. A. (1978) *Biochem. 17,* 5394–5398.
Spierer, P. & Zimmermann, R. A. (1978) *Biochem. 17,* 2474–2479.
Stahl, D. A., Luehrsen, K. R., Woese, C. R. & Pace, N. R. (1981) *Nucleic Acids Res. 9,* 6129–6137.
Stark, M. J. R., Gourse, R. L. & Dahlberg, A. E. (1982) *J. Mol. Biol. 159,* 417–439.
Sutcliffe, J. G., Shinnick, T. M., Verma, I. M. & Lerner, R. A. (1980) *Proc. Natl. Acad. Sci. USA 77,* 3302–3306.
Thompson, J., Schmidt, F. & Cundliffe, E. (1982) *J. Biol. Chem. 257,* 7915–7917.
Ungewickell, E., Garrett, R. A. & Le Bret, M. (1977) *FEBS Lett. 84,* 37–42.
Ungewickell, E., Garrett, R. A., Ehresmann, C., Stiegler, P. & Fellner, P. (1975) *Eur. J. Biochem. 51,* 165–180.
Woese, C. (1980) In: *Ribosomes* (Ed. G. Chambliss *et al.*), pp. 357–373, University Park Press, Baltimore.
Wower, I. & Brimacombe, R. (1983) *Nucleic Acids Res. 11,* 1419–1437
Zimmermann, R. A. (1980) In: *Ribosomes* (Ed. G. Chambliss *et al.*), pp. 135–169, University Park Press, Baltimore.
Zimmermann, R. A. & Singh-Bergmann, K. (1979) *Biochim. Biophys. Acta. 563,* 422–431.
Zimmern, D. (1977) *Cell 11,* 463–482.

DISCUSSION

MAALØE: When you build models, how does the looping out of an adenosine affect the stacking in a double helix, and is there a "hinge" effect?

GARRETT: Model building with space-filling atoms indicates that the protruding nucleotide perturbs the helix very little; a slight kink is formed. Possibly, 2 adjacent looped-out nucleotides would produce a "hinge".

UHLENBECK: We have investigated the RNA binding site of the R17 coat-protein which acts as a translational repressor. Many variants of the binding site were synthesized and binding constants were measured for protein-RNA complex formation using the Millipore filter binding assay. The binding constant for the wild-type sequence to the protein was about 3×10^8. If the single looped out adenosine (see Fig. 5A in the chapter) is either deleted, or substituted by a cytidine, no complex formation is detected. However, this is not the only site in the binding sequence that is important for the association. Each of the 7 single-stranded nucleotides (Fig. 5A) has been changed and 4 of them have a large effect on the binding constant. We also have evidence that one of the uridines in the loop is forming a transient covalent bond between the RNA and protein molecule. The protein-RNA interaction is, indeed, highly specific, and involved extensive intimate contact between this "hairpin" loop and the protein.

GARRETT: Have you investigated whether ribosomal protein S1 binds to the fragment and recognizes the looped out adenosine that lies in the "Shine & Dalgarno" sequence involved in ribosome recognition?

UHLENBECK: No.

GARRETT: I agree that the RNA binding site of a protein involved more than a looped-out nucleotide. We also have evidence for extensive interactions with the double helical segment and with structures adjacent to the double-helix (see for example the site of L18 on 5S RNA in Fig. 2). We are only implying that the bulged nucleotide participates in the protein-RNA interaction, possibly at the recognition state; it could, for example, act as a marker enabling the protein to find the correct helix out of the 150 or so possibilities in the ribosomal RNAs.

KLUG: Although the examples of the looped-out nucleotides that we have just seen are very impressive, showing that they are all needed for recognition by a protein, I think that a word of warning is appropriate: the looped-out nucleotide may have another function, which is namely to destabilize a helix. In the tobacco mosaic viral RNA, there is a "hairpin" stem at the origin of assembly with 2 looped-out nucleotides separated by 7 base pairs (Fig. 5C in the chapter). There, the double helix is temporary and while the secondary structures is recognized by the side chains of the protein it must then be melted-out during virus assembly. It looks, therefore, as though the purpose of the looped-out nucleotides is to weaken the helix.

ROSENBERG: Just to add to Dr. Klug's comment: in termination of transcription a stem and loop structure forms in the RNA and caused termination of the RNA polymerase. It has also been shown that many mutations, which are simply single base pair additions or deletions in that stem structure, lead to non-functioning of the termination events. So that again a single nucleotide change is enough to destabilize that structure.

UHLENBECK: A weakening of the helix cannot be the only explanation for the looped-out nucleotides though. Some years ago, it was shown fairly clearly, by spectroscopy, that the base pairs in the R17 RNA "hairpin" do not open on protein binding.

GARRETT: Similarly, with ribosomal protein-RNA complexes there is no evidence, as yet, for protein-induced opening of secondary structure. On the contrary, for the protein L18-5S RNA complex circular dichroism evidence suggests there may be an increase in secondary structure.

ERDMANN: But your ethidium bromide chasing experiments indicated that although there was double-stranded structure prior to L18 binding, dye binding decreased on complex formation. Might this not indicate that the double helix has been broken?

GARRETT: It could. However, we never knew where the ethidium bromide was binding. It could have been in any pocket in the tertiary structure or around the looped-out nucleotide. All we showed was the 2 or 3 dye molecules were displaced by the protein. The only strong evidence for the occurrence of the helix within the L18-S5 RNA complex derives from the circular dichroism data, and I concede this is not proof since one helix could have broken and another formed.

UHLENBECK: Patel has NMR data on defined DNA helices which demonstrate that an extra adenosine residue in a helix is stacked in the helix in the way that an ethidium molecule might intercalate between adjacent base pairs. In contrast, Tinoco has produced a similar DNA helix with an extra cytidine residue and it is looped-out of the helix by the same NMR criterion. Granted this is in DNA but, nevertheless, it seems difficult to predict what the structure of these extra helical residues will have.

KLUG: It is a kinetic effect. You are both right. If you have a single looped-out base, I am pretty sure it will spend a good deal of its time outside the helix, as the molecule is breathing. We know that rather large motions occur in the DNA structure; there are rotations of a base pair of about 10° in angle which are greater than the average deviation between the angle of the base pairs, but on the average there is a good structure. Although diethyl pyrocarbonate won't attack the N-7 of an adenosine involved in a base pair it might well attack it, slowly, when the adenosine spends part of its time in and part of its time out of the helix.

GARRETT: Nevertheless, the looped-out adenosines in the ribosomal RNA structure are exceptional. They are all hyper-reactive with diethyl pyrocarbonate and are more reactive than the so-called single-stranded adenosines.

KLUG: That would be conclusive then, if, on the same gel, you have some calibration of the degrees of reactivity.

GARRETT: Yes. A possible explanation for this effect is provided by our magnesium-depletion studies which show that looped-out adenosines are only hyper-reactive in the presence of magnesium (10–20 mM). This suggests that a magnesium ion, strongly bound in the neighbourhood of the looped-out base, may fix it in a protruding position.

UHLENBECK: I would add that in the R17 fragment there are 5 single-stranded adenosines including the looped-out nucleotide. They are all equally reactive to chloracetaldehyde.

The Structure and Function of Ribosomal 5S rRNAs

Tomas Pieler, Martin Digweed & Volker A. Erdmann

INTRODUCTION

The large ribosomal subunits of pro- and eukaryotic organisms contain a small ribosomal RNA designated as 5S rRNA. This small ribosomal RNA is 120 nucleotides long, contains in prokaryotic species no, and in eukaryotic species few, modified nucleotides (Erdmann 1976, Erdmann *et al.* 1983). The eukaryotic 60S ribosomal subunits contain a second small-sized 5.8S ribosomal RNA, which is 155 nucleotides in length and hardly modified (Erdmann 1976, Erdmann *et al.* 1983). Sequence comparison shows, that this RNA contains significant homologies to the 5'end of prokaryotic 23S rRNAs.

The third small-sized ribosomal RNA found, is that of the large subunit of chloroplasts. It is approximately 105 nucleotides in length and does not contain modified nucleotides (Erdmann *et al.* 1983). Again sequence comparisons have established that the 4.5S rRNA is 60% homogeneous in sequence to the 3'end of prokaryotic 23S rRNAs (Kumagai *et al.* 1983).

Little is known about the precise function of these small ribosomal RNAs, and it is apparent, that this information can only be obtained after the detailed knowledge of their structures. Of the 3 different small ribosomal RNAs, most is known about the 5S rRNA of prokaryotes, in particular of *E.coli.* We will limit this summary of results primarily to those obtained for *E.coli* and discuss the 5S rRNA structure on the basis of physical measurements and enzymatic and chemical accessibilities. Concerning the 5S rRNA function, we will discuss its well-documented interaction with ribosomal proteins, and, more speculatively, its possible interaction with tRNAs, 16S rRNA and 23S rRNA.

Institut für Biochemie, FB Chemie, Freie Universität Berlin, Thielallee 69–73, D-1000 Berlin 33, West Germany.

GENE EXPRESSION, Alfred Benzon Symposium 19.
Editors: Brian F. C. Clark & Hans Uffe Petersen, Munksgaard, Copenhagen 1984.

PHYSICAL CHARACTERIZATION OF 5S rRNA STRUCTURE

Until recently, most reviews of the application of physical techniques to the analysis of 5S rRNA structure began with the apologia that crystallization and subsequent X-ray diffraction analysis had not yet been achieved. This is no longer the case, as Morikawa and co-workers have recently succeeded in crystallizing this second natural RNA molecule (Morikawa *et al.* 1982). Although the present crystal of 5S rRNA from the thermophilic bacterium *Thermus thermophilus* does not diffract X-rays well enough for elucidation of the detailed three-dimensional structure of the molecule, a decisive step towards this goal has now been made. In the meantime, the use of physical techniques has continued, and considerable progress has been made. The approach of most researchers has been to accumulate data on either the overal shape and size of the molecule (e.g., X-ray scattering, hydrodynamic measurements), or on the extent of base pairing (e.g., infra-red and NMR spectroscopy), and, subsequently, construct models which accommodate these measurements and a maximum of those previously published.

The use of thermal-melting analysis to examine base pairing in nucleic acids is well documented. Applied to 5S rRNA, Fox and Wong (1979) were able to determine 38 base pairs (63%) after correction for base-stacking contributions. UV-absorption difference spectra indicated a higher amount of base pairing, namely 75% (46 base pairs), of which 70% are G-C and 30% A-U pairs. We have carried out IR-spectroscopy on eukaryotic (Stulz *et al.* 1981), and prokaryotic (Appel *et al.* 1979) 5S rRNAs. Spectra taken at 52° C, at which temperature only secondary structure may be assumed to persist, allowed prediction of 30 (+/−2) G-C and G-U base pairs and 20 (+/−2) A-U base pairs in *E. coli* 5S rRNA. Spectra were simulated for several proposed models by summation of spectra from simpler model compounds. No single model spectrum was identical to the observed spectrum. A reasonable fit was found with 6 models, including Österberg's from X-ray scattering (Österberg *et al.* 1976), Luoma & Marshall's (1978) from Raman spectroscopy, and a hypothetical model composed of 70% paired guanines and 35% paired adenines. Raman spectroscopy applied to 5S rRNA predicted a total of 36 base pairs of which 28 are G-C and 8 A-U pairs (Chen *et al.* 1978). NMR analysis conducted by Kearns and Wong (1974) suggested 28±2 base pairs in *E.coli* 5S rRNA, whilst a later study (Burns *et al.* 1980) recorded 33 base pairs. Using IR-spectroscopy, Böhm and co-workers determined 25±3 G-C and 10±1 A-U base pairs in *E. coli* 5S rRNA, a total of 58% base pairing (Böhm *et al.* 1981).

Table I
Estimated base pairing for E. coli 5S rRNA

Model proposed by:	Total base pairs:	G–C	A–U base pairs	G–U
Fox & Woese (1975)	25	17	5	3
Luoma & Marshall (1978)	36	21	10	5
Hancock & Wagner (1982)	34	22	7	5
Pieler & Erdmann (1982)	39	26	6	7
Böhm *et al.* (1981)	43	27	9	7
Methods				
NMR (Kearns & Wong 1974)	28	20	7	1
NMR (Burns *et al.* 1980)	33	–	–	–
IR (Böhm *et al.* 1981)	35	25	10	–
IR (Appel *et al.* 1979)	50	30	20	–
RAMAN (Chen *et al.* 1978)	36	28	8	–
UV (Fox & Wong 1979)	46	32	14	–
UV (Fox & Wong 1978)	38	–	–	–

As is clear from Table I, the extent of base pairing predicted from spectroscopic techniques varies between, and even within, the various methods. All methods, however, predict higher levels of base pairing than biochemical investigations e.g., 7 A-U base pairs based on adenine modification (Cramer & Erdmann 1968). The reasons for this large variation remain unclear. Differences in the ionic conditions and temperature prevailing during the measurements may explain much of this variation. Similarly, the pre-treatment of the 5S rRNA (e.g., heat renaturation) can have dramatic effects on the conformation of the molecule.

E.coli 5S rRNA is known to exist in 2 conformations which may be separated from one another by chromatographic and electrophoretic methods (Aubert *et al.* 1968, Lecandiou & Richards 1975) and interconverted by heat or urea treatment. A possible functional significance for the A-form to B-form conversion has been suggested by Weidner and co-workers (Weidner *et al.* 1977), who suggest a switching mechanism between the 2 forms. Their model requires considerable structural rearrangement between the 2 forms.

Böhm and co-workers (1981) measured 35 base pairs in both A-form and B-form 5S rRNA. IR-melting profiles suggested that the stable structures of these 2 forms are identical at high temperature, and that the observed differences in electrophoretic mobilitiy are due to assumption of different low-temperature-stable structures. Particularly, helices I, II and IV (compare Fig. 2) are to be found in both molecules, whilst a proposed fifth helix and the large hairpin loop formed by helix III are suggested as critical for the A-form to B-form conversion.

A recent NMR analysis by Kime and Moore (1982) has added a further dimension to the question of A-form 5S rRNA structure. This study identified 2 conformations of native 5S rRNA which possess distinct NMR spectra, neither of which is identical to that of B-form 5S rRNA. The forms have been designated H (for high temperature and high Mg^{++} ion concentration), and L (for low temperature and absence of Mg^{++} ions). The involvement of Mg^{++} ions in a conformational change in 5S rRNA had previously been suggested by the tritium exchange data of Ramstein and Erdmann (1981). The transition between H-form and L-form may be induced at room temperature by Mg^{++} ions and is very rapid. A low-temperature melting transition was also detected under near-physiological conditions from temperature-jump and light-scattering experiments by Kao and Crothers (1980), and this is assumed to be the same transition, and which is speculated to be related to 5S rRNA function. The 10% difference in diffusion constant represents a 15 Å change in the molecular dimensions of 5S rRNA, sufficient for extensive movement in the ribosome during protein synthesis. Models based on the extent of base paring, invariably take the secondary structure model of Fox and Woese (1975) as their basis. Most physical measurements, however, would imply that this model must be considered as representing the minimum extent of base pairing (Table I). In contrast, tritium exchange data on the accessibility of nucleotides in the molecule (Farber & Cantor 1981 b) support a relatively open structure at 37° C, which could be satisfied by the Fox and Woese model with minor extensions to the "common-arm base" involving the looping out of 2 adenines (positions 52 and 53). Furthermore, a tentative tertiary interaction was proposed based solely on the exchange rates of nucleotides in the sequences C_{37}CAU and A_{73}UGG. This interaction is supported by the phenyl-diglyoxal crosslink reported between G_{41} and G_{72}, since these 2 guanines would be located opposite one another at the end of the duplex of the tertiary interaction. A model based largely on this crosslink (and one further crosslink between G_2 and G_{112}) has been suggested (Hancock & Wagner 1982) with dimensions 115 Å×75Å×30Å.

From their IR-spectroscopy data, Böhm and co-workers suggested a model based on that of Fox and Woese, but with considerable extensions to the helical regions, and a parallel base pairing interaction between C_{37}CAU and G_{75}GUA (Böhm *et al.* 1981). This interaction, however, might be expected to considerably separate the two guanines crosslinked with 1.4-phenyl-diglyoxal (Hancock & Wagner 1982).

A further tertiary interaction has recently been proposed by ourselves (Pieler & Erdmann 1982) for eubacterial 5S rRNA. The base pairing involved is also between the large hairpin loop closed by helix III and bases in the multibranched loop, namely G_{41}CCG with U_{77}GGU, and was identified by single-strand-specific S1 nuclease digestion: it is universally applicable to unbacterial 5S rRNA sequences. A molecular model including this tertiary interaction has the dimensions 80 Å×140 Å, and is in good agreement with the physical data concerning overall shape and size and, as shown in Table I, estimates of base pairing.

Data on the overall shape/size of 5S rRNA are quite consistent. Thus, the sedimentation behaviour was found to be typical of a prolate ellipsoid with an axial ratio of 1:5 and dimensions of 160 Å×32 Å (Fox & Wong 1979). The calculated friction coefficient for such a structure was found to be in excellent agreement with the observed value. This form for 5S rRNA is further supported by X-ray scattering (Österberg *et al.* 1976) which measured a radius of gyration of 36 Å for native 5S rRNA from *E. coli.* This measurement does not accomodate compact or spherical models, whilst the arrangement of 2 short and one longer helix in a "Y" configuration fits well. Such a model, with overall dimensions of 125 Å×80 Å×40 Å was proposed (Österberg *et al.* 1976). Denatured 5S rRNA showed a radius of gyration of 27 Å, which might reflect the disruption of one of the shorter helices, thus, producing a more open structure. Our small-angle neutron-scattering experiments (Lorenz *et al.* 1980) have determined a radius of gyration of 42 Å for native *E. coli* 5S rRNA, in good agreement with the result from X-ray-scattering experiments. Electron microscope images of 5S rRNA after freeze-drying indicate an elongated ellipsoid containing defined helical regions and with overall dimensions of 135 Å×75 Å (Tesche *et al.* 1980). An examination by Kime *et al.* (Kime & Moore 1983) of the NMR spectrum of a 61-nucleotide-long fragment of 5S rRNA showed it to be similar to that of native 5S rRNA. The conclusion is that the base pairing in the fragment is identical to that of the same nucleotide stretch in the intact molecule and, in analogy to protein structure, represents a discrete structural domain of the molecule.

In summary, the physical measurements support an elongated 5S rRNA structure with the following approximate dimensions (115–160 Å×75 Å×35Å).

The estimated extent of base pairing also varies significantly (between 58 and 75%), which may partially be due to the fact that the 5S rRNA structure is sensitive to temperature and Mg^{++} concentrations.

THE GENERAL SECONDARY STRUCTURE OF 5S RIBOSOMAL RNA

The nucleotide sequences of 93 5S ribosomal RNAs of eukaryotic, eubacterial, archaebacterial and organelle origin have been determined (Erdmann *et al.* 1983). Comparative analysis of these 5S rRNA primary structures has been used for the construction of phylognetic trees (Hori & Osawa 1979, Küntzel *et al.* 1981). A deeper insight into their secondary structure organization and classification has been gained by means of comparative secondary structure analysis (Fox & Woese 1975, Studnicka *et al.* 1981, Luehrsen & Fox 1981, Pieler & Erdmann 1982, Böhm *et al.* 1982, Delihas & Andersen 1982, De Wachter *et al.* 1982, MacKay *et al.* 1982). The first, classical report on a consensus 5S rRNA secondary structure was presented by Fox and Woese (1975) based on a systematic sequence comparison of 10 5S rRNA sequences which were available

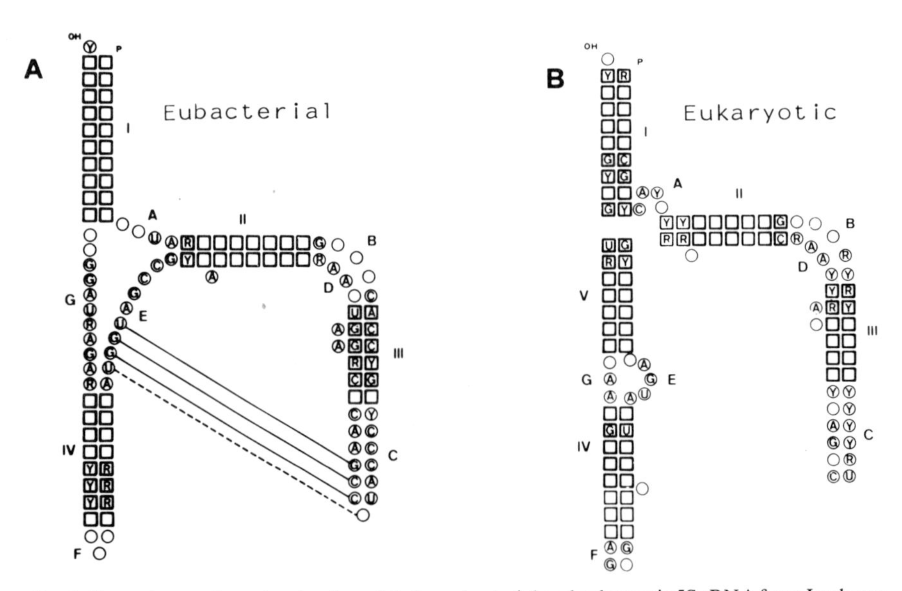

Fig. 1. General secondary structural models for eubacterial and eukaryotic 5S rRNA from Luehrsen & Fox (1981) with extended base pairing in helices II and IV as suggested by Delihas & Andersen (1982). Boxes indicate base-paired and circles single-stranded regions. Conserved and semi-conserved nucleotides (R, purine; Y, pyrimidine) are indicated.

at that time. All the later investigations have confirmed this basic structure as determining the minimal number of base pairs in any 5S rRNA. Studnicka *et al.* (1981) involved the calculation of the minimum energy of homologous structures in their analysis, creating general 5S rRNA models for gram-negative and gram-positive bacteria.

The same approach has been improved (De Wachter *et al.* 1982) using a set of energy values and topological rules proposed by Ninio (1979), after adjusting the existing thermodynamical rules (Tinoco *et al.* 1973) empirically to the predictability of the tRNA cloverleaf. In principal agreement with these studies a universal eukaryotic (Luehrsen & Fox 1981), and a universal eubacterial 5S rRNA model (Pieler & Erdmann 1982), including a tertiary interaction, have been proposed (Fig. 1).

Length and arrangement of helices and single-stranded loops are very similar in these 2 structures. However, there are significant differences, characterizing a given 5S rRNA molecule as belonging to one of the 2 major phylogenetic groups. The general eukaryotic model contains a fifth helical segment (helix V in Fig. 1) of considerable stability, helix IV contains a single base-bulge loop, and loop C consists of 12 nucleotides. The same region contains 13 nucleotides in the eubacterial model and part of this loop is complementary to the conserved single strand E, giving rise to the possibility of an antiparallel base pairing, which is not given for the eukaryotic one.

There are considerable differences in the sequences of conserved nucleotides which may be used as marker sequences for the classification of the eubacterial and eukaryotic structures (Luehrsen & Fox 1981, Pieler & Erdmann 1982).

In a detailed study Delihas and Andersen (1982) precisely defined the number of nucleotide positions between several of the universal nucleotides, found in eubacterial as well as in eukaryotic 5S rRNAs. This number appears to be an additional major distinguishing feature between the 5S rRNAs of eubacteria and eukaryotes. Following this definition and the classification characters listed above, organelle 5S rRNAs may be classified as members of the universal eubacterial structural group (Delihas & Andersen 1982, Pieler *et al.* 1982, Pieler *et al.* 1983 a), whereas the archaebacterial 5S rRNAs have properties of both the prokaryotic and eukaryotic 5S rRNAs, thus establishing a third, distinct structural group (Delihas & Andersen 1982, Pieler *et al.* 1982).

Most models agree in that the amount of conserved nucleotides in single-stranded regions is higher than in double-stranded regions, and that the number of adenines in those conserved single-stranded regions is significantly higher

Table II
Absolute and relative amounts of adenines in the universal eukaryotic and eubacterial 5S rRNAs. Adenines in single- and double-stranded regions of E.coli, spinach cytoplasma and Xenopus laevis oocyte *5S rRNAS.*

Conserved Adenines in:	Total number of adenines in: single strands	double strands	% of total conserved nucleotides in: single strands	double strands	References
General eukaryotic 5S rRNA	9	0	50	0	Luehrsen & Fox 1981
General eubacterial 5S rRNA	14	1	38	13	Pieler & Erdmann 1982

Organism	Adenines in: single strands	double strands	% in conserved positions of: single strands	double strands	References
E.coli	17	6	82	17	Pieler & Erdmann 1982
Spinach cytoplasma	17	6	53	0	Pieler *et al.* 1983a
Xenopus laevis oocyte	17	7	53	0	Delihas & Andersen 1982

than that of the other nucleotides, as illustrated in Table II. In the examples listed herein, the amount of adenine in single-stranded regions is about twice that of those in double-stranded regions, and nearly all of the conserved adenines are located in loops. In addition 5S rRNA exhibits unique, conserved structural features never observed in tRNA secondary structure. Helices II and III (Fig. 1) and the eukaryotic helix IV contain single – or double-base bulge loops. These structural elements as well as the conserved single-stranded regions have been suggested to be important for the recognition of, and for interaction with, other ribosomal components or molecules involved in protein biosynthesis (Pieler & Erdmann 1982, Delihas & Andersen 1982, MacKay *et al.* 1982, Peattie *et al.* 1981).

E. COLI 5S rRNA – rPROTEIN INTERACTION

E.coli 5S rRNA binds to the large ribosomal subunit proteins E-L5, E-L18 and E-L25 (Horne & Erdmann 1972). Each of these proteins was shown to bind independently and stoichiometrically to the 5S rRNA (Spierer & Zimmermann 1978, Spierer *et al.* 1978). Co-operative binding has been observed for EL5 in the presence of E-L18 (Spierer & Zimmermann 1978). Recently, a fourth protein E-L31a was found to be associated with the quaternary complex prepared *in vitro* using partially purified 5S rRNA and total 70S ribosomal proteins (Fanning & Traut 1981). By means of immune electron microscopy E-L5, E-L18, E-L25 (Noa *et al.* 1982), and the 3′ end of *E. coli* 5S rRNA (Stöffler-Meilicke *et al.* 1981, Skatsky *et al.* 1980), have been localized near to, or on, the central protuberance of the large ribosomal subunit. The proteins at the interface between the 30S and 50S subunits have been identified by crosslinking (Lambert & Traut 1981); they include the 5S rRNA-binding proteins. Various isolated *E.coli* 5S rRNA fragments have been bound to E-L18 and E-L25 (Zimmermann & Erdmann 1978, Speek & Lind 1982, Kime & Moore 1983). These data suggest that the 5S rRNA tertiary structure is organized in domains (Kime & Moore 1983) which retain their structure, or at least those structural properties required for protein interaction even in 5S rRNA fragments. Ribosomal RNA-protein-binding studies have established that homologous and heterologous 5S rRNA protein complexes can be reconstituted from different eubacterial sources (Erdmann *et al.* 1980), whereas eubacterial ribosomal proteins are not recognized by

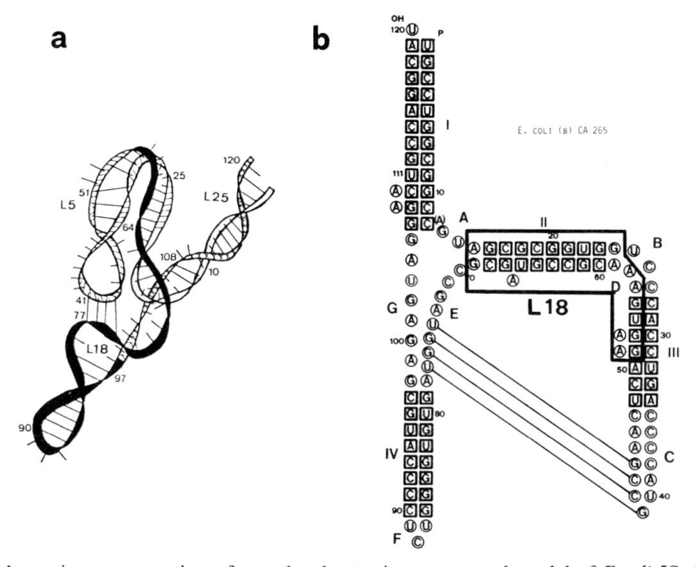

Fig. 2. Schematic representation of a molecular tertiary structural model of *E.coli* 5S rRNA; the ribonuclease T_1 resistant fragments upon binding of E-L5, E-L18 and E-L25 (Zimmermann & Erdmann 1978) are indicated. Secondary structure of *E.coli* 5S rRNA (Pieler & Erdmann 1982); the primary binding region of E-L18 as suggested by Garrett *et al.* (1981) is shown.

eukaryotic 5S rRNAs. This suggests that the structural features of eubacterial 5S rRNA required for protein recognition and interaction have been evolutionarily conserved.

Reconstituted *E.coli* 5 rRNA-protein complexes have been analyzed by means of nuclease digestion (Pieler & Erdmann 1982, Zimmermann & Erdmann 1978, Douthwaite *et al.* 1979, Douthwaite *et al.* 1982, Gray *et al.* 1973), chemical modification (Peattie *et al.* 1981, Garrett & Noller 1979), and oligonucleotide-binding studies (Erdmann *et al.* 1980).

In early attempts (Gray *et al.* 1973, Zimmermann & Erdmann 1978, Douthwaite *et al.* 1979) to define the protein-binding regions of 5S rRNA the ribonuclease-resistant regions in 5S rRNA-protein complexes have been determined, although they do not necessarily constitute those parts of the RNA interacting with the proteins (Fig. 2A). However, the new, rapid sequencing techniques have allowed the identification of particular nucleotides directly protected against nuclease digestion in the presence of individual proteins (Douthwaite *et al.* 1982). Our most detailed knowledge (Garrett *et al.* 1981) results from experiments on the binding of E-L18 to the 5S rRNA including chemical modification of guanines with kethoxal (Garrett & Noller 1979), as well as with dimethylsulfate and of adenines with diethylpyrocarbonate (Peattie *et al.* 1981), (Fig. 2B). We notice that the bulged adenines of helices II and III as well as the conserved adenines in loop D (Figs. 1, 2) are thought to be directly involved in the interaction.

THE TERTIARY STRUCTURE OF *E. COLI* 5S rRNA

Limited digestion of 5S rRNA molecules with nucleases is thought to yield fragments generated from cuts in accessible single-stranded regions. Ribonucleases A, T_1, T_2 (Jordan 1971, Vigne *et al.* 1973, Vigne & Jordan 1977, Douthwaite & Garrett 1981, Speek *et al.* 1980), and nuclease S_1 (Speek *et al.* 1980, Pieler & Erdmann 1982) have been used for this purpose. In a similar approach, double-stranded regions have been identified by cleavage with a double-strand-specific ribonuclease from *Naja naja oxiana* venom (Erdmann *et al.* 1980, Speek *et al.* 1980, Douthwaite & Garrett 1981). Special emphasis has been given to the discrimination between primary and secondary cuts (Douthwaite & Garrett 1981). Thus, a detailed picture of the enzymatically accessible regions in *E. coli* 5S rRNA has emerged (Fig. 3). Double-strand-specific hydrolysis was found in helices I, II and IV which is in perfect agreement with the structure shown in Fig. 3. Loops A, D and F as well as the bulge loops in helices I, II and III are entirely cleaved in the single-strand-specific reactions. The partial or complete inaccessibility of the other loops may be due to several reasons: a) these loops may be internally "buried" within the tertiary structure and, thereby, inaccessible for the relatively large enzymes, b) they may be involved in tertiary interactions comparable to those found in tRNA (Rich & Kim 1978), and c) partial stacking of single strands may reduce the reactivity in single-strand-specific reactions. Several tertiary interactions on the basis of parallel or anti-parallel base pairing between the partially accessible loop C and the inaccessible loop E have been proposed (discussed above). It is most likely that additional tertiary interactions involving the phosphate and sugar residues of the nucleotides take place, in particular between loops C and E. Partial stacking of region G may be the reason for the reduced accessibility of single-stranded region C. This conserved sequence is strikingly rich in purine residues, which favours stacking. In our molecular model (Fig. 2A) we have arranged helices I and IV co-axially, whereby, a highly ordered conformation of loop G is favoured.

Inaccessibility of loop B may be explained by either "internal burying" of this

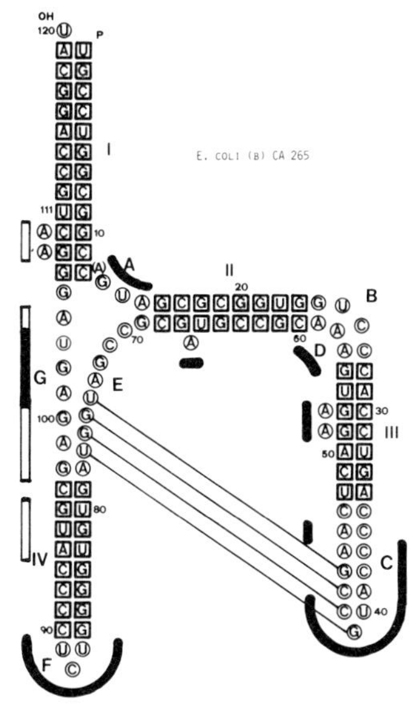

Fig. 3. Strong (filled bars) and reduced (open bars) accessibility of *E.coli* 5S rRNA for limited, single-strand-specific nuclease digestion (Jordan 1971, Vigne *et al.* 1973, Vigne & Jordan 1977, Speek *et al.* 1980, Douthwaite & Garrett 1981, Pieler & Erdmann 1982, Pieler *et al.* 1983, unpublished).

part of the molecule, or involvement in tertiary interactions of unknown nature. Partial reactivity of basepaired region IV may be due to a dynamic folding, unfolding in this unstable part of the helix, which consists of two G-U and one A-U base pair.

Base-specific chemical modification studies have been used in order to discriminate between free nucleotides and those which are involved in secondary or tertiary interactions. Early studies (Bellemare *et al.* 1972, Erdmann 1976) revealed accessibility of cytosine residues for methoxamine, which belong to sequences 34–41 and 45–51. Nucleotides of the same regions could be deaminated with nitrous acid. G_{41} was shown to be most readily accessible for glyoxal (Aubert *et al.* 1973), and kethoxal (Noller & Garrett 1979), which also modifies G_{13} (Noller & Garrett 1979), and G_{44} (Larrinua & Delihas 1979). Establishment of the rapid gel-sequencing methods (Donis-Keller *et al.* 1977, Peattie 1979) has facilitated the exact identification and quantification of modified nucleotides. Base-specific modification can either lead to a loss of reactivity in an enzymatic sequencing reaction (Göringer *et al.* 1983, Pieler *et al.* 1983b), or it may itself be part of the chemical sequencing procedure (Peattie & Gilbert 1980). Depending on the particular groups of the nucleobases involved,

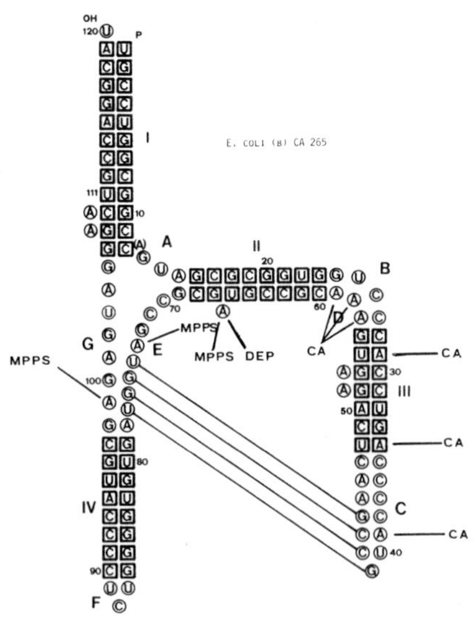

Fig. 4. Chemical modification of adenines in *E.coli* 5S rRNA with diethylpyrocarbonate (Peattie *et al.* 1981), (DEP), chloracetaldehyde (Digweed *et al.* 1981), (CA) and monoperphthalic acid (Pieler *et al.* 1983b, Silberklang *et al.* 1983), (MPPS).

every reagent exhibits different stereospecifity. Dimethyl sulfate monitors the N-7 of guanosines, thus sensing tertiary interactions, and diethylpyrocarbonate has been shown to detect stacking of adenines via its N-7 (Peattie & Gilbert 1980). Examination of the N-3 of cytidines with dimethyl sulfate, probes base pairing (Peattie & Gilbert 1980), as does C to U conversion with bisulfite (Mashkova *et al.* 1979), and the reaction of adenines at the N-1 with monoperphthalic acid (Silberklang *et al.* 1983), or chloracetaldehyde (Digweed *et al.* 1981). The most detailed knowledge exists about the adenines in *E. coli* 5S rRNA making use of the experiments listed above (Fig. 4). Obviously, the 3 reagents differ in their sterical specificity, the only adenine being modified by 2 reagents found in the single base-bulge loop of helix II, indicating that this nucleotide is neither stacked nor base-paired, and probably not involved in tertiary interactions. A detailed picture including the other reagents listed above remains to be worked out.

A third experimental approach to gain information about the tertiary structural organization of 5S rRNA is the use of bifunctional reagents in order to crosslink adjoining nucleotides. Reactivity in helix I, the 3′-5′end base-paired stem region, has been observed with phenyldiaglyoxal (Wagner & Garrett 1978) and aminomethyltrioxsalen (Rabin & Crothers 1979). In addition, crosslinking between G_{41} and G_{72} with phenyldiglyoxal, once again suggesting tertiary interactions of loops C and E, has been reported (Hancock & Wagner 1982).

Apart from protection against nuclease attack and chemical modification, induced reactivity was observed as a result of protein binding (Pieler & Erdmann 1982, Douthwaite & Garrett 1982). In particular, enhanced reactivity in loop C was detected upon binding of E-L25. This observation is consistent with the proposal that the binding of ribosomal proteins to 5S rRNA results in a conformational rearrangement of the RNA molecule. Additional, experimental support comes from circular dichroism spectra (Bear *et al.* 1977, Fox & Wong 1978), and, in particular, from oligonucleotide-binding studies (Erdmann *et al.* 1980).

Again, accessibility of loop C and other regions in the RNA for complementary oligonucleotides is induced in the 5S rRNA upon binding of ribosomal proteins including E-L25. These findings might reveal functional implications for the 5S rRNA molecule, as will be discussed below.

Little is known about the structure of 5S rRNA *in situ,* in the 50S or 70S particle. The experimental data from chemical modification of adenines in the 50S ribosomal subunit with monoperphthalic acid (Silberklang *et al.* 1983), and modification of guanines in the 50S subunit and the 70S ribosome with kethoxal (Brow & Noller 1983, Delihas *et al.* 1975, Raué *et al.* 1976), are summarized in Fig. 5. In the most recent investigation (Göringer *et al.* 1983) protection of G_{41} and G_{44} against kethoxal modification has been observed upon binding of tRNA to the 70S ribosome.

Finally, an intra-RNA cross-link between G_{69} and G_{107} has been reported for *E. coli* 5S rRNA in the 50S ribosomal subunit (Stiege *et al.* 1982).

THE FUNCTION OF 5S rRNA

The importance of 5S rRNA for the function of the ribosome can be demonstrated by total reconstitution experiments involving *B. stearothermophilus* (Nomura & Erdmann 1970, Erdmann *et al.* 1971b), and *E. coli* (Nierhaus & Dohme 1974) 50S ribosomal subunits. Reconstitution of ribosomes in the absence of 5S rRNA, yields particles which have lost their activity in *in vitro*

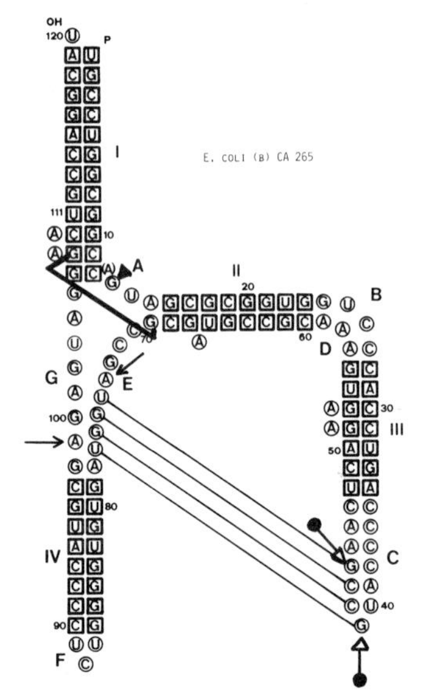

Fig. 5. Chemical modification of *E.coli* 5S rRNA *in situ.* Adenines modified with monoperphthalic acid in the 50S ribosomal subunit (→; Silberklang *et al.* 1983), guanines modified with kethoxal in the 50S or 70S ribosome (△ Delihas *et al.* 1975, Herr & Noller 1979, Brow & Noller 1983), partially protected against kethoxal modification in the 70S ribosome (▲ Brow & Noller 1983), and protected upon binding of tRNA (●—; Göringer *et al.* 1983). The intramolecular crosslink performed on the 50S subunit between G_{69} and G_{107} (Stiege *et al.* 1982) is also indicated.

assays for biological function (Erdmann *et al.* 1971b, Nierhaus & Dohme 1974). In agreement with the results from partial reconstitution experiments (as discussed in the previous paragraph) none of the eukaryotic 5S rRNAs tested could be incorporated into biologically active *B. stearothermophilus* 50S subunits (Table III A), (Wrede & Erdmann 1973), whereas all the eubacterial 5S rRNAs could replace the *B. stearothermophilus* 5S rRNA. The activity of spinach chloroplast 5S rRNA (Vogel *et al.* 1983), as well as that of archaebacterial 5S rRNA (Pieler *et al.* 1982), in this assay further confirms the phylogenetic classification discussed above.

Reconstitution of chemically modified 5S rRNAs (Table III B), (Silberklang *et al.* 1983, Delihas *et al.* 1975, Digweed *et al.* 1981, Erdmann *et al.* 1971a), and artificial 5S rRNA constructs (Table 3 c), (Raué *et al.* 1981) into *B. stearothermophilus* 50S ribosomal subunits did not reduce the biological activity drastically, except N-oxidation of A_{73} and A_{99} (Silberklang *et al.* 1983). Observation of base complementability of 5S rRNA with 16S rRNA (Azad 1979), and 23S rRNA (Herr & Noller 1975), suggests the possibility of base

Table III

Reconstitution of various 5S rRNAs into Bac. stearothermophilus *50S ribosomal subunits:*

Source	Incorporation into 50S subunits	Biological activity
A) intact 5S rRNA from different sources		
eubacterial (Wrede & Erdmann 1973)	+	+
eukaryotic (Wrede & Erdmann 1973)	–	–
archaebacterial (Pieler *et al.* 1982)	+	60–70%
chloroplast (Vogel *et al.* 1983)	+	+
B) chemically modified *E.coli* 5S rRNA		
Kethoxal $G_{13}/_{14}$ (Delihas *et al* 1975)	+	+
Monoperphthalic Acid A_{73}/A_{99} (Silberklang *et al.* 1983)	+	50%
Chloracetaldehyde ($A_{29}/A_{34}/A_{39}/A_{57-59}$ (Digweed *et al.* 1981)	80%	85%
Partial removal of 3′ terminal nucleoside (Erdmann *et al.* 1971b)	+	+
C) artifical *Bac. licheniformus* 5S rRNA molecules constructed combining parts of major and minor type (Raué *et al.* 1981) 5S rRNA.		
Disturbance of base pairing in helix I	reduced	+

pairing between these ribosomal constituents (Fig. 6). A fragment of *E.coli* 5S rRNA including A_{73} was found to be associated with fragments of 23S rRNA after RNase T_1 digestion of 50S subunits (Glotz *et al.* 1981). Modification of A_{73} might interfere with a dynamic interaction of this kind. Base pairing with 16S rRNA implies a possible function of 5S rRNA in the initiation of protein biosynthesis. More recently, it has been shown that addition of 30S ribosomal subunits to a tRNA-TP50-5S rRNA complex protects 5S rRNA from nuclease degradation (Metspalu *et al.* 1983). The proposed interaction (Azad 1979) of *E.coli* 5S rRNA with a region near to the 3′ end of 16S rRNA involves A_{99}. The 3′end of 16S rRNA is also thought to be involved in mRNA binding (Shine & Dalgarno 1974), and in the interaction with IF3 (Wickstrom 1983). Thus, the reduction of biological activity observed after N-oxidation of A_{73} and A_{99} in *E.coli* 5S rRNA might be explained by involvement of these nucleotides in the dynamic interaction of 5S rRNA with other ribosomal components or molecules involved in protein biosynthesis.

It has been proposed that 5S rRNA is directly involved in tRNA binding, namely that the conserved 5S rRNA sequence $C_{43}GAAC_{47}$ interacts with the conserved tRNA sequence GTΨCG (loop IV), (Brownlee *et al.* 1968, Forget & Weissmann 1967). This hypothesis was indirectly supported by the observation

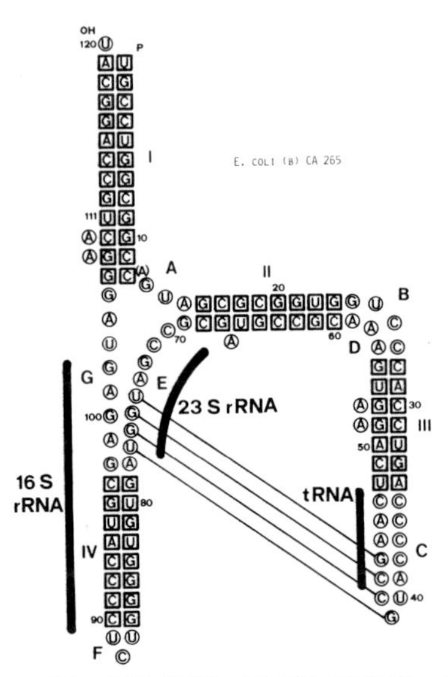

Fig. 6. Hypothetical interaction of *E.coli* 5S rRNA with 16S rRNA (Azad 1979), 23S rRNA (Glotz *et al.* 1981), and tRNA (Brownlee *et al.* 1968, Forget & Weissmann 1967).

that the rRNA fragment TΨCG inhibits binding of aminoacyl tRNA (Ofengand & Henes 1969, Shimizu *et al.* 1970, Sprinzl *et al.* 1976), and AcPhe-$tRNA^{Phe}$ (Ivanov *et al.* 1981) to the ribosome. It is important to note that the binding of fmet $tRNA_f^{Met}$ was not inhibited but stimulated by TψCG (Sprinzl *et al.* 1976). The binding site for TΨCG is located on the 50S ribosomal subunit (Richter *et al.* 1973). Furthermore, it was shown that the 5S rRNA ~ E-L5 ~ E-L18 ~ E-L25 complex is able to bind TΨCG or UUCG (as a synthetic analogue), whereas the free 5S rRNA is not (Erdmann *et al.* 1973). Finally, binding of tRNA to the ribosome leads to a strong protection of G_{41} and G_{44}, which is part of the complementary sequence in the 5S rRNA (Göringer *et al.* 1983).

On the other hand, there are several experimental results arguing against this hypothesis. A tRNA with a C to A change in the TΨCG sequence has been shown to be biologically active, suggesting that this sequence is not essential for ribosomal function (Yarus & Breeden 1981). In addition, the binding of tRNA to the P-site did not have any significant effect on the (^{3}H) exchange of 5S rRNA (Farber & Cantor 1981b). Finally, reconstitution of an artifical *E.coli* 5S rRNA construct with a deletion of $CGAA_{46}$ (Table IV), (Pace *et al.* 1982) did not eliminate 5S rRNA-dependant biological activity, but merely reduced it to 52%. In conclusion, we may say that the conserved complementary sequences, TΨCG in tRNA and CGAA in 5S rRNA, are not essential for ribosomal function, although this does not necessarily exclude the possibility of such a base pairing. It seems more likely that apart from this putative binding to 5S rRNA other,

Table IV

Reconstitution of artificial E. coli *5S rRNA-constructs carrying nicks and defined deletions (modified from Pace* et al. 1982*) *into* E. coli *50S ribosomal subunits.*

5S rRNA constructs	nucleotides deleted	% activity
intact 5S	–	100
1–41/42–120	–	75
1–52/53–120	–	27
1–40/42–120	G_{41}	69
1–41/47–120	$CCGAA_{46}$	52
1–52/57–120	$AGUG_{56}$	9
No 5S rRNA	–	0

* We have recalculated 5S rRNA-dependant ribosome activity, taking the amount of polyphenylalanine synthesized by ribosomes containing intact 5S rRNA as 100% and the residual activity of particles containing no 5S rRNA as 0%, in order to facilitate the comparison with data from Table III.

additional interactions of the tRNA with ribosomal components in the vicinity of 5S rRNA such as the 5S rRNA-binding proteins E-L5, E-L18 and E-L25 might occur. In this respect it is interesting to note that E-L18 could be crosslinked to tRNA (Wickstrom *et al.* 1981), and bound to immobilized tRNA from a mixture of ribosomal proteins (Yukioka & Omori 1977, Ustav *et al.* 1978) with other proteins.

Furthermore, 5S rRNA has been considered to play a role in stringent factor-dependent "magic spot" synthesis (Richter *et al.* 1974), and the 5S rRNA-protein complex was shown to exhibit ATP- and GTPase activity (Gaunt-Klöpfer & Erdmann 1975). In summary, it becomes clear that the precise function of 5S rRNA remains to be determined, and that one requirement for this goal is the precise determination of its structure.

ACKNOWLEDGEMENTS

We would like to thank H. Mentzel and A. Schreiber for drawing and photographing the figures and the Deutsche Forschungsgemeinschaft (Sfb 9/B5) and the Fonds der Chemischen Industrie for financial support.

REFERENCES

Appel, B., Erdmann, V. A., Stulz, J. & Ackermann, Th. (1979) *Nucleic Acids Res. 7,* 1043–1057.

Aubert, M., Scott, J. F., Reynier, M. & Monier, R. (1968) *Proc. Natl. Acad. Sci. (USA) 61,* 292–299.

Aubert, M., Bellemare, G. & Monier, R. (1973) *Biochimie 55,* 135–142.

Azad, A. A. (1979) *Nucleic Acids Res. 7,* 1913–1929.

Bear, D. G., Schleich, T., Noller, H. F. & Garrett, R. A. (1977) *Nucleic Acids Res. 4,* 2511–2526.

Bellemare, G., Jordan, B. R., Rocca Serva, J. & Monier, R. (1972) *Biochimie 54,* 1453–1466.

Böhm, S., Fabian, H. & Welfle, H. (1982) *Acta biol. med. germ. 41,* 1–16.

Böhm, S., Fabian, H., Venyaminov, S., Matveev, S., Lucius, H., Welfle, H. & Filimonov, V. V. (1981) *FEBS Letters 132,* 357–361.

Brow, D. A. & Noller, H. F. (1983) *J. Mol. Biol. 163,* 27–46.

Brownlee, G. G., Sanger, F. & Barrell, B. G. (1968) *J. Mol. Biol. 34,* 379–412.

Burns, P. D., Luoma, G. A. & Marshall, A. G. (1980) *Biochem. Biophys. Res. Comm. 96,* 805–811.

Chen, M. C., Giegé, R., Lord, R. C. & Rich, A. (1978) *Biochemistry 17,* 3134–3138.

Cramer, F. & Erdmann, V. A. (1981) *Nature 218,* 92–93.

De Wachter, R. Chen, M.-W. & Vandenberghe, A. (1982) *Biochimie 64,* 311–329.

Digweed, M., Erdmann, V. A., Odom, O. W. & Hardesty, B. (1981) *Nucleic Acids Res. 9,* 3187–3198.

Delihas, N. & Andersen, J. (1982) *Nucleic Acids Res. 10,* 7323–7344.

Delihas, N., Dunn, J. J. & Erdmann, V. A. (1975) *FEBS Letters 58,* 76–80.

Donis-Keller, H., Maxam, A. M. & Gilbert, W. (1980) *Nucleic Acids Res. 4,* 2527–2538.

Douthwaite, S., Christensen, A. & Garrett, R. A. (1982) *Biochemistry 27,* 2313–2320.

Douthwaite, S. & Garrett, R. A. (1981) *Biochemistry 20,* 7301–7307.

Douthwaite, S., Garrett, R. A., Wagner, R. & Fenntenn, J. (1979) *Nucleic Acids Res. 6,* 2453–2470.

Erdmann, V. A. (1976) *Prog. Nucleic Acid Mol. Biol. 18,* 45–90.

Erdmann, V. A., Appel, B., Digweed, M., Kluwe, D., Lorenz, S., Lück, A., Schreiber, A. & Schuster, L. (1980) in "*Genetics and Evolution of RNA Polymerase, tRNA and Ribosomes*", eds. Osawa, S. Ozeki, M., Uchida, H. & Yura, T. pp. 553–568. Tokyo University Press.

Erdmann, V. A., Doberer, H. G. & Sprinzl, M. (1971a) *Molec. Gen. Genetics 114,* 89–94.

Erdmann, V. A., Fahnestock, S., Higo, K. & Nomura, M. (1971b) *Proc. Natl. Acad. Sci. USA 68,* 2932–2936.

Erdmann, V. A., Huysmans, E., Vandenberghe, A. & De Wachter, R. (1983) *Nucleic Acids Res. 11,* r105–r133.

Erdmann, V. A., Sprinzl, M. & Pongs, O. (1973) *Biochem. Biophys. Res. Commun. 54,* 942–948.

Fanning, T. G. & Traut, R. R. (1981) *Nucleic Acids Res. 9,* 933–1004.

Farber, N. M. & Cantor, C. R. (1981a) *J. Mol. Biol. 146,* 233–239.

Farber, N. N. & Cantor, C. R. (1981b) *J. Mol. Biol. 146,* 241–257.

Forget, B. G. & Weissmann, S. M. (1967) *Science 158,* 1695–1699.

Fox, G. E. & Woese, C. R. (1975) *Nature 256,* 505–507.

Fox, J. W. & Wong, K.-P. (1978) *J. Biol. Chem. 253,* 13–20.

Fox, J. W. & Wong, K.-P. (1979) *J. Biol. Chem. 254,* 10139–10144.

Gaunt-Klopfer, M. & Erdmann, V. A. (1975) *Biochem. Biophys. Acta 390,* 226–233.

Garrett, R. A., Douthwaite, S. & Noller, H. F. (1981) *Trends Biol. Sci. 6,* 137–139.

Garrett, R. A. & Noller, H. F. (1979) *J. Mol. Biol. 132,* 637–648.

Glotz, C., Zwieb, C., Brimacombe, R., Edwards, K. & Kössel, H. (1981) *Nucleic Acids Res. 9,* 3287–3306.

Göringer, H. U., Bertram, S. & Wagner, R. (1983) (submitted).

Gray, P. N., Bellemare, G., Monier, R., Garrett, R. A. & Stöffler, G. (1973) *J. Mol. Biol. 77,* 133–152.

Hancock, J. & Wagner, R. (1982) *Nucleic Acids Res. 10,* 1257–1269.

Herr, W. & Noller, H. F. (1975) *FEBS Lett. 53,* 248–252.

Herr., W. & Noller, H. F. (1979) *J. Mol. Biol. 130,* 421–432.

Hori, H. & Osawa, S. (1979) *Proc. Natl. Acad. Sci. USA 76,* 381–385.

Horne, J. R. & Erdmann, V. A. (1972) *Molec. gen. Genet. 119,* 337–344.

Ivanov, G. V., Grajevskaja, R. A. & Saminsky, E. M. (1981) *Eur. J. Biochem. 113,* 457–461.

Jordan, B. R. (1971) *J. Mol. Biol. 55,* 423–439.

Kao, T. H. & Crothers, D. M. (1980) *Proc. Natl. Natl. Acad. Sci. (USA) 77,* 3360–3364.

Kearns, D. R. & Wong, K.-P. (1974) *J. Mol. Biol. 87,* 755–774.

Kime, M. J. & Moore, P. B. (1982) *Nucleic Acids Res. 10,* 4973–4983.

Küntzel, H., Heidrich, N. & Piechulla, B. (1981) *Nucleic Acids Res. 9,* 1451–1461.

Kumagai, I., Bartsch, M. Subramanian, A. R. & Erdmann, V. A. (1983) *Nucleic Acids Res. 4,* 961–970.

Lambert, J. M. & Traut, R. R. (1981) *J. Mol. Biol. 149,* 451–476.

Larrinua, J. & Delihas, N. (1979) *FEBS Lett. 108,* 181–184.

Lecandiou, R. & Richards, E. G. (1975) *Eur. J. Biochem. 57,* 127–133.

Lorenz, S., Erdmann, V. A., May, R., Stöckel, P., Strell, I. & Hoppe, W. (1980) *Eur. J. Cell Biol. 22,* 134.

Luehrsen, K. R. & Fox, G. E. (1981) *Proc. Natl. Acad. Sci. (USA) 78,* 2150–2154.

Luoma, G. A. & Marshall, A. G. (1978) *Proc. Natl. Acad. Sci. (USA) 75,* 4901–4905.
MacKay, R. M., Spencer, D. F., Schnare, N. N., Doolittle, W. F. & Gray, M. W. (1982) *Can. J. Biochem. 60,* 480–489.
Mashkova, T. D., Mazo, A. M., Scheinker, U. S., Beresten, S. F., Bogdanova, S. L., Avdommia, T. A., T. A. & Kisselev, L. L. (1979) *Molec. Biol. Res. 6,* 83–87.
Metspalu, E., Ustav, M. & Villems, R. (1983) *FEBS Lett. 153,* 125–127.
Morikawa, K., Kawakami, M. & Takemura, S. (1982) *FEBS Lett. 145,* 194–196.
Nierhaus, K. H. & Dohme, F. (1974) *Proc. Natl. Acad. Sci. (USA) 71,* 4713–4717.
Noa, M., Stöffler-Meilicke, M. & Stöffler, G. (1982) *Proc. 10th int. congress of electr. microscopy, Hamburg, 3,* 101–102.
Noller, H. F. & Garrett, R. A. (1979) *Nature 228,* 744–748.
Österberg, R., Sjöberg, B. & Garrett, R. A. (1976) *Eur. J. Biochem. 68,* 481–487.
Ofengand, J. & Henes, C. (1969) *J. Biol. Chem. 244,* 6241–6253.
Pace, B., Matthews, E. A., Johnson, K. D., Cantor, C. R. & Pace, N. R. (1982) *Proc. Natl. Acad. Sci. (USA) 79,* 36–40.
Peattie, D. A. (1979) *Proc. Natl. Acad. Sci. (USA) 76,* 1760–1764.
Peattie, D. A., Douthwaite, S., Garrett, R. A. & Noller, H. F. (1981) *Proc. Natl. Acad. Sci. (USA) 78,* 7331–7335.
Peattie, D. A. & Gilbert, W. (1980) *Proc. Natl. Acad. Sci. (USA) 77,* 4679–4682.
Pieler, T., Digweed, M., Bartsch, M. & Erdmann, V. A. (1983a) *Nucleic Acids Res. 11,* 591–604.
Pieler, T., Schreiber, A. & Erdmann, V. A. (1983b) (in preparation).
Pieler, T. & Erdmann, V. A. (1982) *Proc. Natl. Acad. Sci. (USA) 79,* 4599–4603.
Pieler, T., Kumagai, I. & Erdmann, V. A. (1982) *Zbl. Bakt. Hyg. I Abt. Orig. C 3,* 69–78.
Rabin, D. & Crothers, D. M. (1979) *Nucleic Acids Res. 7,* 689–703.
Ramstein, J. & Erdmann, V. A. (1981) *Nucleic Acids Res. 9,* 4081–4086.
Raué, H. A., Lorenz, S., Erdmann, V. A. & Planta, R. J. (1981) *Nucleic Acids Res. 9,* 1263–1269.
Raué, H. A., Heerschap, A. & Planta, R. J. (1976) *Europ. J. Biochem. 68,* 169–176.
Rich, A. & Kim, S.-H. (1978) *Science 238,* 52–62.
Richter, D., Erdmann, V. A. & Sprinzl, M. (1973) *Nature 246,* 132–135.
Richter, D., Erdmann, V. A. & Sprinzl, M. (1974) *Proc. Natl. Sci. (USA) 71,* 3226–3229.
Shimizu, N., Hayaski, H. & Miura, K. (1970) *J. Biochem. (Japan) 67,* 373–397.
Shine, J. & Dalgarno, L. (1974) *Proc. Natl. Acad. Sci. (USA) 71,* 1342–1346.
Silberklang, M., RajBhandary, U. L., Lück, A. & Erdmann, V. A. (1983) *Nucleic Acids Res. 11,* 605–617.
Skatsky, N., Evstafieva, A. G., Gystrova, T. F., Bogdanov, A. A. & Vasiliev, V. D. (1980) *FEBS Lett. 121,* 97–100.
Speek, M. & Lind, A. (1982) *Nucleic Acids Res. 10,* 947–965.
Speek, M., Ustav, M. B., Lind, A. J. & Saarma, M. J. (1980) *Bioorg. Khim. 6,* 1877–1880.
Spierer, P., Boganov, A. A. & Zimmermann, R. A. (1978) *Biochemistry 17,* 5394–5398.
Spierer, P. & Zimmermann, R. A. (1978) *Biochemistry 17,* 2474–2479.
Sprinzl, M., Wagner, T., Lorenz, S. & Erdmann, V. A. (1976) *Biochem. 15,* 3031–3039.
Stiege, W., Zwieb, C. Brimacombe, R. (1982) *Nucl. Acids Res. 10,* 7211–7229.
Stöffler-Meilicke, M., Stöffler, G., Odom, O. W., Zinn, A., Kramer, G. & Hardesty, B. (1981) *Proc. Natl. Acad. Sci. (USA) 78,* 5538–5542.
Studnicka, G. M., Eiserling, F. A. & Lake, J. A. (1981) *Nucleic Acids Res. 9,* 1885–1904.
Stulz, J., Ackermann, Th., Appel, B. & Erdmann, V. A. (1981) *Nucleic Acids Res. 9,* 3851–3861.
Tesche, B., Schmiady, H., Lorenz, S. & Erdmann, V. A. (1980) *Eur. J. Cell Biol. 22,* 131.

Tinocco, J., Borer, P. N., Dengler, B., Levine, M. D., Uhlenbeck, O. C., Crothers, D. M. & Gralla, J. (1973) *Nature 246,* 40–41.

Ustav, M., Saarma, M., Lind, A. & Villems, R. (1978) *FEBS Lett. 87,* 315–317.

Vigne, R. & Jordan, B. R. (1977) *J. Mol. Evol. 10,* 77–86.

Vigne, R., Jordan, B. R. & Monier, R. (1973) *J. Mol. Biol. 76,* 303–311.

Vogel, D., Hartmann, R. & Erdmann, V. A. (1983) *FEBS Lett.,* (in preparation).

Wagner, R. & Garrett, R. A. (1978) *Nucleic Acids Res. 5,* 4065–4075.

Weidner, H., Yuan, R. & Crothers, D. M. (1977) *Nature 266,* 193–194.

Wickstrom, E. (1983) *Nucleic Acids Res. 11,* 2035–2052.

Wickstrom, E., Parker, K. K., Hursh, D. A. & Newton, R. L. (1981) *FEBS Lett. 123,* 273–276.

Wrede, P. & Erdmann, V. A. (1973) *FEBS Lett. 33,* 315–319.

Yarus, M. & Breeden, L. (1981) *Cell 25,* 815–823.

Yukioka, M. & Omori, K. (1977) *FEBS Lett. 75,* 217–220.

Zimmermann, J. & Erdmann, V. A. (1978) *Molec. gen. Genet. 160,* 247–257.

DISCUSSION

OFENGAND: Thinking about the structure of the 50S subunit and the localizaton at the 3′-end of the 5S rRNA, I would like to ask if the dimensions of your 5S rRNA model are sufficient to reach as far as the base of the stalk of protein L7/L12?

ERDMANN: Yes, the dimensions of the 5S rRNA in the model are reasonable enough to fit into the 50S subunit as you have described.

SCHATZ: I have a question concerning the function of 5S rRNA. As you undoubtedly know, mitochondrial ribosomes from most species don't seem to have such an RNA. Why? Are there sequences in the other RNAs from these ribosomes which are homologous to 5S rRNA or do mitochondrial ribosomes lack certain functions, which you find in other ribosomes?

ERDMANN: First of all, it has been shown that *Triticum aestivum* mitochondrial ribosomes do have a 5S rRNA, so we can only say that some mitochondrial ribosomes lack this RNA species. Indeed, I agree with you that in those cases, the important sequences may be incorporated in one of the other 2 ribosomal RNAs. In this context I would like to remind you that, for example, sequence homologies have been found between the 3′ and 5′ ends of *E. coli* 23S RNA and chloroplast 4.5S RNA and eukaryotic 5.8S RNA.

SCHATZ: The sequences of the large and small RNAs of many mitochondrial ribosomes are now known through the sequence of the corresponding mitochondrial DNA. Have you in fact screened the available DNA sequences to find such homologies.

ERDMANN: We have not done it, but Hans Kuntzel in Göttingen has done these computer studies and found no evidence for such a homology. One of the reasons why we have not done such a study is that we feel strongly that the 3-dimensional arrangement of the RNA is of great importance, and that we currently do not know enough to validate such a comparative analysis.

KURLAND: One possible thought that comes from your question, Dr. Schatz, is that the mitochondrial ribosomes, from animals at any rate, are functioning with truncated tRNA and it is conceivable, without wishing to raise the ghost of other models, that there is a functional correlation between the absence of the arm from the tRNA and the absence of the 5S RNA.

ERDMANN: It may well be.

GARRETT: Isn't it quite likely that the proteins have taken over the role of the 5S rRNA, whatever it might be, in mitochondria? I mean, you have got an excess of ribosomal proteins. You lose a lot of ribosomal RNA compared with eubacteria. How do you explain that? I don't see it as any different from the problem of explaining the absence of 5S RNA.

ERDMANN: I think that it is possible that proteins have taken the role of some ribosomal RNA function. Unfortunately, there is no evidence for this interesting hypothesis.

GARRETT: I would like to raise one objection to the tertiary interaction you propose in the 5S RNA. One of your 4 base pairs involves guanine-41, which is the most reactive guanine in the free 5S RNA molecule, in protein-5S RNA complexes, and in the ribosome.

ERDMANN: I agree. The problem is the following: this tertiary interaction which we are proposing is based on the comparative study which shows that the S1 nuclease stops at the sequences prior to the interaction and also that complementary oligonucleotides do not bind to these sequences. If one looks closer at these so-called tertiary interactions it turns out that they consist of 2 GC base pairs. Our feeling is, that these tertiary interactions are not really the driving force behind the 3-dimensional structure of a molecule. They may be just reflecting interactions which are permissive and are taking place to a certain degree.

CLARK: I would like to re-emphasize what Dr. Erdmann said. Kjeld Marcker and I, earlier on, soon discovered the dangers of comparing primary structures to determine functional sites on large molecules. It is clearly possible that different primary structure can give rise to the same functional tertiary structure, both in the case of proteins and nucleic acid. So there is a big danger in looking at functional activities compared to evolutionary mapping. The question I would like to ask you is what you feel about the ribosome crystals described in Hoppe-Seyler's report a long time ago. Can you tell us what the current state of affairs is for ribosome crystals?

ERDMANN: Well, I was not involved in that earlier report on ribosome crystallization, so I cannot really comment on it. In our case, I think, it is indeed very straight forward, there is nothing unusual about the crystallization conditions, and nothing unusual about the crystals we are getting. One trick was to use *B. stearothermophilus* 50S ribosomal subunits and old vintage ribosomes which we had stored in the +80°C deep freeze for 6–7 years. As we know now, it is apparently the growth conditions of the bacterial cells, which may influence the ability to crystallize the ribosomes.

KURLAND: Are you sure that they are ribosomes?

ERDMANN: Oh, yes.

KURLAND: The reason for the question is not simply to express sarcasm.

ERDMANN: No, of course, I think it is a very basic question which asks if the appropriate controls were made. What we have done is the following: we have isolated the crystals, washed them, and then dissolved them. The redissolved ribosome still showed, mind you after 3–4 months at 4°C, 60% of their biological activity. Normally, ribosomes would lose all of their activity after 1–2 weeks of storage at 4°C. In addition, the crystalized 50S ribosomal subunits co-migrated with radioactive control subunits in a sucrose gradient.

KURLAND: Those are very convincing observations.

PESTKA: You stated that the ribosome crystals were not giving a good resolution. Could you tell us what the resolution is that you are getting?

ERDMANN: We found that the crystals, after embedding and after cutting, gave a resolution of about 45Å, so it is not very good, but a start.

VI. Control systems

Translational Regulation during Bacteriophage T4 Development

L. Gold, M. Inman, E. Miller, D. Pribnow, T. D. Schneider, S. Shinedling & G. Stormo

INTRODUCTION

The bacteriophage T4 infects *Escherichia coli.* After infection, host macromolecular syntheses quickly cease, being replaced by phage-directed reactions. The T4 genome is large enough to encode 200 proteins, many of which have been identified, either directly or by genetic analysis (Wood & Revel 1976, Mathews *et al.* 1983). As do many large, lytic bacteriophages, T4 regulates its gene expression both quantitatively and temporally. Much of this regulation occurs at the level of transcription (Rabussay *et al.* 1977). The major transcriptional regulation is designed so as to first yield T4 mRNAs that encode the proteins that participate in the construction of the T4 "replisome", the macromolecular complex that produces deoxyribonucleotide triphosphates and carries out DNA replication (Mathews *et al.* 1983). Once DNA synthesis has begun, late T4 transcription leads to the expression of genes encoding the structural proteins of the phage itself.

The people in our laboratory at first worked on problems related to early transcriptional regulation during T4 development (O'Farrell & Gold 1973a). We began, inevitably, to wonder about translational yields from various of the early T4 messages. We elected to study the ribosome-binding site of one early T4 gene that had been extensively studied using genetics. At the time we initiated this work, one possibility was that T4 would utilize translational initiation "factors" different from those used by *E.coli,* thus partially accounting for the shut-off of host translation after infection (Wiberg & Karam 1983). The work we

Department of Molecular, Cellular and Developmental Biology, University of Colorado, Boulder, CO 80309, U.S.A.

GENE EXPRESSION, Alfred Benzon Symposium 19.
Editors: Brian F. C. Clark & Hans Uffe Petersen, Munksgaard, Copenhagen 1984.

did would, therefore, either overlap and extend the work of others on the general properties of *E. coli* ribosome-binding sites, or would serve to identify initiation determinants on an mRNA recognized by altered ribosomes.

SPECIFICS

The rIIB ribosome-binding site

We chose the rIIB gene of bacteriophage T4. The genetics of rII are very extensive (Singer *et al.* 1983) and, furthermore, a good selection for translational defectives was available to us (Nelson *et al.* 1981). The selection was very simple. A frameshift mutation (FC238) in the first third of the rIIB cistron directs the synthesis of a toxic polypeptide. Because of this toxic polypeptide, phages carrying the FC238 mutation produce tiny plaques on a host that normally does not require any rIIB protein at all. We began with a phage carrying the frameshift mutation, picked large plaques, sorted out the interesting mutations (Nelson *et al.* 1981), and eventually mapped the translational defectives (Singer *et al.* 1981). We have now sequenced (Pribnow *et al.*, in preparation) most of the translational mutations from this selection, as well as other mutations isolated via different selection schemes (Fig. 1).

Some of the mutations show changes in the Shine and Dalgarno region of the rIIB message, and others are in the initiation codon. In 1974, the Shine and Dalgarno hypothesis was put forth (Shine & Dalgarno 1974), and all of the

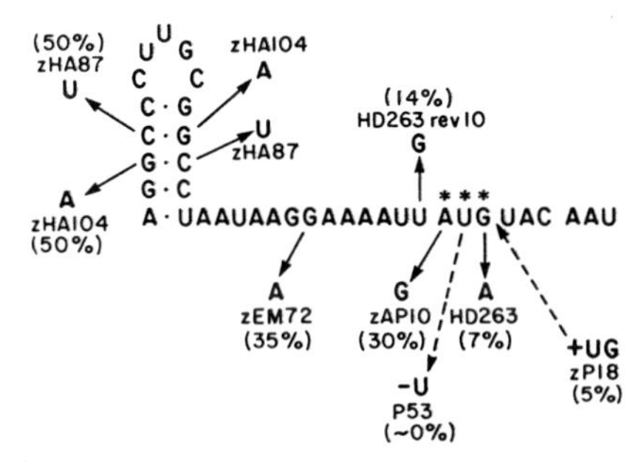

Fig. 1. Ribosome-binding site mutations in rIIB. The rIIB ribosome-binding site is shown with the initiation codon marked (*). The mutations have been described earlier (Singer *et al.* 1981); the sequences of the mutations will be documented elsewhere (Pribnow *et al.* in preparation). The data in parentheses are for rIIB expression relative to wild type, for infections at 30°, 5–8 min after infection (Singer *et al.* 1981). HD263 directs temperature-sensitive rIIB expression; HD263rev10 was selected as a pseudo-revertant whose rIIB expression was below wild type (Singer & Gold 1976). zP18 creates an in-frame GUG just 3′ to the AUG that is no longer in-frame.

biochemistry and genetics that have been done in the ensuing 10 years support the idea that the Shine and Dalgarno element and the initiation codon are major determinants for translational initiation in *E.coli* (Gold *et al.* 1981). Thus, at least one T4 mRNA shares determinants with other *E.coli* mRNAs. We did see one surprise. A predicted hairpin structure 5′ to the Shine and Dalgarno element of the rIIB message is formed by a domain at which 2 mutations (zHA104 and zHA87) were collected. In fact, each isolate is a double mutation, and each isolate, by inspection, would have no hairpin (Fig. 1). We have scanned about 20 T4 ribosome-binding sites, seeking similar hairpins 5′ to the Shine and Dalgarno domains. It is clear from such searching that hairpins 5′ to the Shine and Dalgarno domain are not conserved in T4 ribosome-binding sites. Also, hairpins are not a common feature of *E.coli* ribosome-binding sites. Lastly, the collection of mutations that we obtained included no hits between the Shine and Dalgarno domain and the initiation codon, and no hits 3′ to the initiation codon (but see below).

Table I

T4 ribosome-binding sites

Gene	−15 * −10 * −5 *	0 *	+5 *		+10 *	
e	A T A C T T A G G A G G T A T T	A T G	A A T	A T A	T T T	G A A
IPI	A A T T A G G A A A A T A A A A	A T G	A A A	A C A	T T T	A A A
IPIII	T A T T T A A A G G A A A C A T	A T G	A A A	A C A	T A T	C A A
rIIB	C C T A A T A A G G A A A A T T	A T G	T A C	A A T	A T T	A A A
1	A T T T G A G G A G A A A C A C	A T G	A A A	C T A	A T C	T T T
23	T T A A A G G T T A A C A C A A	A T G	A C T	A T C	A A A	A C T
30	C T C T A A A G G A T G A A C A	A T G	A T T	C T T	A A A	A T T
32	T A A A A A G G A A A T A A A A	A T G	T T T	A A A	C G T	A A A
36	A A T A A A G G G G C A T A C A	A T G	G C T	G A T	T T A	A A A
37	A T T A T T A A G A G G A C T T	A T G	G C T	A C T	T T A	A A A
38	T T C G G C C C T T C T A A A T	A T G	A A A	A T A	T A T	C A T
45	A A T T G A A G G A A A T T A C	A T G	A A A	C T G	T C T	A A A
57	C A C T A A A G G T A C T A T A	A T G	T C T	G A A	C A A	A C T
67	A A C A A G A G G A T T T T T A	A T G	G A A	G G T	T T A	A T T

Ribosome-binding sites from −16 to +14 are shown for 14 T4 genes. References to all available T4 sequences as of this writing are provided in *The Bacteriophage T4,* edited by C. K. Mathews, P. B. Berget, G. Mosig and E. Kutter. The nucleotides from −12 to −1 are 49% A's; the nucleotides from +3 to +14 are also 49% A's. T4 contains only 33% A's, so T4 ribosome-binding sites are rich in A's, relative to the overall genomic abundance. These ribosome-binding sites contain only 6% G's from +3 to +14. For this small data set of 14 sequences, when inspected from −60 to +40, the highest density of non-random positions is between −15 and +14. Complete statistical analyses of T4 ribosome-binding sites will be provided when the data set is large enough to use reliably.

Other T4 ribosome-binding sites

We began anecdotal molecular biology. By reading the literature, and thanks to the generosity of others who work on T4 (who sent us sequences as they were obtained), we made a catalogue of T4 ribosome-binding sites (more precisely, we made a catalogue of sequences flanking specific T4 initiation codons) (Table I). Some interesting members of the set are shown in Fig. 2. We found an example of a sequence whose secondary structure may facilitate initiation by bringing together a Shine and Dalgarno element and an initiation codon. This is the

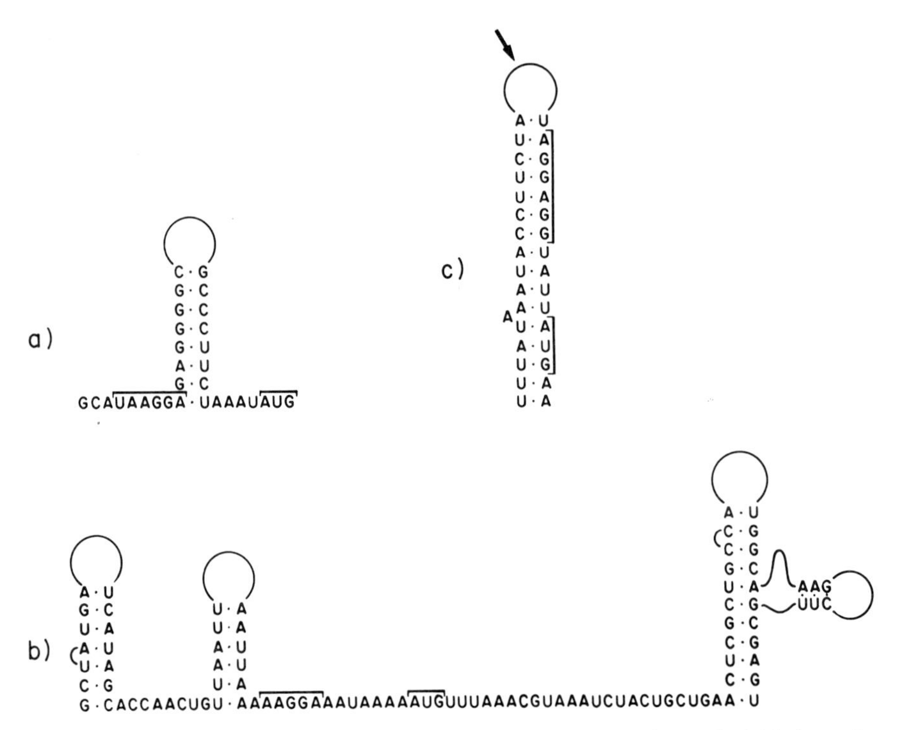

Fig. 2. Some interesting T4 ribosome-binding sites. For each sequence shown, the initiation codon and Shine and Dalgarno domain are in brackets.

a) Gene 38 mRNA is shown. Note that the adjacent gene terminates at the UAA of the Shine and Dalgarno.

b) Gene 32 mRNA is shown. The structure was predicted as described earlier (von Hippel *et al.* 1982). Repression by the gene 32 protein is thought to utilize up to 9 monomers of protein, co-operatively bound along the nucleotides between the 2 helices at the ends of the diagram. The hairpin in the middle, which is not very stable, is thought to denature when repression occurs (von Hippel *et al.* 1982).

c) The T4 lysozyme mRNA is shown. The arrow notes the location of the 5′ end of the late T4 transcript, as demonstrated directly by S1 nuclease mapping (Christensen *et al.* 1983).

messenger RNA for T4 gene 38 (Fig. 2a). We found a ribosome-binding site with almost no potential structure around the initiation codon. This is the gene 32 mRNA, an interesting ribosome-binding site for other reasons (Fig. 2b, and

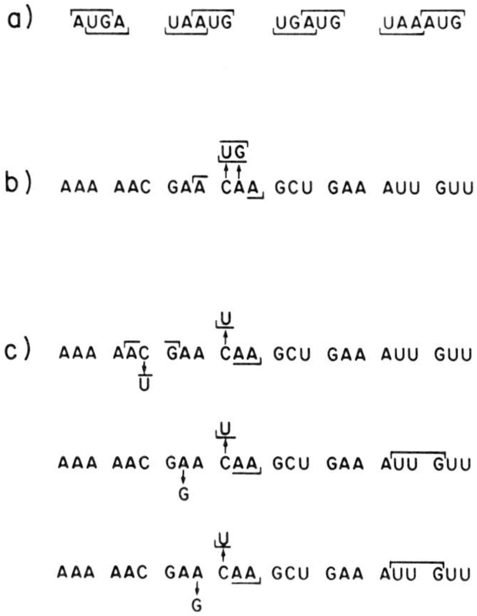

a) AUGA UAAUG UGAUG UAAAUG

b) AAA AAC GAA CAA GCU GAA AUU GUU

c) AAA AAC GAA CAA GCU GAA AUU GUU

AAA AAC GAA CAA GCU GAA AUU GUU

AAA AAC GAA CAA GCU GAA AUU GUU

Fig. 3. Re-initiation configurations in T4.

a) Sequences from Broida and Abelson are shown (personal communication). Open-reading frames terminate (brackets underneath) close by initiation codons (brackets on top) that begin the subsequent long open-reading frames.

b) An rIIB mRNA sequence, including the CAA at codon 12 (Pribnow *et al.* 1981), is shown. Codon 12 is the site of the ochre mutation r360, which was converted to a UGA. The UGA can be coupled to a minus frame-shift mutation 3′ to codon 12 to give a weak r^+ phenotype; the ochre at codon 12 does not behave similarly (Napoli *et al.* 1981). We infer that re-initiation occurs at the AUG in the plus frame (bracket on top) after termination at the UGA (bracket underneath). Note that the UGA in the sequence *GCUGA*A was changed to sense in this experiment.

c) The same rIIB sequence as in *b* is shown. Ochre mutations at codon 12 (the CAA), when coupled to plus frame-shifts 38 nucleotides 3′ to codon 12 (or other plus frame-shifts), revert to r^+ phenotypes after new transitions are introduced into strains carrying the 2 mutations (Sarabhai & Brenner 1967). We have shown that the triple mutants are r^+, because they yield a truncated rIIB polypeptide via translational re-initiation (Napoli *et al.* 1981). We have sequenced 3 different transitions that lead to re-initiation (Pribnow *et al.*, in preparation); the re-initiations must be in the minus frame (Sarabhai & Brenner 1967, Napoli *et al.* 1981). The 3 sequenced transition mutants are shown. In each case, the termination codon has a bracket underneath. The presumed re-initiation codons are shown with brackets on top. In one case, the transition creates a minus-frame re-initiation codon; in the other 2 cases, a new Shine and Dalgarno domain probably allows re-initiation at a minus-frame UUG. Re-initiation at that UUG was actually predicted after a deep look at the genetic analysis done earlier by Sarabhai and Brenner (Napoli *et al.* 1981).

below). We found a T4 message whose structure seemed likely to *eliminate* translation; both the Shine and Dalgarno domain and the initiation codon are buried in a stable secondary structure. This is the messenger RNA for the lysozyme gene (Fig. 2c). We remembered that the T4 lysozyme gene is transcribed both early and late, but is only translated late (Kasai & Bautz 1969, Salser *et al.* 1967). In fact, we know that the early lysozyme mRNA has a long 5′ leader, whereas the late transcript initiates within the loop of the hairpin (Christensen *et al.* 1983). Therefore, the late lysozyme message does not have the structure shown in Fig. 2c, but has instead an exposed Shine and Dalgarno domain and initiation codon.

A long T4 sequence determined by Joel Broida and John Abelson (personal communication) from a region surrounding the T4 tRNA genes, displays another kind of translational initiation arrangement. This sequence has a number of long open-reading frames that terminate in the immediate vicinity of initiators for the next open-reading frame (Fig. 3a). That is, these T4 genes could be expressed via translational *re-initiation* within a polycistronic mRNA. We have studied translational re-initiation within rIIB (Napoli *et al.* 1981); a few sequences (Figs. 3b, 3c) suggest that translational termination very close to an initiation codon leads to re-initiation at sites normally closed to direct initiation. Translational re-initiation plays a major role in so-called "translational coupling" within some of the ribosomal protein cistrons of *E.coli* (Nomura *et al.* 1982a, b).

T4 versus E.coli *ribosome-binding sites: a simplistic comparison*

T4 shuts off host translation (Wiberg & Karam 1983). No phage mutations have been found that are defective in the shut-off event, although some clever mutant hunts are now underway (D. Shub, personal communication). At least 2 general hypotheses may be proposed to account for host shut-off. T4 ribosome binding sites may be "better" than those on *E.coli* mRNAs, and simply "win"; alternatively, T4 infection may rapidly alter the ribosomes so as to facilitate their capture by T4 mRNAs. Data exist in weak support of both models, which are not mutually exclusive. One experiment suggested that T4 mRNAs will compete well for *E.coli* ribosomes, *in vitro,* if the system is challenged simultaneously with both T4 and *E.coli* mRNAs (Goldman & Lodish 1975).

We have published some statistical evaluations of *E.coli* ribosome-binding sites (Gold *et al.* 1981, Stormo *et al.* 1982). The data suggested that the sequences of ribosome-binding sites of *E.coli* are not random within the portion of the messenger RNA that is protected against RNase by ribosomes poised in an initiation complex. That is, *E.coli* messenger RNAs are not random for at least the first 4 codons 3′ to the initiation codon, and for more than 20 nucleotides 5′ to the initiation codon. We do not know if the non-randomness is related to the function of these regions during translational initiation, although that is our intuition. The codon GCU is extremely common as a second codon; recently Dunn and Studier (1983) have noted that *all* of the most abundant bacteriophage T7 proteins are translated from messenger RNAs whose sequences begin AUG GCU. We await biochemistry and genetics to ascertain the role of the GCU in this striking correlation.

Among the non-random nucleotide distributions in *E.coli* mRNAs is an abundance of A's throughout the region and a relatively conserved TTAA around +9 to +13 (Gold *et al.* 1981, Stormo *et al.* 1982). Both features are preserved in the small set of T4 ribosome-binding sites (Table I). In fact, the T4 catalogue is nearly half A's both 5′ and 3′ to the initiation codon, whereas T4 DNA is overall one-third A's. The information (or non-randomness) within *E. coli* and T4 ribosome-binding sites is similarly placed.

Regulation of T4 mRNAs by repressor proteins

The discussion, thus far, has been centered on constitutive translational yields from different T4 transcripts. However, as is now well known, during T4 infection some ribosome-binding sites are regulated by proteins which act upon them. That is, T4 encodes translational repressors. The first well-documented example of a translational repressor (in a system with a DNA genome) was described by Krisch *et al.* (1974) and by our laboratory (Gold *et al.* 1973, Gold *et al.* 1976, Russel *et al.* 1976, Gold *et al.* 1977).

The major T4 single-stranded DNA-binding protein is encoded by gene 32. This protein participates in DNA repair, recombination, and replication (Doherty *et al.* 1983). Synthesis of the T4 gene 32 protein is autoregulated at the translational level. The gene 32 messenger RNA is stable *in vivo,* and is repressed by excess gene 32 protein. Repression was obtained *in vitro* using cell-free systems programmed by mRNA (Lemaire *et al.* 1978). The present model for the mechanism by which T4 gene 32 protein turns off its own translation is quite simple. The protein has a capacity to bind single-stranded DNA cooperatively, but without sequence specificity. A secondary ligand for gene 32 protein is messenger RNA. The sequence suggestes that the initiation domain of gene 32 mRNA is remarkably unstructured (Fig. 2b and von Hippel *et al.* 1982, Krisch &

Allet 1982). Once single-stranded DNA has been saturated with gene 32 protein, excess molecules of gene 32 protein are free to seek the alternative ligand that most closely mimics its primary ligand. By hypothesis, the cooperative binding of gene 32 protein to its own unstructured message prevents translational initiation (Lemaire *et al.* 1978, von Hippel *et al.* 1982). One must still ask if the non-specific model, which utilizes the unstructured gene 32 message as the secondary ligand, is correct. Supporting, but indirect, data come from the large number of missense mutations that have been collected in gene 32; in all cases overproduction of the protein is observed (Doherty *et al.* 1982, Nelson *et al.* in preparation). We believe that the mutations yield proteins containing aberrant single-stranded DNA-binding sites which are directly correlated with a failure to repress translation. That is, the data suggest that the gene 32 protein has only one nucleic acid-binding site. However, the real test of the model will include repression of other (synthetic) ribosome-binding sites that have no sequence similarity to the gene 32 mRNA, but share its lack of structure.

The second T4 translational repressor which has been identified is a small protein encoded by the *regA* gene. This gene was uncovered in several mutant hunts, some of which demanded overproduction of specific early T4 proteins that are needed for DNA replication (Karam & Bowles 1974, Wiberg & Karam 1983). When T4 strains carrying *regA* mutations were compared with wild type, a large number of differences were noted in the patterns of early gene expression. Many early proteins were strongly overproduced late in infection by the *regA* mutants, as though the *regA* gene product is a negative regulator of gene expression. [A presumption is that the available *regA* mutations are nulls or have at least suffered partial loss of function; the *regA* mutations are recessive, and they are isolated at frequencies consistent with the rate at which other T4 genes are inactivated. No amber mutations in *regA* have been identified as yet.]

Many experiments have demonstrated that *regA*-mediated repression *in vivo* is a post-transcriptional phenomenon. The most important observation is that repressed genes are well expressed *in vitro* from template RNAs prepared from T4 *regA*$^+$-infected cells (Trimble & Maley 1976). Although many target genes have been identified, only one "operator" site has been carefully defined. The rIIB gene is under *regA* control; rIIB has been partially sequenced, and the mutational density for rIIB is so high that we were able to search for translational constitutives among existing rIIB mutations (Karam *et al.* 1981). The constitutive mutations all fell within the sequence ATGTACAAT, the first 3 codons of rIIB. As rIIB transcription is initiated substantially upstream from this site (Sweeney & Gold in preparation), the constitutive mutations are consistent with a post-transcriptional mechanism for *regA* activity. The constitutive mutations fall within the rIIB ribosome binding site, therefore, a simple model was proposed in which the *regA* protein competes with ribosomes for the initiation domains of target messengers. That is, we think that the *regA* protein is a direct translational repressor.

Note that our selection for rIIB ribosome-binding site mutations did not yield hits 3′ to the initiation codon (Singer *et al.* 1981), even though our investigation of all ribosome-binding sites suggested that some contact between ribosomes and mRNA uses information from this region (Gold *et al.* 1981, Stormo *et al.* 1982). We think that rIIB initiation defectives 3′ to the initiation codon might have been phenotypically altered by the failure of *regA*-mediated repression. That is, a trade-off (poor translational initiation, combined with poor repression) might have caused some mutations to be lost during our selection. Mutations in this region should be directly made by site-specific oligonucleotide mutagenesis, and tested directly for regulated and unregulated initiation rates.

Given the great interest in all nucleic acid-binding proteins, the renewed interest in RNA-binding proteins, and the ease with which T4 early genes may be transcribed and translated *in vitro* (O'Farrell & Gold 1973b), we set out to purify the *regA* protein to test the model that the *regA* protein is a translational repressor. We did establish a successful purification scheme using T4-infected cells as the starting material, but the yields were too low to be useful (Campbell & Gold unpublished). We recently placed the *regA* under the control of the leftward lambda promoter, using a plasmid that also includes a temperature sensitive variant of the lambda repressor. Cells carrying this recombinant plasmid, when shifted to temperatures that inactivate the lambda repressor, yield enormous quantities (more than 200,000 molecules per cell) of a protein with the same apparent size as the regA protein. That protein has been purified to homogeneity; both peptide sequencing [to confirm the identity of the protein, since the regA DNA sequence now exists (Karam, personal communication)] and direct biochemical experimentation are now underway.

CONCLUSIONS

T4 ribosome-binding sites

Sequence data suggest that T4 ribosome-binding sites utilize features similar to those found on other *E.coli* mRNAs. Although no critical data exist, we believe that T4 ribosome-binding sites may be stronger than most *E.coli* ribosome-binding sites, and that host shut-off could partly be due to competition for limiting ribosomes. We have also noted that T4 ribosome-binding sites have individual characteristics that are utilized by the phage; some sites contain structural information with regulatory significance. We are just beginning to appreciate that T4, like T7 (Dunn & Studier 1983) and lambda (Sanger *et al.* 1982), utilizes nearly all of its genome to encode proteins. The high gene density automatically yields situations where translational termination codons fall close to initiation codons. Some of these arrangements may facilitate co-ordinate regulation via "translational coupling", or re-initiation (Sarabhai & Brenner 1967). In fact, a cluster containing genes 45, 44, 62 and *regA* exists on the T4 genome (Wood & Revel 1976, Mathews *et al.* 1983); all 4 genes are negatively regulated by the *regA* protein (Karam *et al.* 1981). The number of *regA*-protein targets in this cluster is not yet known; if the genes are set up similarly to the ribosomal protein clusters in *E.coli,* the target could be one site at the 5′ end of a polycistronic transcript.

Translational regulation and phage development

Ribosome-binding sites determine translational yields that, to a first approximation, maximize phage yields, or provide reasonable phage yields under nutritionally disastrous conditions. Simple mutations (basepair substitutions, small insertions and deletions) will be frequently tested, and optimal solutions will evolve [we have commented previously on the concept of optimal solutions, using as a starting point the essays on "tinkering" by Jacob (Campbell *et al.* 1983)]. Similarly, translational repressors have evolved through mutations that confer a growth advantage on the organism. While we are fascinated with the repressor activities of the gene 32 protein and the *regA* protein (direct-binding experiments, footprintings, competitions with alternative ligands, and ribosome competitions are all underway in our lab and elsewhere), important *biology* is concerned with the regulatory networks that include the 2 translational repressors. Organisms, including T4, survive by integrating disparate activities into coherence; if a cell is a system that can not use more energy than its fixed energy-generating capacity, all intracellular systems compete with each other for that energy. For *E. coli* and its phages, yields of infective particles must be dependent on the nutritional state at the time of infection. Lytic phages must measure that nutritional state so that their development can be "aimed" at an appropriate burst. It follows, then, that lytic phages may make decisions comparable to the lambda lysogenic/lytic decision (Herskowitz & Hagen 1980).

T4 development is intricately regulated during the early period. The *regA* protein and the gene 32 protein are negative regulators that act at the translational level. The T4-encoded DNA polymerase is an autogenous negative regulator that acts at the transcriptional level (Krisch *et al.* 1977, Russel 1973). Lastly, T4 expresses an early protein (encoded by the *mot* gene) that activates, at the transcriptional level, expression of many of the proteins that are negatively regulated by the *regA*-protein (Mattson *et al.* 1974, Mattson *et al.* 1978). The early T4 period culminates in DNA replication, at rates that must be set by the integration of (at least) the 4 loops described above, plus the intrinsic promoter and ribosome-binding site strengths of the early genes. The responsiveness of the 4 regulatory proteins to physiological conditions has not been studied. In addition, important regulatory circuits couple DNA replication to the quantity of late transcription (Rabussay *et al.* 1977); T4 infection must provide the structural proteins in amounts appropriate for the level of newly replicated, packageable DNA. Thus, T4 gene regulation, including translational regulation, provides an opportunity to study diverse circuits, and to understand the mechanisms that couple growth conditions to altered genomic replication rates and burst sizes.

ACKNOWLEDGEMENTS

We thank Kathy Piekarski for her help with manuscript preparation. Our work has been supported by the National Institutes of Health.

REFERENCES

Campbell, K. M., Stormo, G. D. & Gold, L. (1983) In: *Prokaryotic Gene Expression* (in press). New York: Cold Spring Harbor Laboratory.

Christensen, A. C., Young, E. T., Stormo, G. D. & Gold, L. M. (1983) *Nature* (submitted).

Doherty, D. H., Gauss, P. & Gold, L. (1982) *Mol. Gen. Genet. 188,* 77–90.

Doherty, D. H., Gauss, P. & Gold, L. (1983) In: *Multifunctional Proteins: Regulatory and Catalytic/Structural* pp. 45–72. CRC Press, Cleveland.

Dunn, J. J. & Studier, F. W. (1983) *J. Mol. Biol. 166,* 477–536.

Gold, L. M., O'Farrell, P. Z., Singer, B. & Stormo, G. (1973) In: *Virus Research,* pp. 205–226. Second ICN-UCLA Symp. on Mol. Biol.

Gold, L., Lemaire, G., Martin, C., Morrissett, H., O'Conner, P., O'Farrell, P., Russel, M. & Shapiro, R. (1976) In: *Proceedings of Physicians and Surgeons Biomedical Sciences Symposium on "Nucleic Acid Protein Recognition"*, pp. 91–113. Academic Press, New York.

Gold, L., O'Farrell, P. Z. & Russel, M. (1976) *J. Biol. Chem. 251,* 7251–7262.

Gold, L., Pribnow, D., Schneider, T., Shinedling, S., Singer, B. S. & Stormo, G. (1981) *Ann. Rev. Microbiol. 35,* 365–403.

Goldman, E. & Lodish, H. F. (1975) *BBRC 64,* 663–672.

Herskowitz, I. & Hagen, D. (1980) *Ann. Rev. Genet. 14,* 399–445.

Karam, J. D. & Bowles, M. G. (1974) *J. Virol. 13,* 428–438.

Karam, J., Gold, L., Singer, B. S. & Dawson, M. (1981) *Proc. Natl. Acad. Sci. USA 78,* 4669–4673.

Kasai, T. & Bautz, E. K. F. (1969) *J. Mol. Biol. 41,* 401–417.

Krisch, H. M., Bolle, A. & Epstein, R. H. (1974) *J. Mol. Biol. 88,* 89–104.

Krisch, H. M., Van Houwe, G., Belin, D., Gibbs, W. & Epstein, R. H. (1977). *Virology 78,* 87–98.

Krisch, H. M. & Allet, B. (1982) *Proc. Natl. Acad. Sci. USA 79,* 4937–4941.

Lemaire, G., Gold, L. & Yarus, M. (1978) *J. Mol. Biol. 125,* 73–90.

Mathews, C. K., Berget, P. B., Mosig, G. & Kutter, E. (1983) *The Bacteriophage T4,* 1–130. Am Soc Microbiol.

Mattson, T., Richardson, J. & Goodin, D. (1974) *Nature 250,* 48–50.

Mattson, T., van Houwe, G. & Epstein, R. H. (1978) *J. Mol. Biol. 126,* 551–570.

Napoli, C., Gold, L. & Singer, B. S. (1981) *J. Mol. Biol. 149,* 433–449.

Nelson, M. A., Singer, B. S., Gold, L. & Pribnow, D. (1981) *J. Mol. Biol. 149,* 377–403.

Nomura, M., Jinks-Robertson, S. & Miura, A. (1982) In: *Interaction of Translational and Transcriptional Controls in the Regulation of Gene Expression,* pp. 91–104, Elsevier Biomedical.

Nomura, M., Dean, D. & Yates, J. L. (1982) *Trends in Biochemical Sciences 7,* 92–95.

O'Farrell, P. Z. & Gold, L. M. (1973a) *J. Biol. Chem. 248,* 5502–5511.

O'Farrell, P. Z. & Gold, L. M. (1973b) *J. Biol. Chem. 248,* 5512–5519.

Pribnow, D., Sigurdson, D. C., Gold, L., Singer, B. S., Napoli, C., Brosius, J., Dull, T. J. & Noller, H. F. (1981) *J. Mol. Biol. 149,* 337–376.

Rabussay, D., Geiduschek, E. P., Weisberg, R. A., Gottesman, S., Gottesman, M. E., Campbell, A., Calendar, R., Geisselsoder, J., Sunshine, M. G., Six, E. W. & Lindqvist, B. H. (1977) *Comprehensive Virology 8,* 1–344.

Russel, M. (1973) *J. Mol. Biol. 79,* 83–94.

Russel, M., Gold, L., Morrissett, H. & O'Farrell, P. Z. (1976) *J. Biol. Chem. 251,* 7263–7270.

Salser, W., Gesteland, R. F. & Bolle, A. (1967) *Nature 215,* 588–591.

Sanger, F., Coulson, A. R., Hong, G. F., Hill, D. F. & Petersen, G. B. (1982) *J. Mol. Biol. 162,* 729–773.

Sarabhai, A. & Brenner, S. (1967) *J. Mol. Biol. 27,* 145–162.

Shine, J. & Dalgarno, L. (1974) *Proc. Natl. Acad. Sci. USA 71,* 1342–1346.

Singer, B. S. & Gold, L. (1976) *J. Mol. Biol. 103,* 627–646.

Singer, B. S., Gold, L., Shinedling, S. T., Colkitt, M., Hunter, L. R., Pribnow, D. & Nelson, M. A. (1981) *J. Mol. Biol. 149,* 405–432.

Singer, B. S., Shinedling, S. T. & Gold, L. (1983) In: *The Bacteriophage T4,* pp. 327–333. Am. Soc. Microbiol.

Stormo, G. D., Schneider, T. D. & Gold, L. M. (1982) *Nucleic Acids Res. 10,* 2971–2996.

Trimble, R. B. & Maley, F. (1976) *J. Virol. 17,* 538–549.

von Hippel, P. H., Kowalczykowski, S. C., Lonberg, N., Newport, J. W., Paul, L. S., Stormo, G. D. & Gold, L. (1982) *J. Mol. Biol. 162,* 795–818.

Wiberg, J. S. & Karam, J. D. (1983) In: *The Bacteriophage T4,* pp. 193–201. Am. Soc. Microbiol.

Wood, W. B. & Revel, H. R. (1976) *Bacteriological Reviews 40,* 847–868.

DISCUSSION

GRUNBERG-MANAGO What was the effect of the AUU mutation from AUG? Did you not have such an initiation codon change?

GOLD: No, there is an AUG to AUA, AUG to GUG, and then there is a UAC to AAA at the second codon; these are regA constitutives. An extra U, to give AUGUUAC, has no effect on regA regulation. There is no conversion of the AUG to an AUU. If there was, I would be happy, because then I could study IF3.

KURLAND: You have this overlap in your cases for the T4 between terminator and initiator, and I am wondering if you know enough about the expression of the proteins involved to know whether there is any regularity to the expressions of the downstream gene. If you have out-of-phase overlap between a terminator and the initiator, is the second protein produced in much lower quantity than the first one, or there is no such regulation?

GOLD: I think there are no data that are good enough to be seriously thought about. What data there are, suggest that most of those things are made at about the same level as each other, and I think that the re-initiation efficiencies will not depend on something that subtle. I think that there will be situations similar to what is studied in the Nomura laboratory. If one wants a re-initiation efficiency of one, one may build a good Shine and Dalgarno and then bury that site so it cannot be seen from the outside.

SCHATZ: What happens to the turnover rate of the message whose translation has been slowed down by a regulatory protein?

GOLD: In the case of the regA regulated genes, many of them are destabilized by repression. One is not, one is completely stable, whether it is repressed or not.

MAALØE: What stability are you talking about?

GOLD: I am talking about functional halflives *in vivo* in rif-type experiments, not chemical stability but functional stability. The gene 32 protein message seems to be stable, whether it is repressed or not. This was one of the first clues that its regulation was at the level of translation. There are lovely older data from Ennis

& Cohen who did some experiments to show that you can get an answer in either direction by attacking cells with different antibiotics. Some mRNAs are stabilized, some are destabilized, and the answer depends on which specific antibiotic is used.

ROSENBERG: I didn't understand your answer to the first question about termination and the overlap with the AUG. We have published data (Schumperli *et al.* 1982) that I think answer the question directly. In the gal operon one gene is coupled to the next by delivering ribosomes to a termination site 3 base pairs upstream of the initiating AUG. Expressions of the downstream gene is enhanced by this coupling. If the ribosome is frameshifted to a UGA codon that overlaps the initiating AUG, then the system is still translationally coupled. However, if you frameshift again, such that the ribosome traverses into the gene 98 base pairs, all coupling is lost, and expression is lowered dramatically.

GOLD: Distances mapped carefully in the case of rIIB cistron show that re-initiation can occur if termination occurs within 20 nucleotides on either side of the re-initiation codon (Napoli *et al.* 1981). All of those situations give approximately the same re-initiation levels, i.e., it does not matter where the ribosome terminates (5′ ot 3′ to the re-initiation codon), as long as it is close. As soon as the ribosome terminates far away, then re-initiation is lost. Far enough in that case, could be message-specific also; *lacI* re-initiation studies by Steege show this beautifully.

KURLAND: What I had in mind was the R17 story of van Duin. There you have frameshift, which is a low-frequency event, and then you re-initiate, and you get a very small copy number.

SPRINZL: In the ribosomal binding site you showed in the second slide, there is a Shine and Dalgarno sequence which is flanking a double helical region. Can the Shine and Dalgarno interaction with the 16S rRNA take place when there is such a helix? Can you show that the binding of the ribosome to this type of mRNA is efficient?

GOLD: That sequence that I showed you was the only sequence of the 200 that we have looked at that has such a structure doing that.

SPRINZL: Is the initiation more efficient in this type of mRNA?

GOLD: The protein is found at reasonable levels *in vivo*. It is like most other T4 proteins, found at 5–10,000 copies per cell. It does not give you a clue. I think you have noted a lovely message to work on.

MAALØE: On Day One of this meeting, we talked about the classical model for initiating translation, where you introduce the 30S subunit as an element. For a long time I thought it was obligatory to initiate in that manner and that you must re-initiate along a muliple message. Now, you are interested and Dr. Nomura is greatly interested in notions of how you make translational coupling, and you have shown us a number. How do you visualize a ribosome continuing into the next cistron? Is it without going back to the classical initiation? What is the mechanism?

GOLD: I have always thought that at the termination codon, if you have built close by a strong Shine and Dalgarno for the next ribosome binding site, the local concentration of ribosomes ready to initiate is now very high, and that one doesn't have to build-in anything special, except that the ribosome is there, ready to go. So, I imagine that annealing between Shine and Dalgarno region and the next cistron and the 3′-end happens very fast. I think the only problem for coupling is to make sure that the ribosome binding site is not available for entry from free ribosomes. I don't think you need anything else special.

SAFER: Could you comment on the number of proteins required to achieve regulation on gene 32 versus the rIIB. You mentioned there are 9 involved in gene 32, is there a single protein involved repressing rIIB?

GOLD: By looking at the site that has been identified for rIIB for the regA protein, I think there is room for a protein of about 12,000 molecular weight, such as a head piece of a repressor, something like that. And that is the size of the regA protein, so I don't think there is going to be a lot of them involved. It could be a dimer, it might be a multimer, nobody knows, but I'm betting on a monomer.

SAFER: Are there 9 sites in gene 32 mRNA or do you have additional proteins coming on by protein/protein interaction?

GOLD: Oh, I did not make that as clear as one might. I believe the gene 32 regulatory system is filling by co-operativity. The target is the longest unstructured region of RNA available to it in a T4-infected cell. It co-operatively loads, in a sequence non-specific fashion. I think in that region of 63 nucleotides (which is 9×7, 7 being the site size for a gene 32 monomer by direct measurement) there is *no* sequence recognition. I think the gene 32 protein is a backbone binder, and all you are doing here is looking at *unstructure.* Our paper with Peter von Hippel tries to persuade the reader (without experiments) that the model is right, based on what is known about the gene 32 protein. There are no data yet that say even that our "operator" is where the protein sits during repression, let alone whether there are contacts that are sequence specific.

NOMURA: You suggested that the regulatory mechanism has evolved in order to save energy after T4 infection. In this connection, I would like to get your comments about the regulation of the synthesis of capsit or other structural proteins. The only example of regulation that I know is the P22 scaffolding protein.

GOLD: T4 is not known to regulate, in a cistron-specific fashion, its late genes. However, late gene transcription in T4 is directly coupled to replication. So, the quantity one gets of late proteins is regulated by how much DNA the infected cells have chosen to make. It may be that all the late genes are transcribed and translated constitutively, simply responding to the number of replisomes. There are no examples where an amber or any other mutation in T4 give gene-specific overproduction of the *late* protein.

Schumperli, D., McKenney, K., Sabietski, O. & Rosenberg, M. (1982) *Cell 30*, 865–871.

Napoli, C., Gold, L. & Singer, B. S. (1981) Translational re-initiation in the rIIB cistron of bacteriophage T4. *J. Mol. Biol. 149*, 433–449.

Regulation of Ribosome Biosynthesis in *E. Coli*

Sue Jinks-Robertson, Gail Baughman & Masayasu Nomura

INTRODUCTION

The rate of ribosome biosynthesis in *Escherichia coli* is precisely controlled (Gausing 1980). Under conditions of medium to fast growth, the number of ribosomes per unit amount of cellular protein is directly proportional to growth rate (i.e., the rate of protein synthesis), and the rate of ribosome biosynthesis responds rapidly to fluctuations in environmental conditions. Cells have, thus, evolved a regulatory mechanism which adjusts the number rather than the translational activity of ribosomes to meet the demand for protein synthesis specified by a given set of growth conditions. Since ribosomes are the essential machinery for protein synthesis, and constitute a major fraction of the cellular mass of *E. coli,* an understanding of the regulation of ribosome biosynthesis is central to understanding the regulation of cellular growth rate.

The *E. coli* ribosome is a complex organelle composed of 52 ribosomal proteins (r-proteins) and 3 species of ribosomal RNA (rRNA). Studies have shown that the synthesis rates of all the ribosomal components are balanced and, like ribosomes, are rapidly adjusted in response to changes in environmental conditions. In view of the complexity of the ribosome particle, one approach to studying the regulation of ribosome biosynthesis has been to examine the regulation of the RNA and protein components. Studies of r-protein regulation conducted in this and other laboratories have demonstrated the existence of a novel translational feedback regulatory mechanism which ensures balanced synthesis of the 52 r-proteins. In this article, we will review the essential features of translational feedback regulation of r-protein gene expression, giving emphasis to our recent *in vitro* data which substantiate some of the previously

Institute for Enzyme Research, and the Departments of Genetics and Biochemistry, University of Wisconsin, Madison, Wisconsin 53706, U.S.A.

GENE EXPRESSION, Alfred Benzon Symposium 19.
Editors: Brian F. C. Clark & Hans Uffe Petersen, Munksgaard, Copenhagen 1984.

proposed regulatory features. We will also describe our recent *in vivo* studies of rRNA regulation which suggest that rRNA synthesis is also feedback-regulated. Finally, the relationship of rRNA regulation to a model for the global regulation of ribosome biosynthesis (the ribosome feedback-regulation model) will be discussed.

TRANSLATIONAL FEEDBACK-REGULATION OF R-PROTEIN SYNTHESIS

The genes for 31 of the 52 r-proteins are found in 2 major clusters on the *E. coli* chromosome: the *str-spc* cluster at 72′ and the *rif* cluster at 89′ (Nomura & Post 1980, Lindahl & Zengel 1982). The *str-spc* cluster is organized into 4 transcription units ("operons") which encode 27 r-proteins, elongation factors G and Tu, and RNA polymerase subunit α. The *rif* cluster contains 2 operons which encode 4 r-proteins, and the *β* and *β′* subunits of RNA polymerase. The organization of genes within the *str-spc* and *rif* region operons is shown in Fig. 1.

One can imagine 2 possible levels at which r-protein gene expression might be regulated: at the level of transcription, or at a post-transcriptional (translational) level. To distinguish between these 2 possibilities, we examined the effect of r-protein gene dosage on the transcription rate of r-protein mRNA and on the synthesis rate of r-proteins. *E. coli* strains were constructed which were diploid with respect to operons within the *str-spc* region, and a subsequent comparison of the haploid *rif* region with the diploid *str-spc* region revealed that r-protein mRNA synthesis is gene-dosage-dependent, whereas the synthesis of r-proteins from the message is gene-dosage-independent (Fallon *et al.* 1979). Similar gene dosage experiments were also performed in other laboratories (Geyl & Bock 1977, Dennis & Fiil 1979, Olsson & Gausing 1980). Based on the results of the gene-dosage experiments, we proposed a translational feedback-regulation model to explain the co-ordinate synthesis of r-proteins. We suggested that r-protein synthesis is coupled to the ribosome-assembly process so that r-proteins synthesized in excess of the amount needed for ribosome assembly interact with their respective mRNA to block further translation of protein from the message (Fallon *et al.* 1979).

Since the early gene-dosage experiments, the translational feedback regulation model has been confirmed and refined by *in vitro* experiments using a DNA-dependent protein synthesis system, and *in vivo* experiments using recombinant plasmids with specific r-protein genes fused to, and under control of, an inducible promoter. Although the gene dosage experiments provided no

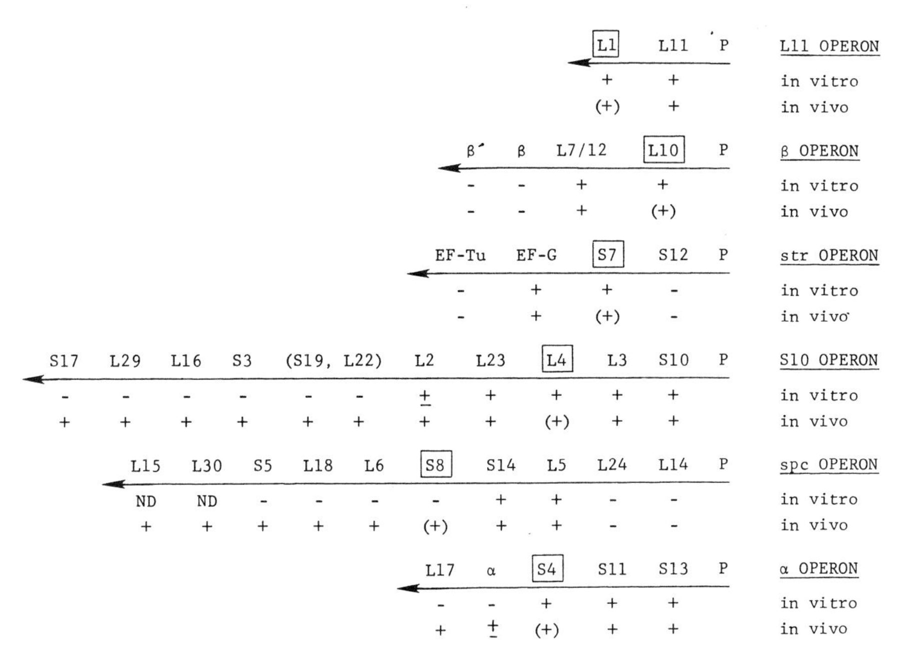

Fig. 1. Organization and regulation of genes within the *str-spc* and *rif* regions of the *E. coli* chromosome. The L11 and β operons are in the *rif* region; all other operons are in the *str-spc* region. Genes are represented by the protein product, and the direction of transcription from the promoter (P) of each operon is indicated by the arrow. It should be noted that *in vivo,* the primary β and α operon transcripts are the result of read-through from the upstream L11 and *spc* operons, respectively (Bruckner & Matzura 1981, Cerretti *et al.* 1983). Regulatory r-proteins are indicated by the boxes, and the effects of the boxed proteins on the *in vitro* or *in vivo* synthesis of proteins in the same operon are indicated. *In vitro* experiments have identified the following r-proteins as translational repressors: L1 (Yates *et al.* 1980, Brot *et al.* 1981), L10 (Brot *et al.* 1980, Fukuda 1980, Yates *et al.* 1981, Johnsen *et al.* 1982), L4 (Yates & Nomura 1980), S7 (Dean *et al.* 1981a) S8 (Dean *et al.* 1981b), and S4 (Yates *et al.* 1980). *In vivo* experiments have identified the following r-proteins as repressors: L1 (Dean & Nomura 1980), L10 (Yates *et al.* 1981), S7 (Dean *et al.* 1981a), L4 (Lindahl & Zengel 1979, Zengel *et al.* 1980), S8 (Dean *et al.* 1981b), and S4 (Dean & Nomura 1980). +, specific inhibition of synthesis; −, no significant effect on synthesis; ±, weak inhibition of synthesis; (+), inhibition assumed to occur *in vivo;* ND, not determined. It has not been established how L14, L24, or S12 are regulated.

information as to whether all or only some r-proteins are capable of inhibiting mRNA translation, the subsequent *in vitro* and *in vivo* experiments demonstrated that only specific r-proteins can function as translational repressors. It was found that each repressor r-protein inhibits its own synthesis and the synthesis of some, or all, of the r-proteins whose genes are in the same transcription unit as the repressor. The results of the *in vitro* and *in vivo* experiments are summarized in Fig. 1.

The experimental evidence for what we believe to be the essential features of translational feedback-regulation of r-protein gene expression has been reviewed previously (Nomura *et al.* 1982a, Nomura *et al.* 1982b), so only a brief summary of these features is given below. Following this account, we will summarize new data which substantiate 2 specific aspects of translational regulation: the notions of a single mRNA target site for translational inhibition and the translational coupling of co-regulated r-proteins. A detailed account of this work will be published elsewhere (Baughman & Nomura submitted). The essential features of translational feedback-regulation of r-protein synthesis are as follows: First, the r-protein genes within the *str-spc* and *rif* regions are organized into discrete regulatory units which are composed of contiguous, co-transcribed genes. A regulatory unit may correspond to an entire operon or only a portion of an operon, and each regulatory unit encodes its own unique translational repressor r-protein (Fig. 1). Second, a translational repressor r-protein acts at a single target site on the polycistronic mRNA to affect the translation of all the proteins encoded within the regulatory unit. The target site is at, or near, the translation initiation site for the first protein in the regulatory unit (Yates *et al.* 1981, Yates & Nomura 1981, Johnsen *et al.* 1982). Third, interaction of the repressor with the target site, directly blocks translation of the first cistron encoded in the regulatory unit. The translation of downstream cistrons is coupled with, and dependent on, that of the first cistron in the unit (Yates & Nomura 1981). This phenomenon has been called "translational coupling" (Oppenheim & Yanofsky 1980), or "sequential translation" (Yates & Nomura 1981). Fourth, the mechanism of translational repression involves competition between structurally similar regions of rRNA and mRNA for the binding of r-proteins (Nomura *et al.* 1980, Branlant *et al.* 1981, Gourse *et al.* 1981, Olins & Nomura 1981a, b, Wirth *et al.* 1981, Johnsen *et al.* 1982, Fig. 2 for example). Repressor r-proteins bind preferentially to rRNA and interact with the mRNA target site only when synthesized in excess of the amount needed for ribosome assembly. Interaction of the repressor r-protein with the mRNA target site presumably "sequesters" the ribosome binding site (or the translation start site) thus, preventing translation of the message (Robakis *et al.* 1981). Finally, it should be noted that translational feedback regulation of r-protein gene expression is physiologically significant. It is important for balancing the synthesis rates of r-proteins during exponential growth (Jinks-Robertson &

Nomura 1981, Stoffler *et al.* 1981, Jinks-Robertson & Nomura 1982), and for the growth-rate-dependent regulation of r-protein synthesis (Miura *et al.* 1981).

Translational repressor r-proteins act at a single target site

One can imagine 2 possible ways that a translational repressor r-protein might inhibit the translation of the co-transcribed cistrons which constitute a regulatory unit. First, the translational repressor might inhibit the translation of each cistron individually. This implies that the message for each regulated cistron would have its own target site (with perhaps a common structural feature shared with other target sites in the unit) for the repressor. A second possibility is that there is a single target site for the repressor r-protein which is located at the beginning of the polycistronic regulatory unit. According to this second model, the translation of the first cistron in the unit would be directly inhibited by the translational repressor, and the inhibition of the translation of downstream cistrons would be an indirect consequence of the repressor interacting with the unique target site. This implies that the translation of the downstream cistrons is coupled with, and dependent on, the translation of the first cistron in the regulatory unit. *In vitro* evidence supporting the "single target site" notion has been obtained for the L11 operon (Yates & Nomura 1981), and the β operon (Yates *et al.* 1981). In both cases, removal of the N-terminal coding sequence of the first cistron in the regulatory unit relieves the normal translational repression of the downstream cistron.

We have recently systematically analysed translational regulation of the L11 operon in order to confirm the existence of only a single target site in the L11-L1 regulatory unit, and to more accurately define this site. It was reasoned that if there is a single target site that depends on structural features of the mRNA near the translation start site of the L11 gene, then it should be possible to disrupt the target site without preventing translation of the message. More specifically, it should be possible to destroy the target site so that the synthesis of L11 and L1 is no longer affected by the repressor r-protein *in vitro* or *in vivo*. The L11 operon was chosen for further study because it is the simplest of the r-protein regulatory units; it encodes only r-proteins L11 and L1, and the synthesis of both proteins in the operon is regulated by L1 (Fig. 1). In addition, the entire operon had been sequenced in this laboratory (Post *et al.* 1979), and a region of structural homology between the L1 binding site on 23S rRNA and the presumptive L1 target site on the L11-L1 mRNA had been identified (Fig. 2).

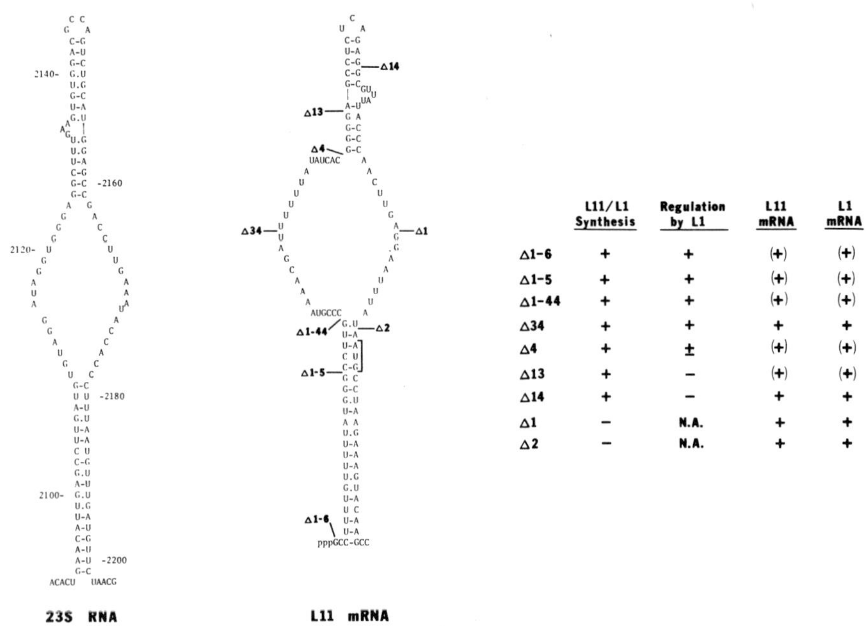

	L11/L1 Synthesis	Regulation by L1	L11 mRNA	L1 mRNA
Δ1-6	+	+	(+)	(+)
Δ1-5	+	+	(+)	(+)
Δ1-44	+	+	(+)	(+)
Δ34	+	+	+	+
Δ4	+	±	(+)	(+)
Δ13	+	−	(+)	(+)
Δ14	+	−	+	+
Δ1	−	N.A.	+	+
Δ2	−	N.A.	+	+

Fig. 2. Model of the L1 binding site on 23S rRNA and L11 mRNA, and the results of *in vitro* analysis of L11 mRNA deletions. The secondary structures for the L1 binding site on 23S RNA and on L11 mRNA are those proposed by Gourse *et al.* (1981). The L1 binding site on 23S RNA was reconstructed from the L1 protection experiments of Branlant *et al.* (1976) using the 23S rDNA sequence of Brosius *et al.* (1981). The sequence of the L11 mRNA leader is from Post *et al.* (1979). The presumptive Shine-Dalgarno sequence (GAGG), the AUG start codon for L11 translation, and the endpoints of the Ba131-generated deletions of the L11 mRNA leader are indicated on the proposed secondary structure. The table to the right of the RNA secondary structures summarizes the results obtained using each of the Ba131 deletions as a template in an *in vitro* DNA-dependent protein synthesis system. The first column indicates whether the L11 and L1 protein products are synthesized; the second column indicates whether the synthesis of L11 and L1 is regulated (i.e., inhibited) by addition of purified L1 to the system; and the third and fourth columns indicate whether L11 and L1 mRNAs, respectively, are transcribed from each of the templates. +, synthesis of protein or mRNA, regulation by L1; −, no synthesis of protein or mRNA, no regulation by L1; N.A., not applicable; ±, probably regulated, but this conclusion is tentative because of weak translation of r-proteins from this plasmid; (+) synthesis assumed to occur, but not measured.

The experimental approach taken to disrupt the L1 target site was to sequentially remove portions of the leader region of the L11 mRNA by Ba131 exonuclease digestion from an upstream restriction enzyme site present on a plasmid containing the entire L11 operon. A set of deletion plasmids was, thus, obtained which lacked the L11 operon promoter (P_{L11}) and varying amounts of the L11 mRNA leader, but left the L11 (and L1) coding sequences intact. These deletions were subsequently subcloned into a vector containing the *lac* operon promoter (P_{lac}) so that the L11 and L1 cistrons could be transcribed from P_{lac}. In Fig. 2, the L11 mRNA secondary structure proposed by Gourse *et al.* (1981) is shown, and the endpoints of the deletions constructed with Ba131 exonuclease are indicated.

The set of P_{lac}/L11 operon fusion plasmids was analyzed in an *in vitro* DNA-dependent protein synthesis system. Each DNA template was examined for its ability to direct synthesis of r-proteins L11 and/or L1, and for the ability of exogenously added purified L1 to inhibit synthesis of L11 and/or L1 from the template. The *in vitro* results from the series of deletions whose endpoints are shown in Fig. 2 are summarized in the accompanying table. All of the Ba131 deletions except for Ba131△1 and Ba131△2 synthesize both L11 and L1. As the L11 Shine-Dalgarno sequence is disrupted in Ba131△1 and deleted in Ba131△2, it is not surprising that the synthesis of L11 and L1 is affected. In 5 of the 7 Ba131 deletions which still direct synthesis of L11 and L1, the synthesis of both proteins is still repressible by addition of purified L1 to the *in vitro* system. In 2 of the 7 deletions (Ba131△13 and Ba131△14), the synthesis of neither L11 nor L1 is affected by addition of purified L1 to the system. Note that if the L1 cistron had its own target site for the repressor, its synthesis should still be repressible even though that of L11 is not. The *in vitro* results obtained with Ba131△13 and Ba131△14 clearly demonstrate that the regulation of r-proteins L11 and L1 is the result of L1 interacting at a single target site which is near the beginning of the L11 structural gene. *In vivo* experiments using the Ba131 deletions are in agreement with the *in vitro* observations; strains harboring the deletion plasmids which still have the regulatory target site do not overproduce L11 or L1, while strains harboring the deletion plasmids which are missing the target site, but still retain the translation initiation site for L11, overproduce L11 and L1 following induction of transcription from P_{lac}.

The *in vitro* experiments with the Ba131 deletion plasmids define a region of L11 mRNA leader sequence that is necessary for translational regulation by r-protein L1. As the first 36 and probably 48 bases of L11-L1 mRNA leader can be deleted without there being any effect on the expression or regulation of L11 or L1 (Fig. 2), it is clear that the bottom stem in the proposed L11 mRNA secondary structure shown in Fig. 2 is not important for the regulation. The absence of regulation with Ba131△13 and Ba131△14 suggests that the top stem in the proposed structure may be important in the regulation, but more studies need to be done to verify this hypothesis.

Translational coupling of r-protein synthesis

The above *in vitro* experiments with the Ba131 deletion plasmids demonstrate that L1 regulates both L11 and L1 synthesis by interacting with a single target site near the beginning of the L11 coding sequence. This finding implies that the regulation of L1 synthesis is a consequence of its synthesis being translationally coupled with that of L11. As was described above, neither Ba131△1 nor Ba131△2 synthesizes r-protein L11 in the *in vitro* system (Fig. 2), presumably because the L11 ribosome-binding site has been affected. It was also noted that neither deletion synthesizes r-protein L1, and this again suggests that the synthesis of the 2 r-proteins is translationally coupled. Another possible explanation for simultaneous loss of L11 and L1 gene expression is that the prevention of L11 translation has a polar effect on the transcription of the downstream L1 gene. To distinguish between the translational coupling and the transcriptional polarity explanations, the synthesis of L11-specific and L1-specific mRNA was examined in the *in vitro* system. As is indicated in Fig. 2, both messages are synthesized. S1 mapping experiments have shown that the *in vitro* polycistronic message is intact, so it can be concluded that the translation of the L1 cistron is coupled with, and dependent on, that of the upstream L11 cistron. *In vivo* experiments with Ba131△1 and Ba131△2 are also consistent with the *in vitro* results; L1 mRNA is synthesized, but not translated, when the translation initiation site for the upstream L11 cistron is disrupted.

The experiments summarized above demonstrate that L1 synthesis is translationally coupled with L11 synthesis. Ribosomes initiate translation of the first cistron in a regulatory unit, and the translation of this cistron presumably opens up the mRNA secondary structure so that the ribosome-binding sites of downstream cistrons become accessible. The subsequent translation of the downstream cistrons could occur by one of two mechanisms. First, the downstream cistrons might be translated by the same ribosomes (or ribosomal subunits, i.e., 30S subunits) that translated the first cistron in the regulatory unit. According to this mechanism, a ribosome (or a 30S subunit) would not completely dissociate from the message when translation is completed, but instead would immediately re-initiate at the now accessible ribosome-binding site for the next cistron. This process would continue until all the cistrons in the unit have been translated. As a special case of translational coupling, we have called this hypothetical process "sequential translation" (Yates & Nomura 1981). The second possible mechanism of translational coupling is more general. One could imagine that ribosomes initiate independently at the downstream cistrons once translation of the first cistron in the unit has begun. While the stoichiometric synthesis of co-translated r-proteins is more easily explained by the specialized process of sequential translation, there is no experimental evidence for this process.

REGULATION OF rRNA SYNTHESIS

Regulation of r-protein gene expression is achieved by a translational feedback mechanism that involves competition between rRNA (or ribosome assembly intermediates) and mRNA for the binding of key repressor r-proteins. The demonstration that this type of regulation is important for balancing the synthesis rates of the 52 r-proteins during exponential growth suggests that the synthesis rates of r-proteins are limited by the availability of a single ribosomal component, presumably rRNA. During the last several years we have been

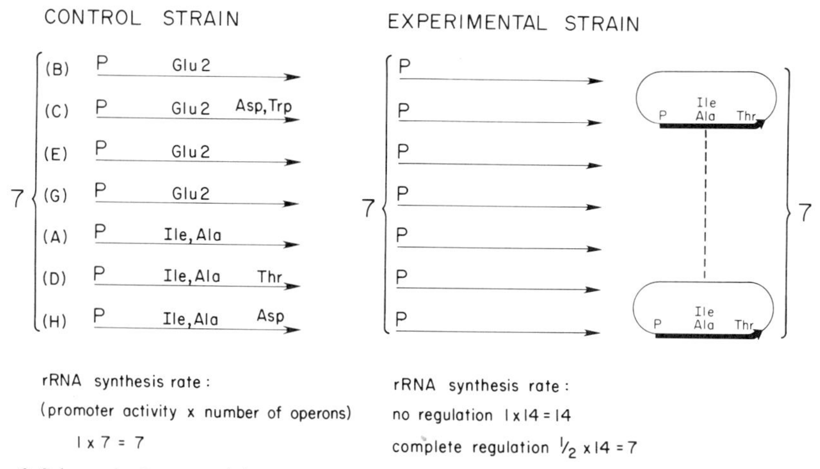

Fig. 3. Schematic diagram of the rRNA gene-dosage experiments. A control strain with the normal number of 7 rRNA operons per haploid chromosome is shown on the left. On the right, an experimental strain with the 7 chromosomal plus an additional 7 plasmid-encoded rRNA operons is shown. the direction of transcription from the promoter (P) is indicated by the arrow, and each operon has the following structure: promotor – 16S RNA coding region – the 16S-23S spacer region encoding tRNA(s) – 23S RNA coding region – 5S RNA coding region – distal tRNA(s). For the control strain, the genetic designation of each operon (*rrnA, rrnB,* etc.) is included parenthetically; for simplicity, only the tRNA(s) encoded within each of the 7 rRNA operons is indicated (Ellwood & Nomura 1982). For the experimental strain, only those tRNAs encoded by the rRNA operon present on the plasmid (*rrnD* in this case) are indicated; the tRNAs encoded by the chromosomal rRNA operons are identical to those shown for the control strain.

studying the regulation of rRNA synthesis in this laboratory, and what follows is a brief description of this work; experimental details described elsewhere (Jinks-Robertson *et al.* 1983).

The experimental approach to studying rRNA regulation has been to examine the effect of gene dosage on the synthesis rate of rRNA. Since *E. coli* has 7 rRNA operons per haploid genome (Kiss *et al.* 1977), multi-copy plasmids carrying intact rRNA operons were used to increase the dosage of rRNA operons 2-fold. Two extreme possibilities regarding gene dosage effects are schematically shown in Fig. 3. One possibility is that rRNA synthesis is gene-dosage-dependent ("no regulation"), and the second possibility is that rRNA synthesis is not affected by an increase in gene dosage ("complete regulation"). The synthesis rate of rRNA in strains harboring the recombinant plasmids was measured using 2 different methods, and was compared to the rate in a control strain with the normal number of rRNA operons. First, rRNA synthesis was directly measured using pulse-labelling and filter hybridization techniques. Second, the synthesis rate of rRNA was indirectly measured by examining the relative accumulations of tRNAs whose genes are co-transcribed with the chromosomal and plasmid-encoded rRNA operons (Fig. 3). Both types of experiments have shown that the synthesis rate of rRNA is not significantly affected by an increase in gene dosage. The analysis of the relative accumulations of the plasmid versus the chromosomally-encoded tRNAs has demonstrated that the regulation of total rRNA synthesis in strains carrying extra rRNA operons is achieved by reducing the transcriptional activity of each individual operon. In cells with 7 extra plasmid-encoded rRNA operons (14 total operons), for example, an analysis of tRNA accumulations has shown that approximately one-half of the rRNA transcripts come from the plasmid and one-half from the chromosome. Relative to a control strain with only 7 operons, the accumulations of those tRNAs which are co-transcribed only with the chromosomal rRNA operons are, thus, reduced by a factor of about 2, while the accumulations of those tRNAs co-transcribed with the plasmid-encoded rRNA operon are increased in the plasmid containing strain (Fig. 3, the operon carried by the plasmid contributes one-seventh of the total rRNA transcripts in a normal strain, and about one-half of the total rRNA transcripts in the plasmid-containing strain).

There are 2 possible ways to explain the observed gene-dosage-independent synthesis of rRNA. One possibility is that a factor necessary for the transcription of rRNA operons (e.g., active polymerase or an anti-termination factor; Travers 1976, Brewster & Morgan 1981) is limiting the total amount of rRNA synthesis.

A second possibility is that rRNA synthesis is feedback-regulated by a product of rRNA operons. A feedback mechanism would presumably prevent the synthesis of excess rRNA (and ribosomes), and would ensure that the amount of rRNA synthesized is the amount needed by the cell. To distinguish between these 2 possibilities, we deleted the central portion of the rRNA operon carried on a recombinant plasmid so that functional transcripts would not be synthesized from the plasmid. If the first possibility is correct (i.e., total rRNA synthesis is limited solely by some unknown positive factor), then the rate of rRNA synthesis should be the same in strains with the intact or the deletion plasmid. If the second possibility is correct (i.e., total rRNA synthesis is regulated by a feedback mechanism), then strains with either the intact or the deletion plasmids should make the same amount of *functional* product. The net result of this would be that the strain with the deletion plasmid should make more total rRNA (functional transcripts from the chromosome plus defective transcripts from the plasmid) than the strain with the intact plasmid (functional transcripts from the plasmid and chromosome). In strains harboring deletion plasmids, it was found that the rate of rRNA synthesis is increased and essentially gene dosage dependent. Based on these experiments, it can be concluded that rRNA transcription is controlled by a feedback mechanism which ensures that the amount of functional product is the amount required by the cell.

THE RIBOSOME FEEDBACK-REGULATION MODEL

As pointed out in the introduction, cells adjust the rate of ribosome biosynthesis to meet the demand for protein synthesis that is specified by a given set of growth conditions. The synthesis of ribosomes is, thus, regulated so that virtually all ribosomes are utilized and the pool of free, non-translating ribosomes is small (an exception is very slow growth; see below). To explain these observations, we previously proposed that ribosome biosynthesis is subject to a feedback-regulatory mechanism which ensures that the "correct" number of ribosomes is maintained (i.e., that the pool of non-translating ribosomes is small). This regulatory model was called the "ribosome feedback-regulation model" and has been described previously (Nomura *et al.* 1982b). Presumably, the regulation of ribosome biosynthesis could be achieved by limiting the expression of a ribosomal component by a feedback loop. Since r-protein synthesis can be regulated by the availability of rRNA, it was suggested that regulation of rRNA synthesis would, in principle, be sufficient to regulate the biosynthetic rate of ribosomes.

If the ribosome feedback-regulation model is correct, and rRNA is the ribosomal component which is regulated by this mechanism, one would predict that rRNA synthesis should not be significantly affected by gene dosage. The rRNA gene dosage experiments described above confirm this prediction, and in addition, demonstrate that rRNA synthesis is feedback-regulated by a functional product of rRNA operons. We suggest that the product responsible for the observed regulation is the ribosome, and that the regulation occurs as follows. When protein synthesis slows down (e.g., as the result of depletion of a nutrient in the growth medium), there would be a transient increase in the amount of free, non-translating ribosomes. These "free" ribosomes would either directly, or indirectly, inhibit the rate of rRNA synthesis, so that the number of ribosomes per unit amount of cell mass is adjusted to the slower growth rate. Conversely, if cells encounter better growth conditions, the pool of non-translating ribosomes would be rapidly depleted as the rate of protein synthesis increases. The result would be derepression of rRNA transcription and an increase in the rate of ribosome biosynthesis. The proposed feedback mechanism, thus, ensures that the ribosome biosynthesis is balanced with cellular growth rate.

The rRNA gene-dosage experiments utilizing the rRNA deletion plasmids demonstrate that a product of rRNA operons is responsible for the observed regulation, but they do not identify the product, which could be a specific rRNA transcript, a ribosome assembly intermediate, or intact ribosomes. The following observations, however, indicate that ribosomes are the product that is responsible for the regulation. First of all, overproduction of ribosomal components has been observed *in vivo* when the rate of protein-chain elongation is reduced without there being a concomitant reduction in the rate of chain initiation (Bennett & Maaløe 1974, Zengel *et al.* 1977). In theory, this would deplete the pool of free ribosomes and, according to our model, should stimulate the transcription of rRNA. Second, our model would predict that if ribosomes are the key factor in the feedback regulation of rRNA synthesis, blocking the assembly of ribosomes without specifically inhibiting protein synthesis should result in stimulation of rRNA synthesis because of depletion of the pool of free ribosomes. In agreement with this prediction, we note that there are several reports of ribosome assembly-defective mutants which overproduce rRNA under non-permissive conditions (MacDonald *et al.* 1967, Tai *et al.* 1969, Buckel *et al.* 1972). A similar phenomenon has been studied in this laboratory using an *rpsE*, cold-sensitive, ribosome assembly-defective mutant. When this mutant is shifted to the non-permissive temperature, the rate of rRNA synthesis is elevated 50–100% relative to the rate in a wild-type control strain within 20 min of the temperature shift (Miura *et al.* unpublished observations). Our interpretation of this is that the rate of ribosome biosynthesis is dependent on the amount of intact, potentially functional ribosomes that are free in the cellular pool rather than on the amount of excess ribosome components. Finally, our model is consistent with the observation that ribosomes are apparently overproduced at very slow growth rates (Alton & Koch 1974). According to the ribosome feedback-regulation model, the pool of free ribosomes should increase as the degree of repression of rRNA operons increases (i.e., as growth rate slows down).

The ribosome feedback-regulation model we have proposed is consistent with our data and with data from the literature. This model accounts for the observed growth-rate-dependent regulation of ribosome biosynthesis by providing a simple explanation for how the proportionality between ribosome content and the rate of protein synthesis is achieved. While the simplest mechanism for the feedback regulation of ribosome biosynthesis would involve non-translating ribosomes as the direct effector of rRNA transcription, we have no evidence to support this. It is difficult to identify the direct effector *in vivo,* and it should be noted that our attempts to inhibit the synthesis of rRNA *in vitro* by the addition of purified ribosomes to the system have been unsuccessful so far. While we believe that ribosomes are the key factor in the feedback regulation, their effect on rRNA transcription might be achieved indirectly.

ACKNOWLEDGEMENTS

The work described in this article was supported in part by the College of Agriculture and Life Sciences, University of Wisconsin-Madison, by grant GM-20427 from the National Institutes of Health, and by grant PCM79-10616 from the National Science Foundation. This is paper number 2643 from the Laboratory of Genetics, University of Wisconsin-Madison.

REFERENCES

Alton, T. H. & Koch, A. L. (1974) *J. Mol. Biol. 86,* 1–9.
Bennett, P. M. & Maaløe, O. (1974) *J. Mol. Biol. 90,* 541–561.
Branlant, C., Korobko, V. & Ebel, J.-P. (1976) *Eur. J. Biochem. 70,* 471–482.
Branlant, L., Krol, A., Machatt, A. & Ebel, J.-P. (1981) *Nucleic Acids Res. 9,* 293–307.

Brewster, J. M. & Morgan, E. A. (1981) *J. Bacteriol. 148,* 897–903.

Brosius, J., Dull, T. J., Sluter, I. I. & Noller, H. F. (1981) *J. Mol. Biol. 148,* 107–127.

Brot, N., Caldwell, P. & Weissbach, H. (1980) *Proc. Natl. Acad. Sci. USA 77,* 2592–2595.

Brot, N., Caldwell, P. & Weissbach, H. (1981) *Arch. Biochem. Biophys. 206,* 51–53.

Bruckner, R. & Matzura, H. (1981) *Mol. Gen. Genet. 183,* 277–282.

Buckel, P., Ruffler, D., Piepersberg, W. & Bock, A. (1972) *Mol. Gen. Genet. 119,* 323–335.

Cerretti, D. P., Dean, D., Davis, G. R., Bedwell, D. M. & Nomura, M. (1983) *Nucleic Acids Res. 11,* 2599–2616.

Dean, D. & Nomura, M. (1980) *Proc. Natl. Acad. Sci. USA 77,* 3590–3594.

Dean, D., Yates, J. L. & Nomura, M. (1981a) *Cell 24,* 413–419.

Dean, D., Yates, J. L. & Nomura, M. (1981b) *Nature 289,* 89–91.

Dennis, P. P. & Fiil, N. P. (1979) *J. Biol. Chem. 254,* 7540–7547.

Ellwood, M. & Nomura, M. (1982) *J. Bacteriol. 149,* 458–468.

Fallon, A. M., Jinks, C. S., Strycharz, G. D. & Nomura, M. (1979) *Proc. Natl. Acad. Sci. USA 76,* 3411–3415.

Fukuda, R. (1980) *Mol. Gen. Genet. 179,* 489–496.

Gausing, K. (1980) In: *Ribosomes, Structure, Function and Genetics,* pp. 693–718, University Park Press, Baltimore.

Geyl, D. & Bock, A. (1977) *Mol. Gen. Genet. 154,* 327–334.

Gourse, R. L., Thurlow, D. L., Gerbi, S. A. & Zimmerman, R. A. (1981) *Proc. Natl. Acad. Sci. USA 78,* 2722–2726.

Jinks-Robertson, S. & Nomura, M. (1981) *J. Bacteriol. 145,* 1445–1447.

Jinks-Robertson, S. & Nomura, M. (1982) *J. Bacteriol. 151,* 193–202.

Jinks-Robertson, S., Gourse, R. L. & Nomura, M. (1983) *Cell 33,* 865–876.

Johnsen, M., Christensen, T., Dennis, P. P. & Fiil, N. P. (1982) *EMBO 1,* 999–1004.

Kiss, A., Sain, B. & Venetianer, P. (1977) *FEBS Letters 79,* 77–79.

Lindahl, L. & Zengel, J. (1979) *Proc. Natl. Acad. Sci. USA 76,* 6542–6546.

Lindahl, L. & Zengel, J. (1982) In: *Advances in Genetics, Vol. 21,* pp. 53–121, Academic Press, New York.

MacDonald, R. E., Turnock, G. & Forchhammer, J. (1967) *Proc. Natl. Acad. Sci. USA 57,* 141–147.

Miura, A., Krueger, J. H., Itoh, S., deBoer. H. & Nomura, M. (1981) *Cell 25,* 773–782.

Nomura, M., Dean, D. & Yates, J. L. (1982a) *Trends in Biochem. Sci. 7,* 92–95.

Nomura, M., Jinks-Robertson, S. & Miura, A. (1982b) In: *Interaction of Translational and Transcriptional Controls in the Regulation of Gene Expression,* pp. 91–104, Elsevier Science Publishing Co, New York.

Nomura, M., Morgan, E. A. & Jaskunas, S. R. (1977) *Ann. Rev. Genet. 11,* 297–347.

Nomura, M. & Post, L. (1980) In: *Ribosomes, Structure, Function and Genetics* pp. 671–691, University Park Press, Baltimore.

Nomura, M., Yates, J. L., Dean, D. & Post, L. (1980) *Proc. Natl. Acad. Sci. USA 77,* 7084–7088.

Olins, P. O. & Nomura, M. (1981a) *Nucleic Acids Res. 9,* 1757–1764.

Olins, P. O. & Nomura, M. (1981b) *Cell 26,* 205–211.

Olsson, P. O. & Gausing, K. (1980) *Nature 283,* 599–600.

Oppenheim, D. S. & Yanofsky, C. (1980) *Genetics 95,* 785–795.

Post, L. E., Strycharz, G. D., Nomura, M., Lewis. H. & Dennis, P. P. (1979) *Proc. Natl. Acad. Sci. USA 76,* 1697–1701.

Robakis, N., Meza-Basso, L., Brot, N. & Weissbach, H. (1981) *Proc. Natl. Acad. Sci USA 78,* 4261–4264.

Stöffler, G., Hasenbank, R. & Dabbs, E. R. (1981) *Mol. Gen. Genet. 181,* 164–168.

Tai, P.-C., Kessler, D. P. & Ingraham, J. (1969) *J. Bacteriol. 97,* 1298–1304.

Travers, A. (1976) *Nature 263,* 641–646.

Wirth, R., Kohles, V. & Bock, A. (1981) *Eur. J. Biochem. 114*, 429–437.

Yates, J. L., Arfsten, A. E. & Nomura, M. (1980) *Proc. Natl. Acad. Sci. USA 77,* 1837–1841.

Yates, J. L., Dean, D., Strycharz, W. A. & Nomura, M. (1981) *Nature 294,* 190–192.

Yates, J. L. & Nomura, M. (1980) *Cell 21,* 517–522.

Yates, J. L. & Nomura, M. (1981) *Cell 24,* 243–249.

Zengel, J. M., Mueckl, D. & Lindahl, L. (1980) *Cell 21,* 523–535.

Zengel, J. M., Young, R., Dennis, P. P. & Nomura, M. (1977) *J. Bacteriol. 129*, 1320–1329.

DISCUSSION

EBEL: In you last scheme, where you show ribosomes, do you think that 30S and 50S subunits act independently or as 70S ribosomes?

NOMURA: We don't know. We have been making different deletions in the plasmid-encoded rRNA operons to test which subunit(s) are required. A deletion that spans both the 16S and 23S gene destroys the effect, but one that is solely in the 16S gene also largely abolishes the effect implying that the 30S subunit is required. However, we cannot be sure because preventing 30S assembly also likely affects 50S subunit assembly because the synthesis of some r-protein of the large subunit are in operons controlled by small subunit proteins.

UHLENBECK: Are all the tRNA genes regulated in this way, or only a subset?

NOMURA: We have analyzed 20–30 tRNA spots from 2-dimensional gels, and every tRNA examined appears to be regulated. The degree of inhibition varied somewhat, but we cannot say whether this variation is significant. It is clear that at least most of them are controlled.

UHLENBECK: Including many which are in multigene operons? I don't know about tRNA genes in *E. coli*, are a lot of them in multigene operons?

NOMURA: Some of them are in multigene operons, and others are monocistronic. We tested whether there is a tRNA gene dosage effect with a plasmid carrying the monocistronic asparagine tRNA gene. In this case, we saw that the asparagine tRNA was strongly overproduced. From this case and from the case of several other tRNAs which have been examined by others, we can infer that *E. coli* cells do not monitor how any tRNA molecules are synthesized, but regulate their synthesis together with rRNA by monitoring free ribosomes. It appears likely that tRNA genes and rRNA genes have a similar target site for regulation and are regulated in the same way by products of rRNA, not tRNA, operons.

MAALØE: The recent findings on the S10 operon suggest that there is an attenuator-type control by L4 similar to that for the *trp* operon.

NOMURA: Yes. However, there is no such evidence for other r-protein operons we have studied.

MAALØE: I am concerned about the mechanism of translational coupling. When operons contain 10 or 11 protein genes and they are spaced in a more conventional way by 40, 50 or 60 nucleotides, the translational coupling mechanism does not seem to be so likely.

NOMURA: I am also concerned, and that is why we started to do experiments with the simplest regulatory unit, that is, the L11-L1 operon. In this case, the 2 genes are separated by 3 nucleotides, and the experimental results demonstrated the existence of translational coupling. Intercistronic regions in the S10 operon have not been sequenced except for the region between the S10 and the L3 gene which are separated by 32 nucleotides. However, our experiments on the L10 operon showed that the second gene in this operon, the L7/L12 gene, is translationally regulated together with the first gene, the L10 gene, at a single site near the translation initiation site of the L10 gene (Yates *et al.* 1981). The results strongly indicate that the L7/L12 is translationally coupled with the L10 gene, even though the 2 are separated by as many as 66 nucleotides.

PESTKA: What happens if free ribosomes are generated with the use of antibiotics that inhibit protein synthesis?

NOMURA: There are experiments using antibiotics which are pertinent to your question. For example Bennett and Maaløe (1974) have shown that fusidic acid at suitable concentrations which decrease the protein elongation rate cause about a 3-fold increase in rRNA synthesis. Several people have also demonstrated that chloramphenicol stimulated rRNA synthesis. One interpretation we have made is that in these situations the free ribosomal pool is temporarily depleted, resulting in derepressed rRNA synthesis.

GRUNBERG-MANAGO: Translational coupling suggests that the Shine and Dalgarno sequence of the AUG initiation codon for the L1 gene is buried in some secondary structure. But is there any actual evidence?

NOMURA: We do not have any direct evidence. We are currently performing a systematic deletion analysis to identify the region in the L11 gene that is responsible for the masking of the translation initiation site for the L1 gene.

KURLAND: I was puzzled by one answer you gave, that *E. coli* does not care too much about the amounts of tRNAs produced. We know that the acceptor levels of tRNAs are regulated with exquisite refinement and that tRNA levels have incredible potential influences on error frequencies.

NOMURA: It is known that the synthesis of tRNA and rRNA in *E. coli* is regulated in the same way as a group by the mechanisms involved in stringent control or growth-rate shift up or shift down. Our model shows that tRNA genes are regulated together with rRNA operons by free ribosomes. The relative levels of tRNAs produces by tRNA genes with different promotor strengths could be maintained in response to environmental changes to keep the required balance, but apparently *not* in response to changes in gene dosages.

SCHATZ: How much repressor r-protein did you add in your *in vitro* experiments? Were these reasonable amounts?

NOMURA: I think that they are reasonable amounts. For example, the concentration of L1 that can cause a near maximal inhibition *in vitro* is about 0.5 μM. If the volume of an *E. coli* cell is 10^{-15} litres, the presence of 0.5 μM which corresponds to only 300 free L1 molecules in a cell or about 0.4% of the L1 present in the ribosomes would inhibit L1 synthesis almost completely.

SCHATZ: Can you titrate the repressors with rRNA?

NOMURA: Yes. The addition of 23S rRNA, but not 16S rRNA, abolished the inhibitory effects of L1.

GALLANT: With regard to free ribosomes as repressors of rRNA transcription, have experiments been done with translation initiation inhibitors, which should produce an accumulation of free ribosomes.?

NOMURA: Some experiments have been done with translation initiation inhibitors such as trimethoprim, and the results were consistent with the model. However, it has been argued that the inhibition by trimethoprim causes a methionine deficiency which could be the real cause of the inhibition of rRNA synthesis in this case.

Studying Eukaryotic Transcription and Translation by Gene Fusion

Hanne Johansen, Mitchell Reff, Daniel Schumperli & Martin Rosenberg

INTRODUCTION

A variety of bacterial genes have now been expressed efficiently in eukaryotic cells (Mulligan & Berg 1980, Colbere-Garapin *et al.* 1981, Schumperli *et al.* 1982, Gorman *et al.* 1982). In each case, expression requires that the prokaryotic gene be fitted with all the regulatory information normally required to direct the steps of transcription, mRNA maturation, and translation of a eukaryotic gene. Generally, the chimeric transcription unit is inserted into a recombinant vector which allows it to be replicated in *E. coli* and then introduced directly into a eukaryotic cell. Expression is monitored either transiently, 24–72 hours after transfection or continuously, after stable transformation of the appropriate cell type. These systems have proven extremely useful in examining the factors which govern the expression of bacterial genes in higher cells, and in particular, the gene-regulatory elements which control their expression. Analogous to the gene-fusion techniques which have been used so successfully to study prokaryotic gene regulation, the expression of an assayable, selectable bacterial function in higher cells allows the same concept to be applied to the study of eukaryotic gene-control elements.

GALACTOKINASE GENE FUSIONS; THE pSVK-gpt VECTOR

Previously, we demonstrated that the *E. coli* galactokinase gene (*gal*K) can be inserted into a plasmid vector such that the expression of *gal*K is controlled by

Laboratory of Biochemistry, National Cancer Institute, Bethesda, MD 20205 and Department of Molecular Genetics, Smith Kline and French Laboratories, Philadelphia, PA 19101 U.S.A.

GENE EXPRESSION, Alfred Benzon Symposium 19.
Editors: Brian F. C. Clark & Hans Uffe Petersen, Munksgaard, Copenhagen 1984.

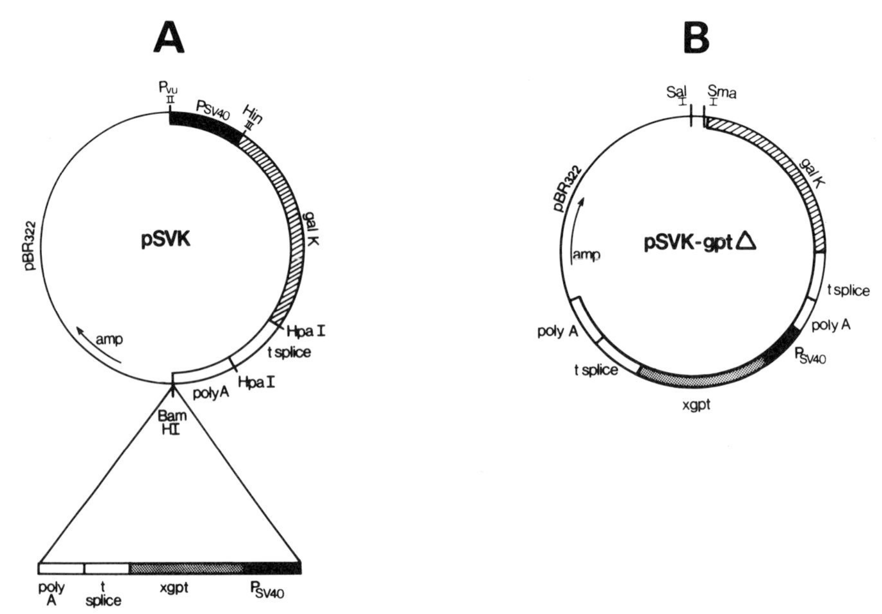

Fig. 1. Schematic representation of the pSVK vector and its derivatives: A) The pSVK vector (Schumperli *et al.* 1982) carries a transcription unit containing the SV40 early-promoter region (P_{SV40}), the *E. coli* galactokinase gene (*gal*K), the SV40 t-antigen intervening sequence (t splice), and the SV40 early-region polyadenylation signal (poly A). This unit is inserted into a pBR322 derivative carrying the *β*-lactamase gene (amp). The pSVK-*gpt* vector is created from pSVK by inserting a second transcription unit, similar to the first, into the unique BamHI site on pSVK. This transcription unit carries the guaninephosphoribosyltransferase gene (*gpt*) (Mulligan & Berg 1980). Important restriction sites are indicated. B) The pSVK-*gpt*∆ vector is derived from pSVK-*gpt* by inserting a SmaI linker into the HindIII site as well as a 34 base-pair fragment obtained from phage M13 into the PvuII site. The M13 fragment contains several unique restriction-enzyme sites, SmaI, SalI, etc. The construction is then recut with SmaI and closed, thereby deleting the SV40 promoter and retaining unique SmaI and SalI sites.

regulatory elements derived from the SV40 viral genome (Schumperli *et al.* 1982). The initial vector, pSVK, carries the SV40 early promoter fused upstream of the *gal*K coding sequence such that the *gal*K initiation codon is the first AUG on the mRNA transcript (Fig. 1). Immediately downstream of *gal*K is a DNA segment carrying the SV40 t antigen intervening sequence and distal to this is a segment carrying the SV40 early mRNA polyadenylation signal. This entire transcription unit is contained within a plasmid pBR322 derivitive. A major feature designed into the pSVK vector is that each of the regulatory elements (i.e., promoter, *gal*K gene, splice signal, poly A signal) is separated by unique

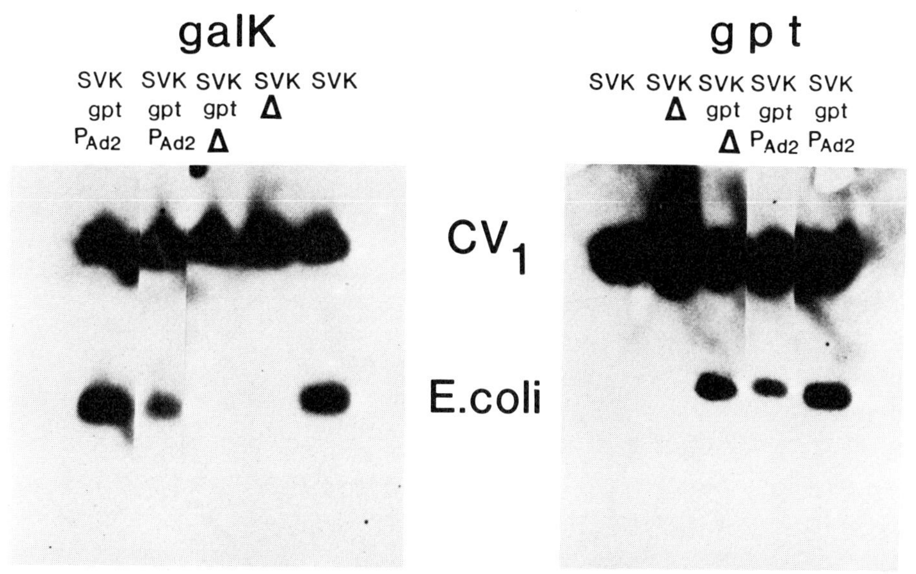

Fig. 2. Autoradiograph showing the *gal*K and *gpt* activities (both endogenous and *E. coli*) assayed in CV_1 monkey cell extracts which have been subjected to starch gel-electrophoresis (Schumperli *et al.* 1983). The CV_1 cells were transfected with the following DNAs: pSVK, pSVKΔ, pSVK-*gpt*Δ, and 2 independent preparations of pSVK-*gpt*-P_{Ad2} (see text in Figs. 1, 3 for details). Transfections and assays were carried out as described by Schumperli *et al.* 1982.

restriction sites appropriately positioned on the vector. This modular arrangement allows each element to be removed selectively and replaced with other DNA segments.

We have used the pSVK vector to obtain expression of the *gal*K gene in a variety of eukaryotic cells (Fig. 2). Expression was shown to be dependent upon the SV40 promoter, but did not require the splice signals (Schumperli *et al.* 1982). Moreover, expression could be obtained either transiently, or continuously after stable transformation. Stable expression was used to complement cells that were genetically deficient in their endogenous *gal*K gene function (i.e., galactosemic). Another important feature of the system is that the *E. coli gal*K activity can be detected and assayed readily even in the presence of endogenous eukaryotic *gal*K activities. This is done using either a starch gel-electrophoretic procedure which resolves galactokinase isozymes or by a recently developed rapid filter assay technique (H. J. & M. R., in preparation). Our ability to assay and select for *gal*K expression suggested the general potential of using this system to study eukaryotic gene-regulatory signals by gene-fusion technology

(Rosenberg *et al.* 1982). It was these considerations which lead to the further development of the vector system described below.

A major difficulty in applying gene-fusion systems to the study of eukaryotic gene-control elements is the large number of experimental variables that will affect quantitative measurements and comparisons of gene expression. For example, variability in DNA preparation, state of the cell culture, transfection conditions, assay conditions, etc., make it extremely difficult to compare accurately, results obtained with even similarly constructed vectors. Even more problematic is the differential gene expression that will result from variations in gene-copy number when these vectors are introduced into cells. There is as yet no easy way to control and/or determine copy number variations in eukaryotic systems. For these reasons, we have inserted a second assayable, selectable gene function contained on its own transcription unit into the pSVK vector system (Fig. 1). This gene, the *E. coli* guaninephosphoribosyltransferase (*gpt*), provides an internal quantitative standard in all transfection experiments. The expression of *gpt* is again controlled entirely by SV40 gene-regulatory signals (Mulligan & Berg 1980). Thus, *gal*K and *gpt* are introduced into cells on a single covalent molecule, pSVK-*gpt*, and expressed from separate transcription units. Both enzymes can be assayed from the same cell extract using either the starch gel procedure and/or the filter assay technique. Most importantly, *gal*K expression is always measured relative to the internal standard, *gpt.* Hence, variations in *gal*K expression resulting from "real" functional changes in gene control can be accurately determined.

In order that any DNA fragment which contains a potential eukaryotic promoter be fused to *gal*K, the SV40 early promoter region was excised from pSVK and replaced with a recombinant linker containing several restriction enzyme sites unique to the vector. It was important to demonstrate that removal of this SV40 promoter fragment selectively eliminated all *gal*K expression, but that the *gpt* transcription unit functioned normally (Fig. 2). The resulting vector, pSVK-*gpt*Δ (Fig. 1B) contains a unique flush-end cloning site (SmaI) positioned 42 base pairs preceding the *gal*K initiation codon. DNA fragments containing transcription start-site information derived from other eukaryotic genes can be inserted at this site, thereby fusing their transcription to *gal*K.

Several important features of the pSVK-*gpt*Δ vector system help insure that accurate comparisons can be made between different promoter-*gal*K fusions. For example, the *gpt* transcription unit is positioned identically on all pSVK-gpt constructs, thereby making its influence on the vector system relatively constant.

As the *gpt* unit is controlled by the SV40 early promoter, it contains the 72 base-pair repeat sequence known to exhibit "enhancer" activity. All pSVK-*gpt*Δ constructs will contain this enhancer, and its position relative to the *gal*K transcription unit remains constant. If desired, the vector is capable of additional flexibility. The *gpt* transcription unit can be placed at other positions on the vector (e.g., at the unique *Sal*I site preceding the *gal*K transcription unit). Moreover, *gpt* expression can be placed under the control of a promoter which does not have enhancer capability. In this way, enhancer-independent expression of different promoter-*gal*K fusions can be studied. It is important, however, that all measurements be made with similar constructs, so as to insure accurate comparisons.

PROMOTER INSERTS

We have inserted a variety of defined DNA segments carrying eukaryotic promoter regions into the pSVK-*gpt*Δ vector system (Fig. 3). In each case, the *gal*K coding sequence was fused with the 5′-non-coding region contained on the

Promoter	Fragment		Percent galK/gpt	
			Monkey CV_1	Hamster R1610
None (pSVK-gptΔ)		cap	0	0
$P_{SV40-EARLY}$	340	↓ 64	100	100
$P_{AD2-LATE}$	867	↓ 33	45	40
$P_{RSV-LTR}$	850	↓ 44	200	90
$P_{\beta-GLOBIN\ (MOUSE)}$	550	↓ 25	15	15
$P_{MT\ (MOUSE)}$	~1900	↓ 68	-metal	20
			+metal	80

Fig. 3. Promoter comparisons. DNA fragments carrying defined eukaryotic promoters were inserted into pSVK-*gpt* vectors (see Fig. 1 and text for details). The size of the inserted promoter fragment is shown, as well as the distance from the cap site to the initiation codon of the *gal*K gene. Each DNA was transfected into monkey kidney CV_1 cells and hamster fibroblast R1610 cells (*gal*K$^-$, *gpt*$^-$; Thirion *et al.* 1976). The cells were harvested 48 and 24 h, respectively, after transfection. The CV_1 cell extract was subjected to starch gel-electrophoresis and the enzyme activity was monitored for both *gal*K and *gpt* by laser scanning of an autoradiograph of the starch gel (Schumperli *et al.* 1982). The enzyme activity in the cell extract from R1610 cells was monitored by filter assay (H. J. & M. R., in preparation). The *gal*K/*gpt* values are given as percent relative to the *gal*K/*gpt* value of the vector containing the SV40 early promoter (P_{SV40}).

promoter fragment. Expression was monitored transiently in both monkey cells (CV_1) and hamster cells (R1610). The values shown in Fig. 3 compare the relative levels of *gal*K gene expression obtained with various constructions ($P_{SV40}=100\%$). These values do not reflect necessarily the relative transcription efficiencies of these promoters. Each promoter construct utilizes a different "cap" site and contains a different 5′-non-coding leader region. These variations, as well as others, may contribute significantly to the overall levels of expression by affecting the maturation, stability, and/or translation efficiency of the *gal*K transcript.

Surprisingly, the promoter which gave the highest level of *gal*K expression was the Rous Sarcoma Virus (RSV) LTR introduced into CV_1 cells. This construct resulted in a 2-fold higher level of *gal*K expression than did the SV40 early promoter, and some 10 to 15-fold higher level expression than did the mouse β-globin promoter. In contrast, the RSV LTR did not express *gal*K with the same high level efficiency in hamster cells. Instead, the LTR gave somewhat less expression than did the SV40 promoter. The LTR was the only promoter examined which exhibited differential expression in the 2 cell types. The mouse β-globin promoter showed the lowest level of *gal*K expression (Fig. 3). Moreover, even this low level was found to be dependent upon the presence of the SV40 enhancer sequence associated with the *gpt* transcription unit. Removal of this enhancer eliminated all *gal*K expression.

The β-globin-*gal*K fusions have been used to study the enhancer dependence of the β-globin promoter as well as the species specificity of 2 different enhancer elements: the SV40 72 base-pair repeat and the Harvey Sarcoma Virus (HSV) 73 base-pair repeat (Berg *et al.* 1983). The results indicate that both enhancer elements activate *gal*K expression however, the 2 enhancers appear to function with some species specificity. That is, the SV40 enhancer resulted in greater *gal*K expression in monkey cells whereas the HSV enhancer worked somewhat better in mouse cells. Among the promoters examined, the globin promoter was the only one which exhibited complete enhancer dependence. The Adenovirus 2 (Ad-2) late promoter, the Rous LTR, and the mouse metallothionein (MT) promoter, all function in the absence of the SV40 enhancer element. However, expression from the Ad2 and MT promoter constructs was stimulated some 3–5 fold by the presence of the enhancer.

We emphasize that the pSVK-*gpt* system can be used to quantitate precisely the changes in gene expression which result from alterations introduced into a particular promoter region. For example, the effects of small deletions or point mutations can be measured accurately and compared. The same is true for induction ratios of those promoters which can be activated to function (e.g., metal induction of the MT promoter, Fig. 3). In these cases, the same RNA transcript is being generated from each construct and, thus, the relative expression levels directly reflect the steady-state mRNA levels in the cell. Carter and Hamer (unpublished data) have used the pSVK vector system to compare the levels of *gal*K expression resulting from a set of deletion mutations introduced into the mouse MT-promoter region. The effects of these mutations on promoter function and metal responsiveness were determined both by monitoring *gal*K expression and by direct measurement of the steady-state levels of initiated transcript (i.e., S_1 nuclease analysis). The results obtained from these 2 types of analyses were consistent. However, the *gal*K analyses were much faster and easier than the S_1 nuclease analyses.

ALTERATIONS IN THE 5′-NON-CODING LEADER REGION; VARYING LENGTH

In addition to characterizing and comparing promoters, the pSVK-*gpt* vector system can be used to study other gene-regulatory elements. In particular, we have used the system to examine the effects on *gal*K expression which result from changes introduced into the 5′-non-coding leader region of the *gal*K transcription unit. Initial experiments involved varying the length of the 5′ leader. pSVK-*gpt* constructs were made which contained increasingly longer segments of 5′-non-coding sequences preceding the *gal*K initiation codon (Fig. 4). These additional sequences derive from the region of the bacterial genome which

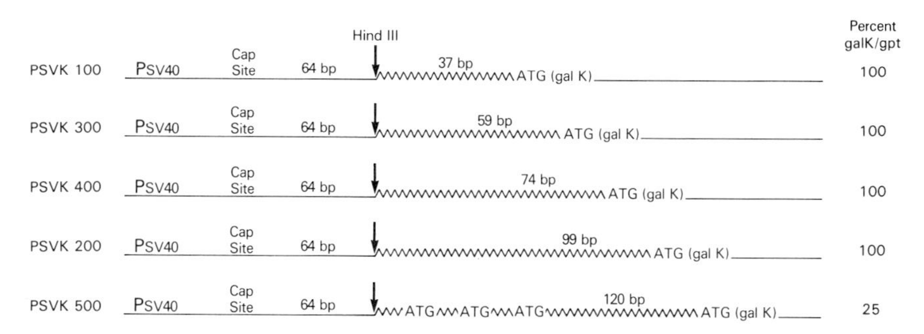

Fig. 4. Schematic representation of the *gal*K transcription unit contained in a set of similarly constructed vectors carrying 5′ non-coding leader regions of different lengths (H. J. *et al.* in preparation). The position of the *gal*K gene is indicated, as is the length of various portions of the leader region and other ATG codons occurring in the leader. See text for details.

normally precedes the *gal*K gene. Constructs were characterized which carry 37 (pSVK100), 59 (pSVK300), 74 (pSVK400), and 99 (pSVK200) base pairs, respectively, of contiguous *gal*K upstream sequence. In each case, these bacterial sequences were fused identically at the HindIII restriction site positioned 64 base pairs downstream from the major transcription start-site of the early SV40 promoter. Thus, the 4 constructions utilize the same promoter, the same cap site, the same initial 64 nucleotides of 5′ leader sequence, the same 37 base pairs immediately preceeding the *gal*K coding region, and the same splice and polyadenylation signals. However, they differ in that the overall length of the 5′ leader sequences vary from 101 nucleotides to 163 nucleotides. Most importantly, the *gal*K initiation codon in these derivatives remains the first AUG sequence occurring downstream from the 5′ mRNA cap site.

We monitored *gal*K expression (relative to *gpt*) from the 4 vectors in both monkey and hamster cells using the transient expression assay. The results (Fig. 4) indicate that the changes made in the length of the 5′ leader have no effect on *gal*K expression in either cell system. Apparently, the *gal*K initiation signal is recognized with the same efficiency when placed at the different distances relative to the 5′ mRNA cap site. One additional vector construction was made in this series (pSVK500, Fig. 4) which contains 120 base pairs of bacterially derived sequence upstream of *gal*K. This segment was again fused at the HindIII site in the SV40 early leader region. Assay of the pSVK500 derivative indicated a significant decrease (approximately 75%) in *gal*K expression. The length of the 5′ leader of pSVK500 is only 21 nucleotides longer than that of pSVK200, however, the additional 21 residues encode 3 separate AUG codons. Our results with the other vectors suggest that it is probably the occurence of the 3 additional AUG codons, and not simply the increased length of the leader, which is responsible for the inhibition. Presumably, one or more of these AUGs is now competing for translational components with the authentic *gal*K initiation codon.*

In the following section, we will describe experiments which utilize the pSVK-gpt system to characterize in detail the differential effects on gene expression resulting from the insertion of specific AUG containing sequences into the 5′-non-coding leader region of the *gal*K transcription unit.

---

*Two codons upstream and in-frame with the *gal*K coding sequence is a chain-terminating codon (TAA). This stop signal prevents any translation initiating upstream of *gal*K from reading through in-frame with the *gal*K gene. Thus, all *gal*K expression observed in any pSVK vector initiates at the authentic *gal*K initiation codon.

INSERTING AUG CONTAINING SEQUENCES

Most eukaryotic mRNA's initiate translation at the first AUG triplet encoded downstream from the 5′ cap site. Based on this observation a scanning model was proposed to explain eukaryotic translation initiation (Kozak 1980). The model suggested that ribosomes first recognize the 5′ end of a mRNA, and then moved to the first AUG codon on the transcript. It was implied that sequences surrounding the initiator AUG would have no role in initiation. Rather, the only feature of import was the position of the appropriate AUG codon 5′ to all other AUG triplets on the transcript. Comparison of additional eukaryotic mRNAs indicated a number of exceptions to the first AUG rule (Kozak 1981, Mulligan & Berg 1981). In order to account for these exceptions, the scanning model was modified (Kozak 1981). It was suggested that certain sequences were preferred around those AUGs that served as functional initiators. The sequence PuXXAUGG emerged as the "favored" sequence for initiation in eukaryotic systems. This sequence was not found among those non-functional AUGs which occurred in various 5′ leader regions. The new model presumes that the ribosome initiates efficiently at the first AUG if its flanking sequence is optimal. The first AUG, however, is ignored if it is flanked by unfavorable sequences, thereby, allowing utilization of downstream positioned AUGs. Some experimental evidence has now accrued which further supports these contentions (Lomedico & McAndrew 1982, H. J. & M. R. in preparation, Peabody & Berg unpublished).

The pSVK-*gpt* vector system is well-suited for examining certain predictions made by the modified-scanning hypothesis. The system allows quantitation of the effects on gene expression which result from introducing AUG codons flanked by different sequences into the 5′-non-coding leader region of a transcription unit. We initiated our studies by obtaining a 16 base-pair synthetic linker (G. Brown, Genetics Institute) which carries a single ATG sequence on each of its DNA strands (Fig. 5). This linker was inserted in both orientations at the unique HindIII restriction site positioned 64 base pairs into the 5′ leader region of the *gal*K transcription unit on the vectors pSVK100, 200, and 300 (Figs. 5, 6). Irrespective of its orientation, this linker adds a single AUG codon into the 5′ leader sequence upstream of *gal*K. In one orientation the inserted AUG is flanked by the favored sequence for eukaryotic translation initiation whereas, in the opposite orientation the AUG is surrounded by unfavorable sequences. In particular, the favored sequence contains the requisite purine 3 base pairs 5′ to the AUG (position-3) whereas, the unfavored sequence contains

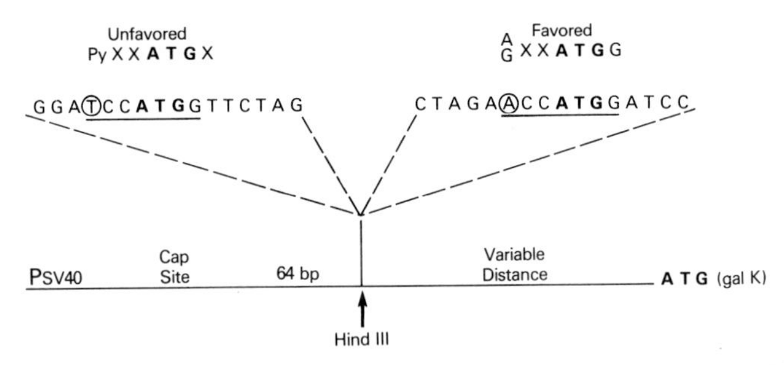

Fig. 5. Schematic representation depicting the insertion of a 16 base-pair synthetic linker (G. Brown, Genetics Inst.) into the unique HindIII site positioned in the leader region of the pSVK100, 200 and 300 vectors shown in Fig. 4 (H. J. *et al.* in preparation). The linker is shown inserting in both possible orientations and ATG codons are noted (bold). Also shown is the consensus "favored" and "unfavored" sequence for eukaryotic translation-initiation sites (Kozak 1981). Py indicates a pyridine residue; x indicates any nucleotide. The circled residue indicates the residue in the linker at the important −3 position (i.e., 3 base pairs upstream of the ATG).

a thymidine residue at this position (Fig. 5). Other residues immediately surrounding the 2 different AUG sequences (e.g., positons −1, −2 and +1) are identical.

Six vector derivatives, 3 containing the favored AUG (pSVK150, 250, and 350; Fig. 6) and 3 containing the unfavored AUG sequence (pSVK160, 260, 360;

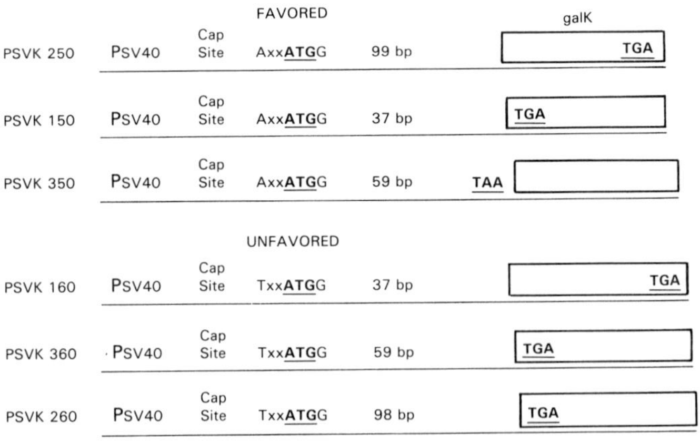

FAVORED
galK
PSVK 250 PSV40 Cap Site AxxATGG 99 bp TGA
PSVK 150 PSV40 Cap Site AxxATGG 37 bp TGA
PSVK 350 PSV40 Cap Site AxxATGG 59 bp TAA

UNFAVORED
PSVK 160 PSV40 Cap Site TxxATGG 37 bp TGA
PSVK 360 PSV40 Cap Site TxxATGG 59 bp TGA
PSVK 260 PSV40 Cap Site TxxATGG 98 bp TGA

Fig. 6. Schematic representation of the 6 vector derivatives which result from the insertion of the synthetic linker, as shown in Fig. 5. The name of each vector, and the position and orientation of the synthetic linker is indicated. Also shown, is the position that translation initiating from the inserted ATG will terminate relative to the *gal*K-coding sequence. All other designations are as in Figs. 3, 4, 5. See text for details.

Fig. 6) were obtained and characterized. Initially, all 6 constructions were assayed for transient expression in CV_1 monkey cells. The results (Fig. 7) indicate that the 2 different AUG sequences had widely different effects on *gal*K

	Position of Stop Codon in Relation to ATG of galK	Percent galK/gpt.	
		CV_1	R1610
PSVK100		100	100
PSVK200		100	100
PSVK300		100	100
PSVK150	Within	~10	~20
PSVK250	Beyond	~10	~5
PSVK350	Before	~10	~20
PSVK160	Beyond	~50	~65
PSVK260	Within	~50	~65
PSVK360	Within	~50	~65

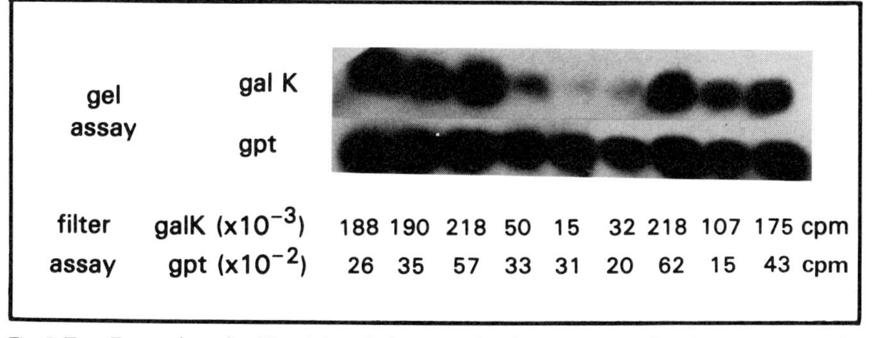

Fig. 7. Top: Expression of *gal*K activity relative to *gpt* for the vectors described in Figs. 4, 5, 6 and in the text. Values are given for transfections carried out both in CV_1 monkey cells and R1610 hamster cells (*gal*K$^-$, *gpt*$^-$). Assays were performed as described in Fig. 3. The 9 lanes shown at the bottom depict one set of representative assays (from *L* to *R*) of the vectors shown above (from top to bottom) in the R1610 hamster cells. The gel assay is a fluorograph of the *E. coli gal*K and *gpt* activities. The filter assay given cpm of C^{14}-galactose and C^{14}-guanine converted to the appropriate phosphorylated derivatives (i.e., a direct measure of the bacterial *gal*K and *gpt* activities).

expression. The 3 constructs carrying the favored AUG expressed *gal*K 10-fold less efficiently than did the parent vectors. In contrast, the 3 vectors carrying the unfavored AUG showed only a 2-fold lower level of *gal*K expression. Assuming that the observed interference with *gal*K expression correlates with the relative ability of the 2 different AUG sequences to be recognized by the translational machinery, then the favored sequence is recognized at least 5 times better than is the unfavored sequence. Apparently, the sequence flanking the AUG codon and in particular, the residue in the −3 position, play a major role in determining the efficiency with which a particular AUG containing sequence is recognized as an initiation codon.

The same experiment was also carried out using hamster cells and qualitatively similar results were obtained. Again, the favored AUG sequence reduced expression of *gal*K far more than did the unfavored AUG-containing sequence (Fig. 7). One interesting difference, however, was noted among the 3 vectors carrying the favored AUG sequence. The pSVK250 derivative reproducibly exhibited some 15% lower *gal*K expression than did the other 2 vectors, pSVK150 and 350. These 3 vectors differ only in that the inserted linker occurs in different positions relative to the *gal*K initiation codon. Most importantly, this places the added AUG codon in the 3 vectors into each of the 3 possible reading frames occurring upstream of the *gal*K gene. As shown in Fig. 6, translation occurring from the various inserted AUG codons terminates before (pSVK350), within (pSVK150), or downstream (pSVK250) of the *gal*K initiation codon. Our results suggest that the *gal*K AUG is used least efficiently when translation initiated upstream traverses out-of-frame well beyond this codon (i.e., terminates 98 codons into the *gal*K gene). In contrast, stopping translation before (as in pSVK350), or even within (as in pSVK150) the *gal*K initiation codon results in increased utilization of this AUG. We did not observe any such reading-frame effect among those vectors which carry the unfavored AUG sequence (pSVK160, 260, 360; Figs. 6, 7). In these vectors, the inserted AUG codon is poorly recognized and, thus, the reading frame is probably unimportant.

FRAME-SHIFT MUTATIONS

The results presented above suggest that translation initiating upstream of a gene can differentially affect its expression depending upon where this translation terminates. To study this further, we utilized the pSVK150 vector, which carries the favored AUG sequence positioned such that its upstream translation reading frame terminates within the *gal*K AUG (Fig. 7). pSVK150 DNA was cleaved at a site between the synthetic linker insert and the *gal*K AUG, then subjected to partial exonucleolytic digestion. Deletion mutations were obtained and characterized (Fig. 8). In particular, 2 derivatives were identified which carried one and 2 nucleotide deletions respectively (pSVK151 and 152, Fig. 8). Most importantly, these deleted residues frame-shift the upstream translation into the 2 other possible reading frames. In pSVK151 translation stops well beyond the *gal*K AUG, whereas in pSVK152 translation stops before *gal*K. Unlike the earlier situation with pSVK250 and 350 (Fig. 6), these vectors are all identical except for the single- and double-nucleotide deletions.

We monitored *gal*K and *gpt* expression of the 3 constructs transiently in the hamster cell line. The results (Fig. 8) confirm those obtained earlier. Apparently, when translation is stopped at or before the *gal*K AUG, this initiation codon is utilized more efficiently (approximately 15%) than when translation is allowed to read-through the *gal*K AUG into the gene. One possible explanation for this effect is that ribosomes terminating upstream, or within the *gal*K AUG, can re-initiate at this AUG, thereby, increasing *gal*K expression. Alternatively, ribosomes may initiate independently at the *gal*K AUG and this interaction is inhibited by the read-through translation. Whatever the mechanism, we emphasize that this effect is secondary in importance to the role played by the sequences surrounding a particular AUG. These sequences help determinine the overall efficiency of translation initiation at the site and in turn, determine its ability to affect the utilization of other initiation codons positioned downstream on the message.

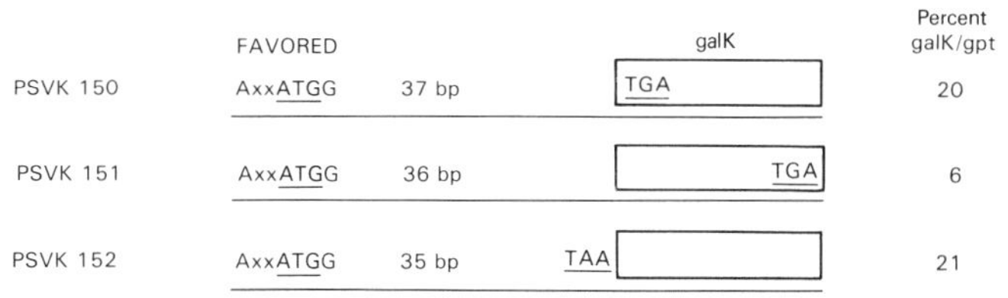

	FAVORED		galK	Percent galK/gpt
PSVK 150	AxxATGG	37 bp	TGA	20
PSVK 151	AxxATGG	36 bp	TGA	6
PSVK 152	AxxATGG	35 bp	TAA	21

Fig. 8. Schematic representation of the *gal*K transcription unit in pSVK150 (as in Fig. 6) and 2 pSVK100 derivatives obtained by deleting one (pSVK151), and 2 (pSVK152) base pairs respectively from the leader region between the synthetic linker insert and the *gal*K gene. pSVK150 DNA was cut at the unique EcoRV restriction site, treated with T4 DNA polymerase, and religated (H. J. & M. R., in preparation). Transformants were obtained and characterized by restriction mapping and DNA sequencing. The importance of the 2 frame-shift mutants are described in detail in the text. The *gal*K/*gpt* values were determined in the R1610 hamster cell line and are given as percent relative to the value obtained for pSVK100.

ACKNOWLEDGEMENT

We thank Linda Hampton for typing and editing the manuscript.

REFERENCES

Berg, P. E., Yu, J-K., Popovic, Z., Schumperli, D., Johansen, H., Rosenberg, M. & Anderson, W. F. (1983) *Mol. & Cell. Biol.* (in press).

Colbere-Garapin, F., Horodniceanu, F., Kourilsky, P. & Garapin, A.-C. (1981) *J. Mol. Biol.* 150, 1–14.

Gorman, C. M., Moffat, L. F. & Howard, B. H. (1982) *Mol. & Cell. Bio. 2,* 1044–1051.

Kozak, M. (1981) *Nucleic Acids Res. 9,* 5233–5252.

Kozak, M. (1980) *Cell 22,* 7–8.

Lomedico, P. T. & McAndrew, S. J. (1982) *Nature 299,* 221–225.

Mulligan, R. C. & Berg, P. (1981) *Mol. and Cell. Biol. 1,* 449–459.

Mulligan, R. C. & Berg, P. (1980) *Science 209,* 1422–1427.

Rosenberg, M., McKenney, K. & Schumperli, D. (1982) In: *Promoters: Structure and Function,* pp. 387–406. Praeger Publishers, New York.

Schumperli, D., Howard, B. H. & Rosenberg, M. (1982) *Proc. Natl. Acad. Sci. USA 79,* 257–261.

Thirion, J.-P., Banville, D. & Noel, H. (1976) *Genetics 83,* 137–147.

Wigler, M., Pellicer, A., Silverstein, S., Axel, R., Vrlaub, G. & Chasin, L. (1979) *Proc. Natl. Acad. Sci. USA 76,* 1373–1376.

DISCUSSION

UHLENBECK: The natural gene, the *gal*K gene, what is the sequence surrounding its ATG?

ROSENBERG: It is the favoured sequence, AXXAUGG. Prokaryotic genes, in general, will have favoured sequences. These same positions are also conserved positions in prokaryotic genes. In retrospect, it was not so surprising that prokaryotic genes have expressed well in eukaryotic systems.

SCHIMMEL: As a point of information, what is the maximum distance ribosomes can travel from the cap site and still properly initiate at this Kozak sequence, that is to say, what is the longest known distance that ribosomes can travel, if not in this system, then in some other system?

ROSENBERG: Some of the distances are pretty long. There are examples where the distances are even greater than 500 nucleotides. For example, certain viral RNAs. However, these distances are measured in nucleotides of primary sequence and may not be relevant in context of a 3-dimensional RNA structure.

SAFER: The correlations you make are made on the basis of *gal*K activity. Is there an internal consistency? Does the amount of message agree with your results, or could these results be from inactive protein, because of fusion of these extra amino acids?

ROSENBERG: There is no fusion.

SAFER: Is there no fusion from the upstream AUG?

ROSENBERG: No, there is a UAA codon 3 base pairs before the *gal*K AUG which would terminate any potential protein fusions. The *gal*K assay is done *in situ* in the starch gel. Thus, if we had fusion protein, we would see the enzyme shift position. Clearly, there are no shifts.

GRUNBERG-MANAGO: Does the position of the translational termination codon upstream of the AUG have an effect. Should it be very close to the AUG?

ROSENBERG: We have no data on that. The termination codon we have in front of *gal*K is very close. To my knowledge, no one has systematically moved the termination codon farther and farther upstream, and then examined quantitatively the relative effects of those changes.

CASKEY: I would like to make a comment to Larry Gold's presentation and also these data, with regard to the peptide chain termination reaction. There are earlier *in vitro* data suggesting that at chain termination there is dissociation of the messenger RNA complex with the ribosome. I have never subscribed to that being authentic *in vivo*, as earlier genetic data, that had been developed with the RII system locus, gave different results. As I recall, Sidney Brenner reported obliteration of polarity if a re-initiation event occurred adjacent to the chain termination codon. He found the genetic distances for re-initiation to be similar to left and right, suggesting that the ribosome, once it recognized a termination signal, wandered and did not have a vectorial direction for re-initiation. I think that the data suggest it is much more likely that you have a re-initiation signal downstream in your terminator codon. The ribosome wanders, once it recognizes its termination signal.

ROSENBERG: That, of course, is a prokaryotic system.

CASKEY: I think that the 2 systems are similar. I think what is being observed here, and what was observed in the obliteration of polarity at the RII locus, really is the same. The second point is that these data, I think, say that re-initiation is a different mechanism than the primary initiation.

HERSHEY: I am wondering how you interpret the position of the termination signal affecting the expression of the *gal*K gene, given that we expect that there is no re-initiation in the eukaryotic cell. The overwhelming evidence in the whole eukaryotic system is for monocistronic messages.

ROSENBERG: Of course, but remember that the restarts observed here are only secondary effects. Keep things in perspective. The major effect we have demonstrated is the 95% down effect seen with the construction containing the favoured AUG which translates into the *gal*K gene. Clearly, it is the sequences surrounding the first AUG, which have the major effect. This restart phenomenon is only a minor effect which is seen because of the sensitivity of the system.

PESTKA: How do you interpret the molecular events involved in the fact that the expression is good in the case where the unfavoured sequence is present and poor when the strong AUG is there? What is happening to the ribosome, with message degradation, etc.?

ROSENBERG: It is a little hard to interpret. I think the basic data are consistent with the Kozak-type scanning model. I think it says that when you have an unfavoured ATG, the ribosome (seeing the message from the 5′-end towards the 3′-end) simply skips that ATG and goes on to the next one. Several years ago, Bruce Paterson and I did some experiments where we enzymatically capped certain prokaryotic messages. For one of those messages, we got a result we did not understand at the time. The capped transcript produced 2 major proteins. One was the full-length gene product, whereas the second initiated at an internal ATG codon that was in frame with the first ATG. The 2 products were made in a ratio of about 60:40. If we look at the sequence surrounding the first ATG, it has an unfavoured sequence, TXXAUGG, however, not the most unfavoured. From our present data, we would now predict that this upstream site should not be efficiently recognized. The ribosome at some frequency should move to the next ATG and sure enough, the smaller product was observed. The smaller protein initiating from this internal ATG is never made in *E. coli.* Clearly, this scanning ability is unique to the eukaryotic translation system and does not occur in *E. coli.*

VII. Maturation of Secreted Proteins, Signal Hypothesis, Posttranslational Modification

Expression of Transforming Gene Products

R. L. Erikson, E. Erikson, Y. Sugimoto, J. Spivack & J. Maller

INTRODUCTION

Malignant transformation by Rous sarcoma virus (RSV) is caused by the expression of its transforming gene product, $pp60^{v-src}$ (Erikson *et al.* 1980). To date, the sole function assigned to $pp60^{v-src}$ that could account for its capacity to alter normal cells is a protein kinase activity specific for tyrosine residues. One limitation in the additional characterization of transforming gene products is the availability of large quantities of highly purified protein. Recently, we have succeeded in developing an ion-exchange protocol that has many of the features we desire (Sugimoto *et al.* in preparation). Some characteristics of $pp60^{v-src}$ purified by this protocol will be described here.

Our experience with $pp60^{v-src}$ indicates that it specifically and exclusively phosphorylates tyrosine residues in protein substrates. However, in transformed cells proteins containing phosphoserine residues also show quantitative increases or decreases in their level of phosphorylation. For example, the phosphorylation of the ribosomal protein S6 appears to be indirectly regulated by $pp60^{v-src}$.

SPECIFICS

Characterization of $pp60^{v-src}$ protein kinase activity

Previous work from this laboratory (Graziani *et al.* 1983) has shown that $pp60^{v-src}$ partially purified by immuno-affinity chromatography is able to utilize

---

Departments of Pathology and Pharmacology, University of Colorado, School of Medicine, Denver, USA.

GENE EXPRESSION, Alfred Benzon Symposium 19.
Editors: Brian F. C. Clark & Hans Uffe Petersen, Munksgaard, Copenhagen 1984.

GTP in addition to ATP as substrate for the phosphorylation of casein. In addition, upon incubation in the protein kinase reaction mixture pp60^{v-src} itself becomes phosphorylated; in the presence of GTP radiolabel is incorporated mostly into the carboxy region of the molecule, whereas in the presence of ATP a greater amount of radioactivity is incorporated into the amino region (Graziani *et al.* 1983). When pp60^{v-src} purified by ion-exchange chromatography was examined in the same manner, we found similar results, as shown in Fig. 1A. Both [γ-^{32}P] ATP and[γ-^{32}P]GTP served as phosphate donor for the phosphorylation of casein. In both cases, the phosphorylation of casein was inhibited by pre-incubation of the enzyme with TBR IgG, and phosphoamino acid analysis revealed radiolabel exclusively in phosphotyrosine residues. In addition, it can be seen that upon incubation of pp60^{v-src} under these conditions radiolabel was incorporated into a protein that migrated at the position of 60,000 daltons.

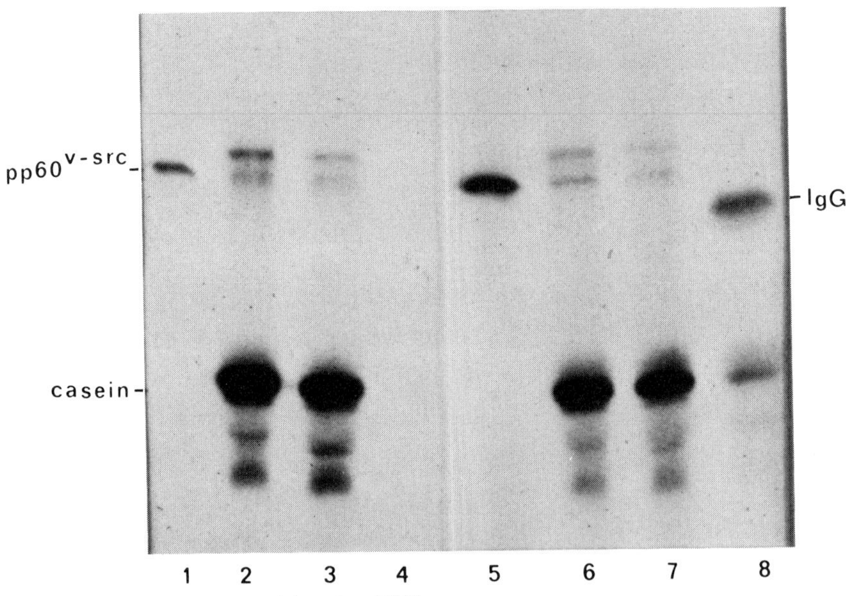

Fig. 1. Phosphotransferase activity of pp60^{v-src}.

Panel A: Samples of pp60^{v-src} were incubated in the presence of [γ-^{32}P]ATP (tracks 1–4) or [γ-^{32}P]GTP (tracks 5–8), and the products of the reaction were resolved by polyacrylamide gel electrophoresis. In addition, prior to the addition of substrates the enzyme was incubated for 15 min at 0°C with non-immune or TBR IgG. Tracks 1 & 5, pp60^{v-src}; tracks 2 & 6, pp60^{v-src} plus casein; tracks 3 & 7, pp60^{v-src} pre-incubated with non-immune IgG, plus casein; tracks 4 & 8, pp60^{v-src} pre-incubated with TBR IgG, plus casein.

Digestion of autophosphorylated pp60^{v-src} with *S. aureus* V8 protease revealed that radiolabel had been incorporated into both the amino and carboxy regions of the molecule in the presence of ATP, and moreover, the amount of radiolabel was much greater in the amino than in the carboxy region (Fig. 1B). In contrast, in the presence of GTP, radiolabel was incorporated mostly into the carboxy region. All of these results are qualitatively similar to those described previously for pp60^{v-src} partially purified by other means (Graziani *et al.* 1983). As the pp60^{v-src} used here is much more highly purified, our data support the previous contention that the results observed are indeed intrinsic properties of the pp60^{v-src} protein.

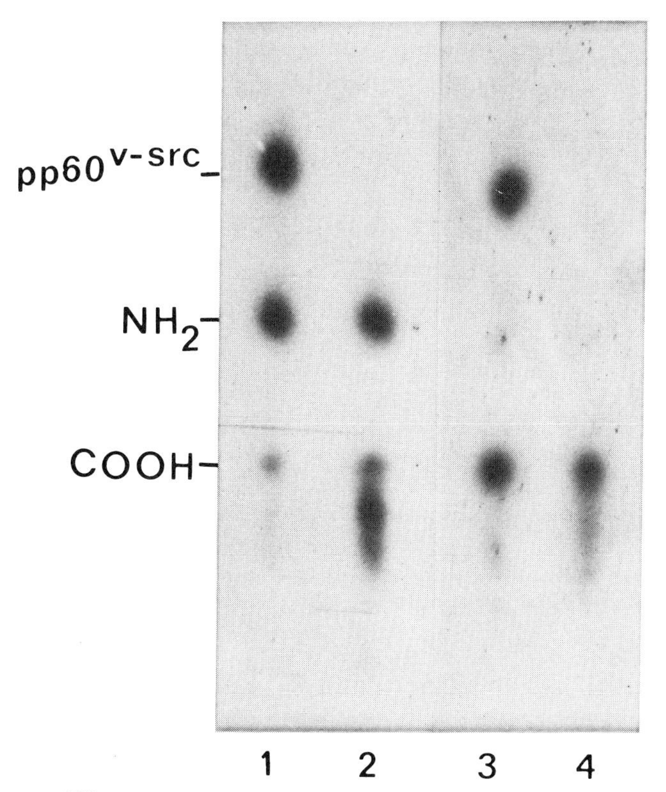

Panel B: pp60^{v-src} that had been incubated in the presence of [γ-^{32}P]ATP or [γ-^{32}P]GTP as shown in panel A, tracks 1 & 5 were subjected to partial proteolytic cleavage during re-electrophoresis. Tracks 1 & 2, ATP-phosphorylated pp60^{v-src}; tracks 3 & 4, GTP-phosphorylated pp60^{v-src}; tracks 1 & 3, digestion with 5 ng *S. aureus* V8 protease; tracks 2 & 4, with 50 ng V8 protease. NH_2, amino terminal region of pp60^{v-src}; COOH, carboxy terminal region of pp60^{v-src}.

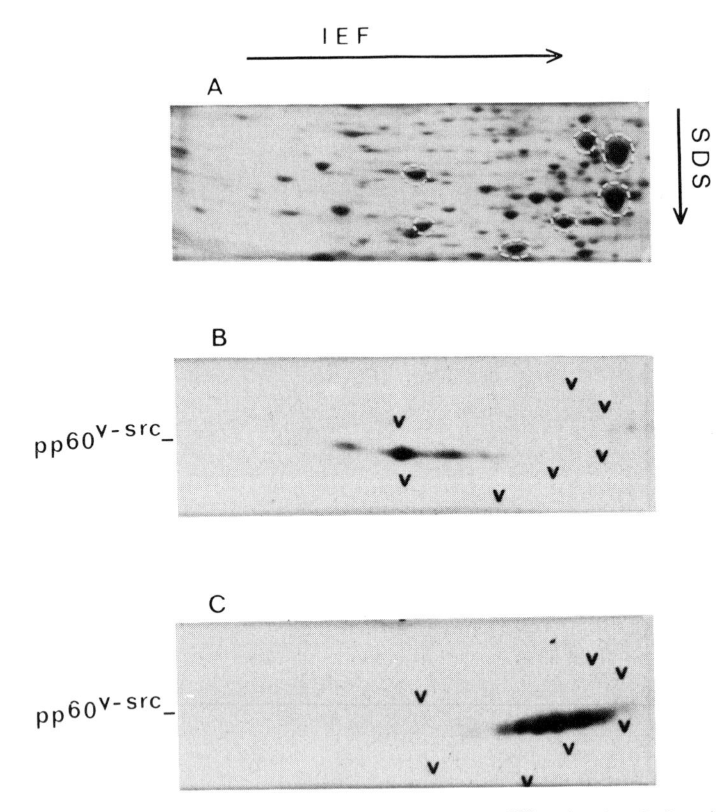

Fig. 2. Separation of the various phosphorylated species of pp60^{v-src} by isoelectric focusing.

pp60^{v-src} was incubated in the presence of 20 μM [γ-^{32}P]ATP or [γ-^{32}P]GTP as shown in Fig. 1A, tracks 1 & 5. The reaction was stopped by the addition of EDTA, SDS and 2-mercaptoethanol (10 mM, 2%, and 5%, final concentrations, respectively) and by treatment at 95° for 2 min. The samples were cooled and then 2 volumes of dilution buffer (0.8% NP-40/9.5 M urea /5% 2-mercaptoethanol) were added. In addition, both unlabelled and [^{35}S]methionine-labelled RSV-transformed cells were lysed by treatment at 95° for 2 min in 2% SDS/5 mM Tris-HCl, pH 8.0/10 mM EDTA/5% 2-mercaptoethanol. After the lysates were cooled, 2 volumes of dilution buffer were added and the lysates clarified by centrifugation at 100,000 g for 30 min. A sample of the unlabelled lysate was added to the pp60^{v-src} samples in order to provide a spectrum of stainable proteins for reference. The proteins were fractionated by isoelectric focusing (1.6% ampholines, pH 5–8; 0.4% ampholines, pH 3–10) in the first dimension (left to right) and by SDS-polyacrylamide gel electrophoresis in the second dimension (top to bottom). The gels were stained with 0.2% Coomassie blue in 50% trichloroacetic acid and destained in 10% acetic acid, 5% methanol. Radiolabelled proteins were visualized by autoradiography. Only the pertinent regions of the gels are shown.

Panel A: Autoradiogram of a lysate of [^{35}S]methionine-labelled cells. Seven prominent proteins, which are used as reference points, are circled. The positions of these proteins are indicated by vs in subsequent panels.

Panel B: [γ-^{32}P]GTP-phosphorylated pp60^{v-src}.

Panel C: [γ-^{32}P]ATP-phosphorylated pp60^{v-src}.

Two-dimensional gel analysis of pp60$^{v\text{-}src}$

From the amount of pp60$^{v\text{-}src}$ protein and the radioactivity incorporated, Graziani *et al.* (1983) estimated that 5–6 tyrosine sites are phosphorylated with ATP and 1–1.5 with GTP. In order to evaluate by another means the number of phosphates incorporated, we subjected autophosphorylated pp60$^{v\text{-}src}$ to isoelectric focusing. As shown in Fig. 2, pp60$^{v\text{-}src}$ that had been phosphorylated in the presence of GTP yielded 1 major and 2 minor species. The majority of radiolabel was located in line with 2 of the "reference proteins." In contrast, a similar preparation that had been phosphorylated in the presence of ATP exhibited a much greater negative charge, focusing in 4 or 5 major spots. These data are consistent with the idea that incubation of pp60$^{v\text{-}src}$ in the presence of ATP results in the generation of very highly phosphorylated molecules.

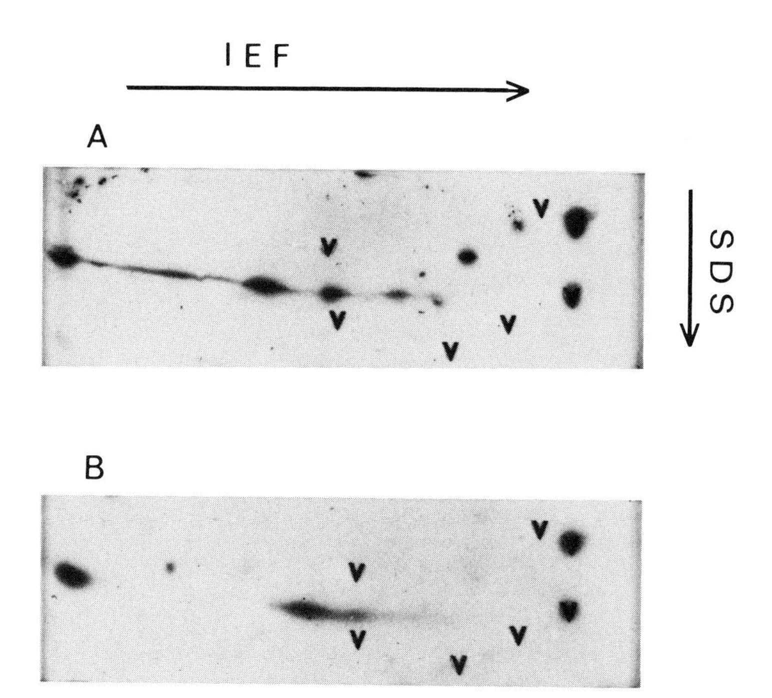

Fig. 3. Species of pp60$^{v\text{-}src}$ present in cultured cells.

Panel A. A lysate of RSV-transformed cells was prepared and fractionated as described in the legend to Fig. 2. The proteins were electrotransferred to nitrocellulose, and pp60$^{v\text{-}src}$ was localized by reaction with anti-p60 serum and radio-iodinated protein A.

Panel B. A sample of purified pp60$^{v\text{-}src}$, approximately 17 ng, was fractionated in the same manner. In each case a small amount of lysate of [^{35}S]methonine-labelled cells was added in order to establish the positions of the reference proteins. As in Fig. 2, these are indicated by vs.

pp60$^{v\text{-}src}$ isolated from the cell contains a phosphorylated serine residue(s) on the amino terminus and a phosphorylated tyrosine residue on the carboxy terminus (Collett *et al.* 1979, Hunter & Sefton 1980). The multiple sites of tyrosine phosphorylation that occur on the amino terminus *in vitro* have not been reported in pp60$^{v\text{-}src}$ isolated from cells that had been biosynthetically radiolabelled in culture. In an effort to ascertain the number of phosphorylated species of pp60$^{v\text{-}src}$ present in cultured cells and the number of species present in purified pp60$^{v\text{-}src}$, we subjected a cell lysate and purified pp60$^{v\text{-}src}$ to isoelectric focusing and then visualized pp60$^{v\text{-}src}$ by "Western blotting." Fig. 3 illustrates that by this procedure 3 major species of pp60$^{v\text{-}src}$ were detectable in the cellular lysate. In contrast, purified pp60$^{v\text{-}src}$ yielded 1 major species; this species migrated to the same position as the least acidic of the pp60^{src} species from the cellular lysate, as determined by their position with respect to the "reference proteins." Therefore, the patterns obtained from autophosphorylated pp60$^{v\text{-}src}$ are consistent with the incorporation of phosphate into one site on the molecule in the case of GTP and into several sites in the case of ATP. The highly phosphorylated species of pp60$^{v\text{-}src}$, as are generated *in vitro* in the presence of ATP, are not detectable in the cell.

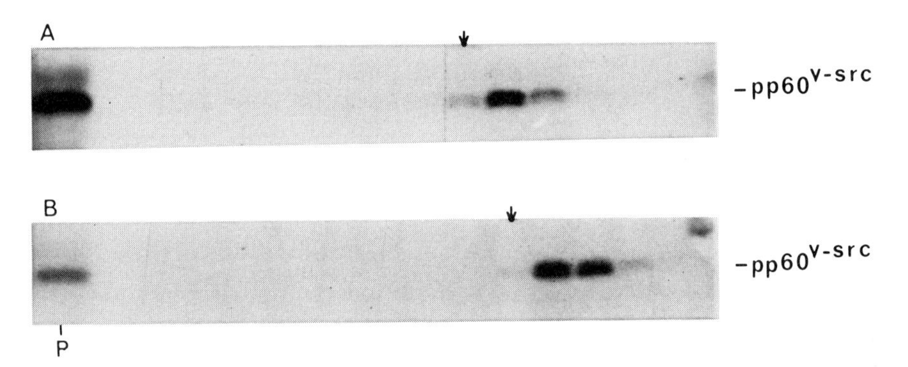

Fig. 4. Sedimentation analysis of pp60$^{v\text{-}src}$.

Samples of pp60$^{v\text{-}src}$ were adjusted to the conditions indicated and sedimented at 49,000 rpm for 17 h through 35–10% glycerol gradients that were also adjusted to the indicated conditions.

Approximately 30 fractions were collected from each gradient, and samples from alternate fractions were assayed for the autophosphorylation of pp60$^{v\text{-}src}$.

Sedimentation was from right to left. The arrow marks the position of hemoglobin sedimented under the same conditions. P indicates the enzymatic activity obtained from the pellet at the bottom of the tube.

A) Buffer contained 0.05% NP40; 20 mM KP_i, pH 7.2.

B) Buffer contained 0.05% NP40; 20 mM KP_i, pH 7.2; 200 mM KCl.

Sedimentation analysis of purified pp60^{v-src}

Preliminary results showed that purified pp60^{v-src} was subject to rapid aggregation. Glycerol gradient centrifugation under various conditions showed that both non-ionic detergent and moderate changes in ionic strength influenced the extent of aggregation. As shown in Fig. 4, 0.05% NP40 and 200 mM KCl provided conditions under which pp60^{v-src} sedimented as a monomer, whereas, lower salt or detergent resulted in aggregation. However, under these conditions the phosphotransferase activity was inhibited by approximately 50% both for autophosphorylation and casein phosphorylation. With these data available additional studies of the interactions of pp60^{v-src} are planned.

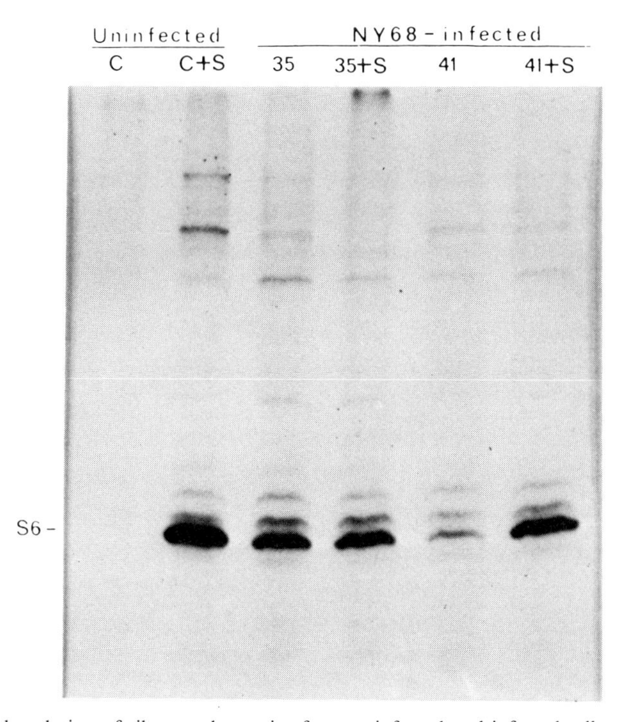

Fig. 5. Phosphorylation of ribosomal proteins from uninfected and infected cells.

Cells were labelled with inorganic $^{32}PO_4^{3-}$, ribosomes were prepared and the ribosomal proteins resolved by polyacrylamide gel electrophoresis. Uninfected cells, serum-starved, C, or after serum stimulation, C+S; NY68-infected chicken embryo fibroblasts grown at 35° in the absence of serum, 35, or after serum stimulation, 35+S; grown at 41° in the absence of serum, 41, or after serum stimulation, 41+S. The position of the ribosomal protein S6 is indicated.

$pp60^{v\text{-}src}$ influences phosphorylation of the ribosomal protein S6

As mentioned above, $pp60^{v\text{-}src}$ phosphorylates specifically tyrosine residues, yet in RSV-transformed cells the phosphorylation of serine residues in a number of proteins is altered. One such example is provided by the work of Decker (1981) which showed that the phosphorylation of the ribosomal protein S6 appears to be controlled by the expression of $pp60^{v\text{-}src}$. As shown in Fig. 5, S6 is dephosphorylated in uninfected cells in the absence of serum, whereas under these conditions it becomes phosphorylated when a functional $pp60^{v\text{-}src}$ is expressed.

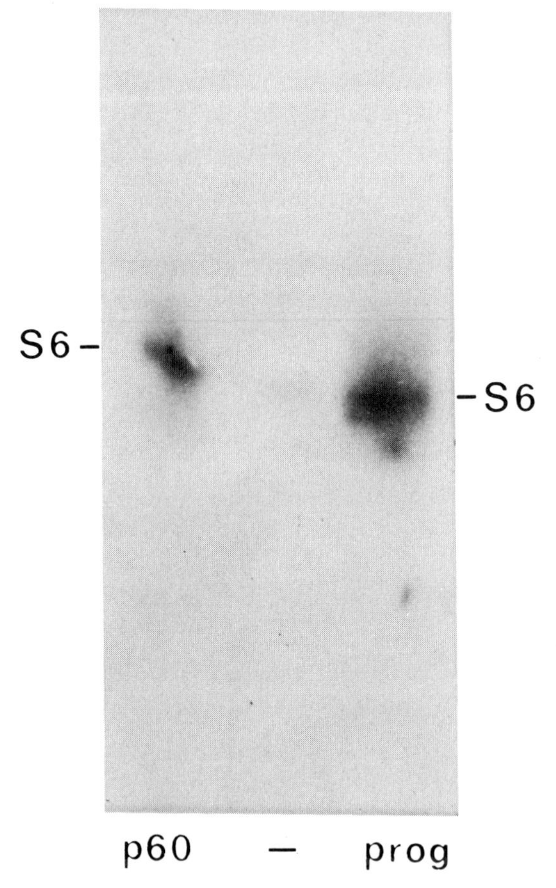

Fig. 6. Autoradiogram of phosphorylated ribosomes from *Xenopus* oocytes.

Oocytes were labelled with $^{32}PO_4^{3-}$ and injected with purified $pp60^{v\text{-}src}$, p60; heated $pp60^{v\text{-}src}$, —; or progesterone, prog. Ribosomes were prepared and the ribosomal proteins resolved by polyacrylamide gel electrophoresis. The position of the ribosomal protein S6 is indicated.

In order to further assess the significance of this phenomenon, $pp60^{v-src}$ was injected into *Xenopus* oocytes. As shown in Fig. 6, 1–2 h after injection of $pp60^{v-src}$, S6 became phosphorylated, whereas, when $pp60^{v-src}$ had been heat inactivated prior to injection no phosphorylation was observed. The phosphorylation was exclusively on serine residues (data not shown).

These data suggest that $pp60^{v-src}$ may activate a protein kinase specific for serine or, alternatively, inactivate a phosphoprotein phosphatase specific for S6 dephosphorylation. Experiments are in progress to test these possibilities.

CONCLUSION

Among the numerous viral transforming gene products that have been identified, the Rous sarcoma virus *src* gene product is, perhaps, the best characterized. That it has the capacity to phosphorylate proteins on tyrosine residues both *in vitro* and *in vivo* seems clear. Whether or not it has other functions is uncertain. Most searches for substrates have utilized techniques for the direct visualization of a newly phosphorylated protein in transformed cells. If one assumes that the transformation-specific phosphorylation observed is physiologically significant, the identification and functional characterization of these substrates is of considerable importance. To date, none of the phosphorylations observed have been shown to be essential to the transformation process. The results previously reported by Decker (1981) and those described here suggest that the assay of a functional activity, such as a kinase that could phosphorylate S6, should be developed. It is of considerable importance at this time to develop new insights concerning the pathways that lead to malignancy and this will necessarily be accomplished by working with the transforming proteins and their substrates.

ACKNOWLEDGEMENTS

This work was supported by grants CA-15823 and CA-21117 from the National Institutes of Health, by an Award from the American Business Cancer Research Foundation and, indirectly, by a gift from R. J. Reynolds Industries, Inc. to the Department of Pathology.

REFERENCES

Collett, M. S., Erikson, E. & Erikson, R. L. (1979) *J. Virol. 29,* 770–781.

Decker, S. (1981) *Proc. Natl. Acad. Sci. USA 78,* 4112–4115.

Erikson, R. L., Purchio, A. F., Erikson, E., Collett, M. S. & Brugge, J. S. (1980) *J. Cell Biol. 87,* 319–325.

Graziani, Y., Erikson, E. & Erikson, R. L. (1983) *J. Biol. Chem. 258,* 6344–6351.

Hunter, T. & Sefton, B. M. (1980) *Proc. Natl. Acad. Sci. USA 77,* 1311–1314.

DISCUSSION

ERDMANN: I was wondering about the phosphorylation on the S6, is there just one phosphate added?

ERIKSON: There are multiple phosphorylations that occur on S6 based on 2-dimensionel gels. George Thomas and co-workers in Basel have shown that when you add serum to these cells a series of phosphorylated S6 appear with perhaps as many as 6 or 7 phosphates in a maximally phosphorylated form.

ERDMANN: If you run a 2-dimensional gel you will see a similar pattern.

ERIKSON: In our 2-dimensional gels I don't believe we see the most maximally phosphorylated forms; but we do see multiple phosphorylations.

KURLAND: This is more directed to Brian Safer, but stimulated by this: is the phosphorylation of the eukaryotic initiation factor 2 a GTP- or an ATP-stimulated phosphorylation?

SAFER: ATP.

SCHIMMEL: Is it known if the phosphorylation of S6 is cyclic AMP-dependent?

ERIKSON: That is an important question. I indicated that the phosphorylation of S6 and understanding of its control regulation is in a very primitive state. One can stimulate, as I understand it, the phosphorylation of S6 with cyclic nucleotides, but I think the general consensus is that that phosphorylation is not carried out by the conventional cyclic AMP-dependent protein kinase. It is probably a different kinase, but that is not to say that the cyclic AMP-dependent protein kinase could not be involved in that pathway.

OFENGAND: Has there been any functional effect ever seen for the S6 phosphorylation?

ERIKSON: No, as I tried to indicate briefly in my closing remark, all of the S6 work, as far as I know, is based on correlation, i.e., it is tightly coupled with the onset of protein synthesis, but no one to my knowledge has gone to the effort of making phosphorylated ribosomes, non-phosphorylated ribosomes, and attempting to do a reconstruction, or if they have, they have not reported it.

KERR: I think the attempts have been made, but failed. Part of the problem is controlling phosphatase.

PESTKA: Do you have any information on the structural differences between the fast and slowly moving monomers?

ERIKSON: We presume that, this is true for molecules other than P60, a more highly phosphorylated form will move more slowly in an SDS gel, and that is what we are seeing. Our calculations would indicate that P60 is able to fill up almost every available tyrosine on the amino terminus in these reactions in 20 micromolar ATP. That amounts to at least 7 or so phosphorylations on tyrosine residues, and that results in a configurational change or else differential binding of SDS in a more slowly moving form on a gel.

KAZIRO: Can P60 phosphorylate protein kinase on tyrosine?

ERIKSON: This protein kinase and this phosphatase are hypothetical phenomena. They are a proposed intermediate. There has been some suggestion that the insulin-stimulated protein kinase called casein kinase 1 or 2. They were purified a number of years ago by Traugh, but I have not seen the recent data, and as far as I know, it is only suggestive.

KAZIRO: What is the Km for GTP and ATP in the phosphorylation reactions?

ERIKSON: For the auto-phosphorylation, the Km for ATP and GTP is about 15–20 micromolar, for phosphorylation of exogenous substrates, for the phosphorylation of an exogenous substrate using ATP, the Km for nucleotide is about 15–20 micromolar. For GTP it is about 200 micromolar. So, exogenous phosphorylation is much higher.

SCHATZ: You mentioned briefly that there is an artefactual tendency for the molecule to lose end termini during work-up. Would it be at all possible that this may be related to some physiological regulation, as there are now increasing numbers of reports on proteases, some of them calcium-activated, which cleave off N-terminal sequences?

ERIKSON: When we saw this we were very excited about that possibility, and I tried very hard to obtain evidence for even a short-lived intermediate in the life of P60, but after a number of experiments, we are really convinced that the cleavage is an artefact that occurs after you break the cell, and that in unbroken cells, or if you use high detergent and you work rapidly, you never see the protease cleavage product.

SCHATZ: By immune blots?

ERIKSON: Yes, by immune blots. I would say in answer to one question, there is a very interesting mutant of RSV that is called 1702. It has a deleted amino terminus. 1702 has a molecular weight of about 56,000, and all the loss occurs at the amino terminus, and although that virus transforms cells very well in culture, the P60 is less tightly associated with the plasma membrane. If one uses that virus to attempt to produce tumours in chickens, nodules appear, begin to grow and then regress, whereas wild-type nodules of course continue to grow. So, this mutation reduces its affinity for the plasma membrane, and also reduces the potential to produce tumours, although the transformed cell in culture is indistinguishable from wild-type, from a morphological standpoint.

Mechanisms for Protein Import into Mitochondria

Gottfried Schatz

INTRODUCTION

A mitochondrion contains several hundred different polypeptides, but only about a dozen of these are made by the mitochondrion itself; all the others are coded by nuclear genes, synthesized in the extramitochondrial cytoplasm, and imported into the mitochondria (Schatz & Mason 1974). There is mounting evidence that this protein import provides not only mitochondrial building blocks, but also signals by which the nucleus regulates the mitochondrial genetic system (Ono *et al.* 1975, Dieckman *et al.* 1982, Faye & Simon 1983, Fox personal communication).

How do the mitochondrial proteins to be imported recognize the mitochondrial surface? How do they penetrate the mitochondrial membrane(s)? How do they become located within the correct intramitochondrial compartment (outer membrane, intermembrane space, inner membrane, matrix)? And how is this entire process regulated?

During the past 5 years, research in several laboratories has shown that different proteins may enter mitochondria by different mechanisms (Schatz & Butow 1983, Hay *et al.* 1983, Reid 1984). Indeed, the multitude of seemingly different pathways came as a major surprise. As the experiments reported so far have only dealt with the import of a relatively small number of polypeptides, additional pathways will probably be found as more proteins are studied. The fact that every new piece of information makes the story more complex need not cause us undue concern; this is quite typical of early phases in biological research. Eventually, we will reach an "integrative" phase in which we will

---

Biocenter, University of Basel, CH-4056 Basel, Switzerland.

GENE EXPRESSION, Alfred Benzon Symposium 19.
Editors: Brian F. C. Clark & Hans Uffe Petersen, Munksgaard, Copenhagen 1984.

recognize that the different import routes are only variations of a few common themes.

IMPORT INTO THE MATRIX AND THE INNER MEMBRANE

Proteins destined to be imported into the matrix are largely synthesized on free cytoplasmic polysomes as precursors. In most cases, the precursors differ from the corresponding mature proteins by an NH_2-terminal extension that contains between about 5 and about 50 amino acids. However, a few precursors have the same molecular weight as the mature proteins, but differ from them in conformation. The finished precursors are discharged into an extramitochondrial (presumably cytosolic) pool whose size may vary depending on the physiological conditions. Next, the precursors bind to a proteinaceous receptor on the mitochondrial surface and are then translocated across both mito-

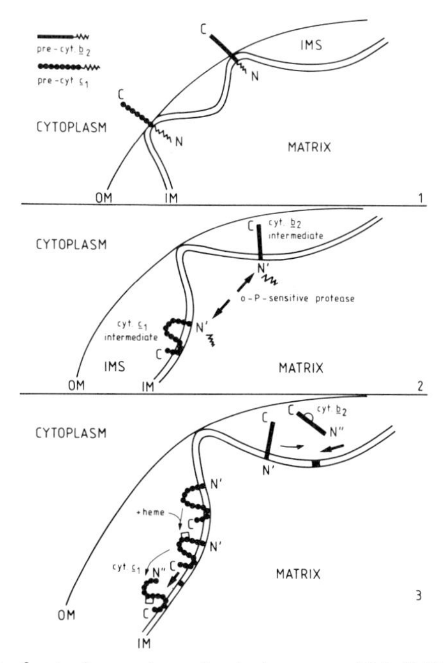

Fig. 1. Two-step import of cytochrome b_2 and cytochrome c_1 OM, IMS, IM: outer membrane, intermembrane space and inner membrane, respectively. C and N denote C-termini and N-termini, N′ and N″ denote N-termini newly created by proteolytic cleavages. The heme group of cytochrome b_2 is indicated by a half-circle, that of cytochrome c_1 by an open square. o-P denotes o-phenanthroline (an inhibitor of the matrix-located processing protease). (From Daum *et al.* 1982).

chondrial membranes. This translocation requires an electrochemical potential across the mitochondrial inner membrane. Precursors containing an NH_2-terminal extension are then cleaved to their mature size by a metallo-endoprotease localized in the soluble matrix.

IMPORT INTO THE INTERMEMBRANE SPACE

Surprisingly, import of several proteins of the soluble intermembrane space is more complex than import into the matrix. Studies on the import of cytochrome b_2, cytochrome c_1 and cytochrome c peroxidase in yeast (Gasser *et al.* 1982) indicate that these proteins initially follow the same import route as proteins destined for the matrix or inner membrane. However, these precursors have unusually long NH_2-terminal extensions that are cleaved in 2 steps: the first cleavage is catalyzed by the matrix metallo-protease mentioned above and results in a transmembrane intermediate whose bulk protrudes into the intermembrane space (Fig. 1). A second cleavage (presumably by a protease located on the outer face of the inner membrane) releases the mature polypeptide into the intermembrane space. With cytochrome c_1, the second cleavage only occurs after heme has been attached to the intermediate form; also, the mature cytochrome c_1 remains attached to the outer face of the inner membrane, probably via its hydrophobic COOH-terminus (Ohashi *et al.* 1982).

Quite a different import route is followed by cytochrome c: studies in *Neurospora crassa* have shown that the heme-free apo-protein without any NH_2-terminal extension is translocated across the outer membrane, and then fitted with its covalently-attached heme group in the intermembrane space (Zimmermann *et al.* 1981). Import does not require an electrochemical potential across the inner membrane nor does it involve proteolytic maturation steps. Similarly, adenylate kinase appears to be imported into the intermembrane space without proteolytic maturation (Watanabe & Kubo 1982). It is not known whether its import requires an energized inner membrane.

IMPORT INTO THE OUTER MEMBRANE

The mitochondrial outer membrane has long been thought to be derived from the endoplasmic reticulum; it was, therefore, quite unexpected to find that its proteins are not imported via the co-translational endoplasmic reticulum pathway, but by a unique mechanism. Insertion of *in vitro*-synthesized outer membrane proteins into isolated mitochondria (Freitag *et al.* 1982, Mihara *et al.* 1982, Gasser & Schatz 1983), or isolated outer membrane vesicles (Gasser & Schatz 1983) can occur post-translationally, is not accompanied by proteolytic processing, and does not require an energized inner membrane. Significantly, no insertion is observed into isolated microsomes. The various import pathways are schematically summarized in Fig. 2.

THE MOLECULES INVOLVED IN PROTEIN IMPORT

Current research attempts to identify, isolate and characterize the molecules that participate in mitochondrial protein import. Some of the recent advances can be summarized as follows:

1) The genes for several imported mitochondrial proteins have been cloned and sequenced. The deduced NH_2-terminal sequence of the precursor proteins appears to be strongly basic, but apart from this, the sequences differ markedly: some are very polar, whereas others exhibit long uninterrupted stretches of uncharged amino acids (Viebrock *et al.* 1982, Kaput *et al.* 1982, Faye & Simon 1983, Hase *et al.,* in press).
2) The precursor of the F_1-ATPase β-subunit from yeast has been isolated in amounts sufficient for protein chemical studies. Under appropriate conditions, the precursor can be processed by the matrix protease and imported into isolated mitochondria (Ohta unpublished).

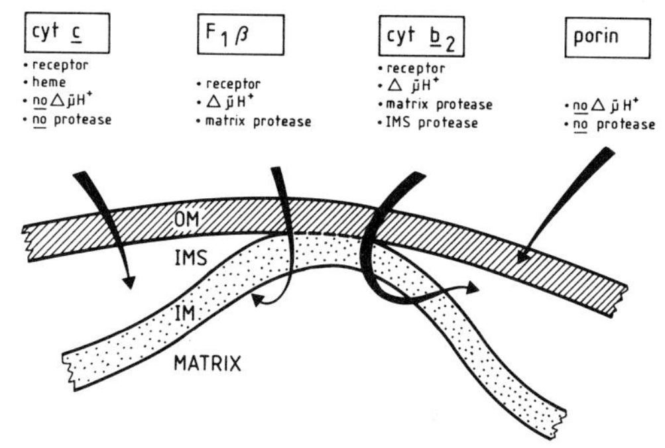

Fig. 2. Various pathways for protein import into mitochondria OM, IMS, IM: outer membrane, intermembrane space and inner membrane, respectively. $\Delta\tilde{\mu}H^+$: proton-motive force (energized inner membrane). cyt *c*, cyt b_2: cytochromes *c* and b_2, respectively. $F_1\beta$: the second largest (β) subunit of F_1-ATPase. porin: the major 29 kDa protein of the outer membrane which forms voltage-controlled pores (From Hay *et al.* 1983).

3) The matrix protease has been purified about 100-fold from yeast (McAda & Douglas 1982, Böhni *et al.* 1983), and shown to correctly cleave the larger precursor of cytochrome oxidase subunit V (Cerletti *et al.* 1983). Interestingly, the protease is itself imported.
4) The receptor for apocytochrome *c* (Hennig & Neupert 1981), and that for cytochrome b_2 (Riezman *et al.* 1983) on the outer membrane have been characterized and are being purified (Neupert personal communication, Hay unpublished). As both cytochromes are imported into yeast mitochondria lacking mitochondrial DNA (Schatz & Mason 1974), the receptors for these proteins are themselves imported.
5) The gene for the 70 kDa outer membrane protein is being altered by selected deletions and by gene fusions to reveal the sequence(s) addressing this protein to the outer membrane (Riezman *et al.* 1983, Hase *et al.* 1983).

CONCLUSIONS

Four major facts have emerged so far:

1) Protein import into mitochondria differs from protein translocation across the endoplasmic reticulum.
2) The molecules catalyzing protein import are themselves imported; mitochondria can, thus, only arise from pre-existing mitochondria.
3) Protein import into the internal mitochondrial compartments requires an energized inner membrane.
4) There exist different import routes for proteins into mitochondria.

REFERENCES

Böhni, P. C., Daum, G. & Schatz, G. (1983) *J. Biol. Chem. 258,* 4937–4943.
Cerletti, N., Suda, K. & Böhni, P. C. (1983) *J. Biol. Chem. 258,* 4944–4949.
Daum, G., Gasser, S. M. & Schatz, G. (1982) *J. Biol. Chem. 257,* 13075–13080.
Dieckman, C. L., Pape, L. K. & Tzagoloff, A. (1982) *Proc. Natl. Acad. Sci. USA 79,* 1805–
Faye, G. & Simon, M. (1983) *Cell 32,* 77–87.
Freitag, H., Janes, M. & Neupert, W. (1982) *Eur. J. Biochem. 126,* 197–202.
Gasser, S. M., Ohashi, A., Daum, G., Böhni, P. C., Gibson, J., Reid, G., Yonetani, T. & Schatz, G. (1982) *Proc. Natl. Acad. Sci. USA 79,* 267–271.
Gasser, S. M. & Schatz, G. (1983) *J. Biol. Chem. 258,* 3427–3430.
Hase, T., Riezman, H., Suda, K. & Schatz, G. (1983) *EMBO J.* (in press).
Hay, R., Böhni, P. & Gasser, S. M. (1983) *BBA Reviews,* (in press).
Henning, B. & Neupert, W. (1981) *Eur. J. Biochem. 121,* 203–212.

Kaput, J., Goltz, S. & Blobel, G. (1982) *J. Biol. Chem. 257,* 15054–15057.

McAda, P. C. & Douglas, M. G. (1982) *J. Biol. Chem. 257,* 3177–3182.

Mihara, K., Blobel, G. & Sato, R. (1982) *Proc. Natl. Acad. Sci. USA 79,* 7102–7106.

Ohashi, A. Gibson, J., Gregor, I. & Schatz, G. (1982) *J. Biol. Chem. 257,* 13042–13047.

Ono, B., Fink, G. R. & Schatz, G. (1975) *J. Biol. Chem. 250,* 775–782.

Reid, G. A. (1984) In: *Current Topics in Membranes and Transport,* Academic Press, New York (in press).

Riezman, H., Hay, R., Witte, C., Nelson, N. & Schatz, G. (1983) *EMBO J. 2,* 1113–1118.

Riezman, H., Hase, T., van Loon, A. P. G. M., Grivell, L. A., Suda, K. & Schatz, G. (1983) *EMBO J.* (in press).

Schatz, G. & Butow, R. A. (1983) *Cell 32,* 316–318.

Schatz, G. & Mason , T. L. (1974) *Ann. Rev. Biochem. 43,* 51–87.

Viebrock, A., Perz, A. & Sebald, W. (1982) *EMBO J. 1,* 565–

Watanabe, K. & Kubo, S. (1982) *Eur. J. Biochem. 123,* 587–592.

Zimmermann, R., Henning, B. & Neupert, W. (1981) *Eur. J. Biochem. 116,* 455–460.

DISCUSSION

ERIKSON: Why don't polysomes making proteins destined to be imported become bound to the mitochondrial surface while these proteins are still being synthetized?

SCHATZ: This is a very good point which has caused a lot of discussion during the past years. Work done by Suissa and Reid in our laboratory suggests that the reason is probably a kinetic one. If protein synthesis is very fast and if mitochondria are poorly developed (as is true for yeast cells growing logarithmically on glucose) the precursor pools are large because precursor polypeptides are made quickly but there are relatively few mitochondrial surface receptors to bind them. On the other hand, if protein synthesis is slow and if mitochondria are well developed (as is true for yeast cells entering stationary phase upon exhaustion of glucose in the medium) precursor pools are small because now binding of nascent precursors to the mitochondrial surface is favoured. This relationship can be shown by progressively inhibiting protein synthesis in yeast cells by titrating in cycloheximide: progressive inhibition of mitochondrial protein synthesis is accompanied by a progressive decrease in the extramitochondrial precursor pools. In fact, I suspect that the long ongoing discussion concerning co- versus post-translational protein movement across bacterial membranes might be explained, at least in part, by a similar relationship.

UHLENBECK: Are there respiration-deficient petite mutants exhibiting Mendelian inheritance?

SCHATZ: Yes, hundreds of them, exactly as you might expect.

UHLENBECK: Can the mutated nuclear genes be assigned to the specific imported mitochondrial proteins?

SCHATZ: In some cases, yes. For example, some nuclear petite mutants lack cytochrome *c*, specific subunits of the mitochondrial ATPase, or components involved in the splicing of mitochondrial pre-mRNA. However, most of the nuclear petite mutants described so far have not been characterized in molecular detail. They still represent a treasure trove for young (or old) investigators.

KURLAND: You mentioned that cytoplasmically-inherited petite mutants invariably lack all proteins coded for by their mitochondrial DNA. This suggests that mitochondrial DNA is not necessary for making mitochondria. I have speculated that it is necessary to synthesize some mitochondrial proteins inside the mitochondria in order to elaborate a new mitochondrion. If this is not so, what is the point of having mitochondrial DNA at all?

SCHATZ: Exactly, there is no obvious reason why higher cells have chosen to retain mitochondrial DNA. It could perhaps be that transfer of genes from the hypothetical endosymbions to the host nucleus has halted once the genetic codes diverged. From this point on, a gene transfer to the host nucleus would have been useless to the cell.

KURLAND: Yet, there must be some additional subtle factor involved since all mitochondrial ribosomes seem to have one protein coded for by mitochondrial DNA. It also seems that this protein is not the same in different species; this in turn suggests that there may be some control which mitochondrial DNA has to exert on the mitochondrial protein synthesizing system or on its interdependence with the nuclear genetic system.

SCHATZ: You could be right, except there is no evidence for such a mitochondrially-coded ribosomal protein in mammalian mitochondria.

MAGNUSSON: Could it be that some mitochondrial precursor polypeptides become glycosylated before entry into the mitochondria? Perhaps such a modification prevents entry of these precursors until the intracellular glucose level drops.

SCHATZ: This is an interesting idea which I have never considered. However, it is probably unlikely to be correct, because the half-time of these precursors in the cytosol is very short, only in the order of seconds or minutes. On the other hand, glucosylation of proteins is a rather slow process, so that there would probably not be enough time for glucose to attach to the precursors while they are still in the cytosol.

EBEL: Do cytoplasmic petite mutants have normal patterns of RNAs?

SCHATZ: No, they don't. Most of them have lost one or more mitochondrial RNA genes. Also, some have lost an, as yet, poorly characterized gene that appears to control the maturation of mitochondrial tRNAs.

SAFER: I have some questions about the transmembrane potential. Is it the same potential that is also used to drive ATP synthesis and ion transport? Also, are there any petite mutants deficient in the ability to generate such a transmembrane potential? If so, these mutants should have an altered protein composition of their mitochondrial inner membrane.

SCHATZ: These are very good and sophisticated questions which I cannot properly answer. I can only give you speculations. First, the potential appears to be of the same type as that driving ATP synthesis. Second, we feel that this transmembrane potential is not a simple electrophoretic driving force for protein import. For example, the polarity of this potential is the same in mitochondria and bacteria (i.e., positive outside) yet proteins move in opposite direction in mitochondria and bacteria. I personally suspect that a transmembrane potential is necessary to destabilize the lipid bilayer. It is not generally appreciated that the fluid mosaic membrane model is great for explaining membrane structure and function, but not nearly as great for explaining membrane biogenesis. By emphasizing bilayer stability, the model makes it difficult to envision how macromolecules move *across* a membrane. It is known, however, that biological membranes can become destabilized by a transmembrane potential. Perhaps the potential induces the transient formation of non-bilayer structures such as inverted micelles which might be involved in transmembrane movement of proteins. If this is true, the direction of such a potential would not be important for facilitating import of proteins into mitochondria or export of proteins from bacteria. Cytoplasmic petite mutants should not be able to set up such a potential across the mitochondrial inner membrane and for this reason we felt initially that the energy for protein import was supplied by ATP itself. However, this is not the case; apparently even cytoplasmic petite mutants can establish a relatively small potential across the mitochondrial inner membrane via the ATP/ADP translocator. Such a potential would be expected to be several fold smaller than the necessary to drive synthesis of ATP.

CRAMER: Does the length of the transient presequence in some of these precursor proteins agree with the length that you would need to span the bilayer?

SCHATZ: In some cases, the presequences of mitochondrial precursor polypeptides are longer than would be necessary for spanning a bilayer; some are as long as 60–70 amino acids. Of course, we really do not know much about the conformation of these presequences even though conformations can easily be calculated from computer programs. I am not sure whether one can trust such predictions, particularly as they are applied to proteins that are certainly not typical globular structures.

KURLAND: I might add that I am not sure what they mean for *any* protein!

VIII. Mechanism of Interferon Action, Control of Interferon Production

The Specific Molecular Activities and Functional Forms of the Human Interferons

Sidney Pestka[1], *Bruce Kelder*[1], *Edward Rehberg*[1], *John R. Ortaldo*[2], *Ronald B. Herberman*[2], *Ellis S. Kempner*[3], *John A. Moschera*[4] & *S. Joseph Tarnowski*[4]

In the past few years, there have been major successes in the purification and the cloning of the human interferons (Pestka 1981a,b, 1983 for reviews and additional citations). With the availability of pure natural and recombinant species of human interferon, we have been able to study in detail the activities of these interferons on various cellular functions. Because many of these interferons were purified to homogeneity, we were able to determine the specific molecular activities of many of them on antiviral action, antiproliferative effects, and stimulation of natural killer cell function.

RELATIVE ACTIVITIES OF INTERFERONS

Our studies on the relative antiviral, antiproliferative, and natural killer stimulatory activity of the native interferons indicated that there is not a direct correlation between these activities (Evinger *et al.* 1981b, Ortaldo *et al.* 1983a, Herberman *et al.* 1982). In some cases, antiviral activity may be high and natural killer cell activity low. In other cases, growth-inhibitory activity may be low and antiviral and natural killer stimulatory activity high. These initial results with the purified natural leukocyte interferons suggested that the activities appeared to be independent of one another. Although the interferons have other biological effects, our initial studies have concentrated on these 3 cellular activities.

[1]Roche Institute of Molecular Biology, Roche Research Center, Nutley, NJ 07110, [2]Biological Therapeutics Branch, National Cancer Institute-Frederick Cancer Research Facility, Frederick, MD 21701, [3]Laboratory of Physical Biology, National Institutes of Health, Bethesda, MD 20205, [4]The Biopolymer Research Department, Hoffmann-La Roche Inc. Nutley, NJ 07110, U.S.A.

GENE EXPRESSION, Alfred Benzon Symposium 19.
Editors: Brian F. C. Clark & Hans Uffe Petersen, Munksgaard, Copenhagen 1984.

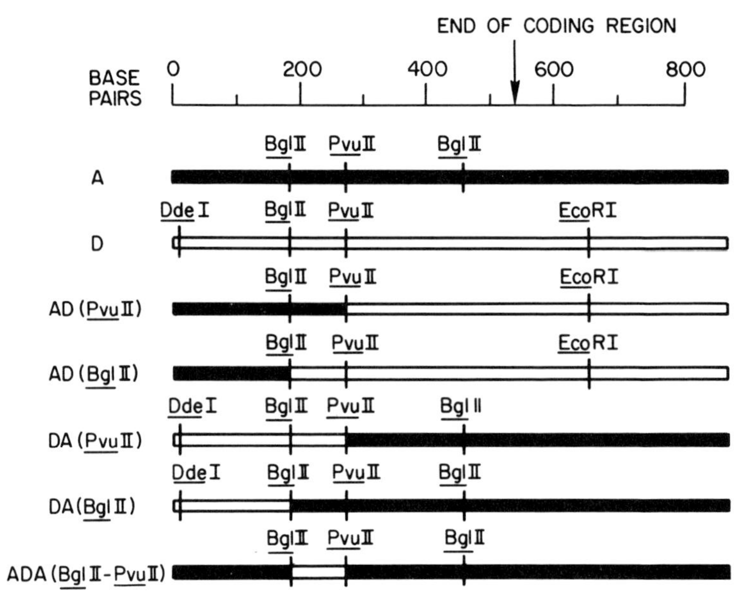

Fig. 1. Schematic illustration and restriction maps of the coding regions for the mature proteins of pIFLrA, pIFLrD, and hybrids constructed from these 2 molecules.

In order to have a large supply of purified leukocyte interferons with different activities for thorough evaluation, hybrid leukocyte interferons (Streuli *et al.* 1981, Weck *et al.* 1981, Rehberg *et al.* 1982) were prepared from recombinant human leukocyte interferons A and D (IFN-αA and IFN-αD). These hybrid interferons are schematically illustrated in Fig. 1 (Rehberg *et al.* 1982).

The specific molecular activity of interferon is defined as the molecules/cell necessary to elicit a specific effect (Rehberg *et al.* 1982). This is an operational definition and is not dependent on knowledge of the specific amount of binding of the interferons to the cells. What is necessary is a supply of pure interferon to determine the concentration in solution of that interferon at the time the experiment is initiated. If binding data were available as well, the operational definition could be refined to include the specific number of molecules bound to cells to elicit a given effect. Comparison of the specific molecular activities of the interferons provides a useful way to compare the relative effects of interferons on different cellular activities. It should be emphasized that the ultimate description of the ability of an interferon to initiate a biological response will be described by the association constant of that interferon to the receptor controlling the specific molecular mechanisms involved as well as the number of

Table I

Molecules/cell for 50% inhibition of viral cytopathic effect for various cell lines

Interferon	AG-1732 (Human)	MDBK (Bovine)	Cell Line L-Cells (Mouse)	Felung (Feline)	Rat C6 (Rat)
A	4,900	1,100	3.1×10^7	8,700	6.1×10^8
D	360,000	6,600	4.3×10^6	3,700	1.1×10^6
A/D (*Bgl*)	5,500	1,200	3,300	450	210,000
A/D (*Pvu*)	4,100	2,400	48,000	1,800	$>1.7\times10^6$
D/A (*Bgl*)	1.5×10^6	1,800	$>7.2\times10^7$	42,000	$>3.8\times10^7$
D/A (*Pvu*)	590,000	5,400	7.1×10^6	12,000	$>3.0\times10^7$
A/D/A	36,000	2,000	1.1×10^6	1,500	$>1.8\times10^7$

The molecular weights of the various species are as follows: IFN-αA, 19,219; IFN-αD, 19,392; IFN-αA/D (*Bgl*), 19,395; IFN-αA/D (*Pvu*), 19,427; IFN-αD/A (*Bgl*), 19,216; IFN-αD/A (*Pvu*), 19,184; IFN-αA/D/A, 19,205. The number of cells in the microtiter well was determined at the time of addition of cells: AG-1732, 12,000; MDBK, 27,000; L-cells, 25,000; Felung, 18,000; Rat C6, 54,000. Data from Rehberg *et al.* (1982).

those receptors on the cell. Until such data are available, the specific molecular activity will provide a useful way to compare various interferons with each other and across species.

Comparison of the antiviral activities of the parental interferons IFN-αA and IFN-αD as well as the hybrid interferons on different species (Table I) showed that new activities of the interferons were generated. Although the antiviral activities were essentially similar for all these molecules on bovine MDBK cells, their activities on human, mouse, feline, and rat cells differed markedly from the parental molecules. For example, the hybrids IFN-αA/D (*Bgl*) and IFN-αA/D (*Pvu*) were not only active on human and bovine cells, but were highly active on mouse and feline cells. The IFN-αA/D (*Bgl*) hybrid was even active on rat cells as well. IFN-αA was highly active on human, bovine, and feline cells, but active on mouse and rat cells only at over 6,000- and 120,000-fold, respectively, the molecular/cellular ratio active on human cells. IFN-αD was most active on bovine and feline cells, slightly active on human cells, but much less active on mouse and rat cells. IFN-αA/D (*Bgl*) was highly active on all but the rat cell lines examined; IFN-αA/D (*Pvu*) was active on all except the rat cell lines, but exhibited a lower specific molecular activity than IFN-αA/D (*Bgl*) on most of these cells.

The most active species at a concentration of 500 to 5000 molecules/cell inhibited the cytopathic effect by 50%. Within a factor of 6, all the species showed a similar specific molecular activity on bovine cells; and most were significantly active on feline cells. The 7 interferons exhibited a range of specific molecular activities on human cells over a factor of 100-fold. Only IFN-αA/D (*BglI*) and IFN-αA/D (*Pvu*) showed activity on mouse cells at 200,000 molecules/cell or less.

These data suggest that the decrease in antiviral activity on human cells appears to be mediated by determinants on the amino-terminal portion of the leukocyte interferon molecule. It is clear that the hybrids designated IFN-αD/A (*BglII*) and IFN-αD/A (*PvuII*) as well as IFN-αD show a much lower activity on human AG-1732 cells than does IFN-αA (Table I). When 3 amino acid substitutions were introduced into the native IFN-αA molecule (as accomplished by construction of the A/D/A hybrid), the specific molecular activity on human cells was reduced about 10-fold. A/D hybrids of both types have approximately equal activity on human, bovine, and feline cell lines. Thus, the amino-terminal half of IFN-αD is responsible for the structural changes that result in a marked reduction in antiviral activity on human cells.

When the antiviral activity of these interferons was examined on L cells, greatest activity was found in all species containing some or all of the 103 carboxy-terminal amino acids of IFN-αD: IFN-αD, IFN-αA/D (*BglII*), IFN-αA/D (*PvuII*), IFN-αD/A (*PvuII*), and IFN-αA/D/A (Rehberg *et al.* 1982). Therefore, it appears that determinants on the carboxy terminus of IFN-αD are responsible for mouse activity.

Table II

Antiproliferative activity: concentrations of interferon at 50% inhibition of control cell growth

Interferon	molecules/cell	Antiviral units/ml
A	13,000	23
D	450,000	12
A/D (*BglI*)	9,300	21
A/D (*Pvu*)	30,000	62
D/A (*BglI*)	910,000	4
D/A (*Pvu*)	4,200,000	73
A/D/A	60,000	17

The values in the Table represent the concentrations of the interferons that inhibited growth of Daudi cells 50%. Antiviral units were determined on AG-1732 cells. Molecules/cell were calculated from the cell number present at the time of addition of the interferons. Data taken from Rehberg *et al.* (1982).

Table III
Ratio of specific molecular activities of interferons for antiproliferative and antiviral activity on human cells

Interferon	AP/AV
A	2.7
D	1.3
A/D (*Bgl*)	1.7
A/D (*Pvu*)	7.3
D/A (*Bgl*)	0.6
D/A (*Pvu*)	7.1
A/D/A	1.7

The specific molecular antiproliferative activity (molecules/cell) for inhibition of growth of human lymphoblastoid Daudi cells (Table II), AP, was divided by the specific molecular antiviral activity on human AG-1732 fibroblasts, AV. The ratio AP/AV is given in the Table. Data from Rehberg *et al.* (1982).

The antiproliferative activities of these interferons were determined on Daudi lymphoblastoid cells (Evinger *et al.* 1981a). The quantity of each interferon that produced 50% inhibition of growth of the cells is shown in Table II. It is evident that the antiviral units necessary to produce this inhibition provide a spurious picture of which interferons are most active on a molecular basis. Comparison of the molecules/cell necessary for 50% inhibition of growth, however, permits an accurate assessment of their relative activities. The ratio of the quantity required for 50% inhibition of viral cytopathic effect (antiviral activity) to the quantity required for antigrowth activity (Table III) varied over a range of 12-fold. In the case of IFN-αA/D (*Pvu*) and IFN-αD/A (*Pvu*), about 7- to 8-fold more interferon was required for 50% inhibition of growth than 50% inhibition of viral cytopathic effect.

Table IV
Molecules/cell for 50% effect

Interferon	AV	AP	NK
A	4,900	13,000	120
D	360,000	450,000	280,000
A/D (*Bgl*)	5,500	9,300	58
A/D (*Pvu*)	4,100	30,000	1300
D/A (*Bgl*)	1.5×10^6	910,000	250,000
D/A (*Pvu*)	590,000	4.2×10^6	46,000
A/D/A	36,000	60,000	20,000

AV, antiviral activity; AP, antiproliferative activity; NK, stimulation of natural killer cell activity of human cells. The data are taken from Rehberg *et al.* (1982) and Ortaldo *et al.* (1983).

Table V
Ratio of specific molecular activities

Interferon	AV/AP	AV/NK
A	0.38	41
D	0.80	1.3
A/D (*Bgl*)	0.59	95
A/D (*Pvu*)	0.14	3.2
D/A (*Bgl*)	1.7	6.0
D/A (*Pvu*)	0.14	13
A/D/A	0.60	1.8

The AV, AP, and NK ratios were calculated from the data of Table IV.

Similar studies were performed to determine the specific molecular activity of these interferons on stimulation of natural killer cell activity (Table IV). These results indicated that, as had been determined with the natural interferons, these recombinant interferons also showed remarkable differences in the number of molecules/cell necessary for the various effects. A more detailed discussion of the natural killer cell stimulatory activity of these interferons is presented by Ortaldo *et al.* (1983b). By comparing the ratio of the specific molecular activities (AV/AP and AV/NK), remarkable differences in these ratios are evident (Table V). In fact, there is approximately a 100-fold difference in the AV/AP and AV/NK ratios for IFN-αA, IFN-αA/D (*Bgl*), and IFN-αD/A (*Pvu*).

These disparities between antiviral, antiproliferative, and natural killer cell stimulatory activities indicate that these activities are mediated by different mechanisms. Similar suggestions were made previously when it was observed that the individual purified human leukocyte interferons exhibited different ratios of antiviral to antiproliferative activity (Evinger *et al.* 1981b). As the effects of interferon are mediated by a number of different mechanisms (Kerr & Brown 1978, Lengyel 1981, Lengyel & Pestka 1981, Maheshwari *et al.* 1981, Sreevalsan *et al.* 1981, Revel *et al.* 1981), it is not surprising that some of the effects can be dissociated. Thus, after comparison of the antiproliferative and antiviral activities of human leukocyte, fibroblast, and immune interferons, Eife *et al.* (1981) observed that the antiproliferative:antiviral activity ratios differed significantly for the 3 interferons. Effects of interferon on lytic viruses such as vesicular stomatitis virus were dissociated from those on Moloney leukemia virus (Epstein *et al.* 1981, Sen & Herz 1983, Herz *et al.* 1983). These results indicate that many of the effects of the interferons can be dissociated and are due to different molecular mechanisms. The individual interferons can apparently turn on several pathways to different degrees.

We have recently shown that recombinant leukocyte interferon J (IFN-αJ) exhibits antiviral and antiproliferative activity, but can hardly stimulate natural killer cell activity (Ortaldo *et al.* 1983b). In addition, preliminary experiments evaluating the antiviral activity of interferons on vesicular stomatitis virus, encephalomyocarditis virus, and retrovirus suggest that the D/A (*Bgl*) hybrid lacks the ability to inhibit encephalomyocarditis virus multiplication of some feline cells whereas its ability to inhibit multiplication of the other 2 viruses remain intact. The parental IFN-αA and IFN-αD molecules as well as the other hybrid interferons inhibited multiplication of all 3 viruses (Sen, Herz, and Pestka, unpublished observations). These results indicate that the mechanisms that generate the various antiviral, antiproliferative, and immunomodulatory activities are distinct. It is, thus, possible that we may be able to tailor the interferons to exhibit a given activity and to fine-tune their effects.

LEUKOCYTE INTERFERONS READILY FORM OLIGOMERS

It was recently observed in purified preparations of human leukocyte interferon that dimers, trimers, and higher oligomers were readily formed (Fig. 2). In order to study this process, we developed a rapid and convenient assay for interferon oligomers or dimers (Pestka *et al.* 1983b). Dimers and oligomers of fibroblast interferon exist as well (Friesen *et al.* 1981, Knight & Fahey 1981).

In addition, human immune interferon appeared to be of high molecular weight ranging between 45,000 to 90,000 daltons when obtained from natural sources (Langford *et al.* 1979, de Ley *et al.* 1980, Yip *et al.* 1981, Von Wussow *et al.* 1982, Wolfe *et al.* in preparation, Braude, personal communication). In contrast, the recombinant immune interferon that was isolated and expressed appeared to code for a protein monomer of molecular weight approximately 17,000 (Gray *et al.* 1982, Devos *et al.* 1982), although Tanaka *et al.* (1983) reported recombinant immune interferon behaved as a dimer on gel filtration.

TARGET SIZE OF LEUKOCYTE, FIBROBLAST AND IMMUNE INTERFERONS

Because of the disparate results for the molecular weights of these interferons, it was of interest to ask what is the actual functional unit of these molecules. Irradiation of macromolecules permits the estimation of their molecular weights

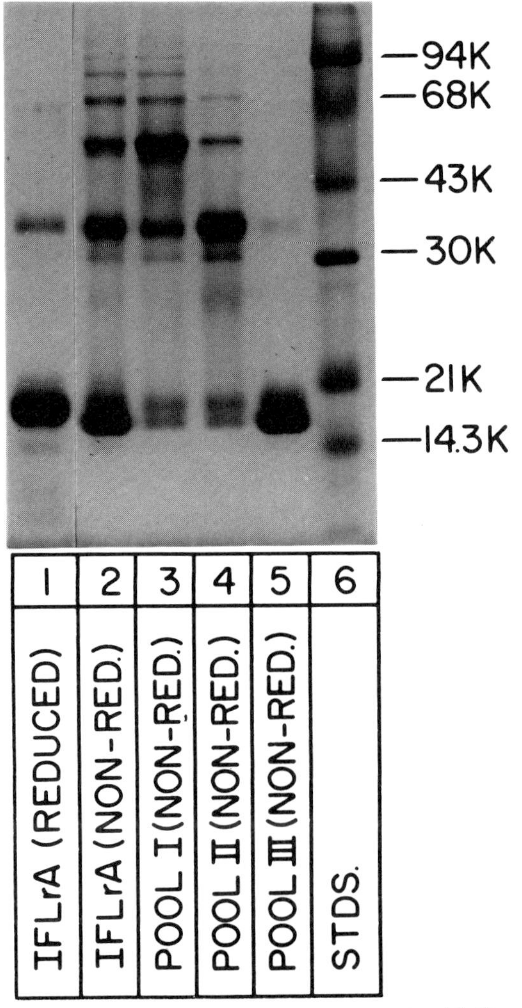

Fig. 2. Sodium dodecyl sulfate polyacrylamide gel electrophoresis of purified interferon separated by Sephadex G-75 chromatography. Five micrograms of each sample of IFN-αA were dried by evaporation and subjected to electrophoresis on a 12.5% polyacrylamide slab gel containing 0.1% sodium dodecyl sulfate. The gel was stained for 1 h at room temperature with 0.2% Coomassie brilliant blue R-250 in methanol/acetic acid/water (25:10:65) and destained with 5% methanol in 10% acetic acid. Electrophoresis of the samples under non-reducing conditions was performed with sample buffer from which *β*-mercaptoethanol had been omitted. Lane 1, CM-52 pool, reduced; lane 2, CM-52 pool, non-reduced; lane 3, Sephadex G-75 pool I, non-reduced (trimer and higher oligomer); lane 4, Sephadex G-75 pool II, non-reduced (dimer); lane 5, Sephadex G-75 pool III, non-reduced (monomer). Lane 6, standard protein markers from BioRad Laboratories (Rockville, NY): 94,000, phosphorylase b; 68,000, bovine serum albumin; 43,000, ovalbumin; 30,000, carbonic anhydrase; 21,000, soybean trypsin inhibitor; 14,300, lysozyme. Standards were run under non-reducing conditions. The data are taken from Pestka *et al.* (1983a). The CM-52 pool represents purified recombinant human leukocyte A interferon purified on carboxymethylcellulose (Kenny *et al.* 1981). The observations that recombinant interferon A can form oligomers were made independently by several groups (unpublished results of Tarnowski *et al.* at Hoffmann-La Roche, Nara *et al.* at Takeda Chemical Industries, and Suhara *et al.* at Nippon Roche, Wetzel *et al.* 1983). IFN-αA was purified as described by Staehelin *et al.* (1981).

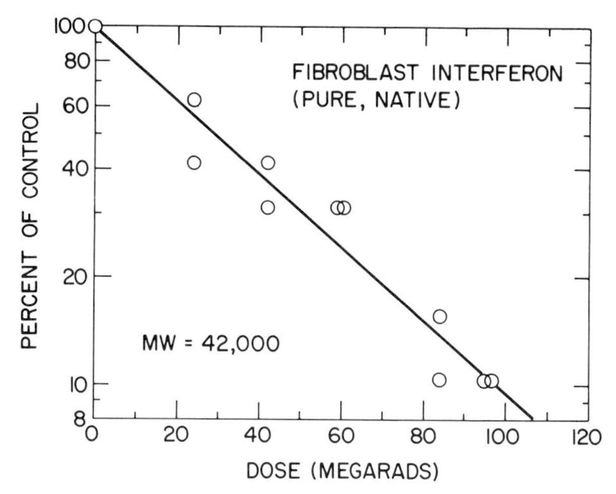

Fig. 3. Loss of antiviral activity of homogeneous human native fibroblast interferon as a function of ionizing radiation. The fibroblast interferon was purified as described (Kenny *et al.* 1981, Stein *et al.* 1980). The procedures are described in detail elsewhere (Pestka *et al.* 1983a).

(target size) independent of their isolation and purity. The target size of proteins has been estimated quite accurately by electron bombardment (Kempner & Schlegel 1979, Kempner *et al.* 1980, Nielsen *et al.* 1981). Thus, we determined the molecular weights of the functional units of the human interferons by determination of their target size (Pestka *et al.* 1983a) for antiviral activity. An example of the target size analysis is shown in Fig. 3. It can be seen that the data appear to represent a mono-exponential inactivation and provide an estimate for the target size of natural fibroblast interferon as 42,000 daltons. In the same manner, we have determined the target sizes of natural and recombinant human leukocyte interferon, natural and recombinant human fibroblast interferon, as well as natural and recombinant human immune interferon. A summary of the inactivation curves for the natural interferons is shown in Fig. 4. It can be seen that the human leukocyte, fibroblast, and immune interferons appear to have target molecular weights of 20,000, 42,000, and 63,000, respectively. Analogous studies with the corresponding recombinant species are shown in Fig. 5. In this case, it can be seen that the target size of the leukocyte, fibroblast, and immune interferons are calculated to be 20,000, 33,000, and 73,000, respectively (Fig. 5).

It appears there is good agreement between the target molecular weight and the known molecular weight of the leukocyte interferons both in crude and pure

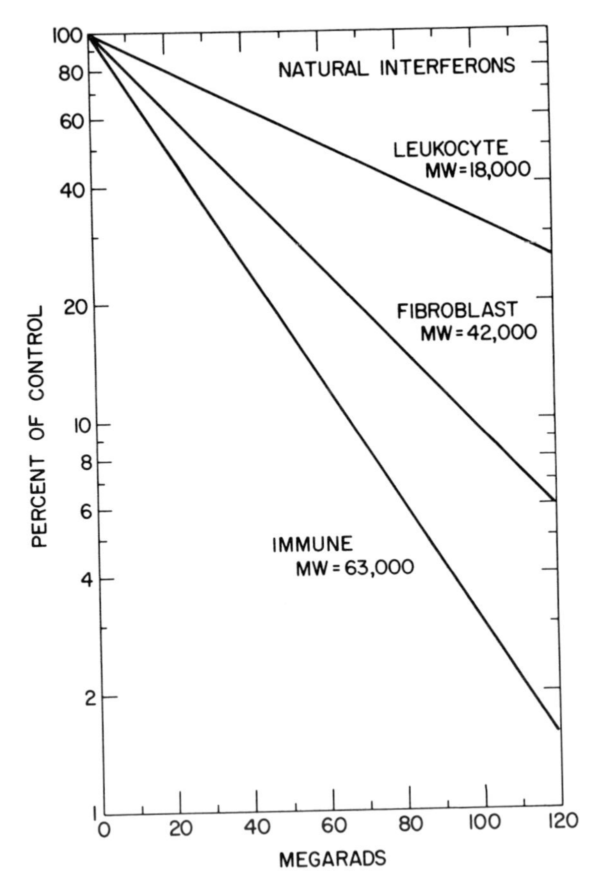

Fig. 4. Summary of the inactivation curves for the natural interferons. Loss of antiviral activity due to increasing doses of ionizing radiation is shown. Data from Pestka *et al.* (1983a).

forms. The leukocyte interferons are molecules that are generally not glycosylated (Rubinstein *et al.* 1981, Allen & Fantes 1980, Pestka 1983), and are present predominantly in the form of a monomer in solution (Pestka *et al.* 1983a).

In the case of fibroblast interferon, the target molecular weight of 33,000 to 42,000 is significantly larger than the experimentally determined molecular weight of about 21,000 to 24,000 (Knight & Fahey 1981, Friesen *et al.* 1981, Pestka 1983). These results indicate that the antiviral activity of fibroblast interferon is due to 2 monomers. Studies of fibroblast interferon have shown that oligomers of the molecule exist together with the monomers (Friesen *et al.*

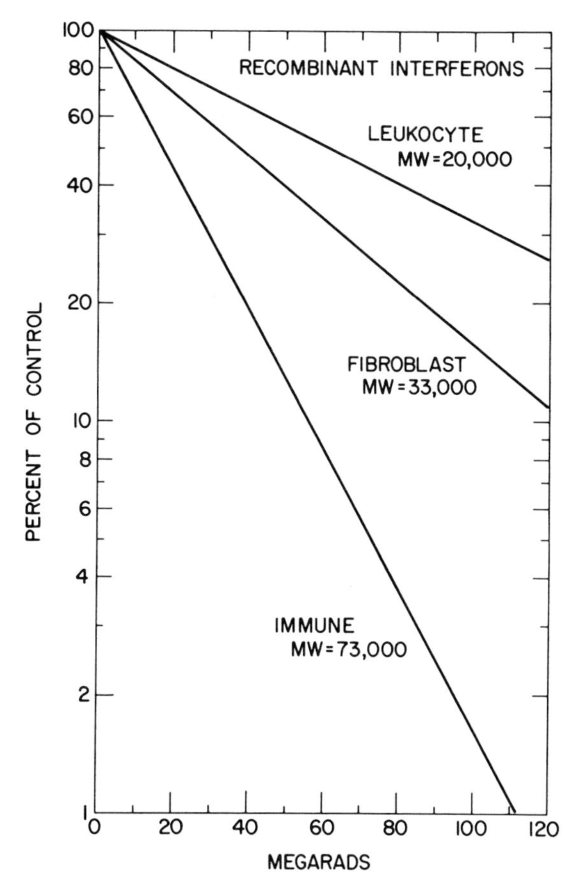

Fig. 5. Summary of the inactivation curves for the recombinant interferons. Loss of antiviral activity due to increasing doses of ionizing radiation is shown. Data from Pestka *et al.* (1983a).

1981, Knight & Fahey 1981). The target size of 33,000 to 42,000 daltons suggests that the dimer may be the predominant active molecular form even in solutions of the crude interferon. Although native fibroblast interferon is glycosylated (Knight 1976, Friesen *et al.* 1981), whereas recombinant fibroblast interferon produced in *Escherichia coli* is not, the glycosylation does not appear to be a significant factor in these determinations. This result is consistent with the observation that the radiation sensitivity of invertase is independent of the presence of attached oligosaccharide (Lowe & Kempner 1982).

In 2 different preparations of natural immune interferon, the target molecular weights ranged from 63,000 to 66,000 (Pestka *et al.* 1983a). The molecular

THE INTERFERON RECEPTOR

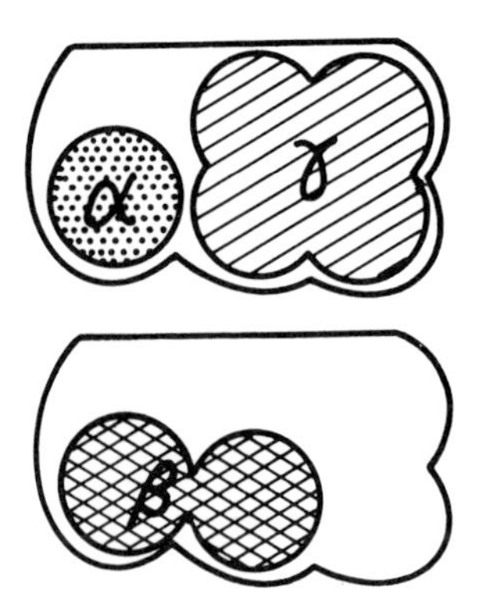

Fig. 6. Schematic illustration of the interferon receptor. IFN-α, -β, and -γ interferons are designated on the figure. As shown, IFN-β overlaps sites occupied by IFN-α and IFN-γ.

weight of these glycosylated molecules in non-denaturing solutions determined by independent means appears to be about 45,000 (Yip *et al.* 1981, Langford *et al.* 1979, de Ley *et al.* 1980). Recombinant immune interferon monomer exhibits a molecular weight of 17,000 (Gray *et al.* 1982, Devos *et al.* 1982, Pestka 1983). The target molecular weight of recombinant immune interferon was determined to be 73,000 (Fig. 5). These results indicated that the functional form of immune interferon may be a tetramer in solution. The difficulty in demonstrating active monomer (Yip *et al.* 1982) seems to reflect the fact that an oligomer is preferred. In solution, there may be very few monomers of immune interferon. These studies indicated that the functional unit of leukocyte interferon is the monomer, that of fibroblast interferon is predominantly a dimer, and that of immune interferon a trimer or tetramer. Although fibroblast and immune interferon monomers may be active as such, their presence as functional dimers or oligomers in solution indicates that the active moiety interacting with the cell receptor is not the monomer species. It would be of interest, indeed, to know the structures of the active moieties interacting with the cell receptor. Nevertheless, it is clear that the high molecular weight oligomers of immune interferon are required for function in crude preparations as well as with the purified recombinant form of the molecule.

Fibroblast interferon competes with the binding of leukocyte (Branca & Baglioni 1981, Joshi *et al.* 1982), and of immune interferon (Anderson *et al.* 1982), to their respective human cell surface receptors. However, leukocyte and immune interferon did not compete with each other for binding to human cells. The conclusion that the target size of fibroblast interferon is a dimer is consistent with the requirement that the fibroblast interferon dimer must bind to both receptors to induce the antiviral state. One monomer binds to the receptor for leukocyte interferon and the other to the receptor for immune interferon (Fig. 6). If, indeed, a structural dimer of fibroblast interferon is the active unit, then the 2 receptors must be adjacent to each other and may, in fact, be portions of a larger integral moiety.

ACKNOWLEDGEMENT

We thank Cynthia Rose, Linda Petervary, Judith Altman for performing some of the interferon assays; and Sophie Cuber, Wendy Ewald, and Pamela Van Wyk for assistance in the preparation of this manuscript.

REFERENCES

Allen, G. & Fantes, K. H. (1980) *Nature (London) 287,* 408–411.

Anderson, P., Yip, Y. K. & Vilček, J. (1982) *J. Biol. Chem. 257,* 11301–11304.

Branca, A. A. & Baglioni, C. (1981) *Nature (London) 294,* 768–770.

de Ley, M., Van Damme, J., Claeys, H., Weening, H., Heine, J. W., Billiau, A., Vermylen, C. & De Somer, P. (1980) *Eur. J. Immunol. 10,* 877–883.

Devos, R., Cheroutre, H., Taya, Y., Degrave, W., Van Heuverswyn, H. & Fiers, W. (1982) *Nucleic Acids Res. 10,* 2487–2501.

Eife, R., Hahn, T., DeTavera, M., Schertel, F., Holtmann, H., Eife, G. & Levin, S. (1981) *J. Immunol. Methods 47,* 339–347.

Epstein, D. A., Czarnieki, W., Jacobsen, H., Friedman, R. M. & Panet, A. (1981) *Eur. J. Biochem. 118,* 9–15.

Evinger, M., Maeda, S. & Pestka, S. (1981a) *J. Biol. Chem. 256,* 2113–2114.

Evinger, M., Rubinstein, M. & Pestka, S. (1981b) *Arch. Biochem. Biophys. 210,* 319–329.

Friesen, H.-J., Stein, S., Evinger, M., Familletti, P. C., Moschera, J., Meienhofer, J., Shively, J. & Pestka, S. (1981) *Arch. Biochem. Biophys. 206,* 432–450.

Gray, P. W., Leung, D. W., Pennica, D., Yelverton, E., Najarian, R., Simonsen, C. C., Derynck, T., Sherwood, P. J., Wallace, D. M., Berger, S. L., Levinson, A. D. & Goeddel, D. V. (1982) *Nature (London) 295,* 503–508.

Herberman, R. B., Ortaldo, J. R., Timonen, T., Reynolds, C. W., Djeu, J. Y., Pestka, S. & Stanton, J. (1982) *Tex. Rep. Biol. Med. 41,* 590–595.

Herz, R. E., Rubin, B. Y. & Sen, G. C. (1983) *Virology 125,* 246–250.

Joshi, A. R., Sarkar, F. H. & Gupta, S. L. (1982) *J. Biol. Chem. 257,* 13884–13887.

Kempner, E. S. & Schlegel, W. (1979) *Anal. Biochem. 92,* 2–10.

Kempner, E. S., Miller, J. H., Schlegel, W. & Hearon, J. Z. (1980) *J. Biol. Chem. 255,* 6826–6831.

Kenny, C., Moschera, J. A. & Stein, S. (1981) In: *Methods in Enzymology,* Vol. 78, Part A, pp. 435–447. Academic Press, New York.

Kerr, I. M. & Brown, R. E. (1978) *Proc. Natl. Acad. Sci. U.S.A. 75,* 256–260.

Knight, E., Jr. (1976) *Proc. Natl. Acad. Sci. U.S.A. 73,* 520–523.

Knight, E., Jr. & Fahey, D. (1981) *J. Biol. Chem. 256,* 3609–3611.

Langford, M. P., Georgiades, J. A., Stanton, G. J., Dianzani, F. & Johnson, H. M. (1979) *Infect. Immun. 26,* 36–41.

Lengyel, P. (1981) In: *Methods in Enzymology,* Vol. 79, Part B, pp. 135–148. Academic Press, New York.

Lengyel, P. & Pestka, S. (1981) In: *Gene Families of Collagen and Other Proteins,* pp. 121–126. Elsevier/North Holland, Amsterdam.

Lowe, M. E. & Kempner, E. S. (1982) *J. Biol. Chem. 257,* 12478–12480.

Maheshwari, R. K. & Friedman, R. M. (1981) In: *Methods in Enzymology,* Vol. 79, Part B, pp. 451–458. Academic Press, New York.

Nielsen, T. B., Lad, P. M., Preston, M. S., Kempner, E., Schlegel, W. & Rodbell, M. (1981) *Proc. Natl. Acad. Sci. U.S.A. 78,* 722–726.

Ortaldo, J. R., Mantovani, A., Hobbs, D., Rubinstein, M., Pestka, S. & Herberman, R. B. (1983a) *Int. J. Cancer 31,* (in press).

Ortaldo, J. R., Mason, A., Rehberg, E. M. Kelder, B., Harvey, C., Oscheroff, P., Pestka, S. & Herberman, R. B. (1983b) In: *Biology of the Interferon System,* 353–358. Elsevier Biomedical Press, Amsterdam.

Pestka, S. (1981a) *Methods in Enzymology,* Vol. 78, Part A. Academic Press, New York.

Pestka, S. (1981b) *Methods in Enzymology,* Vol. 79, Part B. Academic Press, New York.

Pestka, S. (1983) *Arch. Biochem. Biophys. 221,* 1–37.

Pestka, S., Kelder, B., Familletti, P. C., Moschera, J. A., Crowl, R. & Kempner, E. S. (1983a) *J. Biol. Chem. 258,* 9706–9709.

Pestka, S., Kelder, B., Tarnowski, D. K. & Tarnowski, S. J. (1983b) *Anal. Biochem. 132,* 328–333.

Rehberg, E., Kelder, B., Hoal, E. G. & Pestka, S. (1982) *J. Biol. Chem. 257,* 11497–11502.

Revel, M., Wallach, D., Merlin, G., Schattner, A., Schmidt, A., Wolf, D., Shulman, L. & Kimchi, A. (1981) In: *Methods in Enzymology,* Vol. 79, Part B., pp. 149–161. Academic Press, New York.

Rubinstein, M., Levy, W. P., Moschera, J. A., Lai, C.-Y., Hershberg, R. D., Bartlett, R. T. & Pestka, S. (1981) *Arch. Biochem. Biophys. 210,* 307–318.

Sen, G. C. & Herz, R. E. (1981) *J. Virol. 45,* 1017–1027.

Sreevalsan, T., Lee, E. & Friedman, R. M. (1981) In: *Methods in Enzymology,* Vol. 79, Part, B, pp. 342–349. Academic Press, New York.

Staehelin, T., Hobbs, D. S., Kung, H.-F., Lai, C.-Y. & Pestka, S. (1981) *J. Biol. Chem. 256,* 9750–9754.

Stein, S., Kenny, C., Friesen, H. J., Shively, J., Del Valle, U. & Pestka, S. (1980) *Proc. Natl. Acad. Sci. U.S.A. 77,* 5716–5719.

Streuli, M., Hall, A., Boll, W., Stewart, W. E., II, Nagata, S. & Weissmann, C. (1981) *Proc. Natl. Acad. Sci. U.S.A. 78,* 2848–2852.

Tanaka, S., Oshima, T., Ohsuye, K., Ono, T., Mizono, A., Ueno, A., Nakazato, H, Tsujimoto, M., Higashi, N. & Noguchi, T. (1983) *Nucleic Acids Research. 11,* 1707–1723.

Von Wussow, P., Chen, Y.-S., Wiranowska-Stewart, M. & Stewart, W. E., II (1982) *J. Interferon Res. 2,* 11–20.

Weck, P. K., Apperson, S., Stebbing, N., Gray, P. W., Leung, D., Shepard, H. M. & Goeddel, D. V. (1981) *Nucleic Acids Res. 9,* 6153–6166.

Wetzel, R. (1983) In: *Antiviral Research Abstr. 1,* 1.

Wolfe, R., Stein, S., Moschera, J., Hobbs, D. S., Familletti, P. C. & Pestka, S. (in preparation).

Yip, Y. K., Barrowclough, B. S., Urban, C. & Vilček, J. (1982) *Science 215,* 411–413.

Yip, Y. K., Pang, R. H. L., Urban, C. & Vilcěk, J. (1981) *Proc. Natl. Acad. Sci. U.S.A. 78,* 1601–1605.

DISCUSSION

SCHATZ: What is the molecular mechanism of interferon action?

PESTKA: Interferon, as I tried to indicate, does not have a single function. It has many actions. I shall not discuss the specific biochemical events that are induced by interferon because Ian Kerr will describe those such as the 2′-5′-oligoadenylate synthetase and the double-stranded RNA-dependent kinase in his talk. However, as an overview, it appears that interferon interacts with a cell surface receptor. The inteferon-receptor complex somehow communicates with the nucleus of the cell which initiates the transcription of new messenger RNAs (mRNA) in response to this interaction. These mRNAs code for proteins that are responsible for biochemical pathways that are present only or predominantly in cells exposed to interferon. How the surface receptor-interferon complex communicates with the nucleus is unknown. I indicated in my talk that individual interferons activate the various biological activities and biochemical pathways to different degrees. Thus, the receptor for interferon must either be a complex one as I suggested in the talk or, conceivably, there may be multiple receptors for interferon, each controlling the biochemical pathways that generate a specific biological activity such as antiviral, antiproliferative, or immunomodulatory activities. It is of interest that even within the set of closely related leukocyte (α) interferons the antiviral, antiproliferative, and immunomodulatory (natural killer cell stimulation, for example) activities can be modulated to different extents. Thus, it is possible that these molecules may be able to be modified so that they can be used to fine-tune the variety of biological activities that interferon can activate.

SCHATZ: Is there any evidence for receptor-mediated endocytosis in interferon function?

PESTKA: It has been shown that interferon can enter cells through the mechanism of receptor-mediated endocytosis by Zoon, Arnheiter, Willingham and colleagues. However, there is no evidence that this entry of interferon is related to its function.

ERIKSON: Can you sum up what is the current status of interferon in the treatment of cancer?

PESTKA: The purification and cloning of the interferons have moved along very rapidly so that at the present time a number of recombinant interferons are in clinical trials concurrently with crude and purified preparations from natural sources. A large number and variety of cancers have been treated with interferon in preliminary studies. Although it is too early to make any definitive conclusion, results look promising in a number of cancers: malignant melanoma, multiple myeloma, non-Hodgkin's lymphoma, Kaposi's sarcoma, renal carcinoma, chronic leukemias, and some other malignancies. However, although there are encouraging clinical results, there are rarely dramatic total remissions; these have been relatively infrequent. We still do not know how to use the interferons most effectively. Nevertheless, as these clinical trials progress and as our fundamental understanding of how interferon works increases, I believe we may see some improvement in the results of anti-tumour therapy with interferon.

EBEL: Is it reasonable to expect to know effects of therapeutical use of interferon in viral diseases in the near future?

PESTKA: Clinical trials with interferon have been underway in a number of diseases in which viruses have been implicated. It appears that interferon is quite effective in treating juvenile laryngeal papillomatosis and warts. Some effectiveness to one degree or another has been seen in varicella (*herpes zoster*), chronic hepatitis B, and in respiratory virus infections. Additional trials are underway in *herpes genitalis*, cytomegalovirus infections, rabies, and many other viral diseases. In an outbreak of simian varicella in a monkey colony, human interferon was quite effective prophylactically in reducing the attack rate. Nevertheless, even though there are some promising areas, I believe it will take some time to develop the proper criteria for choosing patients and developing appropriate treatment regimens in these diseases. In addition, when considering therapy, the physician must balance the results of treatment with the side effects of the agent. The side effects of interferon are significant so that these may mitigate against its general use in prophylaxis of essentially benign and self-limiting conditions such as the common cold. In any case, I believe the results of these initial clinical trials will begin to accumulate within the next year or two and that the national regulatory agencies will start making decisions about approving interferon for specific indications within the next year.

MAGNUSSON: Did the interferon go in alone or together with another protein in the experiments to which we were referring on receptor-mediated endocytosis?

PESTKA: The interferon was labelled with colloidal gold. It apparently went in without any other protein. However, it is possible that the receptor-interferon complex entered as a unit.

MAGNUSSON: Were these *in vitro* experiments on cell cultures?

PESTKA: Yes.

REHFELD: You mentioned cimetidine together with interferon had an effect on melanoma, was that correct?

PESTKA: That's correct.

REHFELD: How did people get the idea of mixing cimetidine with interferon?

PESTKA: Borgström and colleagues (1982) were treating patients with malignant melanoma with interferon. Several of the patients were simultaneously being treated for symptoms of peptic ulcer with cimetidine. After analysis of their results, they found the best therapeutic responses in those patient given interferon and cimetidine. This initial study is being re-evaluated in many other laboratories. It is of note that patients with malignant melanoma seem to respond to interferon even without cimetidine. Thus, it will be necessary to have the results of further trials before any definitive conclusions can be reached.

Borgström, S., von Euben, F. E., Flodgren, P., Axelson, B. & Sjögren, H. O. (1982) *N. Engl. J. Med. 307,* 1080–1081.

Regulation of Interferon Gene Expression

P. M. Pitha, N. B. K. Raj, K. A. Kelley, M. Kellum, J. D. Mosca, C. H. Riggin & K. Berg

INTRODUCTION

Interferons are inducible peptides with pleiotropic biological activities. Type 1 interferons are induced as a host response to viral infection, whereas Type 2 interferons are induced during the T lymphocyte recognition of foreign histocompatibility antigens together with the other lymphokines. Although interferons were originally described as proteins which induce an antiviral state in cells, it became obvious recently that they also have an anti-proliferative effect.

Interferon induction and action are 2 separate processes regulated in human cells by different sets of chromosomes. The Type 1 interferon genes which are represented by a family of multiple α (13) (Goeddel *et al.* 1981, Brack *et al.* 1981), and β_1 (one) interferon genes are clustered on chromosome 9. The α genes show a high degree of homology in the coding region and diversity in the 3' non-coding region. The human β_1 interferon gene is present in genomic DNA in a single copy (Derynck *et al.* 1980), and shows only 45% sequence homology with the α genes (Taniguchi *et al.* 1980). Both α and β_1 interferon genes belong to a rare group of eukaryotic genes which are not spliced. The molecular events leading to the induction of interferon-gene expression by viral infection or by dsRNA are not known. It is assumed that a dsRNA intermediate which is formed during the virus replicative cycle is the inducing agent. The induction of expression of human α and β_1 interferon genes shows some tissue and inducer specificity. The α interferon genes can be induced in lymphocytes or lymphoblastoid cell lines by viral infection (e.g., Sendai virus), while in cultured fibroblast cells, viral infection or treatment with dsRNA (e.g., poly rI · rC) induces predominantly the synthesis of β_1 interferon.

Oncology Center, The Johns Hopkins University School of Medicine Baltimore, Maryland 21205 USA, and Aarhus University, DK-8000 Aarhus C, Denmark

GENE EXPRESSION, Alfred Benzon Symposium 19.
Editors: Brian F. C. Clark & Hans Uffe Petersen, Munksgaard, Copenhagen 1984.

We have cloned the α and β_1, interferon cDNA by reverse transcription of poly(A^+) RNA isolated both from human fibroblast and lymphoblastoid cells and used the cDNAs obtained as specific probes to examine the molecular mechanisms involved the inducibility and regulation of expression of interferon genes. The rate of transcription was examined in isolated nuclei and compared to the relative levels of interferon mRNA present in the cells and with the amounts of interferon synthesized.

RESULTS AND DISCUSSION

Expression of β_1 interferon genes in human fibroblast cells

Treatment of human fibroblast cells with poly rI·rC leads to the synthesis of human interferon and several other proteins of unknown biological function (Raj & Pitha 1980). Interferon synthesized in these cells was characterized by neutralization with appropriate antiserum as β_1 type and the analysis of the induced mRNA indicates the presence of only one species of translatable β_1 mRNA, identical to the cloned β_1 cDNA (Raj & Pitha 1981). Interferon synthesis under these conditions is maximal at 4–6 h after the induction, and falls to undetectible levels by 8–10 h. The levels of interferon production in induced cells correlates with the amount of interferon β_1 mRNA present in these cells detected both by translability of the β_1 mRNA in *Xenopus* oocytes (Raj & Pitha 1977, Cavalieri *et al.* 1977), and by hybridization with cDNA probe (Raj & Pitha 1981). Thus, the shutoff in interferon synthesis is due to the disappearance of interferon mRNA from the cells.

The shutoff of interferon synthesis requires cellular RNA and protein synthesis. When the induction is done in the presence of clycoheximide, cellular protein synthesis is inhibited by 99%, and β_1 mRNA accumulates in the cells. Upon clycoheximide removal, when the synthesis of cellular proteins and of interferon are restored, β_1 mRNA is degraded at the rate comparable to that in poly rI·rC induced cells; e.g., with a half-life of 30 min. These results indicate that the activation of transcription of the β_1 interferon gene by dsRNA does not require cellular protein synthesis and that the degradation of β_1 mRNA in the cells induced with dsRNA is coupled to cellular protein synthesis (Raj & Pitha 1981).

To determine whether the β_1 mRNA accumulation in the poly rI·rC-induced cells is due to the transcriptional activation of the interferon gene and whether the absence of β_1 mRNA in cells at later times of induction is due to the termination of transcription of interferon gene, we measured the rate of transcription of the β_1 interferon gene. A short pulse-labelling of the cells does not allow enough radioactivity to be incorporated into β_1 mRNA, since the β_1 mRNA represents only less than 0.1% of the total poly(A^+) RNA population. We measured, therefore, the comparative transcription rate in isolated nuclei, where the *in vivo*-initiated RNA chains are faithfully elongated *in vitro*. The nuclei were isolated from poly rI·rC-induced cells at different times after induction, and the growing RNA molecules previously initiated by RNA polymerase II were elongated in the presence of (^{32}P)-UTP. To measure the relative rate of transcription of the β_1 interferon gene, the labelled RNA was hybridized to β_1 cDNA immobilized on nitrocellulose paper. The data show (Fig. 1) that the relative transcription of the β_1 interferon gene was increased in nuclei isolated from induced cells; thus the accumulation of β_1 mRNA in human cells in response to poly rI·rC is due to the transcriptional activation of β_1 interferon gene. There is a good correlation between onset of transcription of β_1 gene in induced nuclei and the detection of β_1 mRNA in the induced cells. However, at times later than 6 h after the induction, there is no correlation between the rate of transcription of the β_1 interferon gene and accumulation of β_1 mRNA in the cells. In nuclei isolated from the poly rI·rC-induced human

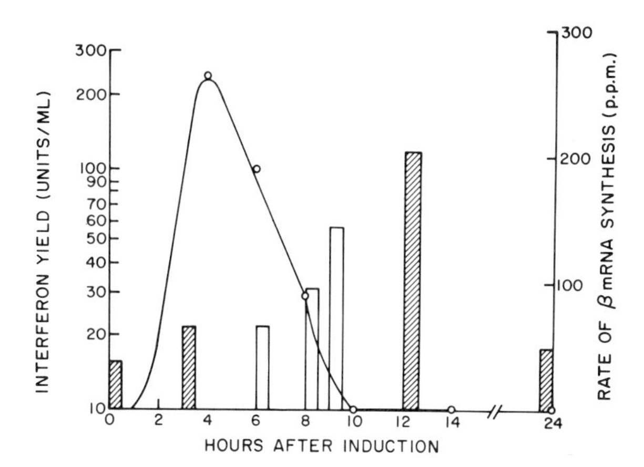

Fig. 1. Time course of β_1 interferon synthesis and relative rate of β_1 interferon mRNA transcription. Human cells were induced with poly rI·rC and medium was collected at times indicated and assayed for interferon activity. Nuclei from the same cells were isolated and transcribed *in vitro*. Relative rates of transcription (histograms) are reported as parts per million; full histograms represent mean values from 2 experiments, the empty ones represent values from a single experiment.

fibroblast cells, the β_1 interferon gene is actively transcribed for at least 12 h after the induction, while in the induced cells β_1 interferon and its mRNA cannot be detected at times later than 7 h after the induction (Fig. 1). These results show that the β_1 interferon gene is transcribed during the shutoff period of interferon synthesis when no β_1 mRNA can be detected in cells, indicating that at later times after induction, the β_1 mRNA is rapidly degraded. The degradation system show some specificity since the half-life of total poly(A^+) RNA in these cells is not greatly affected (Raj & Pitha 1981, Sehgal & Gupta 1980).

Expression of α *and* β_1 *interferon genes in Namalva cells*

In a human lymphoblastoid cell line, (Namalva) infection with Sendai virus leads to the synthesis of both α and β_1 interferons (Paucker *et al.* 1976). The

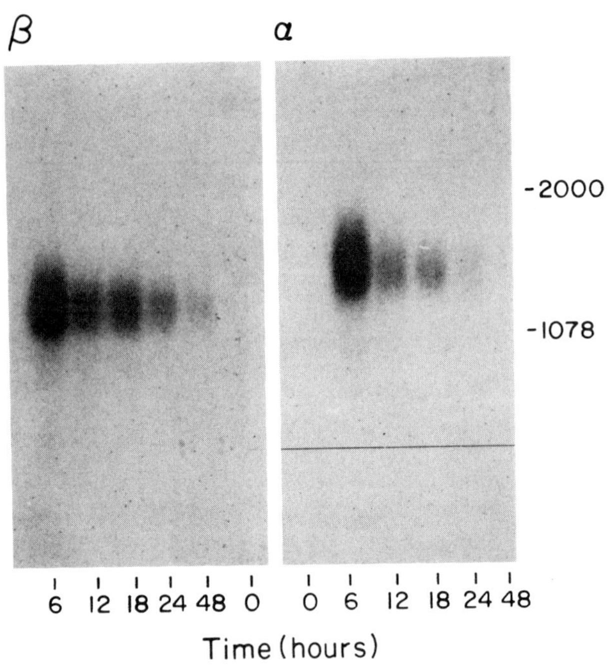

Fig. 2. Kinetics of interferon mRNA expression in Namalva cells induced with Sendai virus. Namalva cells were induced with Sendai virus and at the times indicated, cells were harvested and poly(A^+) RNA isolated. 2 μg of poly(A^+) RNA was denatured with formamide, electrophoresed through 1.1% agarose, 2.2M formaldehyde gels by the method of Lehrach *et al.* (1977), and interferon mRNA detected by Northern hybridization as described by Thomas (1980). Identical filters were hybridized with either β_1 or α cDNA probes.

amount of β_1 interferon synthesized comprises only 10–15% of the total biological activity and the rest is represented by several α interferons. To determine whether the α and β_1 interferon genes are induced and expressed in these cells co-ordinately, we examined the kinetics of interferon mRNA accumulation and the rate of interferon-gene transcription at different times after induction. The relative levels of interferon mRNA present in the induced cells at different times after induction were analyzed by Northern hybridization with ^{32}P labelled β_1 and αA cDNA probes. As the homology at the DNA level between the α and β_1 interferon genes is only 45% these 2 probes do not cross-hybridize under stringent conditions of hybridization employed; however, there is cross-hybridization between αA cDNA and the other α interferon genes and α RNAs. It can be seen (Fig. 2) that maximal amounts of α and β_1 interferon mRNA can be detected in the cells as soon as 6 h after induction, and that both α and β_1 mRNA gradually disappear from the cells within 24 h after induction. The relative levels of α and β_1 interferon mRNAs detected in the induced cells were comparable, although the amounts of biologically active β_1 interferon represented only 10–15% of total interferons synthesized.

The rate of interferon gene transcription was examined in detail in nuclei isolated from induced Namalva cells. RNA pulse-labelled *in vitro* was hybridized to immobilized αA and β_1 cDNA and to the DNA fragments containing the 5' and 3' flanking regions of both αA and β_1 interferon genes. The results show (Fig. 3) that the transcription of α and β_1 interferon genes occurs only after induction, and that both α and β_1 genes are transcribed co-ordinately and with comparable efficiency. In Namalva cells several types of α interferons are synthesized (Allen 1982), and since our α cDNA probe cross-reacts with all of the α interferon genes identified, these data indicate that after induction, the β_1 interferon gene is transcribed more efficiently than the individual α interferon genes. The rate of transcription of α and β_1 interferon genes reached the maximum at 9 h after induction, and no decrease in the rate of transcription was observed up to 15 h after induction. As the levels of both α and β_1 interferon mRNAs are gradually decreased between 9 and 24 h after induction, these data indicate that the clearance of both α and β_1 interferon mRNAs from the cells occurs under conditions when the interferon genes are still actively transcribed, which indicates the presence of the post-transcriptional regulatory shutoff mechanism. In Namalva cells induced with Sendai virus, the shutoff mechanism is, however, less effective than in poly rI · rC-induced human fibroblast, and the half-life of interferon mRNAs is approximately 3 h.

To examine the molecular nature of the transcription of interferon genes in detail, RNA pulse-labelled *in vitro* was hybridized to the 5' and 3' flanking regions of the β_1 and αA interferon genes. No hybridization was detected to the DNA fragment containing the 5' flanking region of either αA or β_1 interferon genes indicating that no transcription occurred in the region upstream from the cap site. However, transcripts hybridizing with the 3' flanking regions of both αA and β_1 interferon genes were detected. The data on the transcription of the β_1 interferon gene indicates that the transcripts extend beyond the 3' polyadenylation of mature β_1 mRNA. Northern blot hybridization revealed the presence of a large (24S) RNA in fibroblast cells induced with Newcastle disease virus, or poly rI·rC which hybridized both with β_1 cDNA and the 3' flanking fragment of human DNA (Fig. 4). The relative levels of this RNA are substantially lower (at least 100-fold) than the levels of the mature β_1 mRNA. Whether the 24S RNA represents primary transcript which is then processed into 11S poly(A^+) RNA, or whether only a small percentage of the β_1 interferon gene transcripts are not properly terminated remains to be determined.

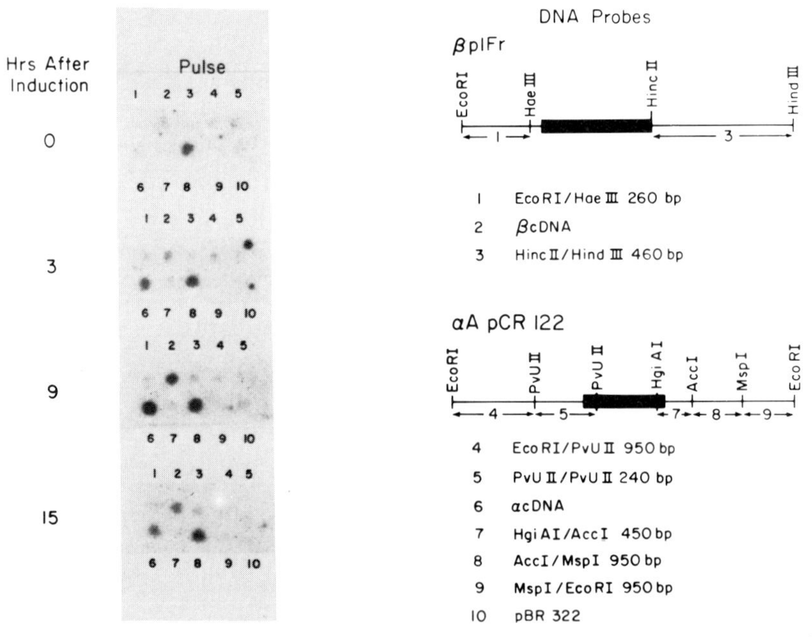

Fig. 3. Relative rate of transcription of αA and β_1 interferon genes in Namalva cells induced with Sendai virus. Nuclei were isolated from induced Namalva cells at different times after induction. *In vitro* transcription, RNA isolation and hybridization were done as described recently (Raj & Pitha 1983).

The transcript of the αA interferon gene seems to have a proper termination, and we did not detect any large molecular size transcripts in the induced lymphoblastoid cells. However, a constitutively transcribed DNA region, in close proximity (~300 nucleotides) to the 3' end of αA interferon gene was identified. This DNA sequence represents a unique region which is present in the genome DNA as a single copy; the sequence analysis of this DNA does not show any open reading frame and the sequence homology was found only in poly(A^-) RNA fraction indicating the lack of a coding potential for protein. The presence of the constitutively expressed DNA sequence in 3' flanking region of the inducible αA gene is of interest. The inducible genes several kb downstream from β_1 interferon gene was previously identifed by (Gross *et al.* 1981), which is in correlation with our previous findings (Raj & Pitha 1980).

Expression of human αA and β_1 genes in mouse Ltk^- cells

With the availability of isolated interferon α and β_1 interferon genes and biological assay for its detection, we have attempted to determine the inducibility and regulation of expression of these genes by transfection into mouse Ltk^- cells using conversion to the Tk^+ phenotype as a selection for transfection (Pitha *et al.* 1982). The DNA sequence present in the upstream 5'

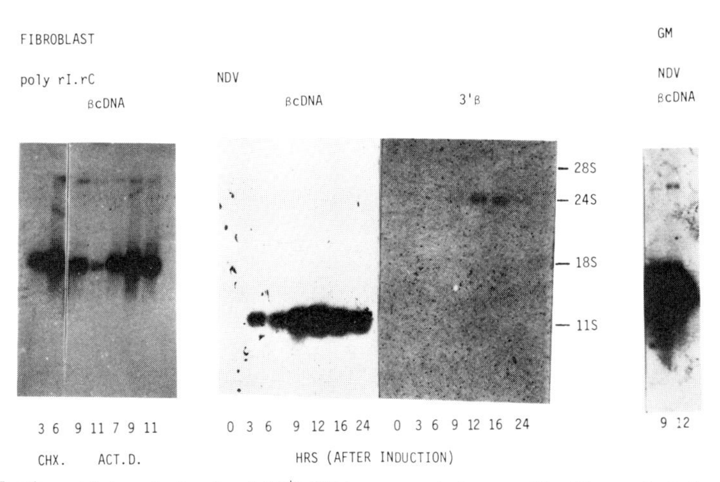

Fig. 4. Northern blot analysis of poly(A^+) RNA present in human fibroblast cells induced with Newcastle disease virus. Poly(A^+) RNA was isolated from the cells at different times after induction and analyzed by Northern hybridization technique as described in Fig. 1. The nitrocellulose filter was hybridize either with the β_1 cDNA probe or the 460 bp HincII/HindIII fragment (Fig. 3) of the 3' flanking region of the β_1 interferon gene.

flanking sequence from the human β_1, shows a sequence homology with the sequence present in the far upstream region of the human αA interferon, and this common sequence has been suggested to have a role in the induction (DeGrave *et al.* 1981, Tavernier *et al.* 1982).

The human αA and β_1 interferon genes cloned into pBR322, were used for transfection. The genomic β_1 interferon clone pIF_R contains interferon gene and 260 and 460 nucleotides of the 5' and 3' flanking sequence respectively. The αA interferon clone contains in addition to the αA interferon gene 1100 and 2300 nucleotides of the 5' and 3' flanking sequences. Thus, both clones contain putative TATAAA transcription-initiation site, and the common sequences which may have a role in the induction of these genes.

Individual colonies from each transfection were isolated, grown into mass cultures and tested for both the constitutive production of human interferon and its inducibility (Table I). When β_1 interferon gene was transfected into Ltk^- cells, the β_1 interferon synthesis could be induced in these cells by viral infection, however, αA interferon gene was not inducible inspite of the fact that it was constitutively expressed at low levels. Since αA interferon gene can be induced in lymphoblastoid, but not in fibroblast cells by viral infection, our results may indicate that the fibroblast cells lack a regulatory factor required for the induction of this gene. The induction of human αD gene in mouse cells was detected on RNA, but not the protein level (Mantei *et al.* 1982), 2 other α interferon genes upon transfection into hamster cells were also poorly inducible (Colby personal communication). Thus, the inducibility of a single α interferon gene seems to be much less efficient, both in homologous and heterologous cells, while compared with the β_1 interferon gene.

CONCLUSIONS

We have cloned the human α and β_1 interferon genes and used these as a specific probe for the study of human α and β_1 interferon-gene expression. Our results indicate that the synthesis of interferon in induced human cells is regulated both by the activation of transcription of interferon genes and by alteration of the interferon mRNA stability. The inducer activates the transcription of interferon genes and the transcription is independent of cellular protein synthesis. The selective nature of inducibility of α interferon gene seems to be reflected even after the transfection of the isolated interferon genes into heterologous cells.

Table I

Synthesis of human α_2 and β_1 interferons in Ltk^+ lines transfected with genomic clones

Gene	Cell lines	Constitutive		Interferon Synthesis $U/10^6$ Cells				
					Induced human		Induced mouse	
		# Lines	U/ml	# Lines	rI·rC	NDV	rI·rC	NDV
4.2 kg α_2	$LH_2pCR122$	20/20	$1\text{–}2.10^3$	0/20	<20	<20	30	1000–4000
1.8 kb β_1	LH_2pIF_g	12/15	10–90	7/15	100	1000–3000	30	1000–4000
Tk-HSV-2	LH_2pGR18	2/2	<5	–	<5	<5	30	3000
–	Ltk	1/1	<5	–	<5	<5	30	6000
–	HF	1/1	<5	–	100–500	1000–3000	<5	<5

Constitutive IFN synthesis was tested on GM2504 cells and the induced IFN was tested on VERO cells (β) or bovine cells (α_2). # number of positive lines from total tested.

REFERENCES

Allen, G. (1982) *Biochem. J. 207,* 397–408.

Brack, C., Nagata, S., Mantei, N. & Weissmann, C. (1981) *Gene 15,* 379–394.

Cavalieri, R. L., Havell, E. A., Vilcek, J. & Pestka, S. (1977) *Proc. Natl. Acad. Sci. 74,* 4415–4419.

DeGrave, W., Derynck, R., Tavernier, J., Haegeman, G. & Fiers, W. (1981) *Gene 14,* 137–143.

Derynck, R., Content, J., DeClercq, E., Volckaert, G., Tavernier, J., Devos, R. & Fiers, W. (1980) *Nature 285,* 285, 542–547.

Goeddel, D. V., Leung, D. W., Dull, T. J., Gross, M., Lawn, R. M., McCandliss, R., Seeburg, P. H., Ullrich, A., Yelverton, E. & Gray, P. W. (1981) *Nature 290,* 20–26.

Gross, G., Mayr, U. & Collins, J. (1981) In: *The Biology of Interferon Systems,* (E. DeMaeyer, E., Gallasso, G. & Schellehens, H., eds.) pp. 85–89.

Lehrach, H., Diamond, D., Wozney, J. M. & Boedtker, H. (1977) *Biochem. 16,* 4743–4751.

Mantei, N. & Weissmann, C. (1982) *Nature 297,* 128–132.

Paucker, K., Dalton, B. J., Ogburn, C. A. & Torma, E. (1976) *Proc. Natl. Acad. Sci. 72,* 4587–4591.

Pitha, P. M., Ciufo, D. M., Kellum, M., Raj, N. B. K., Reyes, G. R. & Hayward, G. S. (1982) *Proc. Natl. Acad. Sci. 79,* 4337–4341.

Raj, N. B. K. & Pitha, P. M. (1977) *Proc. Natl. Acad. Sci. 74,* 1483–1487.

Raj, N. B. K. & Pitha, P. M. (1980) *Proc. Natl. Acad. Sci. 77,* 4918–4922.

Raj, N. B. K. & Pitha, P. M. (1981) *Proc. Natl. Acad. Sci. 78,* 7426–7430.

Sehgal, P. B. & Gupta, S. L. (1980) *Proc. Natl. Acad. Sci. 77,* 3489–3493.

Taniguchi, T., Ohno, S., Fujji-Kuriyama, Y. & Maramatsu, M. (1980) *Gene 10,* 11–15.

Tavernier, J., Gheysen, D., Duerinck, F., Van der Heyden, J. & Fiers, W. (1983) *Nature 301,* 634–636.

Thomas, P. S. (1980) *Proc. Natl. Acad. Sci. 77,* 5201–520505.

The 2-5A and Protein Kinase Systems in Interferon-treated and Control Cells

Ian M. Kerr, A. Rice, W. K. Roberts, P. J. Cayley, A. Reid, C. Hersh* & G. R. Stark**

INTRODUCTION

I propose first to summarise the current state of knowledge concerning the mechanisms of action of interferon (Lengyel 1982). The major part of this presentation will, however, be concerned with the interferon and double-stranded RNA (dsRNA)-mediated protein kinase and 2-5A (ppp(A2′p)$_n$A; n≥2) systems, particularly with the latter, its role in the interferon-treated cell, and possible wider significance.

The interferons are a family of proteins, each of which is capable of inhibiting cell proliferation and virus growth and of inducing a complex response in appropriate cells. There are at least 3 antigenic types of human interferon, α, β and γ in the new terminology, leukocyte and fibroblast (Type I) and immune (Type II), respectively, in the old. A similar situation exists for the mouse interferons. Indeed, it has now become clear that there are at least 16 human α interferon genes, the situation with β is controversial and for the moment only one γ interferon gene has been reported. Do all of these interferons have the same mechanism of action? Certainly they all induce the 2-5A synthetase and protein kinase. In more general terms the α and β interferons share a common receptor (Branca & Baglioni 1981), and induce the same polypeptides (Weil *et al.* 1983). On the other hand γ has different receptors, induces additional polypeptides and shows differences in the kinetics of induction of the antiviral state (Branca & Baglioni 1981, Weil *et al.* 1983, Dianzani *et al.* 1980). Even with the α interferons, however, we have to conclude either that there are subtly

ICRF Laboratories, Lincoln's Inn Fields, London, England, and *Biochemistry Dept., Stanford University, California, U.S.A.

GENE EXPRESSION, Alfred Benzon Symposium 19.
Editors: Brian F. C. Clark & Hans Uffe Petersen, Munksgaard, Copenhagen 1984.

different receptors or that the different interferons can play different tunes on the same receptor. Despite this, for the moment the general concensus remains that all of the interferons are likely to have the same fundamental biochemical mechanisms.

The interferons are not directly antiviral: interferon treatment is always of the cell. Only a few molecules per cell appear to be required to trigger a response. It is not yet certain whether interferon has to get into the cell, but the current concensus is that it works from the membrane. It takes time and there is a requirement for RNA and protein synthesis for the antiviral state to develop. In accord with this, interferon induces the cell to produce a number of polypeptides which are assumed to be responsible for the antiviral and growth inhibitory effects. The outstanding question remains the nature of the second message. In this regard, nothing is known of the events between the interaction of interferon with the membrane receptor and the appearance of a number of mRNAs and polypeptides in response to it, although it is possible that cyclic AMP or cyclic GMP may be in some way involved in some systems.

Over the years the work of a number of groups has indicated that in different cell-virus systems, the antiviral effects of interferon can be expressed at the level of viral RNA synthesis or methylation, viral protein synthesis or at the level of virus maturation or release. Effects on the entry and uncoating of some viruses have also been reported. Multisite models for interferon action are, therefore, accepted, in which any one or more than one of these different possible mechanisms may be involved in a given cell-virus situation. In accord with this, a cell can be in an antiviral state for one virus but not another (Nilsen *et al.* 1980).

It is now accepted that the interferons are not just antiviral agents: they can inhibit the growth of both normal and transformed cells, have profound effects on the cytoskeleton and the cell membrane and either inhibit or enhance cell function (Taylor-Papadimitriou 1980). This last is particularly true in relation to the complex interaction of the interferons with the immune system. The interferons are predominantly cytostatic rather than cytotoxic. Depending on the cell type and conditions, it appears that the effect can be either to elongate all stages of the cell cycle (G1 and G2 normally being more affected than S), or to arrest the cells predominantly in G1. Pure human β-IFN inhibits the proliferation of human diploid fibroblasts while only marginally affecting RNA, DNA, and protein synthesis. In accord with this, there is an increase in cell volume and the number of binucleate cells, suggesting that cell division is in some way inhibited. An increase in actin fibres and in the fibronectin network and a decrease in cell motility and ruffling are also observed (Pfeffer *et al.* 1980). Interferons also induce a multiplicity of changes in the cell membrane (Taylor-Papadimitriou 1980, Friedman 1979). These include a decrease in membrane fluidity, an increase in saturated fatty acids, an inhibition of cap formation and an increase in surface antigen expression.

Effects on translation

We know that protein synthesis is inhibited in a variety of interferon-treated, virus-infected cells, and the cell-free systems isolated from them. Depending on the conditions of cell growth, interferon dose and virus multiplicity, both host and viral or exclusively viral protein synthesis is inhibited. We also know that protein synthesis in cell-free systems from interferon-treated cells is exceptionally sensitive to doublestranded RNA (Kerr *et al.* 1974). This reflects the activity of 2 interferon-induced, dsRNA-dependent enzymes: a cAMP-independent protein kinase and the 2-5A synthetase. Both enzymes are present constitutively in a wide variety of cells and tissues, but their level increases in response to interferon. The kinase and 2-5A systems have been extensively reviewed (Baglioni 1979, Williams & Kerr 1980, Lengyel 1982).

The protein kinase

The dsRNA-dependent protein kinase appears analogous to that first described in rabbit reticulocytes (Farrel *et al.* 1977). It has been observed by a number of groups in a variety of cell-free systems in which it phosphorylates, an endogenous 67,000 (mouse) or 69,000 (human) dalton polypeptide. It also phosphorylates the α subunit of eukaryotic protein synthesis initiation factor 2 (eIF2α) which results in an inhibition of protein synthesis. In addition to being required by the kinase, the dsRNA may also augment its effect by inhibiting an eIF2α protein phosphatase (Epstein *et al.* 1980). It has not been possible, as yet, to establish the significance of the kinase in the antiviral action of interferon in the intact cell. It has, however, recently been shown to be active in interferon-treated, reovirus-infected cells (Gupta *et al.* 1982, Nilsen *et al.* 1982a). It is also worth emphasising that from what we know of the requirements for activation of the 2 systems, evidence for the operation of the 2-5A system makes it probable that the kinase will also be active. This cannot, however, be taken for granted; we have recently obtained evidence for example, that the kinase but not the 2-5A synthetase, is inhibited in response to vaccinia virus infection of mouse L and human HeLa cells. This inhibition was not prevented by interferon pretreatment of the cells (Rice & Kerr, unpublished). It would appear, therefore, that the kinase is not necessarily involved in the effect of interferon on vaccinia replication. However, no very dramatic inhibition of protein synthesis, such as has been reported previously, was observed in these experiments. It remains perfectly possible, therefore, that under different conditions of virus growth, the kinase may indeed be involved. Certainly our current results imply that if not inhibited, activation of the kinase by vaccinia dsRNA, which does appear to be present late in infection (see below), would occur and affect virus growth.

The 2-5A system

Three enzymes are involved: (1) the 2-5A synthetase(s) which synthesises 2-5A, (2) a 2′,5′-phosphodiesterase which degrades 2-5A, and (3) the 2-5A-dependent RNase (RNase L or RNase F). Constitutive levels of all of these enzymes are present in a variety of cells and tissues. The level of the synthetase usually increases substantially (10- to 10,000-fold) in response to interferon. Only relatively small (<5-fold) changes in the other 2 enzymes are normally observed although the 2-5A-dependent nuclease has recently been reported to increase by up to 20-fold in some cell lines (Jacobsen *et al.* 1983). 2-5A is not a single molecule but an oligomeric series $ppp(A2'p)_nA$ ($n=2$ to >4) (Kerr & Brown 1978). Oligomers up to the dodecamer at least have been reported to be synthesised enzymically in progressively decreasing amounts. The trimer, tetramer and pentamer are the predominant products. They are equally active at nanomolar concentrations in cell-free systems and have been found to occur naturally in interferon-treated, encephalomyocarditis virus (EMC)-infected cells (Williams *et al.* 1979a, Knight *et al.* 1980). When introduced into intact cells they inhibit protein and DNA synthesis and, hardly surprisingly therefore, virus growth (Williams *et al.* 1979b, Hovanessian & Wood 1980). When activated by 2-5A, the 2-5A-dependent RNase cleaves both cellular and viral RNA on the 3′ side of UN doublets with a preference for UU or UA, to yield UpNp-terminated products: with rRNA in intact ribosomes a limited and highly characteristic pattern of cleavage products is obtained (Wreschner *et al.* 1981, Floyd-Smith *et al.* 1981). The significance of this specificity is not known. For activity the enzyme requires the continued presence of 2-5A. Since 2-5A is unstable in cells and cell-free systems, the activation of the RNase is transient in the absence of a 2-5A regenerating system (Williams *et al.* 1978). Cell extracts contain a protein which binds 2-5A with high affinity and specificity which is thought to be the 2-5A dependent RNase (Slattery *et al.* 1979, Wreschner *et al.* 1982).

The 2-5A system in interferon-treated, virus-infected cells

RNA viruses. 2-5A is present and the 2-5A-dependent RNase is active in interferon-treated, EMC virus-infected mouse-L and human HeLa cells and in interferon-treated, reovirus-infected HeLa cells. 2-5A *per se* has been detected in TCA extracts from the former by a combination of HPLC analysis with both radiobinding (RB) and biological assays (Williams *et al.* 1979a, Knight *et al.* 1980, Silverman *et al.* 1982a). Evidence for the activation of the RNase has been provided by the detection of a characteristic pattern of 2-5A-mediated rRNA cleavages in such cells (Wreschner *et al.* 1981, Silverman *et al.* 1982a, Nilsen *et al.* 1982b). Moreover, EMC virus growth is resistant to interferon in NIH/3T3 clone 1 cells which are thought to be deficient in the 2-5A-dependent RNase (Epstein *et al.* 1981). It must be remembered, however, that other functions may be deficient in these cells.

DNA viruses. 2-5A and 2-5A-related material has been detected by a combination of HPLC with RB, radioimmune (RI) and RNase activation assays in TCA extracts from SV40-infected, interferon-treated monkey (CV1) cells at late times (>40h) post infection (Hersh, G.R.S., W.K.R. & I.M.K.). 2-5 A is also present and the 2-5A-dependent RNase is ultimately activated in interferon-treated, vaccinia virus-infected mouse L and human HeLa cells (A. Rice, W.K.R. & I.M.K.). Interestingly this implies the natural occurrence of dsRNA in these systems. The significance of the 2-5A in any effect of interferon on the replication of these viruses, however, is uncertain. In the SV40 system, treated with interferon after virus infection there is little, if any, inhibition of virus growth and very little authentic 2-5A capable of the activation of the 2-5A-dependent RNAse has been detected. The nature and function of the additional 2-5A-related material detected in this system remains to be established. The vaccinia system currently presents us with a paradox: high titres of authentic 2-5A capable of activating the RNase *in vitro* are observed, but rRNA cleavage is delayed and there is little obvious overall inhibition of protein synthesis in the intact cell. It will be interesting to determine the basis for this apparent resistance to high levels of 2-5A.

Control of the 2-5A system at the level of the 2-5A-dependent RNase

In the absence of interferon treatment the 2-5A-dependent RNase is lost or inactivated in response to EMC infection (Cayley *et al.* 1981a, Silverman *et al.* 1982a). Interferon pre-treatment prevents this. In the EMC-HeLa cells system in particular, it seems that the crucial aspect of interferon action may be the prevention of the virus-mediated inactivation of the RNase rather than the induction of the 2-5A synthetase (Silverman *et al.* 1982a). The mechanism of the inactivation of the RNase is not known.

The 2-5A system and the selective inhibition of viral protein synthesis

Cells treated with low concentrations of interferon and high multiplicites of EMC virus die, but cells treated with high concentrations of interferon and low multiplicities of virus survive (Vaquero *et al.* 1981, Munoz & Carrasco 1981). Under the latter set of conditions viral, but not host, protein synthesis is inhibited. By the criterion of 2-5A-mediated rRNA cleavage, however, the 2-5A system operates under both sets of conditions (Cayley *et al.* 1982b). A possible model for how such a selective inhibition of viral protein synthesis might operate, involves a localised activation of the 2-5A system in the neighbourhood of partially double-stranded, replicating viral RNA. Evidence suggesting that such localised activation can indeed occur in extracts from interferon-treated cells has been provided by Nilsen & Baglioni (1979). Irrespective of the mechanism, the rRNA cleavage data show that a selective inhibition of protein synthesis can occur under conditions in which the 2-5A system is active. It will be intriguing to determine whether or not the 2-5A or rRNA cleavage is directly involved.

From these data it is reasonable to conclude that the 2-5A system is both present and active in at least some virus-infected cells and that, in the absence of interferon, it can be switched off in response to infection.

The 2-5A system and the effects of interferon on cell growth

The non-phosphorylated 'core' of 2-5A i.e. $(A2'p)_nA$, is inactive in cell extracts but it has been found in intact interferon-treated cells (Williams *et al.* 1979a) and treatment of mitogenically-stimulated lymphocytes or serum-stimulated 3T3 cells with 'core' can inhibit the mitogenic response (Kimchi *et al.* 1981a, b). In addition, the NIH/3T3 cell line mentioned above, which is thought to be deficient in the 2-5A-dependent RNase, is resistant to the growth inhibitory effects of interferon (Kimchi *et al.* 1981b).

The Daudi line of human lymphoblastoid cells is exceptionally sensitive to the growth inhibitory effects of interferon. To date, however, we have failed to detect significant levels (>1nM) of the 5′di- or triphosphorylated components of 2-5A, or any evidence for 2-5A-mediated rRNA cleavage in interferon-treated

Daudi cells (Silverman *et al.* 1982b). On the other hand, material which may correspond to the non-phosphorylated 'cores' of 2-5A has recently been detected (2–15nM) in such cells (Cailla & Kerr).

It may be that the 2-5A system is involved in interferon's effect(s) on growth in some cells but not others. For the moment it is perhaps best to conclude that a role for the 2-5A system in the cell growth inhibitory effects of interferon remains to be firmly established.

Alternative products of the 2-5A synthetase

An enzyme which is almost certainly the 2-5A synthetase will add AMP in 2′–5′ linkage to such interesting metabolites as NAD^+, ADP-ribose and A5′$p_4$5′A (Ball & White 1979, Ferbus *et al.* 1981). Indeed, in our experience, NAD^+ appears to be the preferred acceptor with the HeLa cell enzyme. Despite this we have been unable to detect the natural occurrence of such modified compounds in extracts from EMC virus-infected cells known to be synthesising 2–5A (Cayley & Kerr 1982). This does not, of course, mean that alternative products may not be of importance in other situations. In fact the complexity of 2-5A-related components being detected in a variety of systems (see below) suggests that this may well be the case. The discovery of 2-5A synthetases of about 30,000 and 100,000 in molecular weight in the nuclear and cytoplasmic fractions respectively, of a number of cell lines is particularly intriguing in this respect (St. Laurent *et al.* 1983, Revel *et al.* personal communication).

Alternative functions of 2-5A

To data the only known function of 2-5A is activation of the RNase. If, however, the system is indeed involved in the inhibition of cell growth, one intuitively suspects that it is likely to be through an alternative mechanism(s) (or products). In this connection Lengyel's group have recently reported the detection of additional polypeptides which can be affinity labelled with ppp(A2′p)$_3$ A(^{32}P)pCp. It is particularly intriguing that these were found in nuclear extracts (St. Laurent *et al.* 1983). In addition, we have obtained evidence suggesting an unusually strong interaction between 2-5A and RNA. The precise nature of this interaction is uncertain. If, however, a tight association of 2-5A with RNA can be rigorously proved, it would indicate an alternative function(s) for 2-5A.

Wider significance of the 2-5A system

The enzymes of the 2-5A system are widely distributed in a variety of reptilian, avian and mammalian cells and tissues (Stark *et al.* 1979, Cayley *et al.* 1982c). The major effect of interferon on these enzymes is to elevate the synthetase. The level of this enzyme can also vary with growth and hormone status (Stark *et al.* 1979, Krishnan & Baglioni 1980), suggesting a wider role in normal cell growth or function. The synthetase, however, requires dsRNA for activity and, although the enzyme is present and varies in amount, it has not been established whether it is active and to what extent 2-5A or related products accumulate in normal cells or tissues. Accordingly we have recently surveyed a variety of systems for 2-5A and related products employing RB, RI and RNase activation assays combined with HPLC analysis. 2-5A or 2-5A-related material has been found in the majority of cells and tissues examined. One such tissue is the chick oviduct. In this tissue the 2-5A synthetase increases 10-fold during regression of the oviduct on withdrawal from hormone stimulation (Stark *et al.* 1979). This, however, occurs slowly over a period of days rather than dramatically in the first few hours when major changes in mRNA are occurring. Its significance is not, therefore, clear. HPLC fractionation combined with RI, RB and RNase activation assays have, however, confirmed that 2-5A occurs naturally in the oviduct. The level observed is variable, but from an analysis of 2 sets of oviducts it appears to peak shortly after oestrogen withdrawal. A detailed analysis of individual samples by HPLC has demonstrated that there is also an interesting variability in the 2-5A components present (A. Reid, G.R.S. & I.M.K.).

In similar, but less extensive, studies 2-5A and 2-5A-related material has also been shown to occur naturally in rat mammary gland and human neuroblastoma T98G cells grown in tissue culture (A. Reid, G.R.S. & I.M.K.). In addition significant concentrations ($\geq$10nM) of the non-phosphorylated "cores" of 2-5A ($(A2'p)_nA$; $n \geq 2$) have routinely been found in all normal mouse tissues examined (Cailla *et al.* 1982). The presence of the corresponding phosphorylated molecules is much more variable (L. Laurence, H. Cailla, R. E. Brown & I. M. K.). Even the presence of 2-5A *per se* does not automatically mean that the system will be active. Regulation can also occur at the level of the 2-5A-dependent RNase (Clayley *et al.* 1982a, Silverman *et al.* 1982a). If active, however, it could be that i it plays a role in the processing of RNA. Nilsen et al. (1982c) have presented RB data for the presence of very low levels ($<$ 1nM) of 2-5A in the nucleus of HeLa cells from which the hnRNA is capable of activating the 2-5A synthetase (Nilsen *et al.* 1982d). Alternatively this system could be involved in the normal turnover of ribosomes and ribosomal RNA. The possibility cannot, however, be excluded that the 2-5A detected here may be present in relatively high concentrations in only a small percentage of the cells.

SUMMARY

Essentially nothing is known of the molecular basis of the effects of interferon on cell growth and function, although it is possible that cAMP, the 2-5A system and/or the control of ornithine decarboxylase (Sreevalsen *et al.* 1979) may be involved in some cells. With respect to the antiviral action, the 2-5A system appears to be important in the effects on picornaviruses and it is clear that the changes in the cell membrane have a profound affect on the maturation and release of the RNA tumour viruses. The relative importance of the interferon-induced and dsRNA-mediated protein kinase remains to be established. In addition, it remains perfectly possible that effects on uncoating, transcription, methylation and other as yet unrecognised mechanisms, will prove to be of equal or greater importance in other cell-virus systems. In general it may be quite normal for more than one of the possible antiviral mechanisms to operate in any given system. Certainly the establishment of a role for one does not necessarily exclude others.

With respect to 2-5A, we know that it is unstable and in the absence of a regenerating system it transiently activates an RNase. The nuclease cleaves RNA on the 3′ side of UpNp doublets and shows a high preference for UpA or UpU sequences. With rRNA in intact ribosomes, a very limited and highly characteristic pattern of products is observed. When introduced into intact cells 2-5A inhibits protein and DNA synthesis and virus growth. It occurs naturally in amounts sufficient for it to play a part in the antiviral activity of interferon, and the 2-5A-dependent RNase is active in interferon treated, RNA and DNA virus-infected mouse and human cells. In contrast, in the absence of interferon treatment, the 2-5A system can be switched off in response to virus infection.

Finally, the wide distribution of the enzymes of the 2-5A system, of 2-5A and 'cores', and the variation in the levels of these different components with growth and hormone status, together with the ability of the synthetase(s) to synthesise alternative products, all point to a wider significance for it.

REFERENCES

Baglioni, C. (1979) *Cell 17,* 255–264.

Ball, L. A. & White, C. N. (1979) in: Koch, G. & Richter, D. (Eds.) *Regulation of Macromolecular Synthesis* pp. 303–317. Academic Press, New York.

Branca, A. A. & Baglioni, C. (1981) *Nature 294,* 768–770.

Cailla, H., Laurence, L., Le Borgne de Kaouel, C., Roux, D. & Marti, J. (1982) in: *Radio-immunoassays & Related Procedures in Medicine,* International Atomic Energy, Vienna, pp. 33–44.

Cayley, P. J. & Kerr, I. M. (1982) *Eur. J. Biochem. 122,* 601–608.

Cayley, P. J., Knight, M. & Kerr, I. M. (1982a) *Biochem. Biophys. Res. Comm. 104,* 376–382.

Cayley, P. J., Silverman, R. H., Balkwill, F. R., McMahon, M., Knight, M. & Kerr, I. M. (1982b) in: Friedman, R. M. & Merigan, T. C. (Eds.), Vol. XXV *UCLA Symp. Mol. & Cell. Biol.* pp. 143–157. Academic Press, New York.

Cayley, P. J., White, R. F., Antoniw, J. F., Walesby, N. J. & Kerr, I. M. (1982c) *Biochem. Biophys. Res. Comm. 108,* 1243–1250.

Dianzani, F., Zucca, M., Scupham, A. & Georgiades, J. A. (1980) *Nature 283,* 400–402.

Epstein, D. A., Czarniecki, C. W., Jacobson, H., Friedman, R. M. & Panet, A. (1981) *Eur. J. Biochem. 118,* 9–15.

Epstein, D. A., Torrence, P. F. & Friedman, R. M. (1980) *Proc. Natl. Acad. Sci. USA 77,* 107–111.

Farrell, P. J., Balkow, K., Hunt, T. & Jackson, R. (1977) *Cell 11,* 187–200.

Ferbus, D., Justesen, J., Besancon, F. & Thang, M. N. (1981) *Biochem. Biophys. Res. Comm. 100,* 847–856.

Floyd-Smith, G., Slattery, E. & Lengyel, P. (1981) *Science 212,* 1030–1031.

Friedman, R. M. (1979) in: "Interferon 1" (I. gresser ed.) pp. 53–74. Academic Press, New York.

Gupta, S. L., Holmes, S. L. & Mehra, L. L. (1982) *Virology 120,* 495–499.

Hovanessian, A. G. & Wood, J. N. (1980) *Virology 101,* 81–90.

Jacobsen, H., Czarniecki, L. W., Krause, D., Friedman, A. M. & Silverman, R. H. (1983) *Virology 125,* 496–501.

Kerr, I. M. & Brown, R. E. (1978) *Proc. Natl. Acad. Sci. USA 75,* 256–260.

Kerr, I. M., Brown, R. A. & Ball, L. A. (1974) *Nature 250,* 57–59.

Kimchi, A., Shure, H. & Revel, M. (1981a) *Eur. J. Biochem. 114,* 5–10.

Kimchi, A., Shure, H., Lapidot, S., Rapoport, S., Panet, A. & Revel, M. (1981b) *FEBS Letters 134,* 212–216.

Knight, M., Cayley, P. J., Silverman, R. H., Wreschner, D. H., Gilbert, C. S., Brown, R. E. & Kerr, I. M. (1980) *Nature 288,* 189–192.

Krishnan, I. & Baglioni, C. (1980) *Proc. Natl. Acad. Sci. USA 77,* 6506–6510.

Lengyel, P. (1982) *Ann. Rev. Biochem. 51,* 251–282.

Munoz, A. & Carrasco, L. (1981) *J. Gen. Virol. 56,* 153–162.

Nilsen, T. W. & Baglioni, C. (1979) *Proc. Natl. Acad. Sci. USA 76,* 2600–2604.

Nilsen, T. W., Maroney, P. A. & Baglioni, C. (1982a) *J. Biol. Chem. 257,* 14593–14596.

Nilsen, T. W., Maroney, P. A. & Baglioni, C. (1982b) *J. Virol. 42,* 1039–1045.

Nilsen, T. W., Maroney, P. A., Robertson, H. D. & Baglioni, C. (1982d) *Mol. Cell. Biol. 2,* 154–160.

Nilsen, T. W., Wood, D. L. & Baglioni, C. (1980) *Nature 286,* 178–180.

Nilsen, T. W., Wood, D. L. & Baglioni, C. (1982c) *J. Biol. Chem. 257,* 1602–1605.

Pfeffer, L. M., Wang, E. & Tamm, I. J. (1980) *J. Cell. Biol. 85,* 9–17.

Silverman, R. H., Cayley, P. J., Knight, M., Gilbert, C. S. & Kerr, I. M. (1982a) *Eur. J. Biochem.* 24, 131–138.

Silverman, R. H., Watling, D., Balkwill, F. R., Trowsdale, J. & Kerr, I. M. (1982b) *Eur. J. Biochem. 126,* 333–341.

Slattery, E., Ghosh, N., Samanta, H. & Lengyel, P. (1979) *Proc. Natl. Acad. Sci. USA 76,* 4778–4782.

Sreevalsan, T., Taylor-Papadimitriou, J. & Rozengurt, E. (1979) *Biochem, Biophys. Res. Comm. 87,* 679–685.

St. Laurent, G., Yoshie, O., Floyd-Smith, G., Samanta, H., Sehgal, P. B. & Lengyel, P. (1983) *Cell 33,* 95–102.

Stark, G. R., Dower, W. J., Schimke, R. T., Brown, R. E. & Kerr, I. M. (1979) *Nature 278,* 471–473.

Taylor-Papadimitriou, J. (1980) in: Gresser, I. (Ed.) *Interferon 2,* pp. 13–46 Academic Press, New York.

Vaquero, C., Aujean-Rigaud, O., Sanceau, J. & Falcoff, R. (1981) *Antiviral Res. 1,* 123–134.

Weil, J., Epstein, C. J., Epstein, L. B., Sednick, J. J., Sabran, J. L. & Grossberg, S. E. (1983) *Nature 301,* 437–439.

Williams, B. R. G. & Kerr, I. M. (1980) *Trends Biochem. Sci. 5,* 138–140.

Williams, B. R. G., Golgher, R. R., Brown, R. E., Gilbert, C. S. & Kerr, I. M. (1979a) *Nature 282,* 582–586.

Williams, B. R. G., Golgher, R. R. & Kerr, I. M. (1979b) *FEBS Letters 105,* 47–52.

Williams, B. R. G., Kerr, I. M., Gilbert, C. S., White, C. N. & Ball, L. A. (1978) *Eur. J. Biochem. 92,* 455–462.

Wreschner, D. H., James, T. C., Silverman, R. H. & Kerr, I. M. (1981) *Nucleic Acids Res. 9,* 1571–1581.

DISCUSSION

ERDMANN: Can you say anything about the activity of the nuclease-treated ribosome?

KERR: We don't know. The only bit of data that we have is not our own. It comes from some work from Michel Revel's group, where they noted that there was ribosomal RNA cleavage in cells late after SV40 infection and interferon treatment. The cleavage pattern they showed looks remarkably like the one generated in response to 2–5A and almost certainly was 2–5A-mediated. They noted that the cleavages were mainly in monosomes and not in polysomes. This suggests that the cleaved ribosomes are impaired in function.

ROTTMAN: What information do you have on the levels of stimulation of the oviducts by hormone. How much of a stimulation do you see there?

KERR: Using a radioimmunoassay one is picking up total 2–5A or 2–5A-related material in the fully hormone-stimulated oviducts varying between about 1 and 10 nM in the tissue. After withdrawal from oestrogen stimulation, the level goes up to 10–20 nM. These numbers are derived from about a dozen fully stimulated oviducts and a further dozen taken at various different times after withdrawal. The peak appears to occur about 6 hours after oestrogen withdrawal. This is about the same time scale that one might have expected on the basis of Dick Palmiter's data, saying that there is a very rapid turnover of the egg-white protein-messenger RNAs occurring 6–18 hours after oestrogen withdrawal. Consistent with this if one looks for biologically active material rather than total 2–5A positive material by radioimmunoassay, so far we have only detected active material in the time points immediately after oestrogen withdrawal. In addition, the hplc patterns that I showed you are the simplest that we have obtained. With many of the samples, taken at other times, one is getting a complex of material which may be related products rather than biologically active 2–5A. Accordingly, detailed analyses will be required to know which components of the system are there, whether they are likely to be active, and then, ideally, to prove that they are active by assaying for nuclease. It is not going to be simple.

ROTTMAN: Do you have any information on relative nuclear and cytoplasmic levels of the 2–5A?

KERR: No, we don't. The only bit of data is from Tim Nilsen's work. Tim found very low (subnanomolar) levels in Hela cell nuclei isolated using organic solvent techniques. Detection was by the radio-binding assay only, however, and the levels were so low that I would prefer to see additional data to really establish the presence of biologically active 2–5A in nuclei. Certainly, we should be looking in the nucleus a lot more than we have been.

KJELDGAARD: With your ribosome assay, did you use purified ribosomes?

KERR: No, we just used crude extracts (S10s). Once you purify the ribosomes you get trouble with endogenous nuclease and it is actually better to do it in a crude extract.

SAFER: Will DNA activate the 2–5A synthetase *in vitro*, and can you correlate the inhibition of translation *in vivo* with either the ribosomal RNA cleavage or destruction of messenger RNA?

KERR: Activation is not seen with DNA or with RNA-DNA hybrids. With respect to the second part of your question, if you deliberately introduce 2–5A into cells then you do indeed get messenger RNA cleavage consistent with that being a major factor in the inhibition of protein synthesis. In addition there is both messenger RNA breakdown and ribosomal RNA cleavage in interferon-treated, EMC virus-infected cells in which translation is inhibited. Whether the former are the cause of the inhibition remains to be established.

CLARK: Can you re-iterate the status concerning the T98G cells in culture?

KERR: This is a limited observation to date. With the T98G cells aged in tissue culture we pick up a low level of the tetramer component of 2–5A. In addition, we have known for a long time that in a variety of different cell systems the 2–5A synthetase increases in confluent cells. For example with HeLa or L cells or MRC5 human diploid fibroblasts in the logarithmic phase of growth they normally have relatively low levels of synthetase which increases 10 fold if you maintain them in a confluent state.

2-5A Synthetases

Just Justesen & Peter Lauridsen

(2′-5′)oligoadenylate synthetase (2-5A synthetase) has been described in a number of species of cells treated with interferon and in certain other cell types (reviewed by I. M. Kerr in the preceeding paper). We have chosen 3 types of cells as sources for purification of 2-5A synthetase: reticulocytes from rabbits, chicken embryo fibroblasts treated with interferon and Namalva cells (human lymphoblastoid cells) treated with interferon. These 3 sources represent different cases. The reticulocytes contain 2-5A synthetase for a yet unknown reason, which is probably not interferon in the blood plasma, as the accompanying leucocytes do not show any elevated level of 2-5A synthetase. In the chicken cells treated with interferon in the blood plasma, nobody has been able to demonstrate the 67K protein kinase (Lab *et al.* 1982) which is found in mammalian cells upon treatment with interferon. The Namalva cells, thus, represent the case, where the antiviral state is as well explained by the 2-5A system as by the 67K protein kinase system (Lengyel 1982, Gordon & Minks 1981).

The 2-5A synthetase was first described as an ATP polymerase (Kerr & Brown 1978) that catalyses the reaction $n+1$ ATP→ppp5′A $(2'p5'\ A)_n + n\ PP_i$ in a dissipative mechanism (Justesen *et al.* 1981, Samanta *et al.* 1980, Minks *et al.* 1980), where the oligoadenylate formed does not rest on the enzyme until the next 2′ adenylation takes place. The direction of chain growth is from 5′ to 2′ so that the triphosphate group from the first ATP is maintained. The reaction equilibrium is close to 100% conversion of ATP into ppp5′ A $(2'p5'\ A)_n$, $n = 1,2,3$ ... (called 2-5A) (Justesen *et al.* 1981). The ATP molecule that is used to form PP_i

Department of Molecular Biology and Plant Physiology University of Aarhus, DK-6000 Aarhus, Denmark

GENE EXPRESSION, Alfred Benzon Symposium 19.
Editors: Brian F. C. Clark & Hans Uffe Petersen, Munksgaard, Copenhagen 1984.

and AMP in the 2′ position is called the donor and the other ATP or the oligonucleotide in the same reaction is called the acceptor. The equilibrium is strongly in favour of 2-5A formation and the back reaction has been below detection (Samanta *et al.* 1980).

All purified 2-5A synthetases require double-stranded RNA (dsRNA) for activity. The nature of the *in vivo* dsRNA that activates the enzyme is completely unknown although it is a widely accepted hypothesis that the dsRNA is provided by replicative forms of a virus. Lately heteronuclear RNA has been shown to be able to activate 2-5A synthetase (Nielsen & Baglioni 1982). The dsRNA required *in vitro* has been studied mainly with analogues to poly(I)·poly(C) that so far is the most efficient activator (Gordon & Minks 1981).

Besides ATP other nucleotides have been described as donors to oligo-adenylates (Justesen *et al.* 1981). All nucleoside triphosphates have been shown to react with ppp5′ A 2′p5′A. After this reaction the chain growth is terminated. Furthermore, the accepting ATP can be replaced with a number of oligo- and polynucleotides that all have a common configuration: the terminal nucleoside is adenosine (RpA) (Ball & White 1979, Ferbus *et al.* 1981, Minks *et al.* 1980). The enzyme could, thus, be called a (2′-5′) nucleotidyl transferase.

These alternative products have not been found *in vivo* so far, but the aim of this article is to demonstrate that under the present conditions for measuring enzyme activity *in vitro*, the alternative reactions compete very well with the formation of 2-5A from ATP, which has so far been the definition of the enzyme. Furthermore, the 2-5A synthetase activity can be regulated by addition of divalent metal ions (Mn^{++}) and by nucleoside tetraphosphates.

MATERIAL AND METHODS

Radioactive materials were obtained from the Radiochemical Centre (Amersham). Poly(I)·poly(C) and poly(I)·poly(C) agarose were from Choay (France). Polyethylene imine was a gift from BASF. The PEI-cellulose plates were prepared as described by Justesen *et al.* (1980b).

Preparation of 2-5A synthetase

Cell extracts were prepared by lysing of cells in buffer D (20 mM Tris-HCl pH 8.0, 5 mM $Mg(OAc)_2$, 25 mM KCl, 0.5 mM DTT, 1 mM EDTA; 10% glycerol (v/v)) supplemented with 0.2% NP40. The 2-5A synthetase (50 ml crude extract) was purified by passage through a DEAE-cellulose (50 ml) column followed by binding to a small poly(I)·poly(C) column (2 ml). The 2-5A synthetase activity was eluted by 0.2 M KCl in buffer D, and for Namalva cells an extra top at 0.8 M KCl in buffer D was eluted (Justesen *et al.* 1980a).

Assays for 2-5A synthetase activity

Samples were incubated with nucleotides as described in the legends, under conditions where the activity is maximal (Justesen *et al.* 1980b). High concentrations of Mg^{++} (20 mM) and a sufficient concentration of poly(I)·poly(C) (200 μg/ml) is essential.

The production of (2′-5′)oligonucleotide was analysed by polyethylene imine-cellulose thin-layer chromatography (Justesen *et al.* 1980b). The (2′-5′)oligoadenylate were separated from ATP by developing with 2 M Tris-HCl pH 8.6 (Fig. 2). 1 unit of 2-5A synthetase converts 1 nanomol into 2′-5′ links.

RESULTS

The 2-5A synthetases have been partly purified from cell extracts after treatment with NP40. The supernatant after centrifugation 100,000 g for 3 h is passed over a

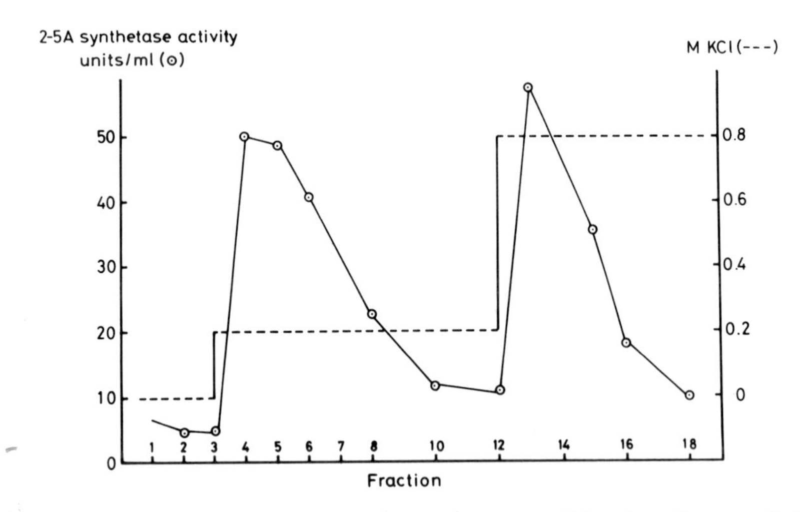

Fig. 1. Affinity chromatography on poly(I)·poly(C)-agarose of Namalva cell extract. Before this chromatography the supernatant after ultracentrifugation (S100) has passed through a DEAE-cellulose column. The elution was performed in 2 steps with 0.2 M and 0.8 M KCl.

DEAE cellulose column and the non-bound proteins are applied directly to a small poly(I)·poly(C) agarose column. The 2-5A synthetase is eluted by KCl at different concentrations. By this procedure the enzyme is not totally pure, but it is free of interfering activities as for example non-specific degradation of ATP.

The Namalva cell extract gave rise to 2 activity tops. The first top elutes at ca. 0.2 M KCl and the second top at ca. 0.8 M KCl. In the following Type I refers to the first top and Type II refers to the second top (Fig. 1).

Acceptors in the (2′-5′) nucleotidyl transferase reaction

The nature of the acceptor in the reaction has so far been described as RpA (Ferbus *et al.* 1981); examples of these acceptors that have been examined are listed in Table I. Several of these acceptors are co-enzymes that prevail in considerable amounts in most cells. Alkaline phosphatase treatment of NADP and co-enzyme A yield compounds that are active as acceptors, presumably because the blocking phosphate in the 2′ position on the ribose has been removed. Of all co-enzymes NAD^+ is a specially good substrate and the kinetic values of reactions with NAD^+ are of the same order of magnitude as the values

Table I
Co-enzymes as acceptors for 2′ adenylation

Co-enzyme 1 mM	% of ATP incorporated into		2′5′ links
	2-5A	co-enz.-2′p5′ A	% of ATP start
no addition	45.7	0.0	
NAD^+	23.7	26.7	15.6
$NADP^+$	37.9	0	28.1
CoA	2.3	0	1.3
FAD	10.5	4.0	2.7
α-NAD^+	19.1	15.0	10.6
NADH	19.4	ca. 1*	10.8
ADP	14.2	0	7.6
$NADP^+$ (a.p.)	21.3	++(n.d.)	12.7
CoA (a.p.)	20.4	10.2	11.3
NAD^+ (a.p.)	20.8	++(n.d.)	11.5

2-5A synthetase was purified from rabbit reticulocyte lysate. Incubations under standard conditions with 1 mM ATP for 2 h. The products were separated by chromatography as described for 2-5A and for NAD.AMP. (Justesen *et al.* 1980b). These separations gave the same per cent conversion.

* probably due to a contamination by NAD^+.

n.d. not determined, but the products was detected qualitatively.

a.p. indicates that the nucleotide has been treated with alkaline phosphatase.

Table II

Kinetic values for ATP and NAD^+

Substrate	Km mM	with alternative substrate	product
ATP	2.4	ATP	ppp5'A(2'p5' A)$_n$
NAD	0.5	ATP	NAD.2'AMP
ATP	0.18	NAD	NAD.2'AMP

Rough estimates from incubations with 2-5A synthetase purified from rabbit reticulocytes to a specific activity of 25 units/mg protein.

for ATP (Table II). The per cent incorporation of ATP into 2-5A is not a linear measure for activity. The enzyme activity is reflected in formation of 2'-5'-links. The first 2'-5' link corresponds to the incorporation of 2 ATP's, while the next link only neads one ATP incorporated. In Table I this is taken into consideration when comparing the co-enzyme-ATP reaction with the ATP-ATP reaction described as 2'-5' link-formation. In fact, it seems that the NAD^+ acts better as acceptor than ATP.

Table III

Km and V_{max} for different acceptors with UTP as donor

Acceptor	Km mM	V_{max} units/mg protein
3'5' CpA	0.92	29.2
NAD^+	0.61	40.0
5'5'' AppppA	0.37	105.4
AMPPCP	ca. 0.4	ca. 0.7

Incubations for 60 min with 2-5A synthetase from rabbit reticulocytes with a specific activity of 150 units/mg protein (2'5' link formation). UTP concentration 2 mM.

Acceptor	Km mM	V_{max} units/mg protein
3'5' CpA	1.32	92.0
NAD^+	1.34	122.0
5'5'' AppppA	0.99	178.0
AMPPCP	0.73	116.0

Incubations with 2-5A synthetase from interferon-treated chick embryo fibroblasts with a specific activity for 2-5A formation (2'5' link formation) of 178 units units/mg protein UTP concentration 2 mM.

The rare nucleotide 5′5″ ApppppA, which is associated with DNA polymerase (Zamecmik *et al.* 1982, Grummt *et al.* 1979) and synthesized by tRNA synthetases, (Plateau *et al.* 1981) is surely a better acceptor than ATP. In incubation, where ATP and 5′5″ ApppppA are present at the same time, 2-5A represents only a minor part of the products (10–25%). These reactions are catalysed by all 3 enzymes. When 5′5″ ApppppA is compared with NAD^+ and 3′5′ CpA as acceptor, this compound is clearly the best substrate, as the Km is lowest and the V_{max} is highest (Table III).

Donors in the (2′-5′) nucleotidyl transferase reaction

As donor nucleotide in the reaction a whole series of normal nucleotides have been examined with NAD^+ as acceptor in addition to the reactions previously shown (Ferbus *et al.* 1981) (Table IV). All natural nucleoside triphosphates do act as donors but with different efficiency when examined at 1 mM concentra-

Table IV

Donor specificity in the reaction $NAD^+ + X$

	Relative activity		Namalva	
X	reticulocytes	chicken	I	II
ATP	57	89	65	40
GTP	20	21	2	7
CTP	=100	33	n.d.	28
UTP	70	=100	10	47
ITP	32	33	=100	=100
2′dATP	33	16	53	51
dGTP	n.d.	n.d.	19	11
dCTP	27	2	30	20
dUTP	23	9	35	29
dITP	21	10	130	73
dTTP	11	2	n.d.	8
εATP	18	2	32	14
3′dATP	3	7	n.d.	n.d.

Compounds without any activity:
5′5″ ApppppA, 3′5′cAMP, 2′3′cAMP, creatine phosphate, AMPPCP, UDPG 5′AMP, 3′AMP.
Partly purified 2-5A synthetase from the sources was incubated for 2 h with 1 mM NAD^+ ((^{3}H-NAD) 70000 cpm/20 μl incubation) under standard conditions and with 1 mM of the donor. The reactions were analysed by PEi-cellose thin-layer chromatography (Justesen *et al.* 1980b). The per cent conversion of $^3H\text{-}NAD^+$ into another compound was calculated after localization by fluorography of the chromatogram.
n.d: not determined.

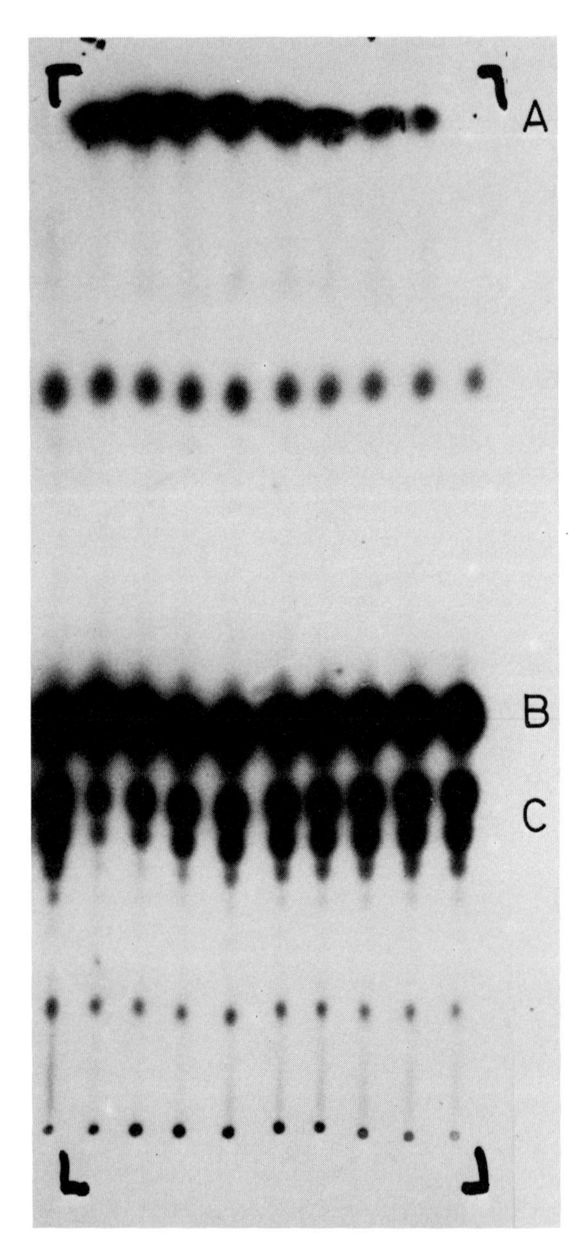

Fig. 2. Incubations of ATP and decreasing amounts of NAD^+ (3 mM to 0.1 mM). $\alpha(^{32}P)$ATP in 2 mM ATP was incubated with NAD^+ in the presence of reticulocyte 2-5A synthetase and the products were chromatographed on PEI-plates with 2 M Tris-HCl pH 8.6. The front spots are NAD.2′AMP (A). The following spots are ATP (B) and the series of ppp5′A $(2'p5'A)_n$(C). The remaining spots are contaminants in the radioactive ATP.

tion. The low activity with GTP and dGTP is, however, a general phenomenon. The very high activity with ITP and dITP with the 2-5A synthetase from Namalva cells is remarkable.

The Km values for ATP alone are much higher than for ATP as donor to NAD^+ (Table II, Fig. 2). This suggests that the Km value for ATP as acceptor is high (2.4 mM), and the Km for ATP as donor is low (ca. 0.2 mM). The demonstration of this in the case of formation of pppA $(2'p5' A)_n$ from ATP is very complex, because the different reactions cannot be separated. Until a way of discriminating between accepting and donating ATP has been found, the kinetics of 2-5A synthetase have to be studied by the alternative reactions that are simple two-substrate reactions, e.g., dATP+NAD.

Requirement for divalent metal ions

The 2-5A synthetase from reticulocyte requires a high concentration of Mg^{++} for full activity. Mn^{++} can partly replace the Mg^{++} but at a concentration higher than 2 mM the Mn^{++} starts to inhibit the 2-5A synthetase activity (D. Ferbus, personal communication). Under certain conditions, the reaction between 5′5″ AppppA and a nucleoside triphosphate requires the presence of a very tiny quantity of manganese ions (Fig. 3).

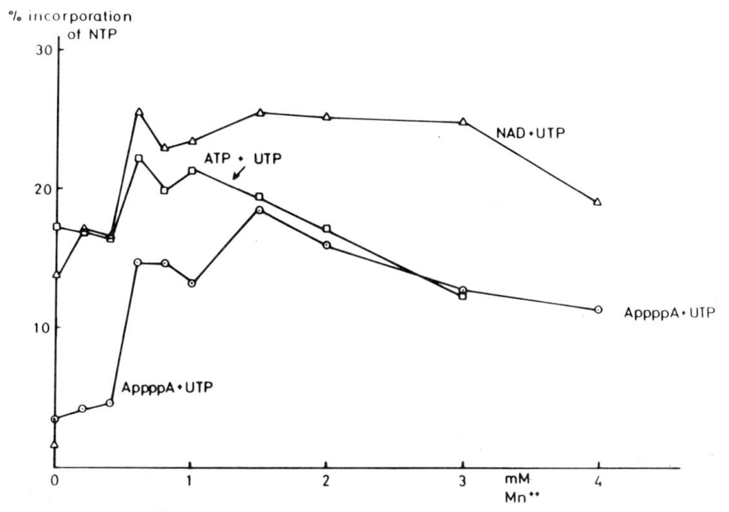

Fig. 3. Dependence on Mn^{++} of 2-5A synthetase activities. The activity is measured by incorporation of $\alpha(^{32}P)UTP$.

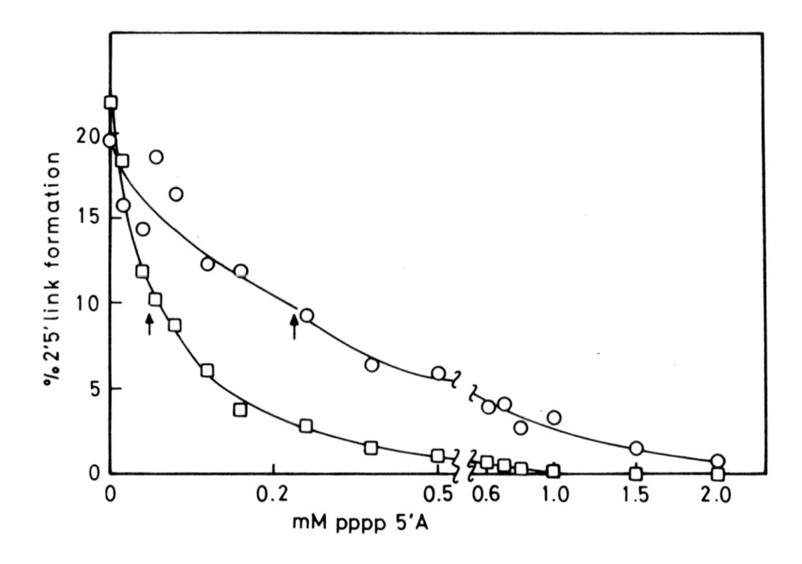

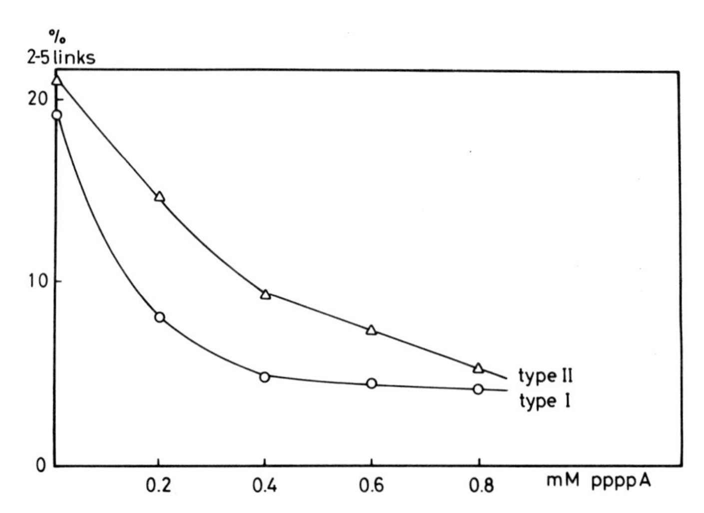

Fig. 4. Inhibitors of 2-5A synthetase activity. The pppp5'A inhibits the formation of 2-5A by reticulocyte (○) and chicken enzyme (□) (panel a) and the Namalva cell enzymes (panel b).

Inhibitors of (2'-5') oligoadenylate synthetase

To distinguish between the different types of 2-5A synthetase, we have looked for characteristics by which the enzymes differ remarkably.

While the compound 5'5'' AppppA is a very good substrate the 5' ppppA is a potent inhibitor of all enzymes, but the degree of inhibition is different (Fig. 4).

It is significant that less than 0.2 mM ppppA inhibits the formation of 2-5A by 50%. We have also examined inorganic triphosphate and tetraphosphate, which showed almost the same capacity of inhibition as ppppA. Under the same conditions inorganic pyrophosphate did not show any effect.

ACKNOWLEDGEMENTS

This work was supported by the Danish National Research Council. We thank Dr. E. Joensen, Heriff, Sdr. Omme, Denmark, for Namalva cells treated with interferon, Dr. M. Lab, Besancon, France for chicken cells treated with interferon. The reticulocyte enzyme was purified in the laboratory of Dr. M. Thang, Institut de Biologie Physico-Chimique, Paris, France, which is gratefully acknowledged.

REFERENCES

Ball, A. L. & White, C. N. (1979) In: *Regulation of Macromolecular Synthesis,* pp. 303–317. Academic Press.

Ferbus, D., Justesen, J., Besançon, F. & Thang, M. N. (1981) *Biochem. Biophys. Res. Commun. 100,* 847–856.

Gordon, J. & Minks, M. A. (1981) *Microbiological Reviews, 45,* 244–266.

Grummt, F., Waltl, G., Jantzen, H.-M., Hamprecht, K., Huebscher, U. & Kuenzle, C. C. (1979) *Proc. Natl. Acad. Sci. USA 76,* 6081–6085.

Justesen, J., Ferbus, D. & Thang, M. N. (1980a) *Proc. Natl. Acad. Sci. USA 77,* 4618–4622.

Justesen, J., Ferbus, D. & Thang, M. N. (1980b) *Nucl. Acids Res. 8,* 510–521.

Kerr, I. M. & Brown, R. E. (1978) *Proc. Natl. Acad. Sci. USA 75,* 256–260.

Lab. M., Thang, M. N., Soteriadou, K., Koehren, F. & Justesen, J. (1982) *Biochem. Biophys. Res. Commun. 105,* 412–418.

Lengyel, P. (1981) *Interferon 3,* 77–99.

Minks, M. A., Benvin, S. & Baglioni, C. (1980) *J. Biol. Chem. 255,*

Nilsen, T. W., Maroney, P. A., Robertson, H. D. & Baglioni, C. (1982) *Mol. Cell. Biol. 2,* 154–160.

Plateau, P., Mayaux, J.-F. & Blanquet, S. (1981) *Biochemistry 20,* 4654–4662.

Samanta, H., Dougherty, J. P. & Lengyel, P. (1980) *J. Biol. Chem. 25,* 9807–9813.

Zamecnik, P. C., Rapaport, E. & Baril, E. F. (1982) *Proc. Natl. Acad. Sci. USA 79,* 1791–1794.

DISCUSSION

SCHIMMEL: You have shown a great diversity of donors and acceptors that can be used. Are we to understand that some of these other products also are formed *in vivo*?

JUSTESEN: We don't know. Dr. Kerr has looked carefully, but so far, I think he has not found any.

SCHIMMEL: Is this 2–5A the major product?

JUSTESEN: Apparently yes, because the 2-5As have actually been found as the only ones in cells.

SCHIMMEL: Is it possible that there are other active species that are made also, because this enzyme apparently has a rather broad spectrum of donors and acceptors.

JUSTESEN: The first problem is that ATP prevails in higher concentrations than the other nucleotides in cells, so that alone might give more products. Another thing is that NAD^+ might be linked as a coenzyme to a lot of proteins, so that although NAD^+ is present in high concentrations in the cell, the amount of free NAD^+ may be very low. This leaves very little NAD^+ as free substrate. But so far this has not been confirmed. The AppppA is present in very low concentrations, so this product is only expected to be seen in lower concentrations. Maybe Dr. Kerr could comment on this.

KERR: So far, we have failed to detect any of the group of NAD-2-5A, AppA-2-5A and ADP ribose-2-5A in cells, specifically in interferon-treated EMC virus infected cells that are known to be synthesizing 2-5A. This does not, however, mean that these products are not going to turn up elsewhere, and, indeed, as I mentioned earlier I think we are possibly now picking up alternative products, although as yet we have not been able to identify any of them and none of them so far correspond to the 3 I have just mentioned.

PETERSEN: In the tRNA tertiary structure there are double-stranded regions. Have you tried tRNA as an activator of 2–5A synthetase?

JUSTESEN: Yes, I have tested several different tRNAs, for instance tRNA-Phe from *E. coli* and yeast, but there was no activation. However, tRNA acts as an acceptor for 2′ adenylation of the adenosine in the 3′-CCA terminal, and this requires double-stranded RNA as activator.

UHLENBECK: There is now another example of the 2′5′ ApA linkage that has been discovered in *E. coli* extracts as the product of the *E. coli* RNA ligase. Have you studied the reversal of this enzyme, and do you have any estimate of what the equilibrium constant of the reaction is?

JUSTESEN: I have studied the reverse, but it is below detection, because the equilibrium constant is much in favour of 2–5A synthesis.

UHLENBECK: When you add high concentrations of pyrophosphate in the reaction, don't you see any reversal?

JUSTESEN: No. I have actually tried that, and it has also been published by Samanta et al. (1980). Energetically, the formation of 2–5A is strongly favourable. The problem is then to have good quantities of pure 2′-5′ oligo As as substrates to push the reaction backwards.

Samanta H., Dougherty, J. P. & Langyel, P. (1980) *J. Biol. Chem.* *255*, 9807–9813.

Summary

David L. Miller

During the past few days we have witnessed the fruition of the renascence of protein biosynthesis research that has occurred in recent years. The early period of investigation, which terminated in about 1976, was characterized by the ingenious application of cell homogenizers, Sephadex chromatography, sucrose gradients and nitrocellulose membrane filtration. That period culminated in the general agreement about the identities of the components of the protein synthesis systems of both eukaryotic and prokaryotic organisms; however, substantial disagreements among various laboratories provoked skepticism about the validity of many mechanism studies.

In recent years, recognizing that mechanism studies can be unambiguously understood only in structural terms, investigators have directed their efforts toward determining the sequences and shapes of most of the proteins and nucleic acids involved in the system. Much of the progress which we have learned about in these reports would have been impossible without the new techniques of nucleic acid sequence determination and molecular cloning. The authors deserve congratulations for their rapid employment of these powerful methods.

A major obstacle to studying the structure and function of initiation and termination factors has been the difficulty in obtaining sufficient amounts of these proteins. The identification of recombinant plasmids bearing sequences coding for IF-2, IF-3, RF-1 and RF-2 offers the possibility for obtaining large quantities of these proteins. This procedure may not be straightforward because of feedback controls on their synthesis. The studies reported here for maximizing interferon synthesis in prokaryotic and eukaryotic hosts could be applied to producing these factors.

New York State Institute for Basic Research in Developmental Disabilities, Staten Island 10314, New York, U.S.A.

GENE EXPRESSION, Alfred Benzon Symposium 19.
Editors: Brian F. C. Clark & Hans Uffe Petersen, Munksgaard, Copenhagen 1984.

It will be interesting to compare the protein sequences of these factors that will be deduced from their DNA sequences. We expect IF-2 to exhibit considerable homology to EF-Tu and EF-G. Already we have seen sufficient homology in EF-Tu and EF-G to suggest an evolutionary and functional relationship between them. Furthermore, considerable sequence homology between EF-Tu and eEF-Tu has been found. At this meeting, we have heard that the yeast mitochondrial EF-Tu gene has been identified and from its deduced amino acid sequence this protein has been found to be 60% homologous to the *E. coli* factor. Maintaining this degree of homology over the eons since mitochondria diverged from bacteria is amazing; it must reflect the severe constraints placed upon the protein's structure by the necessity to interact with several different substances.

The major results of nucleotide sequencing, the primary structure determinations of the ribosomal RNAs, are facilitating the understanding of ribosomal structure and function. Regions containing secondary structure have been determined by nuclease susceptibility. Similarities in the structures of 16S rRNA, free and bound to ribosomal proteins, suggests that 16S rRNA predated the proteins and may have been the primordial ribosome. From studies of the photochemical cross-linking of P-site-bound tRNAs cytidine-1400 appears to be crucial for the decoding process. It occurs near the 3′ end of several small subunit RNAs, and may stick in between adjacent codons like a ratchet. A looped-out base is also important for the function of 5S rRNA, where an unpaired A residue forms part of the binding site for ribosomal protein L 18.

Although biopolymers seem to have no difficulty in correctly folding themselves, the shapes of macromolecules and their complexes have been notoriously difficult to deduce from their primary sequences. We have heard 3 reports on methods to determine the shape of tRNAs and their complexes with proteins. By a careful sequence analysis of the sites of dimethyl sulfate and diethylpyrocarbonate modifications at increasing temperatures the structure of mammalian mitochondrial $tRNA^{Ser}$ was deduced. This tRNA, which lacks the D stem and extra arm, folds up into a structure bearing some resemblance to cytoplasmic tRNAs.

The binding sites on tRNAs for synthetases and elongation factors cannot be reliably determined by chemical modification, because essential protein residues react faster than the nucleotides, causing the complexes to dissociate. Studies of the nuclease digestion of the $Met\text{-}tRNA_m^{Met} \cdot EF\text{-}Tu \cdot GTP$ complex revealed a distinct sidedness to the protein-tRNA binding, a result which agrees with other data on the cross-linking of Phe-tRNA to the protein.

The traditional method for determining the structural basis of reactivity in biopolymers has been to modify specific residues by selective chemical reagents. The limitations of this procedure are apparent: no reagent is specific for a single residue in a single site, and most residues are not uniquely reactive; a glutamyl residue cannot be distinguished from an aspartyl residue, and so forth. Furthermore, the functions of subdomains of proteins and nucleic acids cannot be ascertained readily unless an especially reactive cleavage site appropriately lies between the domains.

We have learned 4 approaches to these problems for modifying specific residues in aminoacyl-tRNA synthetases, interferons, and tRNAs by site directed mutagenesis and by chemical reconstruction. In α-interferon a specific cysteinyl residue was converted to serine, by engineering a base-sequence change in the expression plasmid. The resultant protein exhibited greatly increased activity. By introducing random base-changes in various regions of the Ala-tRNA synthetase gene it was possible to identify domains of the resultant protein responsible for the various functions of the enzyme. In addition deletion mutants were constructed which encoded proteins lacking specific domains and these enzymes lacked some of the complete enzyme's activities.

In both eukaryotes and prokaryotes, the translation process is subject to multiple homeostatic and exogenous controls. In prokaryotes the translational and transcriptional processes cannot be considered to be separable, not only because a sizeable fraction of polysomes form on nascent mRNA chains. In these rapidly growing organisms regulation of the rates of synthesis of the translation system components may actually control translation rates. Therefore, studies of the transcriptional regulation of initiation and elongation factor synthesis are crucial to understanding translational control. These studies have been facilitated by the isolation of recombinants containing the genes for most of the components of the *E. coli* translation system.

Studies of the isolated genes have identified promotor regions, which in the case of IF-3 is located in the preceding gene for threonyl synthetase. There is evidence for autoregulation of IF-2 at the transcriptional level; the level of IF-2 mRNA does not increase a much as the gene dosage in cells containing multicopy plasmids bearing the IF-2 gene. ppGpp seems to regulate synthesis of most of the translational components at the transcriptional level. The genes for IF-2 and EF-Tu were shown to be so regulated. The DNA sequence between the Pribnow box and the transcription-start site is involved in ppGpp control of *tuf* B transcrpition.

The levels of the ribosomal proteins and elongation factors whose genes are located in each of the *str-spc* operons are regulated by feedback control of translation by one of the proteins coded by each polycistronic mRNA. Each control protein inhibits translation by binding to a specific region of the polycistronic mRNA that resembles the structure of its binding site on rRNA. The proteins bind preferentially to rRNA; therefore, translation slows when excess ribosomal proteins appear. The levels of rRNA itself appear to be feedback-regulated by a mechanism in which excess rRNA inhibits its own transcription. The beauty of this mechanism, its logic and economy, is awe-inspiring.

A similar mechanism of translational repression has been shown to regulate expression of 2 bacteriophage T_4 proteins. One of these, the gene 32 protein, binds non-specifically to single-stranded DNA. Excess protein binds to the initiation domain of its own mRNA. *RegA* protein represses the rIIB protein of T_4 phage by a post-transcriptional process. Sequence analysis of translational control mutants revealed that the sites of mutation clustered in the ribosome-binding site. Again, the regulatory mechanism involved inhibition of initiation.

The structural requirements for initiation of translation and for the binding of repressors can now be determined by altering the 5′ non-coding region of the gene through substitution of unrelated DNA sequences. This approach was demonstrated by the systematic manipulation of the initiation region of the galactokinase gene.

Other mechanisms in addition to repression by mRNA-binding proteins control translation in eukaryotes. Poliovirus shuts down host-capped mRNA translation by cleaving the cap-binding protein using the same protease that processes the viral polyprotein. As one of its actions, interferon induces a dsRNA-dependent protein kinase that phosphorylates eIF2α and renders it inactive in catalyzing repeated rounds of initiation.

Eukaryotic IF-2α binds GTP. During the subsequent binding of met-tRNA to ribosomes GTP is hydrolyzed and eIF-2α · GDP is formed. To recycle the factor GDP must be exchanged for GTP. The rate of exchange of GDP from eIF-2α is slower than the rate of initiation of newer peptides in active cells. The exchange is catalyzed by eIF-2β, via an intermediate eIF-2α · eIF-2β complex.

The most widely studied example of translational control is the regulation of globin synthesis by hemin. Hemin deficiency promotes the phosphorylation of eIF-2α; phosphorylated eIF-2α binds eIF-2β, forming a complex which resists dissociation and, thereby, sequesters the limited amount of eIF-2β and prevents it from regenerating non-phosphorylated eIF-2$\alpha \cdot$ GDP.

Proteins that catalyze the exchange of protein-bound GDP (EFT_s and EF-1β) are involved in the regeneration of EF-Tu and EF-1α. In other systems the activation of adenylate cyclase by hormones and neurotransmitters is accomplished through a GTP-activated and GDP-inhibited regulatory subunit. GDP-exchange enzymes will be involved in processes where the reaction rate exceeds the uncatalyzed exchange rate. For other processes requiring GTP, prokaryotic initiation, translocation, and termination, not enough data on the exchange rates are available to allow us to decide whether an exchange enzyme is needed. In the absence of an exchange enzyme, nucleoside diphosphate dissociation may become rate-determining. As an example, ADP dissociation from actomyosin may limit the rate of muscle contraction.

Whereas the recent advances in understanding protein biosynthesis have resulted from structural analysis of the purified components and the kinetic analysis of their reactions, certain properties of the system cannot be studied in the components. One of these phenomena is translational accuracy. Error rates are closely related to elongation rates; therefore, until recently, all definitive studies of translational accuracy were performed *in vivo* by genetic techniques. The recently developed *in vitro* system possessing an elongation rate approaching that of intact *E. coli* has been used to re-investigate the accuracy problem. Using this system kinetic proof-reading can be demonstrated.

In vivo studies provide the ultimate criterion for the validity of conclusions derived from *in vitro* experiments. From such studies we have learned that we still do not know how ppGpp prevents frame-shifting; however, frame-shifting, in general, occurs through incorrect aa-tRNA selection resulting from limitation of the cognate species.

The means for maintaining fidelity depends upon the standard of living of the organism. The view was expressed that a busy organism like *E. coli* can't afford the time and energy necessary to conduct kinetic proof-reading and must get its amino acids and tRNAs right the first time using stereoselective mechanisms. In contrast, eukaryotic cells have the leisure and energy to proof-read. This conclusion was supported by results of the study of aminoacyl-tRNA synthetases from various organisms, which showed that kinetic (pretransfer) proof-reading was a property of eukaryotic enzymes. (If this view is correct, it should be much easier to demonstrate kinetic proof-reading in a eukaryotic protein synthesis system.)

The advances reported here give everyone the satisfaction of participating in a major contribution to knowledge. The information will be valued as long as the human species cares about biology. Nevertheless, many difficult problems remain. Determining the tertiary structures and interactions of proteins continues to be an arduous, risky undertaking. Detailed mechanism studies performed in the light of the known structures of the components must be performed, and the processes controlling the synthesis of the eukaryotic translational apparatus must be identified. As more gene products are identified, one expects that new mechanisms of translational control will appear. Statistically speaking, this enterprise has just begun.

Subject Index